Lecture Notes in Computer Science 8786

Commenced Publication in 1973
Founding and Former Series Editors:
Gerhard Goos, Juris Hartmanis, and Jan van Leeuwen

Editorial Board

David Hutchison
  Lancaster University, UK
Takeo Kanade
  Carnegie Mellon University, Pittsburgh, PA, USA
Josef Kittler
  University of Surrey, Guildford, UK
Jon M. Kleinberg
  Cornell University, Ithaca, NY, USA
Alfred Kobsa
  University of California, Irvine, CA, USA
Friedemann Mattern
  ETH Zurich, Switzerland
John C. Mitchell
  Stanford University, CA, USA
Moni Naor
  Weizmann Institute of Science, Rehovot, Israel
Oscar Nierstrasz
  University of Bern, Switzerland
C. Pandu Rangan
  Indian Institute of Technology, Madras, India
Bernhard Steffen
  TU Dortmund University, Germany
Demetri Terzopoulos
  University of California, Los Angeles, CA, USA
Doug Tygar
  University of California, Berkeley, CA, USA
Gerhard Weikum
  Max Planck Institute for Informatics, Saarbruecken, Germany

T0235739

Boualem Benatallah   Azer Bestavros
Yannis Manolopoulos   Athena Vakali
Yanchun Zhang (Eds.)

Web Information Systems Engineering – WISE 2014

15th International Conference
Thessaloniki, Greece, October 12-14, 2014
Proceedings, Part I

 Springer

Volume Editors

Boualem Benatallah
University of New South Wales, Sydney, NSW, Australia,
E-mail: boualem@cse.unsw.edu.au

Azer Bestavros
Boston University, Boston, MA, USA
E-mail: best@bu.edu

Yannis Manolopoulos
Aristotle University of Thessaloniki, Thessaloniki, Greece
E-mail: manolopo@csd.auth.gr

Athena Vakali
Aristotle University of Thessaloniki, Thessaloniki, Greece
E-mail: avakali@csd.auth.gr

Yanchun Zhang
Victoria University, Footscray, VIC, Australia
E-mail: yanchun.zhang@vu.edu.au

ISSN 0302-9743   e-ISSN 1611-3349
ISBN 978-3-319-11748-5   e-ISBN 978-3-319-11749-2
DOI 10.1007/978-3-319-11749-2
Springer Cham Heidelberg New York Dordrecht London

Library of Congress Control Number: 2014949242

LNCS Sublibrary: SL 3 – Information Systems and Application, incl. Internet/Web
and HCI

© Springer International Publishing Switzerland 2014

This work is subject to copyright. All rights are reserved by the Publisher, whether the whole or part of
the material is concerned, specifically the rights of translation, reprinting, reuse of illustrations, recitation,
broadcasting, reproduction on microfilms or in any other physical way, and transmission or information
storage and retrieval, electronic adaptation, computer software, or by similar or dissimilar methodology
now known or hereafter developed. Exempted from this legal reservation are brief excerpts in connection
with reviews or scholarly analysis or material supplied specifically for the purpose of being entered and
executed on a computer system, for exclusive use by the purchaser of the work. Duplication of this publication
or parts thereof is permitted only under the provisions of the Copyright Law of the Publisher's location,
in its current version, and permission for use must always be obtained from Springer. Permissions for use
may be obtained through RightsLink at the Copyright Clearance Center. Violations are liable to prosecution
under the respective Copyright Law.
The use of general descriptive names, registered names, trademarks, service marks, etc. in this publication
does not imply, even in the absence of a specific statement, that such names are exempt from the relevant
protective laws and regulations and therefore free for general use.
While the advice and information in this book are believed to be true and accurate at the date of publication,
neither the authors nor the editors nor the publisher can accept any legal responsibility for any errors or
omissions that may be made. The publisher makes no warranty, express or implied, with respect to the
material contained herein.

Typesetting: Camera-ready by author, data conversion by Scientific Publishing Services, Chennai, India

Printed on acid-free paper

Springer is part of Springer Science+Business Media (www.springer.com)

Preface

Welcome to the proceedings of the 15th International Conference on Web Information Systems Engineering (WISE 2014), held in Thessaloniki, Greece in October 2014. The series of WISE conferences aims to provide an international forum for researchers, professionals, and industrial practitioners to share their knowledge in the rapidly growing area of Web technologies, methodologies, and applications. The first WISE event took place in Hong Kong, China (2000). Then the trip continued to Kyoto, Japan (2001); Singapore (2002); Rome, Italy (2003); Brisbane, Australia (2004); New York, USA (2005); Wuhan, China (2006); Nancy, France (2007); Auckland, New Zealand (2008); Poznan, Poland (2009); Hong Kong, China (2010); Sydney, Australia (2011); Paphos, Cyprus (2012), and Nanjing, China (2013). This year, for a fifth time, WISE was held in Europe, in Thessaloniki, supported by Aristotle University, the largest University in Greece.

WISE 2014 hosted three well-known keynote and invited speakers: Prof. Krishna P. Gummadi, Head of the Networked Systems Research Group, at Max Planck Institute for Software Systems, Germany who gave a talk on "Understanding Information Exchange in Social Media Systems", Dr Mike Fisher, Chief Researcher in the Research and Innovation Department of British Telekom, UK who gave an industrial focus lecture on "Connected Communities"; and Prof. Santo Fortunato faculty of Complex Systems at the Department of Biomedical Engineering and Computational Science of Aalto University, Finland who gave a talk on "Detecting Communities in Networks". Moreover, four tutorials were presented on the topics : "Blocking Techniques for Web-Scale Entity Resolution", "Community Detection and Evaluation in Social and Information Networks", "Navigating the choices for Similarity Operators", and "Extensions on Map-Reduce".

A total of 196 research papers were submitted to the conference for consideration, and each paper was reviewed by at least two reviewers. Finally, 52 submissions were selected as full papers (with an acceptance rate of 26% approximately), plus 16 as short papers. The program also featured 14 poster papers and 1 WISE challenge summary paper which outlines the WISE challenge succeeded submissions, presented in the WISE challenge separate Workshop. The research papers cover the areas of semantic Web; Web mining, modeling, and classification; Web querying and searching; Web recommendation and personalization; social online networks; Web technologies and frameworks; Software Architectures, techniques, and platforms; and Web innovation and applications.

We wish to take this opportunity to thank the industry program co-chairs, Dr. Shengbo Guo, Dr. Nikos Laoutaris, and Dr. Hamid Motahari; the tutorial and panel co-chairs, Prof. Evimaria Terzi and Prof. Ernestina Menasalvas; the WISE challenge program co-chairs Prof. Grigoris Tsoumakas, Prof. Apostolos

Papadopoulos and Prof. Weining Qian; the Workshop co-Chairs, Prof. Armin Haller and Prof. Barbara Catania; the publication chair Prof. Xue Li; the Local Organizing Committee chairs Prof. Nick Bassiliades and Prof. Eleftherios Angelis; the publicity co-chairs, Prof. Fang Li, the Prof. Roger Whitaker and the Prof. George Pallis; and the WISE society representative, Xiaofang Zhou. The editors and chairs are grateful to the Web site and social media master, Dr. Ioannis Karydis for his continuous active support and commitment.

In addition, special thanks are due to the members of the International Program Committee and the external reviewers for a rigorous and robust reviewing process. We are also grateful to the Department of Informatics of the Aristotle University, and the International WISE Society for supporting this Conference.

We expect that the ideas that have emerged in WISE 2014 will result in the development of further innovations for the benefit of scientific, industrial and societal communities.

October 2014

Azer Bestavros
Boualem Benatallah
Yannis Manolopoulos
Athena Vakali
Yanchun Zhang

Organization

General Chairs

Yannis Manolopoulos   Aristotle University, Greece
Yanchun Zhang   Victoria University, Australia

Program Committee Chairs

Boualem Benatallah   University of New South Wales, Australia
Azer Bestavros   Boston University, USA
Athena Vakali   Aristotle University, Greece

Industry Program Chairs

Shengbo Guo   Samsung Information Systems America, USA
Nikos Laoutaris   Telefonica Research, Spain
Hamid Motahari   IBM, USA

Demo Co-chairs

Srdjan Krco   Ericsson/University of Belgrade, Serbia
Joao Fernandes   Alexandra Institute, Denmark

Tutorial and Panel Co-chairs

Ernestina Menasalvas   Universidad Politecnica de Madrid, Spain
Evimaria Terzi   Boston University, USA

WISE Challenge Program Chairs

Weining Qian   East China Normal University, China
Apostolos Papadopoulos   Aristotle University, Greece
Grigorios Tsoumakas   Aristotle University, Greece

Workshops Co-chairs

Barbara Catania   University of Genoa, Italy
Armin Haller   CSIRO, Australia

Publication Chair

Xue Li   University of Queensland, Australia

Local Organizing Committee Chairs

Nick Bassiliades   Aristotle University, Greece
Eleftherios Angelis   Aristotle University, Greece

Publicity Co-chairs

Fang Li   Shanghai Jiaotong University, China
Roger Whitaker   Cardiff University, UK
George Pallis   University of Cyprus, Cyprus

WISE Society Representative

Xiaofang Zhou   University of Queensland, Australia

Web and Social Media Master

Ioannis Karydis   Ionian University, Greece

Program Committee

Karl Aberer   Ecole Polytechnique Federale de Lausanne,
                                  Switzerland
Divy Agrawal   University of California at Santa Barbara, USA
Marco Aiello   University of Groningen, The Netherlands
Anastasia Ailamaki   Ecole Polytechnique Federale de Lausanne,
                                  Switzerland
Virgilio Almeida   Federal University of Minas Gerais, Brazil
Eleftherios Angelis   Aristotle University of Thessaloniki, Greece
Leonidas Anthopoulos   TEI of Thessaly, Greece
Demetrios Antoniades   Georgia Institute of Technology, USA
Costin Badica   University of Craiova, Romania
Ricardo Baeza-Yates   Yahoo! Labs, Spain
Alistair Barros   Queensland University of Technology, Australia
Ladjel Bellatreche   Laboratoire d'Informatique et d'Automatique
                                  pour les Systemes, France
Boualem Benatallah   University of New South Wales, Australia
Salima Benbernou   University Paris Descartes, France
Christos Berberidis   International Hellenic University, Greece

Elisa Bertino	Purdue University, USA
Azer Bestavros	Boston University, USA
Alex Beutel	Carnegie Mellon University, USA
Antonis Bikakis	University College London, UK
Mahdi Bohlouli	University Siegen, Germany
David Camacho	Autonomous University of Madrid, Spain
Barbara Catania	University of Genoa, Italy
Despoina Chatzakou	Aristotle University of Thessaloniki, Greece
Panagiota Chatzipetrou	Aristotle University of Thessaloniki, Greece
Sotirios Chatzis	Cyprus University of Technology, Cyprus
Richard Chbeir	University de Pau et des Pays de l'Adour, France
Fei Chen	HP Labs, USA
Hong Cheng	The Chinese University of Hong Kong, China
Alexandra Cristea	University of Warwick, UK
Alfredo Cuzzocrea	University of Calabria, Italy
Alex Delis	University of Athens, Greece
Marios Dikaiakos	University of Cyprus, Cyprus
Marlon Dumas	University of Tartu, Estonia
Ahmed Elmagarmid	Qatar Computing Research Institute, Qatar
Christos Faloutsos	Carnegie Mellon University, USA
Marie Christine Fauvet	LIG, France
Joao Fernandes	The Alexandra Institute, Denmark
Joao Eduardo Ferreira	University of Sao Paulo, Brazil
Panos Fitsilis	TEI of Thessaly, Greece
Wen Gao	Peking University, China
John Garofalakis	University of Patras, Greece
Christos Georgiadis	University of Macedonia, Greece
Chryssis Georgiou	University of Cyprus, Cyprus
Vasilis Gerogiannis	TEI of Thessaly, Greece
Aditya Ghose	University of Wollongong, Australia
Maria Giatsoglou	Aristotle University of Thessaloniki, Greece
Alex Gluhak	University of Surrey, UK
Daniela Grigori	Dauphine University, France
Shengbo Guo	Australian National University, Australia
Hakim Hacid	Bell Labs, France
Mohand-Said Hacid	University Claude Bernard Lyon 1, France
Armin Haller	Commonwealth Scientific and Industrial Research Organisation, Australia
Ourania Hatzi	Harokopio University, Greece
Katja Hose	Aalborg University, Denmark
Yuh-Jong Hu	National Chengchi University, Taiwan
Zhisheng Huang	Vrije University of Amsterdam, The Netherlands
Yoshiharu Ishikawa	Nagoya University, Japan

Arun Iyengar	IBM TJ Watson, USA
Peiquan Jin	University of Science and Technology of China, China
George Kakarontzas	TEI of Thessaly, Greece
Georgia Kapitsaki	University of Cyprus, Cyprus
Helen Karatza	Aristotle University of Thessaloniki, Greece
Ioannis Katakis	University of Athens, Greece
Dimitrios Katsaros	University of Thessaly, Greece
Vasiliki Kazantzi	TEI of Thessaly, Greece
Anastasios Kementsietsidis	IBM T.J. Watson Research Center, USA
Fotis Kokkoras	TEI of Thessaly, Greece
Yiannis Kompatsiaris	CERTH, ITI, Greece
Ioannis Kompatsiaris	CERTH, ITI, Greece
Efstratios Kontopoulos	Centre of Research & Technology, Greece
Makrina Viola Kosti	Aristotle University of Thessaloniki, Greece
Manolis Koubarakis	University of Athens, Greece
Srdjan Krco	TEI of Thessaly, Greece
Fang Li	Jiao Tong University, China
Xue Li	School of ITEE, Australia
Xue Li	The University of Queensland, Australia
Xuemin Lin	The University of New South Wales, Australia
Georgios Meditskos	Centre of Research & Technology, Greece
Ernestina Menasalvas	Technical University of Madrid, Spain
Natwar Modani	IBM India Research Lab, India
Hamid Motahari	University of New South Wales, Australia
Hamid Motahari	IBM Almaden, USA
Luis Munoz	University of Cantabria, Spain
Miyuki Nakano	University of Tokyo, Japan
Wilfred Ng	The Hong Kong University of Science & Technology, China
Mara Nikolaidou	Harokopio University, Greece
Kjetil Norvag	Norwegian University of Science and Technology, Norway
Alexandros Ntoulas	Zynga, USA
George Pallis	University of Cyprus, Cyprus
Apostolos Papadopoulos	Aristotle University of Thessaloniki, Greece
Stelios Paparizos	Microsoft Research, USA
Josiane Xavier Parreira	DERI - National University of Ireland, Ireland
Sophia Petridou	Aristotle University of Thessaloniki, Greece
Evaggelia Pitoura	University of Ioannina, Greece
Dimitris Plexousakis	University of Crete, Greece
Weining Qian	East China Normal University, China
Misha Rabinovich	Case Western Reserve University, USA
Sandra Patricia Rojas Berrio	National University of Colombia, Colombia

Marek Rusinkiewicz	New Jersey Institute of Technology, USA
Mohand Said Hacid	LIRIS, France
Rizos Sakellariou	University of Manchester, UK
Neil Shah	Carnegie Mellon University, USA
John Shepherd	The University of New South Wales, Australia
Michalis Sirivianos	Cyprus University of Technology, Cyprus
Georgios Spanos	Aristotle University of Thessaloniki, Greece
Myra Spiliopoulou	University of Magdeburg, Germany
Divesh Srivastava	AT&T Labs-Research, USA
Panagiotis Symeonidis	Aristotle University of Thessaloniki, Greece
Stefan Tai	KSRI, Germany
Leandros Tassiulas	University of Thessaly, Greece
Joe Tekli	Lebanese American University, Lebanon
Evimaria Terzi	Boston University, USA
Christos Tjortjis	International Hellenic University, Greece
Ismail Toroslu	Middle East Technical University, Turkey
Farouk Toumani	Blaise Pascal University, France
Peter Triantafillou	University of Patras, Greece
Athanasios Tsadiras	Aristotle University of Thessaloniki, Greece
Grigorios Tsoumakas	Aristotle University of Thessaloniki, Greece
Iraklis Varlamis	Harokopio University, Greece
Nikolaos Vasileiadis	Aristotle University of Thessaloniki, Greece
Kunal Verma	Accenture Technology Labs, USA
Ray Walshe	Dublin City University, Ireland
Hua Wang	University of Southern Queensland, Australia
Ingmar Weber	Qatar Computing Research Institute, Qatar
Roger Whitaker	Cardiff University, UK
Josiane Xavier Parreira	DERI - National University of Ireland, Ireland
Hao Yan	LinkedIn Corp, USA
Tetsuya Yoshida	Nara Women's University, Japan
Leon Zhang	Fudan, China
Wenjie Zhang	University of New South Wales, Australia
Yanchun Zhang	Victoria University, Australia
Xiaofang Zhou	The University of Queensland, Australia
Rui Zhou	University of Science and Technology of China, China
Xiangmin Zhou	Victoria University, Australia

Keynote Lectures

Prof. Krishna Gummadi

Krishna Gummadi is a tenured faculy member and head of the Networked Systems research group at the Max Planck Institute for Software Systems (MPI-SWS) in Germany. He received his Ph.D. (2005) and M.S. (2002) degrees in Computer Science and Engineering from the University of Washington. He also holds a B.Tech (2000) degree in Computer Science and Engineering from the Indian Institute of Technology, Madras. Krishna's research interests are in the measurement, analysis, design, and evaluation of complex Internet-scale systems. His current projects focus on understanding and building social Web systems. Specifically, they tackle the challenges associated with protecting the privacy of users sharing personal data, understanding and leveraging word-of-mouth exchanges to spread information virally, and finding relevant and trustworthy sources of information in crowds. Krishna's work on online social networks, Internet access networks, and peer-to-peer systems has led to a number of widely cited papers and award (best) papers at ACM/Usenix's SOUPS, AAAI's ICWSM, Usenix's OSDI, ACM's SIGCOMM IMW, and SPIE's MMCN conferences.

"Understanding Information Exchange in Social Media Systems", Monday Oct. 13th, 2014

The functioning of our modern knowledge-based societies depends crucially on how individuals, organizations, and governments exchange information. Today, much of this information exchange is happening over the Internet. Recently, social media systems like Twitter and Facebook have become tremendously popular, bringing with them profound changes in the way information is being exchanged online. In this talk, focus is on understanding the processes by which social media users generate, disseminate, and consume information. Specifically, the trade-offs between relying on the information generated by (i.e., wisdom of) crowds versus experts and the effects of information overload on how users consume and disseminate information are investigated. Limitations of our current understanding is highlighted and arguing that an improved understanding of information exchange processes is the necessary first step towards designing better information retrieval (search or recommender) systems for social media.

Dr. Mike Fisher

Mike Fisher is a Chief Researcher in the Research and Innovation Department of British Telekom, UK. Following a PhD in Physics from University of Surrey, he joined BT and worked on "blue sky" research projects investigating semiconductor optical materials and devices. He later moved into distributed systems where his research interests have included policy-based management, active networks, Grid computing, Cloud computing and most recently the Internet of Things. Mike has had a strong involvement in collaborative projects on these topics at national and European level. He was involved in the establishment of the NESSI European Technology Platform and was the Chairman of the ETSI Technical Committee responsible for Grid and Cloud. His current focus is on information-centric network services that can enable improved exchange of information, and the value that these can deliver.

"Connected Communities", Monday Oct. 13th, 2014

Any process or activity can be improved by timely access to better information. The long-held vision of a connected world is now becoming a reality as technological advances make it increasingly cost-effective to publish, find and use a huge variety of data. New ways of managing information offer the potential for transformational change, with the network as the natural point of integration. In this talk some of BT recent work exploring technologies to promote sharing in communities unified by an interest in similar information is highlighted. This includes experiences in a number of sectors including transport, supply chain and future cities.

 **Prof. Santo Fortunato**

Santo Fortunato is Professor of Complex Systems at the Department of Biomedical Engineering and Computational Science of Aalto University, Finland. Previously he was director of the Sociophysics Laboratory at the Institute for Scientific Interchange in Turin, Italy. Prof. Fortunato got his PhD in Theoretical Particle Physics at the University of Bielefeld In Germany. He then moved to the field of complex systems. His current focus areas are network science, especially community detection in graphs, computational social science and science of science. His research has been published in leading journals, including Nature, PNAS, Physical Review Letters, Reviews of Modern Physics, Physics Reports and has collected over 10,000 citations (Google Scholar). His review article Community detection in graphs (Physics Reports 486, 75-174, 2010) is the most cited paper on networks of the last years. He is the recipient of the Young Scientist Award in Socio- And Econophysics 2011 from the German Physical Society.

"Detecting Communities in Networks", Tuesday Oct. 14th, 2014

Finding communities in networks is crucial to understand their structure and function, as well as to identify the role of the nodes and uncover hidden relationships between nodes. In this talk I will briefly introduce the problem and then focus on algorithms based on optimization. I will discuss the limits of global optimization approaches and the potential advantages of local techniques. Finally I will assess the delicate issue of testing the performance of methods.

Tutorials (Abstracts)

Blocking Techniques for Web-Scale Entity Resolution

George Papadakis and Themis Palpanas

Institute for the Management of Information Systems
- Athena Research Center, Greece
gpapadis@imis.athena-innovation.gr
Paris Descartes University, France
themis@mi.parisdescartes.fr

Abstract. Entity Resolution constitutes one of the cornerstone tasks for the integration of overlapping information sources. Due to its quadratic complexity, a bulk of research has focused on improving its efficiency so that it can be applied to Web Data collections, which are inherently voluminous and highly heterogeneous. The most common approach for this purpose is blocking, which clusters similar entities into blocks so that the pair-wise comparisons are restricted to the entities contained within each block.

In this tutorial, we elaborate on blocking techniques, starting from the early, schema-based ones that were crafted for database integration. We highlight the challenges posed by today's heterogeneous, noisy, voluminous Web Data and explain why they render inapplicable the early blocking methods. We continue with the presentation of the latest blocking methods that are crafted for Web-scale data. We also explain how their efficiency can be improved by meta-blocking and parallelization techniques.

We conclude with a hands-on session that demonstrates the relative performance of several, state-of-the-art techniques, and enables the participants of the tutorial to put in practice all the topics discussed in the theory.

Community Detection and Evaluation in Social and Information Networks

Christos Giatsidis, Fragkiskos D. Malliaros, and Michalis Vazirgiannis

Ecole Polytechnique, France
http://www.lix.polytechnique.fr/~giatsidis
Ecole Polytechnique, France
http://www.lix.polytechnique.fr/~fmalliaros
Ecole Polytechnique, France
http://www.lix.polytechnique.fr/~mvazirg
Athens University of Economics and Business, Greece

Abstract. Graphs (or networks) constitute a dominant data structure and appear essentially in all forms of information (e.g., social and information networks, technological networks and networks from the areas of biology and neuroscience). A cornerstone issue in the analysis of such graphs is the detection and evaluation of communities (or clusters) - bearing multiple and diverse semantics. Typically, the communities correspond to groups of nodes that tend to be highly similar sharing common features, while nodes of different communities show low similarity. Detecting and evaluating the community structure of real-world graphs constitutes an essential task in several areas, with many important applications. For example, communities in a social network (e.g., Facebook, Twitter) correspond to individuals with increased social ties (e.g., friendship relationships, common interests). The goal of this tutorial is to present community detection and evaluation techniques as mining tools for real graphs. We present a thorough review of graph clustering and community detection methods, demonstrating their basic methodological principles. Special mention is made to the degeneracy (k-cores and extensions) approach for community evaluation, presenting also several case studies on real-world networks.

Extensions on Map-Reduce

Himanshu Gupta, L. Venkata Subramaniam,
and Sriram Raghavan

IBM Research, India
http://researcher.watson.ibm.com/researcher/view.php?person=in-higupta8
http://researcher.watson.ibm.com/researcher/view.php?person=in-lvsubram
http://researcher.watson.ibm.com/researcher/view.php?person=
in-sriramraghavan

Abstract. This tutorial will present an overview of various systems and algorithms which have extended the map-reduce framework to improve its performance. The tutorial will consist of four parts. The tutorial will start with an introduction of the map-reduce framework along-with its strengths and limitations. The goal of this tutorial is to explain how research has attempted to overcome these limitations.

The first part will look at systems which focus on (1) processing relational data on map-reduce and (2) on providing indexing support on map-reduce. The second part will then move on to systems which provide support for various classes of queries like join-processing, incremental computation, iterative and recurring queries etc. The third part will present an overview of systems which improve the performance of map-reduce framework in a variety of ways like skew-management, data-placement, reusing the results of a computation etc. The fourth part will finally look at various initiatives in this space currently underway within IBM across all global labs.

Similarity Search: Navigating the Choices for Similarity Operators

Deepak S. Padmanabhan and Prasad M. Deshpande

http://researcher.watson.ibm.com/researcher/view.php?person=
in-deepak.s.p
http://researcher.watson.ibm.com/researcher/view.php?person=
in-prasdesh

Abstract. With the growing variety of entities that have their presence on the web, increasingly sophisticated data representation and indexing mechanisms to retrieve relevant entities to a query are being devised. Though relatively less discussed, another dimension in retrieval that has recorded tremendous progress over the years has been the development of mechanisms to enhance expressivity in specifying information needs; this has been affected by the advancements in research on similarity operators. In this tutorial, we focus on the vocabulary of similarity operators that has grown from just a set of two operators, top-k and skyline search, as it stood in the early 2000s. Today, there are efficient algorithms to process complicated needs such as finding the top-k customers for a product wherein the customers are to be sorted based on the rank of the chosen product in their preference list. Arguably due to the complexity in the specification of new operators such as the above, uptake of such similarity operators has been low even though emergence of complex entities such as social media profiles warrant significant expansion in query expressivity. In this tutorial, we systematically survey the set of similarity operators and mechanisms to process them effectively. We believe that the importance of similarity search operators is immense in an era of when the web is populated with increasingly complex objects spanning the entire spectrum, though mostly pronounced in the social and e-commerce web.

Table of Contents – Part I

Web Mining, Modeling and Classification

Coupled Item-Based Matrix Factorization 1
    Fangfang Li, Guandong Xu, and Longbing Cao

A Lot of Slots – Outliers Confinement in Review-Based System 15
    Roberto Di Pietro, Marinella Petrocchi, and Angelo Spognardi

A Unified Model for Community Detection of Multiplex Networks 31
    Guangyao Zhu and Kan Li

Mining Domain-Specific Dictionaries of Opinion Words 47
    Pantelis Agathangelou, Ioannis Katakis, Fotios Kokkoras,
    and Konstantinos Ntonas

A Community Detection Algorithm Based on the Similarity
Sequence... 63
    Hongwei Lu, Qian Zhao, and Zaobin Gan

A Self-learning Clustering Algorithm Based on Clustering Coefficient ... 79
    MingJie Zhong, ZhiJun Ding, HaiChun Sun, and PengWei Wang

Detecting Hierarchical Structure of Community Members by Link
Pattern Expansion Method.. 95
    Fengjiao Chen and Kan Li

An Effective TF/IDF-Based Text-to-Text Semantic Similarity Measure
for Text Classification 105
    Shereen Albitar, Sébastien Fournier, and Bernard Espinasse

Automatically Annotating Structured Web Data Using a SVM-Based
Multiclass Classifier 115
    Daiyue Weng, Jun Hong, and David A. Bell

Mining Discriminative Itemsets in Data Streams 125
    Majid Seyfi, Shlomo Geva, and Richi Nayak

Modelling Visit Similarity Using Click-Stream Data:
A Supervised Approach... 135
    Deepak Pai, Abhijit Sharang, Meghanath Macha,
    and Shradha Agrawal

BOSTER: An Efficient Algorithm for Mining Frequent Unordered
Induced Subtrees 146
    Israt J. Chowdhury and Richi Nayak

Web Querying and Searching

Phrase Queries with Inverted + Direct Indexes 156
  Kiril Panev and Klaus Berberich

Ranking Based Activity Trajectory Search 170
  Wei Chen, Lei Zhao, Xu Jiajie, Kai Zheng, and Xiaofang Zhou

Topical Pattern Based Document Modelling and Relevance Ranking 186
  Yang Gao, Yue Xu, and Yuefeng Li

A Decremental Search Approach for Large Scale Dynamic
Ridesharing 202
  Ali Shemshadi, Quan Z. Sheng, and Wei Emma Zhang

Model-Based Search and Ranking of Web APIs across Multiple
Repositories 218
  Devis Bianchini, Valeria De Antonellis, and Michele Melchiori

Common Neighbor Query-Friendly Triangulation-Based Large-Scale
Graph Compression 234
  Liang Zhang, Chen Xu, Weining Qian, and Aoying Zhou

Continuous Monitoring of Top-k Dominating Queries over Uncertain
Data Streams.. 244
  Guohui Li, Changyin Luo, and Jianjun Li

Keyword Search over Web Documents Based on Earth Mover's
Distance 256
  Jiangang Ma, Quan Z. Sheng, Lina Yao, Yong Xu,
  and Ali Shemshadi

iPoll: Automatic Polling Using Online Search 266
  Thin Nguyen, Dinh Phung, Wei Luo, Truyen Tran,
  and Svetha Venkatesh

Web Recommendation and Personalization

Comparing the Predictive Capability of Social and Interest Affinity
for Recommendations... 276
  Alexandra Olteanu, Anne-Marie Kermarrec, and Karl Aberer

End-User Browser-Side Modification of Web Pages 293
  Oscar Díaz, Cristóbal Arellano, Iñigo Aldalur, Haritz Medina,
  and Sergio Firmenich

Mobile Phone Recommendation Based on Phone Interest 308
  Bozhi Yuan, Bin Xu, Tonglee Chung, Kaiyan Shuai, and Yongbin Liu

Two Approaches to the Dataset Interlinking Recommendation
Problem 324
    Giseli Rabello Lopes, Luiz André P. Paes Leme,
    Bernardo Pereira Nunes, Marco Antonio Casanova,
    and Stefan Dietze

Exploiting Perceptual Similarity: Privacy-Preserving Cooperative
Query Personalization 340
    Christoph Lofi and Christian Nieke

Identifying Explicit Features for Sentiment Analysis in Consumer
Reviews 357
    Nienke de Boer, Marijtje van Leeuwen, Ruud van Luijk,
    Kim Schouten, Flavius Frasincar, and Damir Vandic

Facet Tree for Personalized Web Documents Organization 372
    Róbert Móro, Mária Bieliková, and Roman Burger

Mobile Web User Behavior Modeling 388
    Bozhi Yuan, Bin Xu, Chao Wu, and Yuanchao Ma

Effect of Mood, Social Connectivity and Age in Online Depression
Community via Topic and Linguistic Analysis 398
    Bo Dao, Thin Nguyen, Dinh Phung, and Svetha Venkatesh

A Review Selection Method Using Product Feature Taxonomy 408
    Nan Tian, Yue Xu, and Yuefeng Li

Semantic Web

A Genetic Programming Approach for Learning Semantic Information
Extraction Rules from News 418
    Wouter IJntema, Frederik Hogenboom, Flavius Frasincar,
    and Damir Vandic

Ontology-Based Management of Conflicting Products in Pixel
Advertising 433
    Ferry Boon, Sabri Bouzidi, Raymond Vermaas, Damir Vandic,
    and Flavius Frasincar

Exploiting Semantic Result Clustering to Support Keyword Search
on Linked Data 448
    Ananya Dass, Cem Aksoy, Aggeliki Dimitriou, and
    Dimitri Theodoratos

Discovering Semantic Mobility Pattern from Check-in Data 464
    Ji Yuan, Xudong Liu, Richong Zhang, Hailong Sun,
    Xiaohui Guo, and Yanghao Wang

An Offline Optimal SPARQL Query Planning Approach to Evaluate
Online Heuristic Planners   480
    Achille Fokoue, Mihaela Bornea, Julian Dolby,
    Anastasios Kementsietsidis, and Kavitha Srinivas

Agents, Models and Semantic Integration in Support of Personal
eHealth Knowledge Spaces   496
    Haridimos Kondylakis, Dimitris Plexousakis, Vedran Hrgovcic,
    Robert Woitsch, Marc Premm, and Michael Schuele

Probabilistic Associations as a Proxy for Semantic Relatedness   512
    Shahida Jabeen, Xiaoying Gao, and Peter Andreae

A Hybrid Model for Learning Semantic Relatedness Using
Wikipedia-Based Features   523
    Shahida Jabeen, Xiaoying Gao, and Peter Andreae

An Ontology-Based Approach for Product Entity Resolution
on the Web   534
    Raymond Vermaas, Damir Vandic, and Flavius Frasincar

Author Index   545

Table of Contents – Part II

Social Online Networks

Predicting Elections from Social Networks Based on Sub-event
Detection and Sentiment Analysis 1
    Sayan Unankard, Xue Li, Mohamed Sharaf, Jiang Zhong,
    and Xueming Li

Sonora: A Prescriptive Model for Message Authoring on Twitter 17
    Pablo N. Mendes, Daniel Gruhl, Clemens Drews, Chris Kau,
    Neal Lewis, Meena Nagarajan, Alfredo Alba, and Steve Welch

A Fuzzy Model for Selecting Social Web Services 32
    Hamdi Yahyaoui, Mohammed Almulla, and Zakaria Maamar

Insights into Entity Name Evolution on Wikipedia 47
    Helge Holzmann and Thomas Risse

Assessing the Credibility of Nodes on Multiple-Relational Social
Networks 62
    Weishu Hu and Zhiguo Gong

Result Diversification for Tweet Search 78
    Makbule Gulcin Ozsoy, Kezban Dilek Onal,
    and Ismail Sengor Altingovde

WikipEvent: Leveraging Wikipedia Edit History for Event Detection ... 90
    Tuan Tran, Andrea Ceroni, Mihai Georgescu,
    Kaweh Djafari Naini, and Marco Fisichella

Feature Based Sentiment Analysis of Tweets in Multiple Languages 109
    Maike Erdmann, Kazushi Ikeda, Hiromi Ishizaki, Gen Hattori,
    and Yasuhiro Takishima

Incorporating the Position of Sharing Action in Predicting Popular
Videos in Online Social Networks 125
    Yi Long, Victor O.K. Li, and Guolin Niu

An Evolution-Based Robust Social Influence Evaluation Method in
Online Social Networks 141
    Feng Zhu, Guanfeng Liu, An Liu, Lei Zhao, and Xiaofang Zhou

A Framework for Linking Educational Medical Objects:
Connecting Web2.0 and Traditional Education 158
    Reem Qadan Al Fayez and Mike Joy

An Ensemble Model for Cross-Domain Polarity Classification
on Twitter 168
    Adam Tsakalidis, Symeon Papadopoulos, and Ioannis Kompatsiaris

A Faceted Crawler for the Twitter Service 178
    George Valkanas, Antonia Saravanou, and Dimitrios Gunopulos

Diversifying Microblog Posts 189
    Marios Koniaris, Giorgos Giannopoulos, Timos Sellis,
    and Yiannis Vassiliou

Software Architectures and Platforms

MultiMasher: Providing Architectural Support and Visual Tools
for Multi-device Mashups 199
    Maria Husmann, Michael Nebeling, Stefano Pongelli,
    and Moira C. Norrie

MindXpres: An Extensible Content-Driven Cross-Media Presentation
Platform 215
    Reinout Roels and Beat Signer

Open Cross-Document Linking and Browsing Based on a Visual Plug-in
Architecture 231
    Ahmed A.O. Tayeh and Beat Signer

Cost-Based Join Algorithm Selection in Hadoop 246
    Jun Gu, Shu Peng, X. Sean Wang, Weixiong Rao, Min Yang,
    and Yu Cao

Consistent Freshness-Aware Caching for Multi-Object Requests 262
    Meena Rajani, Uwe Röhm, and Akon Dey

ϵ-*Controlled-Replicate*: An Improved *Controlled-Replicate* Algorithm
for Multi-way Spatial Join Processing On Map-Reduce 278
    Himanshu Gupta and Bhupesh Chawda

REST as an Alternative to WSRF: A Comparison Based
on the WS-Agreement Standard 294
    Florian Feigenbutz, Alexander Stanik, and Andreas Kliem

Web Technologies and Frameworks

Enabling Cross-Platform Mobile Application Development:
A Context-Aware Middleware 304
    Achilleas P. Achilleos and Georgia M. Kapitsaki

GEAP: A Generic Approach to Predicting Workload Bursts
for Web Hosted Events 319
    Matthew Sladescu, Alan Fekete, Kevin Lee, and Anna Liu

High-Payload Image-Hiding Scheme Based on Best-Block Matching
and Multi-layered Syndrome-Trellis Codes......................... 336
    Tao Han, Jinlong Fei, Shengli Liu, Xi Chen, and Zhu Yuefei

Educational Forums at a Glance: Topic Extraction and Selection 351
    Bernardo Pereira Nunes, Ricardo Kawase, Besnik Fetahu,
    Marco Antonio Casanova,
    and Gilda Helena Bernardino B. de Campos

PDist-RIA Crawler: A Peer-to-Peer Distributed Crawler for Rich
Internet Applications 365
    Seyed M. Mirtaheri, Gregor V. Bochmann,
    Guy-Vincent Jourdan, and Iosif-Viorel Onut

Understand the City Better: Multimodal Aspect-Opinion
Summarization for Travel 381
    Ting Wang and Changqing Bai

Event Processing over a Distributed JSON Store: Design and
Performance.. 395
    Miki Enoki, Jérôme Siméon, Hiroshi Horii, and Martin Hirzel

Cleaning Environmental Sensing Data Streams Based on Individual
Sensor Reliability 405
    Yihong Zhang, Claudia Szabo, and Quan Z. Sheng

Managing Incentives in Social Computing Systems with PRINGL 415
    Ognjen Scekic, Hong-Linh Truong, and Schahram Dustdar

Consumer Monitoring of Infrastructure Performance in a Public
Cloud 425
    Rabia Chaudry, Adnene Guabtni, Alan Fekete, Len Bass,
    and Anna Liu

Business Export Orientation Detection through Web Content
Analysis 435
    Desamparados Blazquez, Josep Domenech, Jose A. Gil,
    and Ana Pont

Web Innovation and Applications

Towards Real Time Contextual Advertising 445
    Abhimanyu Panwar, Iosif-Viorel Onut, and James Miller

On String Prioritization in Web-Based User Interface Localization   460
    Luis A. Leiva and Vicent Alabau

Affective, Linguistic and Topic Patterns in Online Autism
Communities   474
    Thin Nguyen, Thi Duong, Dinh Phung, and Svetha Venkatesh

A Product-Customer Matching Framework for Web 2.0 Applications . . .   489
    Qiangqiang Kang, Zhao Zhang, Cheqing Jin, and Aoying Zhou

Rapid Development of Interactive Applications Based on Online Social
Networks   505
    Ángel Mora Segura, Juan de Lara, and Jesús Sánchez Cuadrado

Introducing the Public Transport Domain to the Web of Data   521
    Christine Keller, Sören Brunk, and Thomas Schlegel

Measuring and Mitigating Product Data Inaccuracy in Online
Retailing   531
    Runhua Xu and Alexander Ilic

Challenge

WISE 2014 Challenge: Multi-label Classification of Print Media Articles
to Topics   541
    Grigorios Tsoumakas, Apostolos Papadopoulos, Weining Qian,
    Stavros Vologiannidis, Alexander D'yakonov, Antti Puurula,
    Jesse Read, Jan Švec, and Stanislav Semenov

Author Index   549

Coupled Item-Based Matrix Factorization

Fangfang Li, Guandong Xu, and Longbing Cao

Advanced Analytics Institute, University of Technology, Sydney, Australia
Fangfang.Li@student.uts.edu.au,
{Guandong.Xu,Longbing.Cao}@uts.edu.au

Abstract. The essence of the challenges *cold start* and *sparsity* in Recommender Systems (RS) is that the extant techniques, such as Collaborative Filtering (CF) and Matrix Factorization (MF), mainly rely on the user-item rating matrix, which sometimes is not informative enough for predicting recommendations. To solve these challenges, the objective item attributes are incorporated as complementary information. However, most of the existing methods for inferring the relationships between items assume that the attributes are "independently and identically distributed (iid)", which does not always hold in reality. In fact, the attributes are more or less coupled with each other by some implicit relationships. Therefore, in this paper we propose an attribute-based coupled similarity measure to capture the implicit relationships between items. We then integrate the implicit item coupling into MF to form the Coupled Item-based Matrix Factorization (CIMF) model. Experimental results on two open data sets demonstrate that CIMF outperforms the benchmark methods.

1 Introduction

Recommender Systems (RS) are proposed to help users tackle information overload by suggesting potentially interesting items to users [15]. The main challenges in RS now are *cold start* and *sparsity* problems. The essence behind these challenges is that traditional recommendation techniques such as Collaborative Filtering (CF) or Matrix Factorization (MF) normally rely on the user-item rating matrix only, which sometimes is not informative enough for making recommendations.

To solve these challenges, researchers attempt to leverage complementary information, such as social friendships, in RS. For instance, many social recommender systems [13] [12] [14] [27] [28] have been proposed utilizing social friendships. To some extent, the social friendships in RS have been well studied but most web sites do not have social mechanisms, therefore, several researchers are currently trying to incorporate users' or items' attributes to improve the performance of recommendation. For example, some have tried to estimate the latent factors through considering the attributes [19] [7] [16] [2] or topic information of users and items [3] [26] [8] for latent factor models. Nevertheless, most of the existing methods assume that the attributes are independent. In reality, however this assumption does not always hold and there exist complex coupling relations between instances and attributes. For example, in Table 1, the "Director", "Actor", and "Genre" attributes in movies are often coupled together and influence each other. Therefore, in this paper, we deeply analyze the couplings between items to capture their implicit relationships, based on the attribute information.

B. Benatallah et al. (Eds.): WISE 2014, Part I, LNCS 8786, pp. 1–14, 2014.
© Springer International Publishing Switzerland 2014

Table 1. A Toy Example

Director	Scorsese	Coppola	Hitchcock	Hitchcock
Actor	De Niro	De Niro	Stewart	Grant
Genre	Crime	Crime	Thriller	Thriller
	God Father	Good Fellas	Vertigo	N by NW
u_1	1	3	5	4
u_2	4	2	1	5
u_3	-	2	-	4

We know that the implicit relationships can be aggregated from the similarity of attribute values for all the attributes. From the perspective of "iid" assumption, different attribute values are independent, and one attribute value will not be influenced by others. However, if we disclose the "iid" assumption, we will observe that one attribute value will also be dependent on other values of the same attribute. Specifically, two attribute values are similar if they present the analogous frequency distribution on one attribute, which leads to another so-called intra-coupled similarity within an attribute. For example, two directors "Scorsese" and "Coppola" are considered similar because they appear with the same frequency. On the other hand, the similarity of two attributes values is dependent on other attribute values from different attributes, for example, two directors' relationship is dependent on "Actor" and "Genre" attributes over all the movies. This dependent relation is called the inter-coupled similarity between attributes. We believe that the intra and inter-coupled similarities disclosing the "iid" assumption should simultaneously contribute to analysing the relationships between items, namely item coupling.

Rating preferences have been well studied but the relationships between items, especially implicit relations, are still far away from being successfully incorporated into the MF model. Therefore, in this paper, we propose a Coupled Item-based MF (CIMF) model incorporating implicit item couplings through a learning algorithm via regularization on implicit and explicit information. After accommodating the implicit relationships between items and users' preferences on items into a unified learning model, we can predict more satisfactory recommendations, even for new users/items or when the rating matrix is very sparse. The motivation for incorporating such couplings into RS is to solve the *cold start* and *sparsity* challenges in RS by leveraging the items' implicit objective couplings and users' subjective rating preferences on items. When we have ample rating data, the user-item rating matrix is mainly applied for recommendation. However, for new users or items of RS or for a very sparse rating matrix, implicit item coupling would be mainly exploited for making recommendations.

The contributions of the paper are as follows:

- We propose a NonIID-based method to capture the implicit relationships between items, namely item coupling, based on the their objective attribute information.
- We propose the CIMF model which integrates the item coupling and users' subjective rating preferences into a matrix factorization learning model.
- We conduct experiments to evaluate the effectiveness of our proposed CIMF model.

The rest of the paper is organized as follows. Section 2 presents the related work. In Section 3, we formally state the recommendation and couplings problems. Section 4 first analyzes the couplings in RS, after which it details the coupled Item-based MF model integrating the couplings together. Experimental results and the analysis are presented in Section 5. The paper is concluded in the final section.

2   Related Work

The approaches related to our work in recommender systems include collaborative filtering and content-based techniques.

2.1   Collaborative Filtering

Collaborative filtering (CF)[22] [21] [6] is one of the most successful approaches, taking advantage of user rating history to predict users' interests. User-based CF and item-based CF are mainly involved in the CF method. The basic idea of user-based CF is to recommend interesting items to the active user according to the interests of the other users with whom they have close relationships. Similarly, item-based CF tries to recommend to the active user potentially interesting items which have close similarities with the historical items that the active user likes. As one of the most accurate single models for collaborative filtering, matrix factorization (MF)[9] [10] is a latent factor model which is generally effective at estimating the overall structure that relates simultaneously to most items. The MF approach tries to decompose the rating matrix to the user intent matrix and the item intent matrix. Then, the estimated rating is predicted by the multiplication of the two decomposed intent matrices.

Although there had been wide adoption of this approach in many real applications, e.g., Amazon, the effect of CF is sharply weakened in the case of new users or items and for a very sparse rating matrix. This is partly because when the rating matrix is very sparse, for new users or items, it is extremely difficult to determine the relationships between users or items. This limitation partly motivates us to consider implicit item coupling to enhance the effectiveness of recommendations.

2.2   Content-Based Methods

Content-based techniques are another successful method by which to recommend relevant items to users by matching the users' personal interests to descriptive item information [1] [17] [18]. Generally, content-based methods are able to cope with the sparsity problem, however, they often assume an item's attributes are "iid" which does not always hold in reality. Actually, several research outcomes [5] [23] [24] [25] have been proposed to handle these challenging issues. To the best of our knowledge, in relation to RS, there is only one paper [29] which applies a coupled clustering method to group the items, then exploits CF to make recommendations. But from the perspective of RS, this paper does not fundamentally disclose the "iid" assumption for items. This motivates us to analyze the intrinsic relationships from different levels to unfold the assumption.

3   Problem Statement

A large number of user and item sets with specific attributes can be organized by a triple $S =< U, S_O, f >$, where $U = \{u_1, u_2, ..., u_n\}$ is a nonempty finite set of users, $S_O =< O, A, V, g >$ describes the items' attribute space. Among S_O, $O = \{o_1, o_2, ..., o_m\}$ is a nonempty finite set of items, $A = \{A_1, ..., A_M\}$ is a finite set of attributes for items; $V = \cup_{j=1}^{J} V_j$ is a set of all attribute values for items, in which V_j is the set of attribute values of attribute $A_j (1 \leq j \leq J)$, V_{ij} is the attribute value of attribute A_j for item o_i, and $g = \wedge_{j=1}^{M} g_j (g_j : U \rightarrow V_j)$ is an mapping function set which describes the relationships between attribute values and items. In the triple $S =< U, S_O, f >$, $f(u_i, o_j) = r_{ij}$ expresses the subjective rating preference on item o_j for user u_i. Through the mapping function f, user rating preferences on items are then converted into a user-item matrix R, with n rows and m columns. Each element r_{ij} of R represents the rating given by user u_i on item o_j. For instance, Table 1 consists of three users $U = \{u_1, u_2, u_3\}$ and four items $O=\{God\ Father,\ Good\ Fellas,\ Vertigo,\ N\ by\ NW\}$. The items have attributes $A=\{Director,\ Actor,\ Genre\}$, and attribute values $V_3 = \{Crime, Thriller\}$. The mapping functions are $g_3(Vertigo) = Thriller$ and $f(u_2, Vertigo) = 1$.

As mentioned, the extant similarity methods for computing the implicit relationships within items assume that the attributes are independent of each other. However, all the attributes should be coupled together and further influence each other. The couplings between items are illustrated in Fig.1, which shows that within an attribute A_j, there is dependence relation between values V_{lj} and V_{mj} ($l \neq m$), while a value V_{li} of an attribute A_i is further influenced by the values of other attributes A_j ($j \neq i$). For example, attributes A_1, A_3, ... to A_J all more or less influence the values of V_{12} to V_{n2} of attribute A_2.

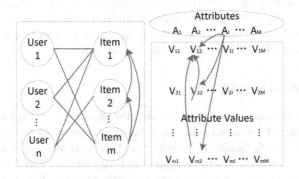

Fig. 1. Item Couplings in Recommender Systems

In order to capture the implicit item coupling and disclose the "iid" assumption, we first introduce several basic concepts as follows.

Given an attribute space $S_O =<O, A, V, g>$, the **Intra-coupled Attribute Value Similarity (IaAVS)** between values x and y of attribute A_j for items is defined as:

$$\delta_j^{Ia}(x,y) = \frac{|g_j(x)|.|g_j(y)|}{|g_j(x)| + |g_j(y)| + |g_j(x)|.|g_j(y)|} \tag{1}$$

where $g_j(x) = \{o_i|V_{ij} = x, 1 \leq j \leq M, 1 \leq i \leq n\}$ is the subset of O with corresponding attribute A_j having attribute value x, and $|g_j(x)|$ is the size of the subset.

The influence of attribute value y of attribute A_k for attribute value x of attribute A_j can be calculated by:

$$P_{j|k}(y|x) = \frac{|g_{j,k}(x,y)|}{|g_j(x)|} \tag{2}$$

where $g_{j,k}(x,y) = \{o_i|(V_{ij} = x) \wedge (V_{ik} = y), 1 \leq j, k \leq M, 1 \leq i \leq n\}$.

Given an attribute space $S_O =<O, A, V, g>$, the **Inter-coupled Relative Similarity (IRS)** between attribute values x and y of attribute A_j based on another attribute A_k is:

$$\delta_{j|k}(x,y) = \sum_{w \in \cap} min\{P_{k|j}(w|x), P_{k|j}(w|y)\} \tag{3}$$

where w is an attribute value for attribute A_k of items with attribute A_j having both values x and y.

Given an attribute space $S_O =<O, A, V, g>$, the **Inter-coupled Attribute Value Similarity (IeAVS)** between attribute values x and y of attribute A_j for item set O is:

$$\delta_j^{Ie}(x,y) = \sum_{k=1,k\neq j}^{n} \gamma_k \delta_{j|k}(x,y) \tag{4}$$

where γ_k is the weight parameter for attribute A_k, $\sum_{k=1,k\neq j}^{n} \gamma_k = 1$, $\gamma_k \in [0,1]$, and $\delta_{j|k}(x,y)$ is inter-coupled relative similarity.

Based on IaAVS and IeAVS, the **Coupled Attribute Value Similarity (CAVS)** between attribute values x and y of attribute A_j is defined as follows.

$$\delta_j^{A}(x,y) = \delta_j^{Ia}(x,y) * \delta_j^{Ie}(x,y) \tag{5}$$

4   Coupled Item-Based MF Model

In this section, we introduce the coupled Item-based MF model as in 2. We first construct the objective attribute spaces of items for computing item coupling which integrates $IaAVS$ and $IeAVS$. Then, we incorporate item coupling and users' rating preferences together into MF model. The CIMF model has the following advantages: (1) the item coupling discloses the common "iid" assumption and consider the real world non-iid characteristic, (2) the item coupling actually reflects the implicit relationships within items, which is able to partly overcome the problem of lacking informative rating knowledge, (3) users' subjective rating preference is taking the leading role in the learning model.

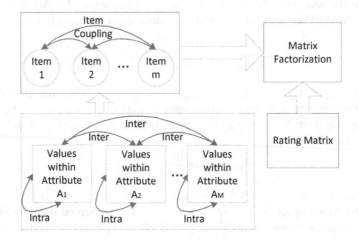

Fig. 2. Coupled Item-based MF Model

4.1   Item Coupling

As mentioned above, item coupling should reflect the implicit relationships between different items. For two items described by the attribute space $S_O = <O, A, V, g>$, the Coupled Item Similarity (*CIS*) is defined to measure the similarity between items.

Formally, given item attribute space $S_O = <O, A, V, g>$, the **Coupled Item Similarity (*CIS*)** between two items o_i and o_j is defined as follows.

$$CIS(o_i, o_j) = \sum_{k=1}^{J} \delta_k^A (V_{ik}, V_{jk}) \qquad (6)$$

where V_{ik} and V_{jk} are the values of attribute j for items o_i and o_j, respectively; and δ_k^A is Coupled Attribute Value Similarity.

From this definition, we clearly see that the intra-couplings between values within an attribute and inter-couplings between attributes are incorporated for measuring item coupling which partly helps to uncover the intrinsic relations within items rather than considering them independently.

4.2   User-Item Coupling

Different from item coupling which is computed from items' objective content, user-item coupling should reflect users' interests on items and disclose an empirical, direct and explicit indication of the couplings between users and items. Simply, based on the rating data, the rating preference $f(u_i, o_j) = r_{ij}$ of user u_i on item o_j can be directly considered as the user-item coupling. As we know, the most existing recommendation methods such as CF and MF are based on the rating preferences and the user-item coupling actually has been well studied.

4.3   Coupled Item-Based MF Model

Traditionally, the matrix of predicted ratings $\hat{R} \in \mathbb{R}^{n \times m}$, where n, m respectively denote the number of users and the number of items, can be modeled as:

$$\hat{R} = r_m + PQ^T \tag{7}$$

with matrices $P \in \mathbb{R}^{n \times d}$ and $Q \in \mathbb{R}^{m \times d}$, where d is the rank (or dimension of the latent space) with $d \leq n, m$, and $r_m \in \mathbb{R}$ is a global offset value. Through Eqn. 7, the prediction task of matrix $\hat{R}$ is transformed to compute the mapping of users and items to factor matrices P and Q. Once this mapping is completed, $\hat{R}$ can be easily reconstructed to predict the rating given by one user to an item by using Eqn. 7. To avoid over-fitting, the regularization factors $\|P\|$ and $\|Q\|$ are added into the loss function to penalize over influence by observations.

On top of the traditional MF method, we propose a novel CIMF model which takes not only the rating preferences, but also item coupling into account. The learning procedure is constrained two-fold: the learned rating values should be as close as possible to the observed rating values, and the predicted item profiles should be similar to their neighbourhoods as well, which are derived from their implicit coupling information. More specifically, item coupling is incorporated by adding an additional regularization factor in the optimization step. Then, the computation of the mapping can be similarly optimized by minimizing the regularized squared error. The objective function is amended as Eqn. 8.

$$
\begin{aligned}
L = \frac{1}{2} \sum_{(u,o_i) \in K} \left(R_{u,o_i} - \hat{R}_{u,o_i} \right)^2 + \frac{\lambda}{2} \left(\|Q_i\|^2 + \|P_u\|^2 \right) \\
+ \frac{\alpha}{2} \sum_{all(o_i)} \left\| Q_i - \sum_{o_j \in \mathbb{N}(o_i)} CIS(o_i, o_j) Q_j \right\|^2
\end{aligned}
\tag{8}
$$

In the objective function, the item coupling and users' rating preferences are integrated together. Specifically, the first part reflects the subjective rating preferences and the last part reflects the item coupling. This means the users' rating preferences and item coupling act jointly to make recommendations. In addition, another distinct advantage is that, when we do not have ample rating data, it is still possible to make satisfactory recommendations via leveraging the implicit coupling information.

We optimize the above objective function by minimizing L through the gradient decent approach:

$$\frac{\partial L}{\partial P_u} = \sum_{o_i} I_{u,o_i} (r_m + P_u Q_i^T - R_{u,o_i}) Q_i + \lambda P_u \tag{9}$$

$$\frac{\partial L}{\partial Q_i} = \sum_u I_{u,o_i}(r_m + P_u Q_i^T - R_{u,o_i})P_u + \lambda Q_i +$$

$$\alpha(Q_i - \sum_{o_j \in \mathbb{N}(o_i)} CIS(o_i, o_j)Q_j) - \tag{10}$$

$$\alpha \sum_{o_j : o_i \in \mathbb{N}(o_j)} CIS(o_j, o_i)(Q_j - \sum_{o_k \in \mathbb{N}(o_j)} CIS(o_j, o_k)Q_k)$$

where I_{u,o_i} is an logical function indicating whether the user has rated item o_i or not. $CIS(o_i, o_j)$ is the coupled similarity of items o_i and o_j. $\mathbb{N}(o_i)$ represent the item neighborhood.

The optimum matrices P and Q can be computed by the above gradient descent approach. Generally, the CIMF model starts by computing item coupling based on the objective content, then commences an iteration process for optimizing P and Q until convergence, according to Eqn. 9 and 10. Once P and Q are learned, the ratings for user-item pairs (u, o_i) can be easily predicted by Eqn. 7.

5 Experiments and Results

In this section, we evaluate our proposed model and compare it to the existing approaches respectively, using the MovieLens and Book-Crossing [30] data sets.

5.1 Data Sets

The MovieLens data set has been widely explored in RS research in the last decade. The MovieLens 1M data set consists of 1,000,209 anonymous ratings of approximately 3,900 movies made by 6,040 MovieLens users who joined MovieLens in 2000. The ratings are made on a 5-star scale and each user has at least 20 ratings. The movies have titles provided by the IMDB (including year of release) and a special "genre" attribute which is applied to compute the item couplings.

Similarly, collected by Cai-Nicolas Ziegler from the Book-Crossing community, the Book-Crossing data set involves 278,858 users with demographic information providing 1,149,780 ratings on 271,379 books. The ratings range from 1 to 10 and the books' "book-author", "year of publication" and "publisher" have been used to form the item couplings.

5.2 Experimental Settings

The 5-fold cross validation is performed in our experiments. In each fold, we have 80% of data as the training set and the remaining 20% as the testing set. Here we use Root Mean Square Error (RMSE) and Mean Absolute Error (MAE) as evaluation metrics:

$$RMSE = \sqrt{\frac{\sum_{(u,o_i)|R_{test}}(r_{u,o_i} - \hat{r}_{u,o_i})^2}{|R_{test}|}} \tag{11}$$

$$MAE = \frac{\sum_{(u,o_i)|R_{test}} |r_{u,o_i} - \hat{r}_{u,o_i}|}{|R_{test}|} \tag{12}$$

where R_{test} is the set of all pairs (u, o_i) in the test set.

To evaluate the performance of our proposed CIMF, we consider five baseline approaches based on a user-item rating matrix: (1) the basic probabilistic matrix factorization (PMF) approach [20]; (2) the singular value decomposition (RSVD) [4] method; (3) the implicit social matrix factorization (ISMF) [11] model which incorporates implicit social relationships between users and between items;(4) user-based CF (UBCF) [22]; and (5) item-based CF (IBCF) [6].

The above five baselines only consider users' rating preferences on items but ignore item coupling. In order to completely evaluate our method, we also compare the following three models PSMF, CSMF and JSMF, which respectively augment MF using the same strategy as in 8 with the Pearson Correlation Coefficient, and the Cosine and Jaccard similarity measures to compute items' implicit relationships, based on the objective attribute information.

5.3   Experimental Analysis

We respectively evaluate the effectiveness of the proposed CIMF model by comparing it with the above methods.

Superiority over MF Methods. It is well known that MF methods are popular and successful in RS, hence, in this experiment, we compare our proposed CIMF model with the existing MF methods. In Table 2, different latent dimensions regarding MAE and RMSE metrics on MovieLens are considered to evaluate the proposed CIMF model. In general, the experiments demonstrate that our proposed CIMF outperforms the other three MF baselines. Specifically, when the latent dimension is set to 10, 50 and 100, in terms of MAE, our proposed CIMF can reach improvements of 16.33%, 17.53% and 27.85% compared with the PMF method. Regarding RMSE, the improvements reach up to 48.35%, 55.00% and 70.53%. Similarly, CIMF can respectively improve by 6.02%, 9.89%, 20.74% regarding MAE, and 26.76%, 32.84%, 57.76% regarding RMSE over the RSVD approach. In addition to basic comparisons, we also compare our CIMF model with the latest research outcome ISMF which utilizes the implicit relationships between users and items based on the rating matrix by Pearson similarity. From the experimental result, we can see that CIMF can respectively improve by 11.55%, 10.89% and 21.23% regarding MAE, and by 41.07%, 35.52%, 58.60% regarding RMSE for different dimensions, 10, 50 and 100.

Similarly, we depict the effectiveness comparisons with respect to different methods on the Book-Crossing data set in Table 2. We can clearly see that our proposed CIMF method outperforms all its counterparts in terms of MAE and RMSE. Specifically, when the latent dimension is set to 10, 50 and 100, in terms of MAE, our proposed CIMF can reach improvements of 3.72%, 3.65% and 3.64% compared with the PMF method. Regarding RMSE, CIMF also improves slightly by 0.85%, 0.80% and 0.69%. Similarly, CIMF can respectively improve by 3.71%, 3.68%, 3.68% regarding MAE, and

Table 2. MF Comparisons on MovieLens and Book-Crossing Data Sets

Data Set	Dim	Metrics	PMF (Improve)	ISMF (Improve)	RSVD (Improve)	CIMF
MovieLens	100D	MAE	1.1787(27.85%)	1.1125 (21.23%)	1.1076 (20.74%)	**0.9002**
		RMSE	1.7111 (70.53%)	1.5918 (58.60%)	1.5834 (57.76%)	**1.0058**
	50D	MAE	1.1852 (17.53%)	1.1188 (10.89%)	1.1088 (9.89%)	**1.0099**
		RMSE	1.8051 (55.00%)	1.6103 (35.52%)	1.5835 (32.84%)	**1.2551**
	10D	MAE	1.2129 (16.33%)	1.1651 (11.55%)	1.1098 (6.02%)	**1.0496**
		RMSE	1.8022 (48.35%)	1.7294 (41.07%)	1.5863 (26.76%)	**1.3187**
Book-Crossing	100D	MAE	1.5127 (3.64%)	1.5102 (3.39%)	1.5131 (3.68%)	**1.4763**
		RMSE	3.7455 (0.69%)	3.7397 (0.11%)	3.7646 (2.60%)	**3.7386**
	50D	MAE	1.5128 (3.65%)	1.5100 (3.37%)	1.5131 (3.68%)	**1.4763**
		RMSE	3.7452 (0.80%)	3.7415 (0.43%)	3.7648 (2.76%)	**3.7372**
	10D	MAE	1.5135 (3.72%)	1.5107 (3.44%)	1.5134 (3.71%)	**1.4763**
		RMSE	3.7483 (0.85%)	3.7440 (0.42%)	3.7659 (2.61%)	**3.7398**

2.61%, 2.76%, 2.60% regarding RMSE over the RSVD approach. In addition, from the experimental results, compared with the latest research outcome ISMF, we can see that CIMF can respectively improve by 3.44%, 3.37%, 3.39% regarding MAE, and by 0.42%, 0.43%, 0.11% regarding RMSE for different dimensions, 10, 50 and 100.

Based on the experimental results on the MovieLens and Book-Crossing data sets, we can conclude that our CIMF method not only outperforms the traditional MF methods PMF and SVD, but also performs better than the state-of-the-art model ISMF in terms of MAE and RMSE metrics. Furthermore, the prominent improvements are the result of considering item couplings.

Table 3. CF Comparisons on MovieLens and Book-Crossing Data Sets

Data Set	Metrics	UBCF (Improve)	IBCF (Improve)	CIMF
MovieLens	MAE	0.9027 (0.25%)	0.9220 (2.18%)	**0.9002**
	RMSE	1.0022 (-0.36%)	1.1958 (19.00%)	**1.0058**
Book-Crossing	MAE	1.8064 (33.01%)	1.7865 (31.02%)	**1.4763**
	RMSE	3.9847 (24.61%)	3.9283 (18.97%)	**3.7386**

Superiority over CF Methods. In addition to the MF methods, we also compare our proposed CIMF model with two different CF methods, UBCF and IBCF. In this experiment, we fix the latent dimension to 100 for our proposed CIMF model. On the MovieLens data set, the results in Table 3 indicate that CIMF can respectively improve by 0.25% and 2.18% regarding MAE. Regarding RMSE, CIMF can improve by 19.00% compared with IBCF, and slightly decreases by 0.36% compared with UBCF but it is still comparable. Similarly, on the Book-Crossing data set, the results show that CIMF can respectively reach improvements of 33.01%, 31.02% regarding MAE, and 24.61%, 18.97% regarding RMSE compared with UBCF and IBCF. Therefore, this experiment clearly demonstrates that our proposed CIMF performs better than the traditional CF methods, UBCF and IBCF. The consideration of item couplings in RS contribute to these improvements.

Superiority over Hybrid Methods. In order to demonstrate the effectiveness of our proposed model, we compare it with three different hybrid methods, PSMF, CSMF and JSMF, which respectively augment MF with the Pearson Correlation Coefficient, and the Cosine and Jaccard similarity measures.

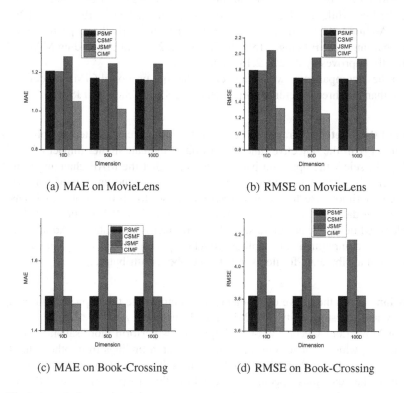

(a) MAE on MovieLens   (b) RMSE on MovieLens

(c) MAE on Book-Crossing   (d) RMSE on Book-Crossing

Fig. 3. Superiority over Hybrid Methods on MovieLens and Book-Crossing Data Sets

From the results shown in Fig. 3 on the MovieLens data set, generally we can clearly see that the coupled similarity method CIMF largely outperforms the three different comparisons with PSMF, CSMF and JSMF in terms of MAE and RMSE. Specifically, on the MovieLens data set for three different dimensions 10, 50 and 100, CIMF can respectively reach an improvement of 47.87%, 44.72% and 68.72% regarding RMSE compared to PSMF. In terms of MAE, CIMF also can increase by 15.79%, 16.14% and 26.48% compared to PSMF. Similarly, compared to CSMF on the MovieLens data set, CIMF can improve by 47.22%, 43.57% and 67.14% regarding RMSE, while for MAE, the improvement can reach up to 15.49%, 15.38% and 26.03%. Additionally, CIMF also performs better than JSMF, the respective improvements regarding RMSE being 74.18%, 70.02% and 93.47%, while regarding MAE, CIMF also improves by 23.23%, 23.63% and 34.48%.

On the Book-Crossing data set, the results in Fig. 3 also indicate that the coupled similarity method CIMF constantly performs better than corresponding comparison methods regarding RMSE and MAE. Specifically, for three different dimensions 10, 50 and 100, CIMF can respectively reach an improvement of 7.91%, 8.10% and 8.01% regarding RMSE compared to PSMF. In terms of MAE, CIMF also can slightly improve by 2.25%, 2.21% and 2.22%, compared to PSMF. Similarly, compared to CSMF, on the MovieLens data set, CIMF can improve by 44.82%, 44.24% and 43.13% regarding RMSE, while for MAE, the improvement can reach up to 19.31%, 19.62% and 19.75%. Additionally, CIMF also performs better than JSMF, the respective improvements regarding RMSE being 8.15%, 8.4% and 8.22%, while regarding MAE, CIMF also slightly improves by 2.22%, 2.18% and 2.16%.

From these comparisons, we can conclude that our proposed CIMF model is more effective than the three different hybrid methods, PSMF, CSMF and JSMF.

Impact of Parameters. Parameter α controls the influence of the item couplings. Bigger value of α in the objective function of Eqn. 8 indicate higher impact of the item coupling. To select the optimum parameters, we depict the MAE changing trends of CIMF methods when α is ranged in [0,1]. Fig. 4 shows the impacts of parameter α when neighborhood size for items is respectively set to 10 or 30 on Movielens and Bookcrossing data sets. Experimental results show that α=0.2 are proper values for Movielens, while α=1.0 are more suitable for Bookcrossing data set. Additionally, in this paper, for computing item coupling, we set the parameter $\gamma_k = \frac{1}{n-1}$ which controls the weight of attribute A_k for items, n is the number of attributes.

Discussion. From the above experiments, we demonstrate the impact of our coupled similarity and the superiority over MF and CF methods. Also, we notice that sometimes on MovieLens CF or SVD-based approaches perform more satisfactorily than PMF method, which violates the well-known research findings from others that PMF usually performs better than SVD or CF methods. This result inspires us to meditate on the data characteristics, specifically we believe that the performance of the recommendation methods might be closely dependent on the data characteristics in addition

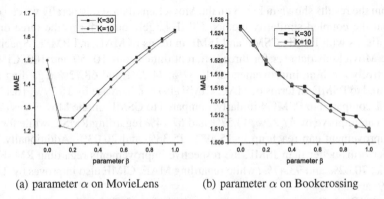

(a) parameter α on MovieLens   (b) parameter α on Bookcrossing

Fig. 4. Parameter Comparison

to approaches themselves. For example, when the data sets largely follow Gaussian distribution, PMF method which assumes that the rating matrix is Gaussian distribution may be more suitable for recommendation. Otherwise, other methods may get a better result. Therefore, we may need to incorporate the data characteristics into our model, we believe that the performance can be further enhanced through this consideration.

6 Conclusion

The significant implicit information within items was studied for solving the *cold start* and data *sparsity* challenges in RS. To capture the implicit information, a new coupled similarity method based on items' subjective attribute spaces is proposed. The coupled similarity method discloses the traditional "iid" assumption and deeply analyzes the intrinsic relationships within items. Furthermore, a coupled item-based matrix factorization model is proposed, which incorporates the implicit relations within items and the explicit rating information. The experiments conducted on the real data sets demonstrate the superiority of the proposed coupled similarity and CIMF model. Moreover, the experiments indicate that the implicit item relationships can be effectively applied in RS. Our work also provides in-depth analysis for the implicit relations within items and greatly extends the previous matrix factorization model for RS. Other aspects for enhancing our recommendation framework such as data characteristics or implicit relationships between users, will be investigated in the future.

Acknowledgments. This work is sponsored in part by Australian Research Council Discovery Grants (DP1096218 and DP130102691) and ARC Linkage Grant (LP100200774).

References

1. Adomavicius, G., Tuzhilin, A.: Toward the next generation of recommender systems: A survey of the state-of-the-art and possible extensions. IEEE Transactions on Knowledge and Data Engineering 17(6), 734–749 (2005)
2. Agarwal, D., Chen, B.-C.: Regression-based latent factor models. In: KDD, pp. 19–28 (2009)
3. Agarwal, D., Chen, B.-C.: fLDA: matrix factorization through latent dirichlet allocation. In: WSDM, pp. 91–100 (2010)
4. Alter, O., Brown, P.O., Botstein, D.: Singular value decomposition for genome-wide expression data processing and modeling. Proceedings of the National Academy of Sciences (PNAS) 97(18) (August 2000)
5. Cao, L., Ou, Y., Yu, P.S.: Coupled behavior analysis with applications. IEEE Trans. Knowl. Data Eng. 24(8), 1378–1392 (2012)
6. Deshpande, M., Karypis, G.: Item-based top-n recommendation algorithms. ACM Trans. Inf. Syst. 22(1), 143–177 (2004)
7. Gantner, Z., Drumond, L., Freudenthaler, C., Rendle, S., Schmidt-Thieme, L.: Learning attribute-to-feature mappings for cold-start recommendations. In: ICDM, pp. 176–185 (2010)
8. Hu, Y., Koren, Y., Volinsky, C.: Collaborative filtering for implicit feedback datasets. In: Proceedings of the 2008 Eighth IEEE International Conference on Data Mining, ICDM 2008, pp. 263–272. IEEE Computer Society, Washington, DC (2008)

9. Koren, Y.: Factorization meets the neighborhood: a multifaceted collaborative filtering model. In: Li, Y., Liu, B., Sarawagi, S. (eds.) KDD, pp. 426–434. ACM (2008)
10. Koren, Y., Bell, R.M., Volinsky, C.: Matrix factorization techniques for recommender systems. IEEE Computer 42(8), 30–37 (2009)
11. Ma, H.: An experimental study on implicit social recommendation. In: SIGIR, pp. 73–82 (2013)
12. Ma, H., King, I., Lyu, M.R.: Learning to recommend with social trust ensemble. In: Allan, J., Aslam, J.A., Sanderson, M., Zhai, C., Zobel, J. (eds.) SIGIR, pp. 203–210. ACM (2009)
13. Ma, H., Yang, H., Lyu, M.R., King, I.: Sorec: social recommendation using probabilistic matrix factorization. In: Shanahan, J.G., Amer-Yahia, S., Manolescu, I., Zhang, Y., Evans, D.A., Kolcz, A., Choi, K.-S., Chowdhury, A. (eds.) CIKM, pp. 931–940. ACM (2008)
14. Ma, H., Zhou, D., Liu, C., Lyu, M.R., King, I.: Recommender systems with social regularization. In: King, I., Nejdl, W., Li, H. (eds.) WSDM, pp. 287–296. ACM (2011)
15. Melville, P., Sindhwani, V.: Recommender systems. In: Encyclopedia of Machine Learning, pp. 829–838 (2010)
16. Menon, A.K., Elkan, C.: A log-linear model with latent features for dyadic prediction. In: ICDM, pp. 364–373 (2010)
17. Mooney, R.J., Roy, L.: Content-based book recommending using learning for text categorization. In: Proceedings of the Fifth ACM Conference on Digital Libraries, DL 2000, pp. 195–204. ACM, New York (2000)
18. Pazzani, M.J., Billsus, D.: Content-based recommendation systems. In: Brusilovsky, P., Kobsa, A., Nejdl, W. (eds.) Adaptive Web 2007. LNCS, vol. 4321, pp. 325–341. Springer, Heidelberg (2007)
19. Rendle, S.: Factorization machines. In: ICDM, pp. 995–1000 (2010)
20. Salakhutdinov, R., Mnih, A.: Probabilistic matrix factorization. In: Advances in Neural Information Processing Systems, vol. 20 (2008)
21. Sarwar, B., Karypis, G., Konstan, J., Riedl, J.: Item-based collaborative filtering recommendation algorithms. In: Proceedings of the 10th International Conference on World Wide Web, WWW 2001, pp. 285–295. ACM, New York (2001)
22. Su, X., Khoshgoftaar, T.M.: A survey of collaborative filtering techniques. Adv. Artificial Intellegence 2009 (2009)
23. Wang, C., Cao, L., Wang, M., Li, J., Wei, W., Ou, Y.: Coupled nominal similarity in unsupervised learning. In: CIKM, pp. 973–978 (2011)
24. Wang, C., She, Z., Cao, L.: Coupled attribute analysis on numerical data. In: IJCAI (2013)
25. Wang, C., She, Z., Cao, L.: Coupled clustering ensemble: Incorporating coupling relationships both between base clusterings and objects. In: ICDE, pp. 374–385 (2013)
26. Wang, C., Blei, D.M.: Collaborative topic modeling for recommending scientific articles. In: KDD, pp. 448–456 (2011)
27. Yang, X., Steck, H., Liu, Y.: Circle-based recommendation in online social networks. In: KDD, pp. 1267–1275 (2012)
28. Ye, M., Liu, X., Lee, W.-C.: Exploring social influence for recommendation: a generative model approach. In: Hersh, W.R., Callan, J., Maarek, Y., Sanderson, M. (eds.) SIGIR, pp. 671–680. ACM (2012)
29. Yu, Y., Wang, C., Gao, Y., Cao, L., Chen, X.: A coupled clustering approach for items recommendation. In: Pei, J., Tseng, V.S., Cao, L., Motoda, H., Xu, G. (eds.) PAKDD 2013, Part II. LNCS, vol. 7819, pp. 365–376. Springer, Heidelberg (2013)
30. Ziegler, C.-N., McNee, S.M., Konstan, J.A., Lausen, G.: Improving recommendation lists through topic diversification. In: Proceedings of the 14th international conference on World Wide Web, WWW 2005, pp. 22–32. ACM, New York (2005)

A Lot of Slots –
Outliers Confinement
in Review-Based Systems*

Roberto Di Pietro[1,2], Marinella Petrocchi[2], and Angelo Spognardi[2]

[1] Bell Laboratories, Paris, France
[2] IIT-CNR, Pisa, Italy
roberto.di_pietro@alcatel-lucent.com,
marinella.petrocchi@iit.cnr.it, angelo.spognardi@iit.cnr.it

Abstract. Review-based websites such as, e.g., Amazon, eBay, TripAdvisor, and Booking have gained an extraordinary popularity, with millions of users daily consulting online reviews to choose the best services and products fitting their needs. Some of the most popular review-based websites rank products by sorting them aggregating the single ratings through their arithmetic mean. In contrast, recent studies have proved that the median is a more robust aggregator, in terms of ad hoc injections of outlier ratings. In this paper, we focus on four different types of ratings aggregators. We propose to the slotted mean and the slotted median, and we compare their mathematical properties with the mean and the median. The results of our experiments highlight advantages and drawbacks of relying on each of these quality indexes. Our experiments have been carried out on a large data set of hotel reviews collected from Booking.com, while our proposed solutions are rooted on sound statistical theory. The results shown in this paper, other than being interesting on their own, also call for further investigations.

1 Introduction

The choice of a service (or a product) on the Web is highly influenced by the reputation they expose on review-based websites. These websites offer to customers to submit a review concerning their experience, as users, for that service, and aggregate the numerical ratings of these reviews in order to calculate a *review score*. One of the most influenced markets is the one of Internet travel booking, whose revenue has grown by more than 73% over the six-year 2007-2012, with an average of 148.3 million of travel bookings per year [1]. A recent survey [3] reveals how review scores impact on lodging performance in terms of bookings, occupancy, and revenue. The analysis finds that, from 2008, the percentage of users consulting review scores on TripAdvisor prior to making their hotel choice is steadily increasing over time. Moreover, the report reveals that those hotels increasing their review scores by 1 point (on a 5-point scale) can afford to increase

---

* This research has been partially funded by the Tuscany region project *MyChoice* and by the Registro.it project *MIB* (My Information Bubble).

B. Benatallah et al. (Eds.): WISE 2014, Part I, LNCS 8786, pp. 15–30, 2014.
© Springer International Publishing Switzerland 2014

their price by 11.2 percent, still maintaining the same booking probability and market share[3]. Those hotel facilities with a 1-percent increase in their online reputation may gain up to a 1.42 percent in revenue per available room. Not surprisingly, fake reviews are widespread and this phenomenon has captured the attention of Academia and News, e.g., [27,29,28]. Posting fake reviews is a way to mislead the users' opinion, to either promote or damage a particular target and, thus, providing strong incentives for opinion spamming. Defining effective techniques to mitigate the effects of fake reviews has become a compelling issue. In this paper, we contribute with solutions that could be easily adopted by online vendors and providers. We concentrate on reviews generated assigning quantitative (numerical) scores.

Analytics say that, in the last few years, Booking.com is the top visited travel site by traffic [2] and, nowadays, it plays a central role in the business of tourism, even raising doubts of running unfair competitions [25]. In this work, we consider the hotels of five major touristic cities throughout the world (New York, Paris, Rome, Tokyo, Rio de Janeiro), advertised on Booking.com. We collect all the scores assigned to these hotels, for a total of about one million of scores. We freshly introduce here two new metrics, the "slotted" mean and "slotted" median, and, based on our score-set, we compare the rankings based on the new metrics with the ones based on the mean and the median. The mean is currently used by the most popular websites for e-advice (including Booking.com). The median has been proposed in the literature as a metric less susceptible than the mean to outliers and bias, see, e.g., [13,19,12]. The slotted mean and the slotted median are computed assigning the ratings received by the hotel to temporal slots of equal length (e.g., weeks or months) and, then, calculating, respectively, the mean and median of the averages over the slots. All the aggregators considered in this paper (mean, median, and slotted aggregators) are compared with respect to different properties, such as the dispersion of the reviewers' scores around each aggregator, the similarity of the rankings obtained sorting the hotels per aggregator, and the robustness of each aggregator to resist to injection of outliers (in terms of degree of ranking alteration).

Supported by both analytic and experimental results, we conclude the following: 1) slotted aggregators substantially preserve the original information on the hotel ranking (currently based on the simple mean of ratings); 2) although the simple median is a quite robust aggregator in terms of number of outliers required to alter the original ranking, the slotted aggregators (particularly, the slotted median) enjoy the appealing property that if the outliers are put within a single time slot, then the ranking alteration is not always possible; 3) they also are more informative than the simple median (e.g., by using the median many hotels in the dataset exhibit the same global score). In all the experiments we have carried out, we demonstrate that the adoption of the slotted aggregators constitute a valid alternative to the mean and median, that strongly reduce the effects of score spamming, still preserving the informative value and similarity to current rankings. Note that the adoption of slotted aggregators can be easily implemented by current review-based Web sites.

Roadmap. This paper is structured as follows. Section 2 recalls existing review systems and focuses on the phenomenon of fake reviews. In Section 3, we introduce both standard and slotted aggregators, analytically illustrating their mathematical properties. Section 4 presents the series of experiments that we ran over the collected data-set and discuss the results derived from those experiments. Finally, Section 5 concludes the paper.

2 Related Work

Online review- and reputation-systems have been adopted in the past few years as a mean for decision making on the Web. Parties in a community may decide to choose a specific item (e.g., a product, a hotel or a service) based on its reputation, usually obtained from the elaboration of the *review scores* assigned by the community to that item. The reputation can be dynamically updated according to various kind of algorithms, that are generally built on the principle that the updated reputation is a function of the old reputation and the most recent feedback [18]. Simple reputation models, as the one adopted by eBay (www.ebay.com) until May 2008, exploit a *plain* combination of past and new feedback. Since May 2008, eBay started considering only the percentage of positive ratings of the last twelve months.

The schemes where the global reputation score is the arithmetic mean of the ratings are the more common and adopted by popular websites for e-commerce, such as Amazon (www.amazon.com) and Epinions (www.epinions.com). In such schemes, past and new ratings contribute in an equal manner to the calculation of the global score [26]. Our work focuses on this kind of schemes.

Works in [12,13,19] focus on recommender systems and users' feedback. The authors show that the arithmetic mean is quite sensitive to outliers and biases, and may not be the most informative rating aggregator. In contrast, they prove that, in most cases, the median is a better choice than the mean, being more robust to the intentional injection of outliers. To evaluate the robustness of different aggregators (mean, median, mode, and weighted ones), the authors count the number of outliers required, for each aggregator, to alter the ranking of a given hotel, from those ranked by a popular travel site that collects hotel reviews.

Work in [9,32,35,10,14,20] focus on numerical scores too. Pioneer work in [9] considers extremely high ratings, aimed at fictitiously increase rates' reputations. The study provides a filtering approach to separate unfairly high ratings from fair ones. Similarly, in [32] the authors observe that unfair ratings have a different statistical pattern than fair ratings. The work considers repeated interactions between a rater and a ratee. The goodness of each interaction is judged by a rating expressing the degree of satisfaction of the rater towards the ratee. By comparing long-run average ratings, a filtering algorithm rejects raters who, over some period, have rated significantly differently from the average. Work in [35] relies on a set of advisor agents to handling unfair ratings within a centralized reputation system. Advisor agents are rated themselves, to build a ranking of most reputable agents to rely on. Authors of [10] study a large data set of eBay transactions, where both partners of a bilateral exchange report their satisfaction towards the transaction. Analyzing the transactions where one or both trading partners choose to remain silent, they notice that eBay traders are more likely to post positive than negative ratings. A study on such reporting bias reveals unbiased estimates that, according to the authors, are more realistic than the surprisingly high percentage of positive feedback advertised, at that time, by eBay. Work in [20] proposes a methodology for empirically detecting review manipulation, by comparing the distribution of reviewer scores of two well-known travel websites, i.e., TripAdvisor and Expedia, on the same set of hotels (around 3k). Considering that Expedia allows only verified hotel customers to post review scores, the authors refer to those scores as genuine ones and use the deviation between the two score distributions to estimate review manipulation possibly introduced in TripAdvisor.

More complex reputation models combine old and new scores to form a new value computed as a weighted mean. Some proposals exist in the literature on how to determine the weights. For example, work in [6,8,33,34,30,7] consider various attributes of the reviewers, such as their reputation, trustworthiness, and expertise of the reviewer, while authors of [15] evaluate the different degrees of the user's satisfaction for a set of parameters characterizing the reviewed item. Work in [4,5] considers the evolution of sellers reputation in electronic marketplaces as an aggregation of past and recent transactions, and proposes an optimization of the *Window Aggregation Mechanism*, in which the seller's score is the average of the last n most recent ratings, and the *Weighted Aggregation Mechanism*, where the seller's score is a weighted average of past ratings, optimal with respect to persuade the seller to be truthful. Fan *et al.* [11] propose to achieve a similar goal by adopting exponential smoothing to aggregate ratings. All these proposals may require substantial modifications to the existing e-advice and e-commerce platforms. Instead, our aim is to propose solutions that, besides mitigating the damages of fake reviews, could be directly implemented on such platforms, without in-depth modifications.

Finally, work in [17,22,23] focus on textual reviews. Authors of [23] study *deceptive opinion spam*, i.e., fictitious opinions that have been deliberately written to sound authentic. They created a *gold standard* of real and deceptive reviews and developed, over this data-set, a classifier to detect fictitious reviews that is nearly 90% accurate. Work in [17] presents pioneer results in the area of spam review detection, and provides a first taxonomy of opinion spam. A large number of (near-)duplicate reviews, almost certainly representing spam reviews, constitute the authors' validation set. Spam review detection models have been built from the analysis of such a set, and results show that those models can provide a good prediction of likely harmful reviews. In [22], investigations are towards detection of opinion spam in a collaborative setting, i.e., to discover fake reviewer groups. On a set of suspicious groups, several behavioral methods have been successfully applied to build a labeled data-set of fake reviewer groups.

Table 1. Data-set from *Booking.com*, updated at July 2013

city	hotels	reviews
New York	489	165253
Paris	1605	447656
Rio de Janeiro	329	53617
Rome	1847	323850
Tokyo	399	33427
total	4669	1023803

3 Ratings Aggregators

In this section, we introduce and discuss the aggregators considered in the paper. For ease of exposition, we quickly describe the dataset we collected for the present work. We gathered the online global scores of all the hotels advertised by Booking.com, for five cities (Rome, Paris, New York, Rio de Janeiro, and Tokyo), obtaining a total of 1,023,806 reviews. Tables 1 summarizes the amount of collected data. The reference period is September 23, 2011 - 11 July, 2013. Reviewers may post numerical sub-scores regarding six hotels parameters (cleanness, comfort, services, staff, value for money, and location). The average of the six sub-scores represents the reviewer score. Figure 1 shows an example of a hotel review, reporting some reviewer data, such as ID, typology, review creation date, the reiewer score, and, possibly, some textual comments.

The totality of reviewer scores for one hotel is then averaged to obtain the *global score* for that hotel: all the collected scores are in the range $[2.5, 10]$. Hotels are ranked by Booking.com according to global scores, from the hotel with the highest score, to the one with the lowest one. Hereafter, we may refer to global scores as simply *scores* or *rating*.

Fig. 1. Review example (*Booking.com*)

3.1 Standard Aggregators

Given a hotel h_j, X_j is the set of ratings for h_j (with $|X_j| = n_j$), μ_j is the mean of such ratings, and M_j is their median. The mean μ_j is the average of the ratings in X_j, namely $\mu_j = \frac{1}{n_j}\sum_{r\in X_j} r$. The *median* M_j is obtained ordering the ratings $r \in X_j$, from the lowest to the highest, in a sequence $x_1, \ldots, x_{n_j}$ and picking

$$M_j = \begin{cases} x_{[\frac{n_j}{2}]} & \text{if } n_j \text{ is odd} \\ \frac{1}{2}\left(x_{[\frac{n_j}{2}]} + x_{[\frac{n_j}{2}]+1}\right) & \text{if } n_j \text{ is even} \end{cases}$$

where $[\frac{n_j}{2}]$ is the greatest integer less than or equal to $\frac{n_j}{2}$. In practice, if the sample size is an odd number, the median is defined to be the middle value of the ordered samples; if the sample size is even, the median is the average of the two middle values [24].

As rating aggregator there is also the *trimmed mean* [24], i.e., the mean obtained discarding a percentage of the lowest and highest values. Formally, the $100\alpha\%$ trimmed average μ_α of our n_j ratings is obtained ordering the values and evaluating:

$$\mu_\alpha = \frac{x_{[n_j\alpha]+1} + \cdots + x_{n_j-[n_j\alpha]}}{n_j - 2[n_j\alpha]}$$

where $[n_j\alpha]$ is the greatest integer less than or equal to $n_j\alpha$. By definition, μ_0 is the mean and μ_{50} is the median. Considered the similarities with the median, we do not considered the trimmed mean in our study.

To evaluate the dispersion of the ratings around the mean and median aggregators, we use the following indices: the *standard deviation* σ_j, defined as the square root of the variance, namely $\sqrt{\sigma_j^2} = \sqrt{\frac{1}{n_j}\sum_{r\in X_j}(r - \mu_j)^2}$ and MAD_j, the *median absolute deviation from the median*, defined as the median of the set $\{|M_j - r| : r \in X_j\}$.

3.2   Slotted Aggregators

A usually public metadata associated to an online rating is the date of the review, namely, when the rating has been casted by the reviewer. This provides another way to sort the ratings and enables the definition of another type of aggregators, which consider all the ratings divided into temporal *slots* of equal length. Here, we define the *slotted mean* $S\mu_s$ as the mean of the means of the ratings over a set S of slots s of a same amount of time. If we denote with a_i the mean of the ratings received by the hotel h_j during slot i, we define

$$S\mu_s = \frac{1}{|S|}\sum_{i\in S} a_i$$

omitting the slots with no ratings. For example, to evaluate the *weekly* slotted mean $S\mu_w$, we consider all the means of the ratings received during all the weeks and we average those means. Similarly, we can consider the *monthly* slotted mean $S\mu_m$, etc. The slotted mean shows many similarities with the "moving average", typical of time series analysis [31]: both the metrics consider the time when the measurements are collected and the average of such measurements. However, the moving average captures the trend of a phenomenon, reducing the effects of fluctuations, whereas the slotted mean gives a representative, but static, value of the phenomenon.

Similarly to above, we may define the *slotted median* SM as the median of the means of the ratings over temporal slots of the same amount of time. In the following, we will consider both the *weekly slotted median* SM_w and the *monthly slotted median* SM_m.

3.3   Observations

The use of the median as aggregator has characteristics that should be considered when applied to our real context, like the ratings casted to the hotels of Booking.com. Indeed, in our scenario, the scores we have collected have a bounded range, namely $[2.5, 10]$, being respectively the minimum and the maximum ratings a reviewer assigned to a hotel.

Score Distribution. First, ranking hotels according to their medians leads to less granularity in the global ranking. Indeed, the median spans only 68 different values, while the mean can assume 4166 distinguished values. This means that, adopting the median, the global hotel ranking will see many hotels with exactly the same position: for example, 335 hotels in Paris have the same median (7.5), and 244 hotels in Rome have the median of 8.8. Sorting also with respect to the number of received reviews is a possible way to sub-rank the hotels with the same score, however, ranking hotels by the median could lead to loose informativeness. Consequently, the median has a less smooth progression when compared to mean, since it increases (or decreases) by steps, only assuming the values of the actual ratings given by reviewers. In particular, over the about one million ratings that we have collected, the median only assumes 37 different values: as an example, the only values used between 6.0 and 7.0 are $6.3, 6.5, 6.7$ and 6.9. This makes the hotels rough jump from a position to another, without the linear progression exposed by the mean.

Moreover, even considering a very large (infinite) number of ratings injections, only the aggregators based on the median can be forced to assume the extreme values, while those based on the mean can only get close to those values (clearly, assuming that there is at least one rating different from the outliers).

Robustness. We argue that the slotted versions of the mean and median aggregators could be, under certain circumstances, more robust against bursts of outliers injections. Indeed, as long as the outliers fall within a single slot, the effects of the attack are mitigated.Furthermore, slotted aggregators do not exhibit the characteristics of the median discussed before: being based on means of ratings, they produce rankings where the hotels have values that better distribute over the rating range. In our dataset, the slotted median assumes more widely spread values (*i.e.*, 1480 distinct values for the weekly, 3379 for the monthly), being based on the averages of the single slots.

To discuss more rigorously the benefits of the slotted aggregators, we consider the *breakdown point* as a measure of their robustness [12]: in robust statistics, the breakdown point is the "smallest fraction of bad observations that may cause an estimator to take on arbitrarily large aberrant values" [16]. Theoretically, the mean has a breakdown point equal to 0 (namely, $\lim_{n \to \infty} \frac{1}{n}$), for the median it is $\frac{1}{2}$ and for the trimmed average it is α. However, when the ranges are bounded (as in the case of hotels ratings), the aggregators can not take arbitrary values, but will only have their limits on their extremal values, namely 2.5 and 10 for the Booking.com.

Garcin *et al.* [12] concluded that the median is the most robust aggregator, compared to the mean and the trimmed mean, based also on their breakdown point. In particular, they empirically considered how many outliers were needed to change the position of a hotel in the global ranking, per aggregator, using a selection of 400 hotels from 4 cities.

In a similar manner, we can consider the number of outliers required in case of slotted aggregators, assuming that the outliers are injected only in one slot (corresponding to the slot where the attack takes place). The injection in different slots, that we do not consider in this paper, would be either time consuming (since the attacker should wait weeks, or even months, to inject scores) or it would require a full access to the ratings already present on the data server of the websites.

The value of the slotted aggregators depends on *when* the ratings are casted and/or injected, namely, it depends on the slot the ratings are casted. Let $X_h = \{r_1^1, r_2^1, \ldots, r_{n_1}^1, r_1^2, \ldots, r_{n_2}^2, \ldots, r_1^s, \ldots, r_{n_s}^s, r_{n_s+1}^s, \ldots, r_{n_s+l}^s\}$ be the set of ratings for a given hotel h, ordered and divided within s slots, where n_i is the number of "regular" ratings for slot i and the l ratings $r_{n_s+i}^s$ $(1 \le i \le l)$ are the "outliers", injected (without loss of generality) into the last slot s. Denoting with $a_j = \frac{1}{n_j} \sum_{i=1}^{n_j} r_i^j$ the average of the ratings within the slot j $(j \ne s)$, and supposing that the outliers in slot s are all equal to r', we can write that

$$S\mu_s(h) = \frac{1}{s} \left(\sum_{j=1}^{s-1} a_j + \frac{1}{n_s+l} \left(\sum_{i=1}^{n_s} r_i^s + lr' \right) \right) = A + \frac{B + lr'}{s(n_s + l)}$$

where $A = \frac{1}{s} \sum_{j=1}^{s-1} a_j$ and $B = \sum_{i=1}^{n_s} r_i^s$. Given another hotel h', with slotted mean equal to $S\mu_s(h')$, the number l of low-score outliers $r' \le \min(X_h)$ required to have a new slotted mean $S\mu_s(h)$ lower than $S\mu_s(h')$ can be evaluated considering the inequality

$$S\mu_s(h) = A + \frac{B + lr'}{s(n_s + l)} \le S\mu_s(h')$$

From the theory of limits for polynomial fractions, we have that $\lim_{l\to\infty} S\mu_s(h) = A + \frac{r'}{s}$. Let us suppose that $S\mu_s(h') < A + \frac{r'}{s}$. This implies that there is no way for the adversary to move hotel h to a position in the global hotel ranking lower than the position of h'. The same consideration holds when looking for the number of l high score injections able to move h to a position higher than that of h'. We conclude that, when only one slot is the target of the attack, even injecting an infinite number of outliers, the breakdown point does not exist.

With a similar reasoning, we can observe that the slotted median SM enjoys the same property:

$$SM_s(h) = median(\{a_1, a_2, \ldots, a_s\})$$

where a_s is the average of ratings in the last slot $\{r_1^s, r_2^s, \ldots, r_{n_s}^s, \underbrace{r', r', \ldots, r'}_{l \text{ times}}\}$,

that is $a_s = \frac{1}{n_s+l} \left(\sum_{i=1}^{n_s} r_i^s + lr' \right)$. To have a new slotted median for h, lower than the one of some hotel h', the number l of low-score outliers $r' \leq \min(X_h)$ to be added can be evaluated considering the inequality $SM_s(h) < SM_s(h')$.

Even considering a possibly infinite number of injections in the last slot, the average of ratings in the last slot will be a value limited in its lower bound:

$$SM_s(h) = median(\{a_1, a_2, \ldots, a_{s-1}, \lim_{l \to \infty} a_s\}) \geq median(\{a_1, a_2, \ldots, a_{s-1}, r'\}) = \rho_h$$

In all cases in which $SM_s(h') < \rho_h \leq SM_s(h)$, it is impossible to lower hotel h more than h' in the global ranking.

4   Experiments

In this section, we detail the experiments we have carried out to evaluate how the slotted aggregators perform (compared to mean and median) and to provide more insights when dealing with a real scenario, as the Booking.com website.

4.1   Dispersion of Reviewers' Scores

Here, we evaluate the dispersion of the reviewers' scores around the following aggregators: the average μ, the median M, the weekly and monthly slotted means $(S\mu_w, S\mu_m)$ and the weekly and monthly slotted medians (SM_w, SM_m). A measure of dispersion provides a numerical indication of the "scatteredness" of a batch of numbers [24], the reviewer scores in our context. The goal is to understand how the scores are distributed around the aggregators, and, possibly, show how good they are in representing the elements of the whole score set. Table 2 reports the average of the dispersions of the scores around the different aggregators, for the five cities in our dataset. For those aggregators based on the mean (i.e., mean and weekly and monthly slotted means), we list the average of the standard deviation, while for the other aggregators, we list both the average MAD (Section 3.1) and the average of the average absolute deviation. We can observe that the aggregators based on the mean have a very similar variance. In particular, the slotted means have an average standard deviation that differs for less than 0.01 from the mean. This means that the slotted means have almost the same representativeness when compared to the simple mean. Considering the slotted aggregators based on the median, we can observe that the dispersion is a bit wider when compared to the simple median. In particular, although the overall distributions of the scores are comparable, the simple median dispersion is slightly smaller for all the considered cities. However, we can still conclude that all the considered aggregators are interchangeably representatives of the whole set of reviewer scores.

4.2   Similarity

We have measured the distance between the positions obtained by the same hotel in the global ranking, using different aggregators. We have used the absolute difference between the hotel positions within two distinct rankings and we report

Table 2. Analysis of rating dispersions around the aggregators

aggregator	average of σ (standard deviation)				
	Rome	New York	Tokyo	Rio d.J.	Paris
μ	1.41375	1.44133	1.39171	1.49676	1.43193
$S\mu_w$	1.41735	1.44471	1.39561	1.50215	1.43477
$S\mu_m$	1.42180	1.44838	1.40366	1.50845	1.43768

aggregator	average of MAD (median absolute deviation)				
	Rome	New York	Tokyo	Rio d.J.	Paris
M	0.92030	0.94519	0.93083	0.98267	0.92491
SM_w	0.95981	0.99089	0.94776	1.00731	0.94786
SM_m	0.97729	1.02045	0.97872	1.05085	0.96001

aggregator	average of the mean absolute deviation				
	Rome	New York	Tokyo	Rio d.J.	Paris
M	1.10526	1.13130	1.09913	1.18496	1.12363
SM_w	1.11735	1.14712	1.11097	1.19687	1.13462
SM_m	1.12728	1.15979	1.12461	1.21233	1.14300

the average results, grouped by cities, in Table 3. The ranking based on the simple median is the most different, compared to the others, while those based on the slotted weekly mean and the mean are really close one to each other. Overall, the rankings based on slotted aggregators (primarily those based on the mean) exhibit more similarity among them, with respect to the simple media. Moreover, the slotted means are the ones that produce the most similar rankings to the one obtained by the simple mean, as highlighted in bold in Table 3.

4.3   Robustness

To evaluate the robustness of the different aggregators for ranking hotels, we performed several experiments, injecting scores at the extremes of the value scale, as in [12]: to help the readability we say that, by injecting the highest value, we perform a *push* attack and, by injecting the lowest value, a *nuke* attack [21]. In our collected dataset, presented in Section 3, the extreme values of the scale are 2.5 and 10: these, indeed, represent the lowest and highest score present in the score-set of Booking.com. The first experiment is similar to the one in [12]: for each hotel, we count the outliers required to alter the ranking of the hotel. In the second experiment, for each hotel with at least 4 ratings, we inject 5%, 10%, and 20% of outer scores (with respect to the number of actual ratings) and count the number of lost/acquired positions. For example, for a hotel with 30 scores, we inject 1, 3, and 6 outer scores, respectively. We exclude those hotels having less than 4 ratings, since even 1 outlier would exceeds the 20% of the actual ratings. For the slotted aggregators, we inject the outliers into the most recent slot, since injecting in multiple slots would be time consuming (spanning on

Table 3. Average ranking difference for the aggregators

	New York	Paris	Rio d.J.	Rome	Tokyo
$\mu - M$	25.4	92.6	17.0	106.6	29.6
$\mu - S\mu_w$	**12.3**	**36.5**	**9.0**	**48.4**	**13.8**
$\mu - SM_w$	19.7	57.5	14.5	74.5	25.4
$\mu - S\mu_m$	**16.1**	**43.6**	**11.5**	**65.2**	**23.0**
$\mu - SM_m$	19.6	56.7	12.2	71.2	25.2
$M - S\mu_w$	29.4	99.3	19.5	115.5	33.1
$M - SM_w$	26.9	89.0	17.9	104.7	27.4
$M - S\mu_m$	31.6	97.7	20.5	126.0	37.9
$M - SM_m$	32.4	103.6	20.7	120.4	35.0
$S\mu_w - SM_w$	20.1	59.7	14.6	75.3	24.1
$S\mu_w - S\mu_m$	15.2	46.5	11.9	61.3	21.0
$S\mu_w - SM_m$	21.2	63.9	13.5	77.6	26.5
$SM_w - S\mu_m$	23.1	65.6	17.4	91.8	31.4
$SM_w - SM_m$	21.7	62.5	14.9	84.4	26.2
$S\mu_m - SM_m$	17.7	54.0	12.5	72.7	25.8

Table 4. *Push* attack: Average number of outliers to perform a ranking alteration (for the hotels for which the alteration is possible) and *no-win* ratio

	New York		Paris		Rio de Janeiro		Rome		Tokyo	
	outliers	*no-win*	outliers	*no-win*	outliers	*no-win*	outliers	*no-win*	outliers	*no-win*
μ	1.99	0	1.14	0	1.76	0	1.08	0	1.30	0
M	43.16	0	34.57	0	16.79	0	21.26	0	11.38	0
$S\mu_w$	2.11	0.01	1.19	0	1.88	0.01	1.13	0	1.31	0.01
SM_w	1.39	0.51	1.34	0.50	1.26	0.55	1.32	0.43	1.25	0.49
$S\mu_m$	2.76	0.00	1.26	0	1.72	0.01	1.15	0	1.40	0
SM_m	2.06	0.25	2.14	0.31	1.70	0.26	2.26	0.23	1.51	0.25

weeks or months) or would mean having access to the system where the ratings are stored.

Altering the Ranking. As discussed in Section 3.3, when using the slotted aggregators it can happen that the ranking alteration is impossible. Then, the results of a *push* attack (injections of the highest value 10) listed in Table 4 show two columns for each city: 1) the average number of outliers required to perform a ranking alteration (for the hotels for which the alteration is possible) and 2) the *no-win* ratio (namely, the fraction of hotels for which that alteration is unfeasible). If the alteration is possible, the number of required outliers for the slotted aggregators is very similar to the one of the mean, while for the median the number is relatively larger: our results confirm the robustness of the simple median with respect to the simple mean, already highlighted in [12]. However, the *no-win* ratio reports the fraction of hotels whose rankings were impossible to alter.For example, in the ranking obtained using the weekly slotted median (SM_w), the alteration was not possible for almost half of the hotels. The ratio decreases to 0.25 considering the monthly slotted median.

Table 5. *Nuke* attack: Average number of outliers to perform a ranking alteration (for the hotels for which the alteration is possible) and *no-win* ratio

	New York		Paris		Rio de Janeiro		Rome		Tokyo	
	outliers	*no-win*	outliers	*no-win*	outliers	*no-win*	outliers	*no-win*	outliers	*no-win*
μ	1.39	0	1.09	0	1.23	0	1.03	0	1.06	0
M	35.80	0	34.19	0	18.59	0	20.14	0	10.20	0
$S\mu_w$	1.15	0.08	1.13	0.05	1.48	0.09	1.06	0.09	1.01	0.11
SM_w	1.03	0.49	1.02	0.40	1.01	0.45	1.01	0.40	1.01	0.42
$S\mu_m$	1.46	0.02	1.18	0.01	1.22	0.04	1.05	0.01	1.13	0.02
SM_m	1.15	0.49	1.12	0.37	1.13	0.46	1.08	0.46	1.06	0.44

Table 5 reports the results of the experiments for the *nuke* attack (injections of the lowest value, namely 2.5). The average number of outliers for the median is considerably larger when compared with the *push* attack, while those required for the slotted aggregators are slightly smaller and, all in all, similar to those of the simple mean (similarly to above, the first column lists the outliers only for those hotels for which it was possible to alter the ranking). The *no-win* ratio, instead, varies noticeably: it is slightly smaller for the weekly slotted median, but almost doubled for the monthly slotted median. Also, the slotted means appreciably increase their *no-win* ratio when compared with the *push* attack.

Burst of Injections. Tables 6 and 7 report the average alteration in the global ranking, experienced by the hotels of a given city when we inject a number of outliers, depending on the whole set of ratings for each hotel. Table 6 considers injection of 10 (*push*). As the injection rate increases, the pace at which the average ranking alteration increases varies considerably for the different aggregators. The mean and the median have the largest ranking alterations, since the effect of the outliers keeps increasing as the injection rate grows. For the slotted aggregators, instead, the effects are clearly mitigated: since the injections are only confined in a single slot (simulating a massive burst of injections), the ranking alteration is lower. The aggregator that better resists against massive injections is the slotted weekly median.

Table 7 considers injection of 2.5 (*nuke*). The alterations experienced with the mean are severely higher, even for the slotted mean aggregators. The median, instead, appreciably reduces the average alterations (with respect to the push attack). As above, the best performing aggregator is the slotted weekly median.

We finally report in Table 8 the ratio of hotels that do not change their position in the global ranking, after the burst of injections, averaged for the five

Table 6. Average ranking alteration for a *push* attack made with burst of injections

city	injection rate	μ	M	$S\mu_w$	SM_w	$S\mu_m$	SM_w
	5%	17.04	14.71	6.66	5.05	11.51	7.86
New York	10%	31.93	31.98	8.31	5.76	16.66	10.69
	20%	60.31	63.20	9.83	7.52	22.66	13.58
	5%	61.88	50.81	24.98	15.77	43.54	34.01
Paris	10%	121.18	107.67	30.38	19.53	65.81	41.42
	20%	226.18	203.50	39.48	27.30	91.00	49.37
	5%	11.92	7.56	5.80	3.32	10.07	6.76
Rio de Janeiro	10%	22.62	18.11	7.44	5.63	15.03	9.24
	20%	42.07	36.12	9.50	6.92	19.54	12.94
	5%	56.43	56.66	23.75	24.36	41.70	37.03
Rome	10%	113.62	109.92	29.85	29.80	63.85	50.10
	20%	220.22	231.22	36.22	33.68	86.97	60.07
	5%	15.82	14.98	8.08	7.21	13.36	9.82
Tokyo	10%	30.85	31.19	10.96	8.76	18.52	12.78
	20%	58.24	59.87	14.40	11.94	25.24	17.44

Table 7. Average ranking alteration for a *nuke* attack made with burst of injections

city	injection rate	μ	M	$S\mu_w$	SM_w	$S\mu_m$	SM_w
	5%	48.74	9.39	19.71	4.97	34.19	11.28
New York	10%	90.67	18.87	24.30	7.22	50.74	14.23
	20%	150.60	38.52	28.76	9.73	67.68	16.22
	5%	139.01	32.05	51.40	13.91	91.89	24.07
Paris	10%	266.25	62.54	62.69	18.81	140.37	29.42
	20%	456.81	123.74	80.66	27.25	192.80	44.13
	5%	24.27	6.09	10.32	7.15	19.11	6.68
Rio de Janeiro	10%	46.08	12.50	13.38	7.81	28.52	9.41
	20%	81.82	28.61	18.30	10.83	39.43	14.09
	5%	165.18	34.79	69.95	25.33	122.18	46.96
Rome	10%	322.67	67.68	87.34	28.68	186.58	54.40
	20%	568.42	147.53	104.38	34.60	252.35	61.10
	5%	41.70	8.68	19.87	7.67	32.49	13.23
Tokyo	10%	79.77	18.24	27.37	11.34	47.29	18.39
	20%	135.42	37.91	35.97	14.84	62.82	24.14

cities (both *push* and *nuke* injections). The rankings based on the mean-based aggregators nearly always experience a position alteration. On the contrary, the rankings that rely on the median are more stable and exhibit significant alterations only with large bursts of injections (*i.e.*, 10% and 20%). Overall, the most stable ranking is obtained using the slotted weekly median.

Table 8. Ratio of hotels whose rank does not change after attacks made with burst of injections. Average over the 5 cities.

attack type	injection rate	μ	M	$S\mu_w$	SM_w	$S\mu_m$	SM_w
push	5%	0	0.75	0.08	0.60	0.01	0.44
	10%	0	0.53	0.07	0.59	0.01	0.41
	20%	0	0.20	0.07	0.58	0.01	0.39
nuke	5%	0	0.74	0	0.58	0	0.48
	10%	0	0.52	0	0.58	0	0.46
	20%	0	0.18	0	0.57	0	0.45

Concluding, our experiments on a real dataset confirm that the slotted aggregators exhibit some interesting properties when compared with the median. In particular 1) they have a more balanced distribution of the hotels on the ranking, 2) their rankings require time (weeks or months) to be severely altered, 3) their rankings are more similar to the one of the simple mean, but more resilient against the outliers.

5 Conclusions

In this paper we introduced two new ratings' synthesis indicators, that we referred to, given their construction, as slotted indicators. We compared the mean, the median, and the slotted indicators with respect to 1) their robustness to injections of outliers, 2) the degree of similarity the global rankings built on such indicators expose, and 3) the informativeness exhibited by the different rankings. The slotted indicators manifest satisfactory outcomes: if we consider the slotted median, it preserves the hotel ranking even after a massive number of outliers injection, for almost half of the hotels under investigation. Moreover, the new indicators would lead to rankings that are more fine grained than the one based on the median (proved, by past work, to be well robust against injections as well). Interestingly, the slotted aggregators could be easily adopted by specialized websites as a patch to mitigate the fake scores phenomenon. The discussed solutions are based on sound statistical theory and supported by an extensive experimental campaign on a real data-set.

Acknowledgements. The authors would like to thank Riccardo Conti and Emanuel Marzini, for the fruitful discussion that made possible the obtained results, and the anonymous reviewers, for their earnest comments and suited suggestions.

References

1. Internet Travel Hotel Statistics (June 18, 2013), http://statisticbrain.com
2. Statistics Summary for Booking.com (May 2013), http://alexa.com
3. Anderson, C.: The impact of social media on lodging performance. Cornell Hospitality Report 12(15) (November 2012)
4. Aperjis, C., Johari, R.: Designing aggregation mechanisms for reputation systems in online marketplaces. SIGecom Exch. 9 (2010)
5. Aperjis, C., Johari, R.: Optimal windows for aggregating ratings in electronic marketplaces. Management Science 56(5) (2010)
6. Buchegger, S., Le Boudec, J.: A robust reputation system for mobile ad-hoc networks. Technical report, IC/2003/50 EPFL-IC-LCA (2003)
7. Chen, W., Zeng, Q., Wenyin, L.: A User Reputation Model for a User-Interactive Question Answering System. In: SKG 2006. IEEE (2006)
8. Cornelli, F., Damiani, E., di Vimercati, S.D.C., Paraboschi, S., Samarati, P.: Choosing reputable servents in a p2p network. In: WWW 2002. ACM (2002)
9. Dellarocas, C.: Immunizing online reputation reporting systems against unfair ratings and discriminatory behavior. In: ACM Conf. on Electronic Commerce (2000)
10. Dellarocas, C., Wood, C.A.: The sound of silence in online feedback: Estimating trading risks in the presence of reporting bias. Management Science 54(3) (2008)
11. Fan, M., Tan, Y., Whinston, A.B.: Evaluation and design of online cooperative feedback mechanisms for reputation management. IEEE Trans. Knowl. Data Eng. 17(2) (2005)
12. Garcin, F., Faltings, B., Jurca, R.: Aggregating reputation feedback. In: In ICORE 2009 (2009)
13. Garcin, F., Faltings, B., Jurca, R., Joswig, N.: Rating aggregation in collaborative filtering systems. In: Proceedings of the Third ACM Conference on Recommender Systems, RecSys 2009. ACM, New York (2009)
14. Gorner, J., Zhang, J., Cohen, R.: Improving the use of advisor networks for multiagent trust modelling. In: Privacy, Security and Trust (2011)
15. Griffiths, N.: Task delegation using experience-based multi-dimensional trust. In: AAMAS 2005. ACM (2005)
16. Huber, P., Ronchetti, E.: Robust statistics, vol. 10(1002). Wiley, NJ (2009)
17. Jindal, N., Liu, B.: Opinion spam and analysis. In: WSDM 2008. ACM (2008)
18. Jøsang, A., Ismail, R., Boyd, C.: A survey of trust and reputation systems for online service provision. Decis. Support Syst. 43 (2007)
19. Jurca, R., Garcin, F., Talwar, A., Faltings, B.: Reporting incentives and biases in online review forums. ACM Trans. Web 4(2) (April 2010)
20. Mayzlin, D., Dover, Y., Chevalier, J.A.: Promotional reviews: An empirical investigation of online review manipulation. Working Paper 18340, National Bureau of Economic Research (August 2012)
21. Mobasher, B., Burke, R., Bhaumik, R., Williams, C.: Toward trustworthy recommender systems: An analysis of attack models and algorithm robustness. ACM Trans. Internet Technol. 7(4) (October 2007)
22. Mukherjee, A., Liu, B., Glance, N.: Spotting fake reviewer groups in consumer reviews. In: WWW 2012. ACM (2012)
23. Ott, M., Choi, Y., Cardie, C., Hancock, J.T.: Finding deceptive opinion spam by any stretch of the imagination. In: Proc. of ACL: HLT, vol. 1 (2011)
24. Rice, J.: Mathematical Statistics And Data Analysis. Number p. 3 in Duxbury Advanced Series. Brooks/Cole CENGAGE Learning (2007)

25. Rubino, M.: Albergatori contro Expedia e Booking. L'Antitrust avvia un'istruttoria. La Repubblica (Italian edition, online) (May 19, 2014), http://goo.gl/U4guRQ
26. Schneider, J., Kortuem, G., Jager, J., Fickas, S., Segall, Z.: Disseminating trust information in wearable communities. Personal Ubiquitous Comput. 4 (2000)
27. Smith, M.D.: Fake reviews plague consumer websites. The Guardian (online ed.) (January 2013), http://gu.com/p/3d9qy
28. Staff writers. Choice warns TripAdvisor users to be aware of fake hotel reviews. Herald Sun (online ed.) (November 2013), http://goo.gl/94Qbdq
29. D. Streitfeld. The best book reviews money can buy. New York Times (online ed.) (August 2012), http://goo.gl/3vuB4
30. van Deursen, T., Koster, P., Petkovic, M.: Hedaquin: A reputation-based health data quality indicator. Electr. Notes Theor. Comput. Sci. 197(2) (2008)
31. Wei, W.W.-S.: Time series analysis, 2nd edn. Pearson Addison-Wesley (2006)
32. Whitby, A., Jøsang, A., Indulska, J.: Filtering out unfair ratings in bayesian reputation systems. In: Workshop on Trust in Agent Societies (2004)
33. Yu, B., Singh, M.P.: Detecting deception in reputation management. In: AAMAS 2003. ACM (2003)
34. Zacharia, G., Moukas, A., Maes, P.: Collaborative reputation mechanisms in electronic marketplaces. In: HICSS (1999)
35. Zhang, J., Cohen, R.: Trusting advice from other buyers in e-marketplaces: the problem of unfair ratings. In: ICEC (2006)

A Unified Model for Community Detection
of Multiplex Networks

Guangyao Zhu and Kan Li

Beijing Institute of Technology, Beijing, China
zhgycolin@gmail.com, likan@bit.edu.cn

Abstract. Multiplex networks contain multiple simplex networks. Community detection of multiplex networks needs to deal with information from all the simplex networks. Most approaches aggregate all the links in different simplex networks treating them as being equivalent. However, such aggregation might ignore information of importance in simplex networks. In addition, for each simplex network, the aggregation only considers adjacency relation among nodes, which can't reflect real closeness among nodes very well. In order to solve the problems above, this paper presents a unified model to detect community structure by grouping the nodes based on a unified matrix transferred from multiplex network. In particular, we define importance and node similarity to describe respectively correlation difference of simplex networks and closeness among nodes in each simplex network. The experiment results show the higher accuracy of our model for community detection compared with competing methods on synthetic datasets and real world datasets.

Keywords: Multiplex networks, Community detection, Information of importance, Node similarity, Unified model.

1 Introduction

Multiplex networks consist of different simplex networks whose links either reflect different kinds of relationships, or represent different values of the same kind of relationships among a same set of elementary components [18,14,15,7]. Multiplex networks are also known as multi-relational, multi-layered or multi-slice networks [4,19,6]. One fundamental task of multiplex network research is to detect community structure [5,8,17]. Community detection of multiplex networks is a clustering problem that clusters nodes into different groups. This work faces a problem that information of all simplex networks should be considered synthetically to get better results of community detection. Most of current approaches to the problem aggregate all the links in different simplex networks together treating them as being equivalent. However, such aggregation will bring about two drawbacks. Firstly, it will lead to lose importance information of original systems, since in many cases the basic components of a system might be connected by relationships which have different correlation and importance. For

B. Benatallah et al. (Eds.): WISE 2014, Part I, LNCS 8786, pp. 31–46, 2014.
© Springer International Publishing Switzerland 2014

instance, a multiplex network (Indonesian terrorists[1]) that includes four simplex networks (T, O, C, B) respectively representing four relationships (trust, operations, communication, business) among 78 Indonesian terrorists. Simplex network T has larger importance than O, C and B, because T relationship makes more influence to other three relationships and the stronger T relation is, the stronger other three relationships become. Aggregation treating all the links of the four relationships as being equivalent will result into that we can't distinguish importance difference between them. In addition, for each simplex network, the aggregation considers only adjacency relation among nodes. But adjacency relation can't reflect very well nodes' closeness, which is important to community detection. For example, for two nodes, even though they are not connected by one edge directly, it doesn't mean that there is no closeness between them. But such two nodes are measured by value 0 in adjacency relation. If such closeness is ignored, the accuracy of community detection for multiplex networks will decrease. Therefore, it is necessary to mine the interaction of simplex networks more precisely.

Aiming at solving the two drawbacks to get better results of community division, we propose a unified model to detect community structure of multiplex networks. The model divides nodes into different communities according to a unified matrix transferred from multiplex networks. Particularly, we define the correlations among simplex networks firstly, and demonstrate that there is difference between simplex networks by analyzing the correlations in Section 3. Then we define importance of simplex networks to describe the difference in multiplex network and define node similarity to measure closeness among nodes in simplex networks precisely in Section 4.The model gets ideal results on synthetic datasets and real world datasets.

2   Related Work

Community detection of multiplex networks needs to consider all the information from different simplex networks at the same time.A better strategy to resolve this problem is to aggregate all the information of different simplex networks together. Several typical methods of information aggregation in multiplex networks are as follows.

In [23], authors defined inter-simplex connections to describe correlations across different simplex networks. Based on the inter-simplex connections, authors extended the definition of modularity to fit multiplex networks. The authors treated each inter-simplex connection as being equivalent and measured them with the same value in the modularity. But it can't reflect importance of simplex networks very well because different simplex networks have different correlations with other simplex networks. In [9], authors presented a method to reduce network size aiming at improving efficiency and then took modularity optimization to detect communities of multiplex network. However, the problem mentioned above in modularity is still unsolved.

---

[1] https://sites.google.com/site/sfeverton18/research/appendix-1

Literatures [22] and [3] both defined an aggregated adjacency matrix obtained from multiplex network. In the matrix, two nodes are joined when they are connected by an edge in at least one simplex network of multiplex network. The method is easy to be understood, but it ignores the difference between edges from different simplex networks. In addition, if the aggregated adjacency is used to detect community of multiplex networks, the results might not be accurate enough, because it can't reflect closeness of nodes in multiplex networks well. In [22], authors only demonstrated that different simplex networks of multiplex network have information difference, but they didn't propose a quantitative measure method for the difference.

In [30], authors described one method, called Network Integration, for information integration by calculating average interactions of nodes in multiplex networks. It considered the fact that interactions come from different simplex networks, but it treats each simplex network as being equivalent that will result into losing difference between different simplex networks. In [29], authors proposed another method for information integration named Feature Integration. The method performs the integration over the structural features extracted from each simplex network. However, when constructing feature matrix, authors didn't distinguish importance among the feature vectors.

Strehl et.al presented Partition Integration [27]. The method detected communities for each simplex networks firstly, and then constructed similarity matrix for each simplex network where the similarity between two nodes is 1 if the two nodes are in the same simplex community, otherwise 0. Nevertheless, the two values are not enough to describe similarity among nodes in simplex networks very well, because the similarities among nodes of a simplex community are different and we should not treat them all as being 1.

Therefore, community detection of multiplex networks should consider closeness of nodes for each simplex network and importance difference among simplex networks.

3   Correlation and Importance

In this section, we will define correlations among simplex networks, and then use the correlations to analyze and define importance of simplex networks.

3.1   Correlations among Simplex Networks

We first define inter-simplex correlations of one node that measure how correlated two simplex networks are for the node. Then we extend it to correlations among simplex networks.

Inspired by [23], we couple arbitrary node with itself between different simplex networks by inter-simplex connections described as Fig. 1. Here, the inter-simplex connections represent correlations for one node between different simplex networks.

One node has interactions with other nodes in different simplex networks. There are relationships between the interactions for the node. To measure the

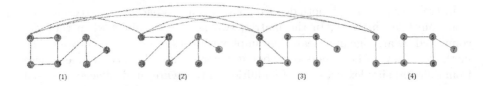

Fig. 1. Schematic of connections among different simplex networks. The multiplex network is composed of four simplex networks s = 1,2,3,4. For each node, there exists one connection between any two simplex networks. For clarity, only connections for node "1" are shown. The connections for other nodes are the same as node "1".

relationships, we define the inter-simplex correlation for one node. The inter-simplex correlation of one node is defined as Eq. (1).

$$C_{isr} = \frac{\sum_j a_{ij}^s a_{ij}^r}{\sum_j a_{ij}^s + \sum_j a_{ij}^r - \sum_j a_{ij}^s a_{ij}^r} \tag{1}$$

where a_{ij}^s represents interaction of node i and node j in simplex network s. It takes value 1, if there is an edge connecting them in simplex network s, otherwise, it takes value 0. Note that the numerator of Eq. (1) is equal to the number of such edges that connect node i and appear in both simplex network s and simplex network r. The denominator is equal to the number of the edges that connect node i and appear in simplex network s or simplex network r, but if one edge of node i appears in both simplex network s and r, it is only counted by once in denominator. The inter-simplex correlation of node i, C_{isr} takes values in [0,1]. If C_{isr} is larger, the interactions of node i between different simplex networks are closer and the inter-simplex correlation of node i is stronger. C_{isr} can get values 1 and 0 respectively under following two cases. The two cases are correspondingly that for one node, if its' interactions in two simplex networks are same or totally different, the node has strongest or weakest inter-simplex correlation between the two simplex networks.

Based on the inter-simplex correlations of one node, we now analyze the correlations among different simple networks. The relation between inter-simplex correlations of one node and correlations among different simple networks is that the stronger the inter-simplex correlations between two simplex networks of nodes are, the stronger the correlations between the two simplex networks is. We thus can utilize inter-simplex correlations to define correlations among simplex networks. The correlations among simplex networks is defined as Eq. (2).

$$C_{sr} = \frac{1}{n} \sum_i C_{isr} \tag{2}$$

where n is the number of nodes in the multiplex network and C_{isr} is inter-simplex correlation of node i between simplex network s and r. C_{sr} takes value in [0,1]. The stronger the correlation among simplex network s and r becomes, the larger the value of C_{sr} is.

3.2   Importance of Simplex Networks

In this section, we verify that different simplex networks have different importance by analyzing correlations among simplex networks, and then define the importance. As a case of study, in this part we focus on the multiplex network of Indonesian terrorists described in Section 1. Especially, T includes three sub-simplex networks, i.e. C (classmates), F (friendship) and K (kinship).

We analyze correlations among the four relationships in Indonesian terrorists according to Eq. (2), and the result is plotted as a symmetric heat-map in Fig. 2 (1). Each color section of the heat-map shows the correlation between two simplex networks. For instance, the first color section in the first row and the last color section in the last row represent the correlation between T and C. The color sections in back diagonal stand for correlations between one simplex network with itself and they are the largest. Besides, the map tells us that T has larger correlations than other three simplex networks. Now that T has three sub-simplex networks, we can consider part of them when computing correlations between T and B, O, C. Fig. 2 (2) shows correlations between T and B, O, C under three cases .The first case described by dark line is that we consider all three sub-simplex networks of T; the second case represented by red line is that we consider sub-simplex networks C and F of T, while the third case described by blue line is that we just consider sub-simplex network C of T. Notice that the more sub-simplex networks of T we consider, the stronger the correlations between T and B, O, C become.

We can explain the phenomenon above by importance of simplex networks in multiplex networks. Because different simplex networks have different importance in multiplex network and T has larger importance in Indonesian terrorists than B, O and C, it can influence the three simplex networks more in multiplex networks and it has larger correlations with them. Besides, if we consider more sub-simplex networks of T, the importance of T becomes larger and its' correlation with others becomes stronger.

To verify the importance of simplex network in multiplex networks, we construct two random networks T1 and T2 that are obtained by randomizing the edges keeping fixed the total number of links and the degree distribution for T respectively. In Fig. 2 (3), we compare the correlations of T, T1 and T2. The results show that the two constructed networks have lower correlations than T. It illustrates that there is importance of simplex networks in original multiplex network indeed.

Correlations among simplex networks have close relationship with importance of simplex networks. If correlations between one simplex network and other simplex networks are larger, the importance of the simplex network is stronger. Therefore, we utilize the correlations to define importance of simplex networks as shown in Eq. (3).

$$I_s = \frac{\sum_r C_{rs}}{\sum_s \sum_r C_{rs}} \tag{3}$$

where C_{rs} represents correlation between simplex network r and s. The numerator of Eq. (3) is correlations between simplex network s and others. The

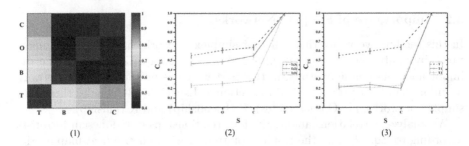

Fig. 2. (1) Heat map of correlations among T, B, O and C. (2) The correlations of T(1), T(2), T(3). T(1), T(2), T(3) stand for Trust relationship respectively considering one, two and three sub-simplex networks. (3) The correlations of T, T1 and T2.

denominator is all correlations among any two simplex networks. I_s takes values in $[0, 1]$.

4   Community Information Extraction

After discussing importance of simplex networks that is a relation between two simplex networks, we focus our attention to closeness that is a relation between two nodes in simplex networks. Closeness measures how close two nodes are, which is the useful simplex community information because two nodes having large closeness should be divided into the same community in simplex networks according to definition of community [13,12,21]. To get better community division of multiplex networks, it is necessary to extract and utilize the closeness of nodes.

Jaccard's coefficient, Salton index, Sorenson index, Common neighbors, Adamic-Adar and LHN (Lei cht-Holme-Newman) [10,25,26,24,1] are five famous methods for node silimarity. Zhou Tao and Liben-Nowell have proved that common neighbors is the simplest method "Common neighbors" which usually performs surprisingly well. However, "Common neighbors" requires that there are common neighbors between two nodes, otherwise the similarity of two nodes will be 0. Literature [28] resolved the problem above and defined the similarity between two nodes connected by only one edge without common neighbors. But similarity of two nodes is 0 if the two nodes are not connected by one edge directly. Here, we propose a new method to compute node similarity between node v_i and v_j described as Eq. (4).

$$Sim(v_i, v_j) = \begin{cases} 1, & \text{if } v_i \text{ and } v_j \text{ are same;} \\ \sum_{B \in P} \prod_{E(u,w) \in B} S(u, w), & \text{if there are paths between } v_i \text{ and } v_j; \\ 0, & \text{otherwise.} \end{cases} \tag{4}$$

$$S(u, w) = \frac{1}{deg(u) + deg(w) - 1} \tag{5}$$

In Eq. (4), P is the set of paths between node v_i and v_j; E(u, w) is one edge belonging to the path B and u, w is two nodes of the edge E(u, w). In Eq. (5), deg(u) and deg(w) are degree of nodes u and w respectively. The node similarity can measure similarity of such two nodes that are not connected by one edge directly or between which there are not common neighbors . To extract node similarity from each simplex network, we construct a similarity matrix A^i that is square matrix for the ith simplex network. $A^{(i)}_{(u,w)}$ is node similarity for node u and node w of the i-th simplex network. The value of $A^{(i)}_{(u,w)}$ is defined as :

$$A^{(i)}_{(u,w)} = Sim(u, w) \tag{6}$$

The node similarity matrix of a multiplex network containing m simplex networks can be represented as

$$A = \left\{ A^{(1)}, A^{(2)}, ..., A^{(m)} \right\} \tag{7}$$

with $A^{(i)}$ satisfying

$$A^{(i)} \epsilon R^{n*n}_+, A^{(i)} = (A^{(i)})^T, i = 1, 2, ..., m \tag{8}$$

where n is the total number of nodes involved in the multiplex network.

5   A Unified Model to Community Structure of Multiplex Networks

In order to detect community structure of multiplex network better, we propose a unified model. In the model, we use importance in Section 3 and node similarity in Section 4 to construct a unified matrix.The matrix L is defined as:

$$L = \sum_{i=1}^{m} I_i A^{(i)} \tag{9}$$

where m is the number of simplex networks; $A^{(i)}$ is node similarity matrix of the ith simplex network and I_i is the importance of the ith simplex network described as Eq. (3). To explain the process of constructing the unified matrix, we construct a multiplex networks having 7 nodes and 3 simplex networks described as Fig. 3 and Fig. 4 shows the schematic illustration of constructing the unified matrix for the multiplex network.

After constructing unified matrix, we describe the unified model as Table1.

Step5 of the model needs to detect community of multiplex networks according to an unified simplex network corresponding to the unified matrix. Therefore, we can detect multiplex community structure utilizing simplex community discovery algorithms in Step5. In this paper, we chose three simplex community discovery algorithms for Step5, i.e. Louvain [11], OSLOM[20] and Infomap[16]. They all can be applied to the unified simplex network. Correspondingly, we call our unified model as OurModel(Louvain), OurModel(OSLOM) and OurModel(Infomap).

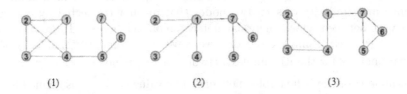

Fig. 3. A multiplex network with 7 nodes and 3 simplex networks.(1),(2),(3) are the 3 simplex networks.

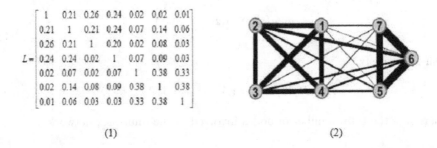

(1)   (2)

Fig. 4. Schematic illustration of constructing a unified matrix for the multi-graph described in Fig. 3. (1) The unified matrix L got from multiplex network in Fig. 3 according to Eq. (9). (2) A simplex network corresponds to the unified matrix L and the thicker the edges are, the larger the corresponding value in unified matrix is.

Table 1. The unified model for multiplex community detection

Input:	$Net = \left\{ N^{(1)}, N^{(2)}, ..., N^{(m)} \right\}$; one multiplex network containing n nodes and m simplex networks, and $N^{(m)} \epsilon R_+^{n*n}$.
Output:	$Community = \left\{ C^{(1)}, C^{(2)}, ..., C^{(s)} \right\}$, where $C^{(s)}$ is the sth community.
Step1:	Compute importance of each simplex network $N^{(i)}$ according to Eq. (3) and get importance vector $I = \{I_1, I_2, ..., I_m\}$.
Step2:	Compute node similarity among n nodes of each simplex networks according to Eq. (4).
Step3:	Construct node similarity matrix for each simplex network according to node similarity got in Step2.
Step4:	Construct unified matrix L utilizing node similarity matrix in Step3 and importance vector in Step1 based on Eq. (9).
Step5:	Use one simplex community detection method to detect communities of multiplex network according to matrix L .
Step6:	Return $Community = \left\{ C^{(1)}, C^{(2)}, ..., C^{(s)} \right\}$.

6   Experiments

In this section, we apply our methods to two synthetic datasets and four real datasets.

The two synthetic datasets [30] have 4 kinds of interactions and 3 communities, having 50, 100, 200 members respectively. We call them S1 and S2. They are shown as Fig. 5 and Fig.6.

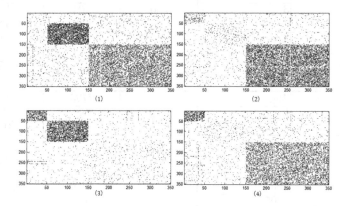

Fig. 5. The first synthetic dataset with (1),(2),(3),(4)representing four simplex networks. Edge distribution of the four simplex networks are relatively average.

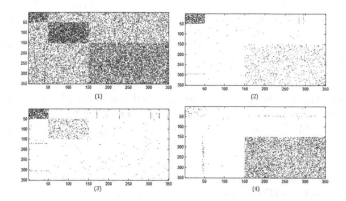

Fig. 6. The second synthetic dataset with (1),(2),(3),(4)representing four simplex networks. Edge distribution of the four simplex networks are not relatively average.

The four real data sets are listed as follow.

Flickr[2]. The multiplex network is a person-person network, having three simplex networks, "Tag", "Favorites" and "Comment", representing respectively

---

[2] http://www.flickr.com

Table 2. Basic characters of six data set for experiments

Multiplex Networks	Simplex Network	Node Number	Edge Number	Average Degree
S1	1	350	3558	20.33
	2		3178	18.16
	3		3015	17.23
	4		3246	18.55
S2	1	350	6326	36.15
	2		555	3.17
	3		306	1.75
	4		670	3.83
Indonesian Terrorists	Trust	78	259	6.64
	Operation		473	12.13
	Communication		200	5.13
	Business		15	0.38
Flickr	Tag	517	7783	30.11
	Favorite		8885	34.37
	Comment		8119	31.41
DBLP	2001	332	1753	10.56
	2002		1589	9.57
	2003		1346	8.11
	2004		1376	8.29
	2005		1678	10.11
	2006		1509	9.09
	2007		1547	9.32
	2008		1472	8.87
	2009		1464	8.82
	2010		1745	10.51
IMDB	Comedy	91	695	15.27
	Romance		2021	44.42
	Crime		243	5.34

that two users tagged the same picture, marked the same picture as a favorite and commented on the same picture. **Indonesian Terrorists**. It is described in Section1, but in this section, we don't consider T's sub-simplex networks. **DBLP**[3]. We extracted author-author relationships if two authors collaborated in writing at least one paper in each year from 2001 to 2010 among 332 authors. In the multiplex network, the author-author collaboration of each year is a simplex network. **IMDB**[4]. We extract a multiplex network of collaborations among 91 actors. The multiplex network includes four simplex networks, each one representing collaboration in one of four movie categories, i.e., "Action", "Comedy", "Romance" and "Crime",if two actors collaborate in at least one movie of corresponding kind.Table2 shows the basic characters of six datasets.

We compare our methods with two methods. One method is based on multiplex modularity [23], and we call it as CUMN (community unfolding in multi-slice

---

[3] http://www.informatik.uni-trier.de/?ley/db
[4] http://www.imdb.com/

networks) [9]. Another method is based on information aggregation, and we call it as CUPI (community detection based on partition aggregation)[27].

6.1 Experiment on Synthetic Datasets

We carry on this experiment with our methods and two other methods, i.e., CUMN and CUPI. Since we know the real communities for the synthetic data, we can use the NMI (Normalized Mutual Information) [2] to measure the accuracy of our methods. NMI takes values in [0, 1]. The higher the value of NMI is, the accurater the result of community detection is.

Table 3 and Table 4 show the importance of simplex networks in synthetic data S1 and S2 respectively. As shown in S1, the importance of different simplex networks is close and there is not large difference between them. Differently, in S2, simplex network 1 plays an absolutely more important role than others. We want to test our methods on such two different synthetic data. In Fig. 7, we report the NMI values for different methods. The dark dot line shows the NMI values for different methods on S1 and the blue solid line shows the results on S2. Comparing the two results, we can find that our methods get better results of multiplex community detection than CUMN and CUPI no matter whether the synthetic data has large importance difference. In particular, if there are distinct importance differences between simplex networks, our model is better, because our methods consider the influence of simplex importance to multiplex community detection. Notice that CUPI that is based on information aggregation performs worse on S2 than on S1, because on S2, there is big difference between the importance of simplex networks, and if it is ignored, there will be big biases. For multiplex network S1, where there are not distinct importance difference between simplex networks, our methods extract useful node similarity information rather than aggregate the simplex networks together to detect multiplex communities, that is why our methods perform better than others.

In addition, comparing OurModel(Louvain), OurModel(OSLOM) and Our-Model(Infomap) on the two synthetic datasets, we notice that the NMI values of the three methods are close. We can conclude that the choice of simplex community detection methods does not significantly affect result of our method.

Table 3. Importance of different simplex networks in multiplex network S1

Simplex networks	1	2	3	4
Importance	0.271	0.243	0.245	0.241

Table 4. Importance of different simplex networks in multiplex network S2

Simplex networks	1	2	3	4
Importance	0.532	0.173	0.145	0.150

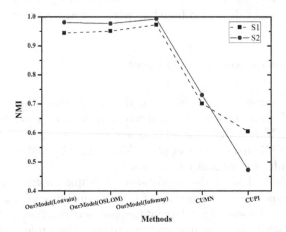

Fig. 7. NMI values of five methods on two synthetic data. The dark dot line describes the NMI values of five methods on S1. The blue solid line describes NMI values on S2.

6.2   Experiment on Real World Datasets

For the real data, we don't have the ground truth and we take modularity [14] to measure the accuracy of our methods. This measure gives a value between 0 and 1. The higher the value of modularity is, the better the result of community detection is. We divide the real data into two groups that are Indonesian Terrorists, Flickr and DBLP, IMDB. Because in Indonesian Terrorists and Flickr, different simplex network represents different relationship for instance T (Trust relationship) and O (Operational relationship) in Indonesian Terrorists, while in DBLP and IMDB, different simplex network represents different value of a same relationship for example Romance (cooperation in romantic movies) and Crime (cooperation in crime movies) in IMDB. If different simplex network stands for different relationship, we can't aggregate the simplex networks together simply without information extraction, while if different simplex network stands for different value of a same relationship, we can aggregate them together. We test our methods on such two groups of real data.

Fig. 8 shows the importance of simplex networks in different multiplex networks. (1)-(4) stand for Indonesian Terrorists, Flickr, IMDB and DBLP respectively. Notice that in different multiplex networks, the importance distributions are different. For instance, in Flickr the importance distribution is homogeneous while in Indonesian Terrorists, Trust relationship takes bigger importance than others. The same situation occurs on DBLP and IMDB. Analyzing the reason why Trust in Indonesian Terrorists and Romance in IMDB are more important, we find that if two members trust each other, they are more possible to operate together, communicate or have common business actually. In other words, Trust relationship have actually caused other three relationships. In IMDB, the

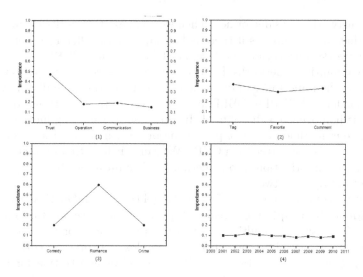

Fig. 8. Importance of simplex networks in different multiplex networks. (1)-(4) stand for Indonesian Terrorists, Flickr, DBLP and IMDB respectively.

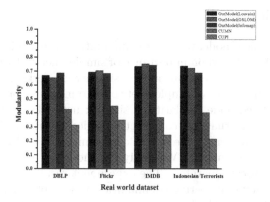

Fig. 9. A multiplex network with 7 nodes and 3 simplex networks.(1),(2),(3) are the 3 simplex networks.

91 actors cooperate each other more in Romance movies. The result indicates that our definition of importance can reflect real importance of simplex networks in real multiplex networks.

In Fig. 9, we show the modularity of five methods in four real multiplex networks. Firstly, we compare the two groups i.e. Indonesian Terrorists, Flickr, and IMDB, DBLP. As show, our methods i.e. OurModel (Louvain), OurModel (OSLOM) and OurModel (Infomap) perform better than CUMN and CUPI in both the two groups. As analyzed above, the difference of the two groups is whether the relationships in multiplex networks are same or not. The result shows that our methods are better no matter whether the relationships in multiplex networks are same or not. Then we compare the results in each group. We find that in both groups if there is big importance differences between the simplex networks, our methods perform much better, while CUPI performs worse on such real multiplex networks. It turns out that it is necessary that our methods consider the importance of different simplex networks. In addition, we compare our model and CUPI in DBLP and Flickr where the importance of simplex networks is homogeneous distributed. In this case, considering importance difference in our methods doesn't play a major role, however, the results show that our methods are better than CUPI. We can conclude that extraction useful node similarity information in our methods is helpful to get better community division of multiplex networks.

Finally, we compare our model i.e. OurModel (Louvain), OurModel(OSLOM) and OurModel(Infomap) in four real multiplex networks, we find that no matter whether relationships in multiplex networks are the same or not and whether there are importance differences between simplex networks or not, the three methods get close modularity values. It is a good demonstration for that the choice of simplex community detection methods will not change the result of our methods significantly.

7   Conclusion

We propose a unified model to detect communities of multiplex networks in the paper. The model is based on importance of simplex networks and node similarity. Firstly, we define correlations among simplex networks, and then we prove and define the importance of simplex networks to measure correlation difference and then we defined node similarity in simplex networks to measure closeness between nodes. After that, we construct a unified matrix utilizing importance and node similarity. Finally, we detect communities of multiplex networks according to the unified matrix. We apply the model on two synthetic datasets and four real datasets separately. The experiment results show that our model has higher accuracy in detecting communities of multiplex networks comparing with the competitive methods.

Acknowledgments. The Research was supported in part by Natural Science Foundation of China (No.60903071), National Basic Research Program of China (973 Program, No.2013CB329605), Specialized Research Fund for the Doctoral Program of Higher Education of China, and Training Program of the Major Project of BIT.

References

1. Adamic, L.A., Adar, E.: Friends and neighbors on the web. Social Networks 25(3), 211–230 (2003)
2. Aldecoa, R., Marín, I.: Deciphering network community structure by surprise. PloS One 6(9), e24195 (2011)

3. Battiston, F., Nicosia, V., Latora, V.: Structural measures for multiplex networks. Physical Review E 89(3), 032804 (2014)
4. Berlingerio, M., Coscia, M., Giannotti, F., Monreale, A., Pedreschi, D.: Multidimensional networks: foundations of structural analysis. World Wide Web 16(5-6), 567–593 (2013)
5. Berlingerio, M., Pinelli, F., Calabrese, F.: Abacus: frequent pattern mining-based community discovery in multidimensional networks. Data Mining and Knowledge Discovery 27(3), 294–320 (2013)
6. Bródka, P., Kazienko, P., Musiał, K., Skibicki, K.: Analysis of neighbourhoods in multi-layered dynamic social networks. International Journal of Computational Intelligence Systems 5(3), 582–596 (2012)
7. Brummitt, C.D., Lee, K.M., Goh, K.I.: Multiplexity-facilitated cascades in networks. Physical Review E 85(4), 045102 (2012)
8. Cai, D., Shao, Z., He, X., Yan, X., Han, J.: Community mining from multi-relational networks. In: Jorge, A.M., Torgo, L., Brazdil, P.B., Camacho, R., Gama, J. (eds.) PKDD 2005. LNCS (LNAI), vol. 3721, pp. 445–452. Springer, Heidelberg (2005)
9. Carchiolo, V., Longheu, A., Malgeri, M., Mangioni, G.: Communities unfolding in multislice networks. In: da F. Costa, L., Evsukoff, A., Mangioni, G., Menezes, R. (eds.) CompleNet 2010. CCIS, vol. 116, pp. 187–195. Springer, Heidelberg (2011)
10. Chowdhury, G.: Introduction to modern information retrieval. Facet Publishing (2010)
11. De Meo, P., Ferrara, E., Fiumara, G., Provetti, A.: Generalized louvain method for community detection in large networks. In: 2011 11th International Conference on Intelligent Systems Design and Applications (ISDA), pp. 88–93. IEEE (2011)
12. Fortunato, S.: Community detection in graphs. Physics Reports 486(3), 75–174 (2010)
13. Girvan, M., Newman, M.E.: Community structure in social and biological networks. Proceedings of the National Academy of Sciences 99(12), 7821–7826 (2002)
14. Gomez, S., Diaz-Guilera, A., Gomez-Gardeñes, J., Perez-Vicente, C.J., Moreno, Y., Arenas, A.: Diffusion dynamics on multiplex networks. Physical Review Letters 110(2), 028701 (2013)
15. Gómez-Gardenes, J., Reinares, I., Arenas, A., Floría, L.M.: Evolution of cooperation in multiplex networks. Scientific Reports 2 (2012)
16. Greene, D., O'Callaghan, D., Cunningham, P.: Identifying topical twitter communities via user list aggregation. arXiv preprint arXiv:1207.0017 (2012)
17. Gregory, S.: Finding overlapping communities using disjoint community detection algorithms. In: Fortunato, S., Mangioni, G., Menezes, R., Nicosia, V. (eds.) Complex Networks. SCI, vol. 207, pp. 47–61. Springer, Heidelberg (2009)
18. Hao, J., Cai, S., He, Q., Liu, Z.: The interaction between multiplex community networks. Chaos: An Interdisciplinary Journal of Nonlinear Science 21(1), 016104 (2011)
19. Harrer, A., Schmidt, A.: An approach for the blockmodeling in multi-relational networks. In: 2012 IEEE/ACM International Conference on Advances in Social Networks Analysis and Mining (ASONAM), pp. 591–598. IEEE (2012)
20. Lancichinetti, A., Fortunato, S.: Community detection algorithms: a comparative analysis. Physical Review E 80(5), 056117 (2009)
21. Leskovec, J., Lang, K.J., Mahoney, M.: Empirical comparison of algorithms for network community detection. In: Proceedings of the 19th International Conference on World Wide Web, pp. 631–640. ACM (2010)
22. Magnani, M., Micenkova, B., Rossi, L.: Combinatorial analysis of multiple networks. Tech. rep. (2013)

23. Mucha, P.J., Richardson, T., Macon, K., Porter, M.A., Onnela, J.P.: Community structure in time-dependent, multiscale, and multiplex networks. Science 328(5980), 876–878 (2010)
24. Salton, G.: Automatic Text Processing: The Transformation, Analysis, and Retrieval of. Addison-Wesley (1989)
25. des Sciences Naturelles, S.V.: Bulletin de la Société vaudoise des sciences naturelles, vol. 7. Impr. F. Blanchard (1864)
26. Sorenson, T.: A method of establishing groups of equal amplitude in plant sociology based on similarity of species content. Kongelige Danske Videnskabernes Selskab 5(1-34), 4–7 (1948)
27. Strehl, A., Ghosh, J.: Cluster ensembles—a knowledge reuse framework for combining multiple partitions. The Journal of Machine Learning Research 3, 583–617 (2003)
28. Symeonidis, P., Tiakas, E., Manolopoulos, Y.: Transitive node similarity for link prediction in social networks with positive and negative links. In: Proceedings of the Fourth ACM Conference on Recommender Systems, pp. 183–190. ACM (2010)
29. Tang, L., Liu, H.: Uncovering cross-dimension group structures in multi-dimensional networks. In: SDM Workshop on Analysis of Dynamic Networks (2009)
30. Tang, L., Wang, X., Liu, H.: Community detection in multi-dimensional networks. Tech. rep., DTIC Document (2010)

Mining Domain-Specific Dictionaries
of Opinion Words

Pantelis Agathangelou[1], Ioannis Katakis[2], Fotios Kokkoras[3],
and Konstantinos Ntonas[4]

[1] Open University of Cyprus, Nicosia, Cyprus
`pandelis.agathangelou@st.ouc.ac.cy`
[2] National and Kapodistrian University of Athens, Athens, Greece
`katak@di.uoa.gr`
[3] Technological Educational Institute of Thessaly, Larisa, Greece
`fkokkoras@teilar.gr`
[4] International Hellenic University, Thessaloniki, Greece
`k.ntonas@ihu.edu.gr`

Abstract. The task of opinion mining has attracted interest during the
last years. This is mainly due to the vast availability and value of opinions
on-line and the easy access of data through conventional or intelligent
crawlers. In order to utilize this information, algorithms make extensive use of word sets with known polarity. This approach is known as
dictionary-based sentiment analysis. Such dictionaries are available for
the English language. Unfortunately, this is not the case for other languages with smaller user bases. Moreover, such generic dictionaries are
not suitable for specific domains. Domain-specific dictionaries are crucial
for domain-specific sentiment analysis tasks. In this paper we alleviate
the above issues by proposing an approach for domain-specific dictionary building. We evaluate our approach on a sentiment analysis task.
Experiments on user reviews on digital devices demonstrate the utility of
the proposed approach. In addition, we present NiosTo, a software that
enables dictionary extraction and sentiment analysis on a given corpus.

1 Introduction

Sentiment analysis is the task of extracting valuable, non-trivial knowledge from
a collection of documents containing opinions. Most of the times, extracted
knowledge represents a summary of the opinions expressed in the collection.
Opinions can refer to products, services or even political figures. Advanced algorithms are able to discover opinions of multiple features for the entity under investigation. These techniques are applied to data extracted from various sources
like discussion boards, social networks, blogs or video sharing networks.

Sentiment analysis is utilized in different levels of granularity: at sentence-
level, document-level or corpus-level. A key-role to all these levels play the so-
called list of "opinion-words". This is a dictionary containing terms of known
polarity. In most cases, the list is dual. A *positive* word-list ("beautiful",

B. Benatallah et al. (Eds.): WISE 2014, Part I, LNCS 8786, pp. 47–62, 2014.
© Springer International Publishing Switzerland 2014

"astonishing") is coupled with a *negative* word-list ("ugly", "slow"). Opinion dictionaries are critical to various steps of sentiment analysis since algorithms depend their initialization, enhancement and operation on them. Therefore, high-quality dictionaries are required for these type of analytics.

Such dictionaries are already available for the English language [4]. Many of them are manually constructed. However, the availability of opinion word-lists for less popular languages is very limited. Another issue with such lists is that they are domain dependent. For example the words 'cool' or 'low', might have different meaning in different domains. Therefore approaches that are able to extract domain-specific opinion dictionaries are necessary.

In this paper, we provide a method that mines domain-specific dictionaries given a corpus of opinions. This is a multiple-stage iterative approach that gets a small seed of generic opinion words as input and extends it with domain-specific words. It utilizes language patterns and takes advantage of a double propagation procedure between opinion words and opinion targets. The effectiveness of the approach is estimated in a sentiment analysis task. Results justify the utility of all steps of the proposed algorithm. In addition, we present a software that enables the use of the above techniques under a user-friendly interface.

The *advantages* of the proposed algorithm are: a) it is domain independent, b) it can operate with a very small initial seed-list, c) it is unsupervised and, d) it can operate in multiple languages provided the proper set of patterns. The *contribution* of this work can be summarized in the following points.

- Introduces a novel resource-efficient approach for building domain-specific opinion dictionaries.
- Provides an experimental study where all steps of the proposed algorithm are evaluated through a sentiment analysis task.
- Provides a dataset of opinions in the Greek language containing user reviews. This dataset can be exploited in various opinion mining tasks.
- Offers `NiosTo`, a free application that integrates opinion dictionary discovery and sentiment analysis tasks.

The rest of the document is structured as follows. Section 2 summarizes and highlights related work. In Section 3, we outline our approach and provide detailed description of all steps. After that, the experimental evaluation is presented (Section 4) followed by results and discussion (Section 4.2). The reader can learn about the basic feature-set of the `NiosTo` software at Section 5, while the last section highlights significant conclusions and suggests future work.

2   Related Work

In [1] a probabilistic method is presented that builds an opinion word lexicon. The method uses a set of opinion documents which is used as a biased sample and a set of relevant documents as a pool of opinions. In order to assess the effectiveness of the algorithm a dictionary made up of $8K$ words is used, built by [9, 10]. Certain probabilistic functions such as Information Content, Opinion

Entropy and Average Opinion Entropy are used as extraction tools. The method is based upon the observation that nouns contain high information value, while adjectives, adverbs and verbs (usually opinion words) provide additional information to the context. Upon these observations and the probabilistic tools they extract the opinion word lexicon.

The authors of [12] tackle the problem of opinion target orientation and summarization. The method uses an opinion lexicon [4] from WordNet. A list of content dependent opinion words such as nouns, verbs and word phrases that are joined together is utilized. The algorithm uses a score function, which is a formula that calculates opinion target orientation, by exploiting coexistence of opinion words and opinion targets in a sentence and the variance of distance among them. Linguistic patterns and syntactic conventions are used in order to boost the efficiency of the proposed method.

[3] proposes an unsupervised lexicon building method for the detection of "polar clauses" (clauses that can be classified as positive or negative) in order to acquire the minimum syntactic structures called "polar atoms" (words or phrases that can be classified as positive or negative opinion modifiers). This part of process includes a list of syntactic patterns that helps the identification of propositional sentences. Moreover the method uses an opinion lexicon and statistical metrics such as coherent precision and coherent density in order to acquire true polar atoms from fake ones.

The authors of [7] exploit a model called partially supervised word alignment, which discovers alignment links between opinion targets and opinion modifiers that are connected in bipartite graph. Initially some high precision low recall syntactic patterns are used as training sets for generating initial partial alignment links. Then these initial links will be fed into the alignment model. The selection of opinion target candidates is based upon a factor called confidence. Candidates with higher confidence will be extracted as the opinion targets.

Our approach combines various components of the above methods, refines them and introduces new processes to overcome their disadvantages. We propose a multi-stage approach that includes conjunction based extraction and double propagation. The latter is applied more than once. However, in parallel with word extraction we employ *polarity disambiguation* after every step in order to identify the sentiment of the newly discovered words. Finally, we utilize a word validation process that takes advantage of opinion-words - opinion-target relationships by introducing parameters that improve retrieval performance.

3 Our Approach

The approach comprises of a series of steps that gradually detect opinion words. Each step creates a pool of opinion words that will constitute the feed for the next detection step. More specifically the construction of the proposed algorithm can be summarized in the following steps: 1) Opinion Preprocessing 2) Auxiliary List Preparation, 3) Seed Import and Filtered Seed Extraction, 4) Conjunction Based Extraction, 5) Double Propagation, and 6) Opinion Word Validation. In

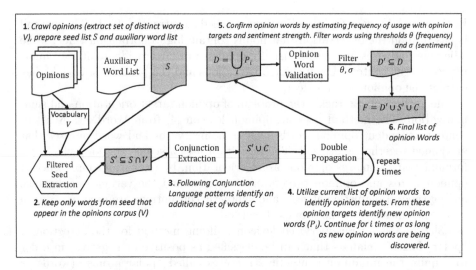

Fig. 1. Overview of the Proposed Approach. Gray shaded items represent the opinion word list that is improved in each step. Please follow the numbers (1-6).

the following subsections, we present the above steps in more detail. Figure 1 presents an overview of the proposed approach.

Opinion Preprocessing. At this step the designed algorithm receives user opinions in raw form. We implement some form of preprocessing in order to filter-out noise. Sentence splitting is a critical step in this module (opinion delimitation) since double propagation takes into account neighbourhood sentences in order to propagate sentiment. Additionally in order to increase the efficiency of the extraction process we have adopted an on-line stemmer engine for the Greek Language[1]. At this step the set of distinct words V is constructed.

Auxiliary List Preparation. This is a particularly crucial step. Auxiliary word list comprises a series of word sets like articles, verbs, comparatives, conjunctions, decreasers (e.g. "less"), increasers (e.g. "extra"), negations (e.g. "not") and pronouns. These words will constitute a main feed of the algorithm. The proposed approach utilizes this seed in the construction of all extraction patterns.

3.1 Seed Import and Filtered Seed Extraction

The initial Seed S of the system is a set opinion words with known polarity (e.g. "bad", "ugly", "wonderful", etc) [4,11]. The Seed is *generic*, meaning that it is not domain-specific. The Filtered Seed S' is the set of Seed words that also appeared in the collection of opinion documents $S' = S \cap V$. In other words we filter-out words from the Seed that don't appear in the corpus. In Seed list, the

---

[1] http://deixto.com/greek-stemmer/

Table 1. Examples of patterns used for sentiment disambiguation

					Pol							Pol
{fut}	{?}	{fut}	{verb}	{pos}	-1		{neg}	{verb}	{conj}	{comp}	{neg}	+1
{neg}	{verb}	{conj}	{comp}	{pos}	-1		{neg}	{fut}	{art}	{?}	{neg}	+1
{neg}	{pron}	{art}	{pos}	{neg}	+1		{art}	{neg}	{verb}	{decr}		+1
{decr}	{comp}	{pos}			-1	e.g.	the	noise	has	depleted		
e.g.	little	more	useful				{neg}	{verb}	{neg}			+1
{neg}	{pos}				-1		{pos}	{neg}				+1
{pos}					+1		{neg}					-1

polarity of each word is provided. However, depending on the way the word is used, it might alter its polarity ("this phone is definitely not lightweight"). Hence, we apply a step of *polarity disambiguation* using a set of language patterns. Some examples of polarity patterns are presented in Table 1. Word abbreviations stand for {fut}: future word, {pos}: positive word, {neg}: negative word, {conj}: conjunction, {art}: article, {decr}: decreaser, {comp}: comparative word, {? }: any word. In total, we've used 21 positive and 12 negative polarity patterns. At the end of this process we have a pool of newly discovered opinion words along with their polarity. Algorithm 1 and 2 provide the logic of this step.

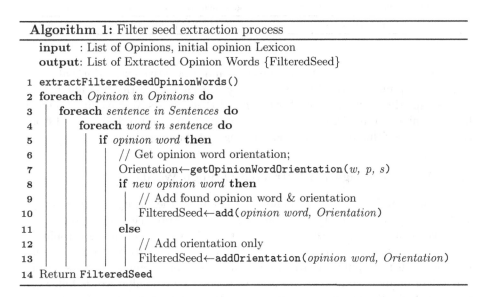

Algorithm 1: Filter seed extraction process

input : List of Opinions, initial opinion Lexicon
output: List of Extracted Opinion Words {FilteredSeed}

```
1  extractFilteredSeedOpinionWords()
2  foreach Opinion in Opinions do
3      foreach sentence in Sentences do
4          foreach word in sentence do
5              if opinion word then
6                  // Get opinion word orientation;
7                  Orientation←getOpinionWordOrientation(w, p, s)
8                  if new opinion word then
9                      // Add found opinion word & orientation
10                     FilteredSeed←add(opinion word, Orientation)
11                 else
12                     // Add orientation only
13                     FilteredSeed←addOrientation(opinion word, Orientation)
14 Return FilteredSeed
```

Algorithm 2: Filter seed & Conjunction based polarity exploration

---

   input : Polarity patterns
   output: Word orientation
1 getOpinionWordOrientation(*word, position, sentence*)
2 **if** *match polarity pattern* **then**
3 | Return **pattern orientation**
4 **else**
5 | Return **default orientation**

3.2 Conjunction-Based Extraction of Opinion Words

At this step we exploit the assumption of *sentiment consistency* [2] that applies in conjuncted words (e.g. "lightweight and well-built device"). That way our algorithm discovers new opinion words by making use of certain conjunction patterns that have been selected to fit the sentiment consistency theories. Table 2 provide examples of such language patterns. Word abbreviations stand for {cpos}: candidate positive word {cneg}: candidate negative word.

Table 2. Examples of positive and negative word conjunction dependencies

	Positive					Negative				
	{pos}	{conj}	{comp}	{art}	{cpos}	{neg}	{conj}	{fut}	{verb}	{cneg}
	{pos}	{conj}	{neg}	{cpos}		{neg}	{conj}	{incr}	{cneg}	
e.g.	thin	and	not	sticky		e.g. expensive	and	too	small	
	{pos}	{conj}	{cpos}			{neg}	{conj}	{cneg}		

For this step we have utilized 6 positive and 4 negative extraction patterns. Candidate opinion words of this step also go through a polarity disambiguation process like the previous step. At the end of this process we have an extended list of opinion words $S' \cup C$ where C is the list of opinion words extracted from this step. Algorithm 3 and 4 provide the logic of this step.

3.3 Double Propagation Extracted Opinion Words

This process of detecting new opinion words follows the theory of double propagation [5, 8]. The assumption is that each opinion word has an opinion target attached to it (e.g. "nice screen"). Here, there is a direct connection (opinion word, 'nice')→(opinion target, 'screen'). Based on double propagation and using the current list of opinion words, we are able to identify opinion targets. Using this set of opinion targets we are able to extract new opinion words following the same logic. So this process is repetitive. It iterates i times or as long as new opinion words are discovered. At each step P_i opinion words are discovered. In the end of this step we end up with list $D = S' \cup C \bigcup_i P_i$. Our experiments

---

Algorithm 3: Conjunction based extraction process

---

input : List of Opinions, filter seed word list, Conjunction list
output: List of Extracted Opinion Words {ConjList}

```
1  extractConjunctionBasedOpinionWords()
2  foreach Opinion in Opinions do
3  |   foreach sentence in Sentences do
4  |   |   foreach word in sentence do
5  |   |   |   if filter seed word then
6  |   |   |   |   if next word conjunction then
7  |   |   |   |   |   // Search the existence of an opinion word
8  |   |   |   |   |   opinion word←getConjunctionBasedOpinionWord(w, i, s)
9  |   |   |   |   |   if opinion word then
10 |   |   |   |   |   |   // Get orientation of opinion word
11 |   |   |   |   |   |   Orientation←getOpinionWordOrientation(w, i, s)
12 |   |   |   |   |   |   if new opinion word then
13 |   |   |   |   |   |   |   // Add found opinion word & orientation
14 |   |   |   |   |   |   |   ConjList←add(opinion word, Orientation)
15 |   |   |   |   |   |   else
16 |   |   |   |   |   |   |   // Add orientation only
17 |   |   |   |   |   |   |   ConjList←addOrientation(opinion word,
                                    Orientation)
18 Return ConjList
```

---

indicate that double propagation is repeated 4 to 8 times. For the reverse step
of double propagation we use a set of patterns that can be seen in 3. The word
abbreviations stand for: {copt}: candidate opinion target, {copw}: candidate
opinion word. In total we have used 6 such word patterns.

The newly discovered opinion words are going through polarity disambiguation. At this step we take advantage of of intra-sentential and inter-sentential
sentiment consistency [3]. The *intra-sentential* consistency suggests that if there
are other opinion words in a sentence with known orientation, then, the newly
found word will get the accumulated sentiment of these words. When there are
no other known opinion words in the sentence, the *inter-sentential* assumption
is applied. It suggests that users tend to follow a certain opinion orientation at
succeeded sentences when forming an opinion. This way if a sentence does not
have an accumulated sentiment we search at nearby sentences (up to a certain
limit) and we assign to the new word, the largest sentiment score found at these
nearby sentences. At the end of this process we have a pool of new opinion words
and their orientation, the double propagation opinion words. Algorithms 5, 6, 7,
8 provide the details of the above processes.

3.4   Opinion Word Validation

The double propagation process makes extensive use of all possible ways to discover new opinion words, but appears to have low precision (see Experimental

---

Algorithm 4: Conjunction based opinion word extraction process

---

  input : Conjunction based extraction patterns
  output: opinion word or null

1 getConjunctionBaseOpinionWord(*opinion word, index, sentence*)
2 **if** *conjunction based pattern* **then**
3 | Return opinion word
4 **else**
5 | Return null

Table 3. Double propagation opinion word dependencies

	{art}	{copt}	{pron}	{verb}	{copw
	{copw}	{conj}	{art}	{copt}	
e.g.	amazing	and	the	cost	
	{copw}	{copt}			

Section). For this reason we apply a filtering procedure, namely *opinion word validation*. We employ two thresholds, the *sentiment threshold (σ)* and the *frequency threshold (θ)*. If a newly found word exceeds these two thresholds then we consider it an opinion word. The intuition behind these thresholds is the following. From the set of candidate opinion words discovered from double propagation, we consider valid opinion words only those that: a) appear more than θ times along with an opinion target *and* b) their sentiment polarity calculated (through polarity disambiguation) is larger than σ (see Equations 1 and 2).

$$w_i = \begin{cases} w_i, \text{ if } \sum_{opinion=0}^{n} (w_i \wedge t_i) \geq \theta \\ \varnothing, \text{ Otherwise} \end{cases} \qquad (1)$$

$$[Sent]_i = \begin{cases} [Sent]_i, \text{ if } \mathrm{Abs}([Sent]_i) \geq \sigma \\ \varnothing, \qquad \text{ Otherwise} \end{cases} \qquad (2)$$

Where w_i stands for double propagation opinion word and t_i for opinion target word. Parameters θ and σ are user defined. The higher these thresholds are, we get higher precision and lower recall. The user can try different values of these parameters through the NiosTo interface.

Algorithm 5: Double propagation extraction process

---

1 DoublePropagation()
2 **while** *new opinion words or new opinion targets* **do**
3 | extractOpinionWordTargets()
4 | extractOpinionWords()

Algorithm 6: Opinion target List extraction process

---

input : List of opinions, explicit opinion word to opinion target rules
output: List of extracted opinion targets {OpTargetList}

1 `extractOpinionWordTargets()`
2 **foreach** *Opinion in Opinions* **do**
3   **foreach** *sentence in Sentences* **do**
4   **foreach** *word in sentence* **do**
5   **if** *opinion word* **then**
6   // Explore opinion target existence
7   opinion target←`getOpinionWordTarget`(w, i, s)
8   **if** *new opinion target* **then**
9   OpTargetList←opinion target
10 Return `OpTargetList`

Algorithm 7: Opinion word to opinion target extraction process

---

input : Explicit opinion target connection rules
output: opinion target or null
1 `getOpinionWordTarget`(*opinion word, index, sentence*)
2 **if** *explicit opinion target connection* **then**
3   Return `opinion target`
4 **else**
5   Return `null`

Algorithm 8: Double propagation opinion word List extraction process

---

input : List of opinions, List of opinion words, List of opinion targets
output: List of Extracted double propagation Opinion words {DoublePropList}

1 `extractOpinionWords()`
2 **foreach** *Opinion in Opinions* **do**
3   **foreach** *sentence in Sentences* **do**
4   **foreach** *word in sentence* **do**
5   **if** *opinion target* **then**
6   // Explore existence of opinion word
7   opinion word←`getDoublePropagationOpinionWord`(w, i, s)
8   **if** *opinion word* **then**
9   // Explore Orientation of opinion word
10   Orientation←`getDoublePropagationOpWordOrientation`(w, i, s)
11   **if** *new opinion word* **then**
12   // Add opinion word & orientation
13   DoublePropList←**add**(*opinion word, Orientation*)
14   **else**
15   // Add orientation only
16   DoublePropList←**addOrientation**(*opinion word, Orientation*)
17 Return `DoublePropList`

Algorithm 9: Double propagation polarity extraction process

---

input : Double propagation extraction patterns
output: opinion word or null

1 `getDoublePropagationOpWordOrientation`(*opinion target, index, sentence*)
2 **if** *opinion words in the sentence* **then**
3   // Use intra-sentential sentiment consistency
4   Orientation←(sum orientation of opinion words in sentence)
5 **else**
6   // Use inter-sentential sentiment consistency
7   Orientation←(strongest orientation from nearby sentences)
8 Return `Orientation`

---

4   Experimental Evaluation

The evaluation aims at answering the following research questions: a) How well does the proposed approach extracts opinion words?, b) What is the added value of each step of the method?, c) How useful are the extracted domain-specific opinion words lists for performing an unsupervised sentiment classification task?

Dataset Creation and Description. The dataset was created by extracting review data from a popular, Greek e-shopping site[2]. A total of 4887 reviews were extracted referencing 1052 different products, belonging to 7 domains: TVs (189 products/322 reviews), Air Conditioners (105/139), Washing Machines (63/83), Cameras (122/166), Refrigerators (77/103), Mobile Phones (339/3626), Tablets (157/448). For the extraction process we used DEiXTo [6], a popular free and open source web content extraction suite. The dataset as well as other assets used in this work, are available at `http://deixto.com/niosto`.

4.1   Evaluation of Opinion Word Retrieval

In this section we evaluate the overall retrieval quality of opinion words and then study the contribution of each individual module of the proposed approach. For the evaluation of this step, we consider the initial seed as the ground truth set of opinion words. We "hide" a percentage of this ground truth from the algorithm and study how well it can discover these words. Since the initial seed is generic, at this step we evaluate the ability of the approach to extract opinion words that are not domain specific. We focus on domain-specific words in the next section.

At Table 4 we present the results of the extraction processes upon various category domains. We present a different set of results with and without using the stemmer. This set of results focuses on the impact of each step at opinion-word discovery. In brief, Conjunction-based extraction is more conservative at finding new opinion words while Double Propagation tends to discover more words. Table 5 presents the results of the algorithm evaluation upon the extraction

---

[2] `http://www.skroutz.gr/`

Table 4. Extracting Words using the proposed approach for various domains

Source Category	Opin	Senten Process	Filtered seed		Conj. extr.		Double prop.	Opin targ	Filtered seed		Conj. extr.		Double prop.	Opin targ
			(Stemmer-out) Extracted words						(Stemmer-in) Extracted words					
			pos	neg	pos	neg	total		pos	neg	pos	neg	total	
Televisions	322	1630	103	28	6	0	370	348	228	60	23	9	409	452
Air Conditioners	139	847	74	18	9	0	173	186	157	35	15	3	242	239
Washing Machines	83	515	46	11	4	0	119	120	103	29	8	3	153	151
Digital Cameras	166	872	79	16	6	0	172	168	170	31	8	1	188	258
Refrigerators	103	539	42	14	6	0	86	109	106	31	8	1	131	142
Mobiles	3626	20284	245	89	131	8	2633	2063	700	231	242	66	2139	2906
Tablets	448	2142	129	23	18	2	378	333	262	60	33	6	424	453

processes in terms of precision (for Conjunction-step) and recall (for the Double Propagation step). Note that this evaluation is only indicative since the approach is evaluated in terms of ability to identify opinion words from the original seed - which is not domain-specific and only a few of them appear in the extracted opinions. Another issue, is that language patterns are not always followed by the users. Reviews are most of the times just a set of keywords put together to describe advantages and disadvantages of the devices. As expected, domains with large number of opinions (mobiles, tablets) present better precision / recall.

Table 5. Average Precision - Recall Metrics

Source Category	Opinions	Average Values	
		Conj. Precision	Double Prop. Recall
Televisions	322	0%	49%
Air Conditioners	139	28%	15%
Washing Machines	83	3%	5%
Cameras	166	0%	46%
Refrigerators	103	35%	10%
Mobiles	3626	19%	38%
Tablets	448	54%	35%

4.2   Evaluating Utility in Sentiment Classification

At this point we discuss the contribution of each extraction step to the sentiment classification task. To create an evaluation set we utilized the "star" ratings of user reviews. We consider an opinion positive when the user assigned 4 or 5 stars. An opinion is considered negative when the user have assigned 1 or 2 stars to the product. Table 6 presents the sentiment classification accuracy for all steps of the approach. Naturally, classification accuracy is calculated as $sent_{acc} = \frac{\#correct\ classifications}{\#opinions}$. In general, we observe that the accuracy of all these dictionary-based approaches can vary from $59,95\%$ to $83,09\%$. Note that these approaches are completely unsupervised, i.e. no labelled data are required. Concerning the contribution of the various steps of the approach, we observe that in most cases the double propagation step leads to an improved classification accuracy. The step of conjunction extraction has a smaller impact. The improvement of Conjunction extraction and Double propagation over the filtered seed (generic opinion words) leads to the conclusion that the algorithm manages to identify effectively domain-specific words that aid in the task of sentiment classification. In most cases, stemming aids in classification accuracy. However, there are some domains, like refrigerators, where stemming has actually a negative impact. This can be explained by the fact that stemming unifies many different words incorrectly and identifies false opinion words.

Next we study another interesting factor in our analysis which is the quality of the expressed opinion. We judge quality of opinions by terms of length (actual word count). We have observed that longer opinions are more carefully written and the 'star' rating corresponds more accurately to the expressed opinion. For example, in short casual written approaches the 'star' rating seem not to correlated well with the actual text. Hence, very short opinions will negatively affect our approach (and, in fact, any language analysis task) and the evaluation as well, since the ground truth labels (positive-negative based on 'star' rating) are inaccurate. In the following figures we present classification accuracy taking into consideration opinions of various length. In Figures 2a, 2b, 2c 2d point x_o in the

Table 6. Sentiment Classification Accuracy of Various Steps of the Approach

Source Category	Opinions	Sentences Processed	Average polarity evaluation (stemmer out)			Classification Accuracy (stemmer in)		
			Filtered Seed	Conj. extr.	Double prop.	Filtered Seed	Conj. extr.	Double prop.
Televisions	322	1630	77,19%	77,19%	80,71%	80,73%	80,73%	75,49%
Air Conditioners	139	847	61,47%	65,14%	72,87%	63,41%	67,08%	69,51%
Washing Machines	83	515	60,48%	60,48%	79,84%	59,13%	59,13%	62,27%
Cameras	166	872	75,75%	75,75%	81,77%	83,09%	83,09%	82,76%
Refrigerators	103	539	63,59%	63,59%	77,85%	59,95%	59,95%	66,97%
Mobiles	3626	20284	70,35%	70,39%	78,05%	71,87%	71,97%	74,96%
Tablets	448	2142	74,60%	74,60%	79,71%	74,16%	74,51%	79,73%
Total Average:			69,06%	69,59%	78,69%	70,33%	70,92%	73,10%
Step Contribution:				1%	12%		1%	3%

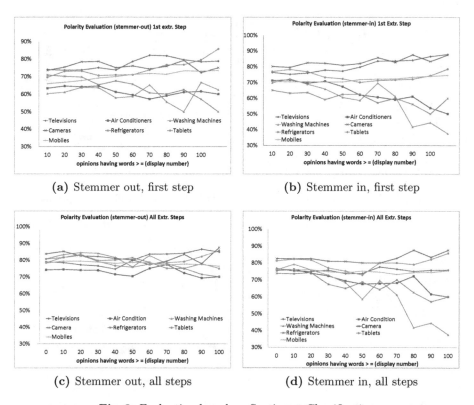

(a) Stemmer out, first step

(b) Stemmer in, first step

(c) Stemmer out, all steps

(d) Stemmer in, all steps

Fig. 2. Evaluation based on Sentiment Classification

x-axis presents the sentiment classification by considering only opinions with more than x_o words. This set of results confirms the results of Table 6: a) stemming has a positive impact, b) double-propagation accuracies outperform the based line (filter seed), c) Sentiment classification is more accurate in domains with more opinions. Values for more opinions with more than 50 words present higher variance since they are very few and this makes the accuracy fluctuate.

5   The NiosTo Tool

We developed a software application (NiosTo) that implements the above opinion word extraction algorithm as well as the dictionary-based sentiment classification. All features are accessible through an easy-to-use Graphical User Interface (see Figures 3a,3b,3c,3d). NiosTo takes as input opinions in csv format and presents the discovered words in each level. In addition, retrieval results as well as sentiment classification are visualized. All parameters can be changed and tuned through the various options available. In Table 7, we observe the text-outcome of the tool provided a set of opinions for mobile devices. The opinions

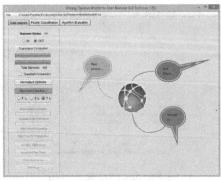

(a) GUI Welcome screen

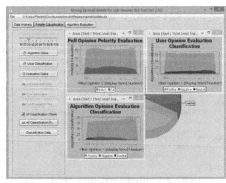

(b) Polarity classification options

(c) Stemmer Opinions process

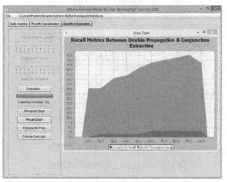

(d) Algorithm evaluation options

Fig. 3. The NiosTo Graphical User Interface

Table 7. Sample of extracted opinion words for various extraction steps from the gui and the mobiles domain source

Positive Filterseed Opinion words and sentiment polarity weights					
good [1064]	protective [4]	well [191]	perfect [248]	functional [147]	fast [392]
exactly [28]	detailed [38]	excellent [75]	honest [8]	serious [3]	Finance [45]
incredible [92]	warm [1]	beautiful [96]	handy [91]	mild [57]	thrilled [7]
Negative Filterseed Opinion Words and sentiment polarity Weights					
bad [-63]	tragic [-5]	negative [-155]	difficulty [-4]	slow [-57]	annoying [-6]
inadequate [-3]	unhappy [-1]	expensive [-28]	Terrible [-8]	radiation [-18]	accurate [-12]
irrelevant [-2]	ridiculous [-2]	useless [-11]	unsavory [-1]	ugly [-8]	bad [-3]
Positive Conjunction extracted opinion words and sentiment weights					
responsive [2]	turns out [4]	hang [10]	practices [2]	inexpensive [1]	enough [7]
covers [1]	single [1]	great [1]	better for [1]	smooth [1]	aesthetically [1]
ease of use [1]	fast [1]	smooth [1]	Unfortunately [1]	rugged [4]	problems [1]
Negative Conjunction extracted opinion words and sentiment weights					
supersaturated [-1]	deprived [-1]	failure [-1]	appearance [-1]	packages [-1]	many [-1]
Double Propagation extracted opinion words and sentiment weights					
costing [6]	exceptional [3]	evaluation [-1]	suggests [9]	monster [3]	works [1]
heavy [21]	heavy [21]	forthcoming [1]	bigger [-1]	relaxed [4]	crazy [10]
authorities [3]	easily [1]	reluctantly [1]	important [1]	Bet [4]	demanding [7]

are originally written in Greek. We have translated the outcome in English. In brackets the system outputs the sentiment strength of each word.

6 Conclusions

In this paper we presented a method for domain-specific opinion word discovery. It consists of a series of steps that complement each other in discovering new words. We follow language patterns and opinion-words opinion-targets relationships to identify new words. Word polarity is calculated automatically by following a set of polarity disambiguation procedures. We evaluated the approach on a set of opinions about digital devices written in the Greek language. The experimental evaluation suggests that we can achieve satisfactory sentiment classification using this completely unsupervised approach. Finally, we presented a software tool, `NiosTo`, that implements the approach and enables the user to experiment with dictionary extraction and apply sentiment classification.

References

1. Amati, G., Ambrosi, E., Bianchi, M., Gaibisso, C., Gambosi, G.: Automatic construction of an opinion-term vocabulary for ad hoc retrieval. In: Macdonald, C., Ounis, I., Plachouras, V., Ruthven, I., White, R.W. (eds.) ECIR 2008. LNCS, vol. 4956, pp. 89–100. Springer, Heidelberg (2008)
2. Hatzivassiloglou, V., McKeown, K.R.: Predicting the semantic orientation of adjectives. In: Proceedings of the 35th Annual Meeting of the Association for Computational Linguistics and Eighth Conference of the European Chapter of the Association for Computational Linguistics, ACL 1998, pp. 174–181. Association for Computational Linguistics, Stroudsburg (1997)
3. Nasukawa, T., Kanayama, H.: Fully automatic lexicon expansion for domain - oriented sentiment analysis (2006)
4. Hu, M., Liu, B.: Mining and summarizing customer reviews (2004)
5. Kanayama, H., NasuKawa, T.: Fully automatic lexicon expansion for domain - oriented sentiment analysis
6. Kokkoras, F., Ntonas, K., Bassiliades, N.: Deixto: A web data extraction suite. In: Proceedings of the 6th Balkan Conference in Informatics, BCI 2013, pp. 9–12. ACM, New York (2013)
7. Liu, K., Xu, L., Liu, Y., Zhao, J.: Opinion target extraction using partially-supervised word alignment model. In: Proceedings of the Twenty-Third International Joint Conference on Artificial Intelligence, IJCAI 2013, pp. 2134–2140. AAAI Press (2013)
8. Qiu, G., Liu, B., Bu, J., Chen, C.: Opinion word expansion and target extraction through double propagation. Comput. Linguist. 37(1), 9–27 (2011)
9. Wiebe, J., Riloff, E.: Learning extraction patterns for subjective expressions. In: Proceedings of the 2003 Conference on Empirical Methods in Natural Language Processing, pp. 105–112 (2003)

10. Skomorowski, J., Vechtomova, O.: Ad hoc retrieval of documents with topical opinion. In: Amati, G., Carpineto, C., Romano, G. (eds.) ECIR 2007. LNCS, vol. 4425, pp. 405–417. Springer, Heidelberg (2007)
11. Wu, Y., Zhang, Q., Huang, X., Wu, L.: Phrase dependency parsing for opinion mining. In: Proceedings of the 2009 Conference on Empirical Methods in Natural Language Processing, vol. 3, pp. 1533–1541. Association for Computational Linguistics, Stroudsburg (2009)
12. Yu, P.S., Ding, X., Liu, B.: A holistic lexicon-based approach to opinion mining (2008)

A Community Detection Algorithm
Based on the Similarity Sequence

Hongwei Lu, Qian Zhao, and Zaobin Gan*

School of Computer Science and Technology
Huazhong University of Science and Technology
Wuhan 430074, PR China
{luhw,zgan}@mail.hust.edu.cn, zqhuster@163.com

Abstract. Community detection is a hot topic in the field of complex social networks. It is of great value to personalized recommendation, protein structure analysis, public opinion analysis, etc. However, most existing algorithms detect communities with misclassified nodes and peripheries, and the clustering accuracy is not high. In this paper, in terms of the agglomerative hierarchical clustering, a community detection algorithm based on the similarity sequence is proposed, named as ACSS (Agglomerative Clustering Algorithm based on the Similarity Sequence). First, similarities of nodes are sorted in descending order to get a sequence. Then pairs of nodes are merged according to the sequence to construct a preliminary community structure. Secondly, the agglomerative clustering process is carried out to get the optimal community structure. The proposed algorithm is tested on real network and computer-generated network data sets. Experimental results show that ACSS can solve the problem of neglecting peripheries. Compared with the existing representative algorithms, it can detect stronger community structure, and improve the clustering accuracy.

Keywords: Community Detection, Similarity, Sequence, Agglomeration, Modularity.

1 Introduction

In the real world, many systems can be abstracted as social networks, such as protein interaction networks, movie actor collaboration networks, viral spreading networks, web linkage networks, transportation networks. Besides the scale-free distribution and the small-world effect, social networks possess another relevant feature, which is the community structure. Generally, the community structure corresponds to the division of nodes into groups within which the network connections are dense, but between which they are sparser[1].

Community detection for large social networks,web graphs, and biological networks is a problem of considerable practical interest that has attracted a great deal of attention[2][3]. For example, in online social network sites, by detecting

---

* Corresponding author.

B. Benatallah et al. (Eds.): WISE 2014, Part I, LNCS 8786, pp. 63–78, 2014.
© Springer International Publishing Switzerland 2014

communities with the same interests or backgrounds, people can do personalized recommendation based on properties. In protein interaction networks, by detecting groups of proteins with special functions in cells, people can find the structure units of similar functions in biological systems. In the World Wide Web, by detecting websites talking about relevant topics, people can do public opinion analysis, hot topic tracking, etc. On the whole, community detection has great significance for social networks' topology analysis, functional property understanding, inherent hidden laws and knowledge discovering, and behavior prediction.

In fact,community detection for social networks is a graph partitioning problem or a clustering problem. Exploring community structure can be regarded as finding sub-graphs where nodes are densely connected among themselves and loosely connected with others. On the one hand, the size and number of communities in social networks are unknown, and the connection relationships between nodes are complex in networks, for a node connecting densely with others in the same community may connects relatively densely with nodes in different communities. In this way, the node will be misclassified at a high probability, which results in low clustering accuracy.

On the other hand, some peripheries connect loosely with nodes in the same community. If agglomerative methods are applied, the core nodes in a community often have strong similarity, and are connected early in the agglomerative process, but peripheral nodes that have no strong similarity to others tend to get neglected[1], leading to the divided structure unreasonable. Therefore, a proper strategy should be chosen to get strong community structure with fewer misclassified nodes and without neglecting peripheries. Besides, the clustering accuracy is improved further.

In order to address the issue, we investigate the current studies of community detection in social networks, and propose an algorithm for community detection named as ACSS. We use a similarity metric of nodes to measure the connectivity of two nodes. The higher the similarity of two nodes is, the higher the probability of the two nodes in the same community is. In light of this idea, a preliminary community structure can be constructed according to the similarity sequence. Finally, the agglomerative clustering process is carried out to get the optimal community structure.

The rest structure of this paper is organized as follows: in Section 2, a brief introduction of related work is presented. Section 3 discusses the approach for community detection based on the similarity sequence in detail. Experimental results and analysis are given in Section 4, followed by conclusions and future work in Section 5.

2   Related Work

So far, many scholars have studied approaches for community detection in complex social networks. The existing approaches can be classified as the spectral method, the optimal method and the heuristic method.

The spectral method means to get networks' hierarchical structure by doing spectrum analysis of the laplacian matrix and its variants in social networks[4,5]. This kind of method depends largely on the eigenvector, thus the time complexity is high, the accumulate error rate is also high for large-scale networks in general.

The optimal method is built on an objective function, and its goal is to maximize the value of the objective function by repeated iterative divisions. Usually, This kind of method has a tendency to produce the local optimal solution.

Based on the *Modularity Q* in [1], Newman proposed a fast algorithm[6]. In [7], Duch et al. proposed an extremal optimization method. Blondel et al. proposed a network modularity maximization strategy called LM for community detection[8]. It consists of two steps and two steps are repeated until an arbitrarily small improvement ΔQ is attained at each iteration. Pasquale et al. used LM to propose a new clustering method called CONCLUDE[9].

The algorithm in [10] is based on the drops of densities between each pair of parent and child nodes in the dendrogram, and communities are formed by making the drop in density high. Zhang et al.[11] proposed a algorithm based on local cores, with which local communities can be constructed by optimizing the local modularity. Nguyen proposed an algorithm based on optimizing local gained modularity[12]. After that, Thang et al. took advantage of power-law distribution, and proposed an adaptive algorithm[13]. Besides, there still exists the optimal method based on the genetic algorithm[14], and so on.

What the heuristic method shares in common is that algorithms are designed by some intuitive assumptions, and a feasible solution of instance is given by computing. So they can't detect strong communities without considering community quality evaluation. They can find the optimal result for most networks, but not for every network strictly[15]. Newman et al.[16] proposed a splitting algorithm based on the edge betweenness, and a division of the graph is got by deleting the edges with the highest betweenness.

In addition, FEC(Finding and Extracting Communities) algorithm based on Markov random walk model was proposed in [17]. Jin et al[18] used an ant colony optimization strategy building on random walks named as RWACO to detect the community structure of complex networks. Huang et al[19] formalized the task of community detection as a multi-objective problem, and proposed a method based on multi-objective particle swarm optimization. Lancichinetti et al proposed a method by finding clusters of statistical feasures in networks[20]. Pan et al.[21] proposed an agglomerative algorithm based on node similarity. But the similarity threshold is different for different network, and it is hard to tune the value.

Although many algorithms have been proposed, it is still an open question how to detect strong community structure with fewer misclassified nodes, and without neglecting peripheries. Therefore, a heuristic community detection algorithm based on the similarity sequence is proposed in the following. We consider the community quality evaluation into consideration to detect a strong community structure.

3   ACSS Algorithm

By means of the idea that the nodes in a community are connected more densely, we first merge nodes based on the similarity of two nodes to get the preliminary community structure, and then merge communities based on the similarity of two communities in the agglomerative clustering process. The community structure with the maximum modularity is the optimal result. First, the metrics of the node similarity and the community similarity are introduced in Section 3.1, followed by the evaluation method of community quality in Section 3.2. Finally, a detailed description of the ACSS algorithm is given in Section 3.3.

3.1   Similarity Metrics

The similarity metrics of nodes in social networks can be divided into two categories, which are the metric based on the network topology structure and the metric based on the local information.

The former constructs a similarity matrix using the topological structure properties of nodes in networks. In [22], each node is assumed to be a system which can send, receive and record signals, then N-dimensional vectors are constructed in the signaling process by initializing each node as the signal source. With N-dimensional vectors, the topological relationship of nodes can be transferred into a geometrical structure of nodes in Eudidean, Manhattan space, and so on. After that, clustering methods like k-means, hierarchical methods can be applied to detect the final community structure[23]. But, that kind of method may lead to some peripheries in the detected communities. Leicht et al[24] proposed a measure of structural similarity for pairs of vertices in networks. The method is fundamentally iterative, with the similarity of a vertex pair being given in terms of the similarity of the vertices' neighbors.

The latter mainly depends on the neighbor information in networks. In the social network analysis, it is reasonable to consider that two nodes in a social network have something in common if they have many same neighbors. The more neighbors that two nodes share, the higher the similarity of the two nodes is. In view of this idea, we use a *Jaccard* metric to compute the similarities of nodes in this paper. The *Jaccard* metric is shown in formula (1). Assuming that N_v is defined as the neighbor set of the node v, namely those nodes directly connected to the node v. And $|x|$ denotes the number of elements in set x.

$$s_{Jaccard}(v, w) = \frac{|N_v \bigcap N_w|}{|N_v \bigcup N_w|} \tag{1}$$

The *Jaccard* metric will get the maximum value 1 when $N_v = N_w$. But it doesn't take the connectivity condition between two nodes into account. It indicates the rate of the neighbors that two nodes share. Apparently, no matter how much the *Jaccard* value of two disconnected nodes is, they are less similar than those connected directly. Considering the connectivity relationship, a new version of the *Jaccard* metric is defined by formula (2).

$$s_{node}(v, w) = \begin{cases} \frac{s_{Jaccard}}{2} & if \quad e(v, w) = 0 \\ \frac{s_{Jaccard}+1}{2} & if \quad e(v, w) = 1 \end{cases} \tag{2}$$

Here, $e(v, w)$ denotes whether an edge exits between the node v and node w or not. $e(v, w)$ equals 0 with no edge, and 1 with an edge. The value range of the $Jaccard$ metric is $[0, 1]$, so the range of the node similarity is also $[0, 1]$. The node similarity gets the maximum value 1 only when the neighbors of two nodes are completely the same and they connect each other at the same time. And it equals 0 when two nodes share no neighbors and they are not connected.

For any two communities c_i and c_j, the community similarity can be measured by the average similarity of nodes between the two communities. It can be defined by formula (3), (4).

$$count(v, w) = \begin{cases} 1 & s_{node} > 0 \\ 0 & s_{node} = 0 \end{cases} \tag{3}$$

$$s_{community}(c_i, c_j) = \frac{\sum\limits_{v \in c_i, w \in c_j} s_{node}(v, w)}{\sum\limits_{v \in c_i, w \in c_j} count(v, w)} \tag{4}$$

Here, if the node similarity between the node v and the node w is more than zero, then $count(v, w) = 1$, otherwise, $count(v, w) = 0$. Because the node similarity varies from 0 to 1, the community similarity also falls in the range from 0 to 1.

3.2 Community Quality Evaluation Method

In general, there are two criteria when thinking about how good of a community is a set of nodes. The first is the number of edges between the members of the community, and the second is the number of edges between the members of the community and the remainder of the network[25]. Many criterion are proposed to evaluate the quality of communities, such as $Conductance$ [26], $Expansion$[27], $Internal \ density$[27], $Modulairty$[1], and the like. $Modularity$ Q is one of the most widely used methods to evaluate the quality of a division of a network into modules or communities[15][18]. For a given division of a network into communities, $Modularity$ measures the number of within-community edges, relative to a null model of a random graph with the same degree distribution. The Q criterion is also applied in this paper.

Assuming that a network with n nodes is divided into k communities. Then a $k \times k$ symmetry matrix $\mathbf{e}$ is defined whose element e_{ij} is one-half of the fraction of edges in the network that connects nodes in the community i to those in the community j. e_{ii} is the fraction of edges that connects nodes in the community i. It can be concluded that the sum of the matrix is $\sum\limits_{i} \sum\limits_{j} e_{ij} = 1$. $a_i = \sum\limits_{j} e_{ij}$ is the fraction of edges that connect to nodes in the community i. The trace of the matrix $\mathbf{e}$ is $Tr(\mathbf{e}) = \sum_i e_{ii}$, denoting the fraction of all edges that connects nodes in the same community. Then Q is measured by formula (5).

$$Q = \sum_i (e_{ii} - a_i^2) = Tr(\mathbf{e}) - \|\mathbf{e}^2\| \tag{5}$$

Where $\|\mathbf{x}\|$ indicates the sum of elements of the matrix $\mathbf{x}$ which reflects the random connections between nodes of the same division. The value range of the *Modularity* Q is $[0, 1]$. The closer the Q value is to 1, the stronger the community structure is. The quality $Q = 0$ will be acquired if the number of edges within communities is no better than random. In practice, Q generally varies from about 0.3 to 0.7.

3.3 Algorithm Description

A network is given with n nodes and m edges. Consider that two nodes v and w have a high node similarity, meaning connecting densely, if the node v belongs to a community c_t, then the node w will belong to the same community c_t at a high probability. In this way, if all the node similarities of connected nodes in the network are sorted in descending order, then pairs of nodes whose node similarities rank in the front of sequence must be in the same community. Therefore, the ACSS algorithm is classified into two steps.

First, preliminary communities are constructed by merging pairs of nodes according to the sequence of the node similarity. Secondly, the agglomerative clustering process is carried out based on the community similarity. Finally, the division which has the maximum Q is the optimal output result.

Step 1 of ACSS is described in Algorithm 1. The node similarities of connected nodes are first calculated according to the formula (1) and (2). Then all the similarity values are sorted in descending order to get a sequence. Each item of the sequence corresponds to a node similarity of two nodes. Pairs of nodes are merged by traversing the sequence from the beginning to the end. If both of the two nodes at an item aren't traversed, they will be merged in a new community. If only one node is traversed, then the other node will be merged in the community that the former node belongs to. The process is repeated until all the n nodes are traversed.

In Algorithm 1, the time complexity of the node similarity calculation is $O(m)$. The merging sort is applied to get the sequence in descending order, whose time complexity is $O(m \log m)$. The next merging process based on the sequence needs n loop operations, thus it requires $O(n)$ running time. As a result, the time complexity of Algorithm 1 is $O(m \log m)$.

After Algorithm 1, the community set C that has $\bar{n} = \log n$ communities is acquired. Step 2 is the agglomerative hierarchical clustering process described in Algorithm 2. First, the community similarities are calculated according to the formula (3) and (4). After two communities with the maximum community similarity are merged to form a new community, Q is calculated according to the formula (5) and similarities between old communities and the new community

---

Algorithm 1. Constructing Preliminary Communities

---

input:$G(V, E)$
1: **for** connected nodes $v, w \in V$
2: **do** calculate $s_{node}(v, w)$
3: **end for**
4: Sort all of the $s_{node}(v, w)$, get F_T //F_T denotes the ordered sequence
5: $P = \phi$ //P denotes the visited nodes set
6: $C = \phi$ //C denotes the divided communities set
7: **while** $P \neq V$
8:　　　**for** t=1 : M in F_T //M denotes the item number in F_T
9:　　　　**if** $v, w \notin P$ **then**
10:　　　　　$c_{new} = \{v, w\}$ //c_{new} denotes a new community
11:　　　　　$C = C \cup c_{new}$
12:　　　　　$P = P \cup \{v, w\}$
13:　　　　**else if** $v \notin P$ && $w \in P$ **then**
14:　　　　　$c_w = c_w \cup \{v\}$ //c_w denotes the community that the node w belongs to
15:　　　　　$P = P \cup \{v\}$
16:　　　　**else if** $w \notin P$ && $v \in P$ **then**
17:　　　　　$c_v = c_v \cup \{w\}$
18:　　　　　$P = P \cup \{w\}$
19:　　　　**else continue**
20:　　　**end for**
21: **end while**
output: the preliminary community set C

Algorithm 2. Agglomerative Clustering

---

input:the community set C
1: **for** two communities $c_i, c_j \in C$
2: **do** calculate $s_{community}(c_i, c_j)$
3: **end for**
4: **while** $\bar{n} \neq 1$
5:　　　choose the maximum $s_{community}(c_i, c_j)$
6:　　　$c_{new} = c_i \cup c_j$
7:　　　$C = (C - c_i - c_j) \cup c_{new}$
8:　　　calculate the Q value
9:　　　$\bar{n} = \bar{n} - 1$
10:　　　**for** $c_r \in C$
11:　　　　recalculate $s_{community}(c_r, c_{new})$
12:　　　**end for**
13: **end while**
output: C set with the maximum Q

---

are recalculated. That process is repeated until all the nodes are in only one community. Finally, the division with the maximum Q is chosen as the optimal result of the whole algorithm.

The process of the community similarity calculation in C requires $O(\overline{n}^2) = O(\log^2 n)$ running time. The agglomerative clustering process needs $\overline{n}-1$ loop operations, so the time complexity is $O(\overline{n}^2)$. Thus, Algorithm 2 requires $O(\log^2 n)$ running time. For the time complexity of Algorithm 1 is $O(m \log m)$, the total time complexity of ACSS is $O(m \log m)$.

4   Experimental Results and Analysis

In this section, the ACSS algorithm is tested on both computer-generated networks and several real networks with known community structures, which are commonly used for evaluating community detection algorithms. For better illustrating ACSS's effectiveness, several other community detection algorithms are also tested on the same data sets respectively.

4.1   Methods for Evaluation

Because the community structures of the tested networks are known, we can evaluate the correctly classified noes for a given division. For a given community c_i, compare it with all the real communities, and find the community c_t with the largest degree of overlapping nodes. The same thing is done for the other detected communities. Then we can get the fraction of nodes classified correctly in a network. It is the criteria $FVIC$(Fraction of Vertices Identified Correctly) in [16], which is applied widely for evaluating a division.

Besides, Danon et al[29] proposed the use of another criteria, names as NMI (Normalized Mutual Information). It is based on a diffusion matrix N. The rows correspond to the real communities, and columns correspond to the detected communities. $N_{i,j}$ denotes the number of nodes in the real community i that appear in the detected community j. R denotes the set of real communities and F denotes the set of detected communities. Then NMI based on information theory is

$$NMI(R,F) = \frac{-2\sum\limits_{i=1}^{c_R}\sum\limits_{j=1}^{c_F} N_{ij}log(\frac{N_{ij}S}{N_{i.}N_{.j}})}{\sum\limits_{i=1}^{c_R} N_{i.}log(\frac{N_{i.}}{S}) + \sum\limits_{j=1}^{c_F} N_{.j}log(\frac{N_{.j}}{S})} \tag{6}$$

When the detected communities are the same with the real communities, NMI=1. When the number of detected communities is more than the number of real communities and all the detected communities belong to the real communities, this division can be regarded a more detailed division of a real network structure. In this way, all the nodes can be considered correctly classified and we hope that it is a right division. But NMI is not equal to 1, and without the effect of a detailed division on $FVIC$, $FVIC = 1$. What's more, the time complexity of NMI is higher than $FVIC$, so we select $FVIC$ as the criteria for evaluating the accuracy of a division.

When the real community structure of a network is not clear, like the American College Football Network, all the teams are divided into 8-12 conferences, so we cannot just use $FVIC$ to measure the accuracy of a particular division. We need another criteria to measure the quality of a division. The $Modularity$ Q proposed by Newman[1] described in section 3.2 can be used to measure the quality for a division. The higher Q is, the stronger the detected community structure is.

4.2 Computer-Generated Networks

Computer-generated networks are small networks generated by computer programs[16]. They have been widely used to test algorithms for community detection[18][19]. The networks are defined as $RN(C, s, d, p_{in})$[15], where C denotes the number of communities, s is the number of nodes in a community, and d presents the average degree of a node. Edges are placed between nodes at random with probability p_{in} for nodes belonging to the same community and p_{out} in different communities. Obviously, the higher p_{in} is, the clearer the community structure is. $p_{in} < p_{out}$ should be set to ensure the community structure in computer-generated networks.

In this paper, the GN network[16] is used, which is $RN(4, 32, 16, p_{in})$. The rule is that each network has 4 communities with 32 nodes each, and the average degree of a node is 16. To evaluate the accuracy of ACSS, the criteria $FVIC$ (Fraction of Vertices Identified Correctly) in [16] is applied which denotes the fraction of nodes classified correctly in a network. As a comparison, FN[6] and GN[16] are tested on the same data sets. The comparison results are shown in Fig. 1. Each data point is the average value on 10 networks.

As shown in Fig. 1, ACSS performs better and get the clustering accuracy more than 50%. When $p_{in} \geq 0.65$, the clustering accuracy of ACSS reaches 95.08%, whereas FN's is 85.94%, and GN's is 75.16%. When $p_{in} = 0.4$, the community structure is not clear, but the clustering accuracy of ACSS is already 51.41%, but FN's is just 23.28% and GN's is 26.88%. What's more, ACSS's performance is stable with p_{in} increasing, but other two algorithms are not stable, having a dependence on the network's real community structure.

4.3 Real Networks

As real networks have different topological features compared with computer-generated networks, in the following, the ACSS algorithm is tested on several real networks with known community structures. And GN, FN, FEC[17] and RWACO[18] are also tested on the same networks for a comparison. To measure the clustering accuracy of the ACSS algorithm, the $FVIC$ metric is applied. The $Modularity$ Q is also used to measure the division quality of the networks.

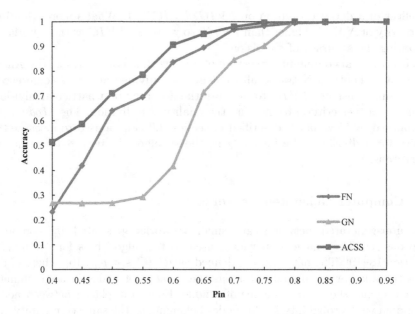

Fig. 1. Comparison with FN, GN on computer-generated networks

Zachary's Karate Club Netwrok. Wayne Zachary[28] observed social interactions between the members of a karate club at an American university over a period of 2 years, and constructed a club network of members in the club. He found that a dispute arose between the principal teacher and the administrator during his study. As a result, the club eventually split in two groups, centered around the administrator and the principal teacher respectively.

Fig. 2 shows the real network. The nodes denote members in the club, and edges between pairs of nodes denote that two members usually appear in the club's activities together. The nodes with the same color and the same shape are in one group, so it is clear that the club is divided into 2 groups. The graph has 34 nodes and 78 edges in total, with node 1 denoting the teacher, and node 34 denoting the administrator.

Fig. 3 shows ACSS's division with different colors and different shapes denoting different communities. The algorithm can not only detect the real community structure successfully but also further divide the real community structure with four communities. It shows strong community structure without neglecting peripheries. The results of the comparison with other algorithms are shown in Table1. RWACO's division is the real community structure.

The division results of FN, GN and FEC all have misclassified nodes, which leads to Q lower. ACSS's division has the value of $Q = 0.4151$, higher than the value 0.3751 of the real community structure, indicating stronger community structure.We can conclude that the real community structure for a given network may not be a strong community structure. By dividing the real community

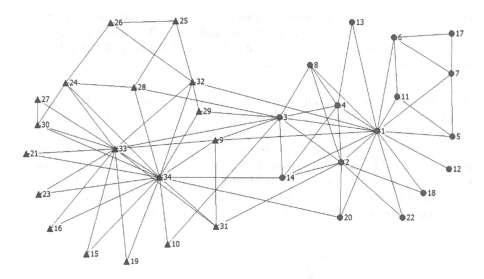

Fig. 2. Real community structure

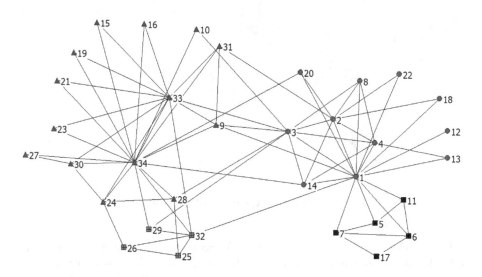

Fig. 3. Result of ACSS on zachary's karate club network

Table 1. Comparison with other algorithms on Zachary's karate club network

Algorithms for Community Detection	GN	FN	FEC	RWACO	ACSS
Q	0.401	0.381	0.3744	0.3715	**0.4151**
$FVIC$	0.971	0.971	0.971	1	**1**
Communities	2	2	3	2	4
Wrong Node	Node 3	Node 10	Node 9	None	**None**

structure into more communities we can get a stronger community structure. Because FN, GN, FEC all misclassifies one node, the $FVIC$ values of them is 0.971. Without considering the effect of a detailed division on $FVIC$, ACSS's $FVIC$ value is 1 without misclassifying one node.

American College Football Network. American college football network is a representation of the schedule of Division I games for the 2000 season by Given and Newman[16]. The nodes in the graph represent teams and edges represent regular-season games between two teams. The graph has 115 nodes and 616 edges. All the 115 teams are divided into 8-12 conferences. When evaluating the accuracy of a division for the football network, people always refer to the structure of 12 conferences. Games are more frequent between teams of the same conference than that between teams of different conferences.

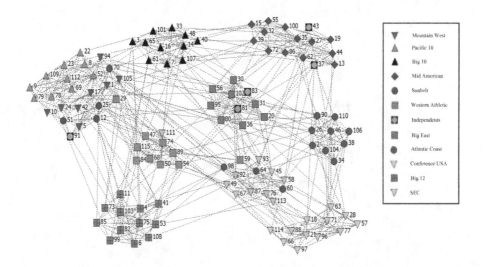

Fig. 4. Result of ACSS on American college football network

The division of the football network by ACSS is shown in Fig. 4. The network is divided into 9 conferences. Most of the teams are correctly grouped with other teams in their conferences. The independent teams including $\{43, 37, 83, 81, 91\}$ do not belong to any conference, and they play more games with teams in other conferences than in themselves, so they tend to be grouped with teams with which they are closely associated.

Pacific 10 and Mountain West are grouped together because games are frequent among teams in the two conferences, so it can be regarded as a correct division. Sunbelt Conference is broken into two pieces with 7 teams, and this happens because they played nearly as many games with teams in other conferences as they did in their own conference. So this will lead to $FVIC$ lower, but it reveals the inner knowledge in Sunbelt Conference.

Table 2. Comparison with other algorithms on American college football network

Algorithms For Community Detection	GN	FN	FEC	RWACO	ACSS
Q	0.546	0.601	0.5697	0.6010	**0.6043**
$FVIC$	0.6261	0.8261	0.7739	**0.9304**	0.8696
Communities	6	11	12	12	9

The comparison results are shown in Table 2. ACSS's $FVIC$ is 0.8696, just lower than RWACO's and higher than other three algorithms. GN detects only 6 communities, resulting in lower $FVIC$ and $Modularity$. ACSS detects 9 communities, and the $Modularity$ Q value of ACSS is 0.6043, higher than other algorithms, showing a stronger community structure.

Other Real Networks. To further analysis the feasibility of the ACSS algorithm, another two real network data sets are also applied: the dolphin network by D. Lusseau[30] and Kreb's network of books on American politics(V. Krebs, http://www.orgnet.com).

In the dolphin network the nodes represent 62 dolphins and edges join pairs of dolphins that are frequently collected. Dolphins are divided into 2 groups according to the frequent association. One group has 20 dolphins, and the other has 42 dolphins.

The network of books on American politics is constructed from the online bookseller Amazon.com. There are 105 nodes representing books about American politics on the online bookseller, and 441 edges between nodes representing that books are frequently purchased by the same buyer. The network is divided into three groups, which represent "Liberal", "Conservative" and "Centrism" respectively. The division is from the analysis of opinions and evaluation on books of Amazon.com by Mark Newman[31]. The comparison results are shown in Table 3. It is clear that the Q value of ACSS is higher both in the dolphin network and the network of books on American Politics.

Table 3. Comparison on the dolphin network and book network

Networks	Size(nodes,edges)	FN	GN	FEC	RWACO	ACSS
Dolphin Network	62,159	0.5104	0.4706	0.4976	0.3774	**0.5132**
Book Network	105,441	0.502	0.5168	0.4904	0.4569	**0.5208**

On the above four real networks, ACSS's Q is higher with high clustering accuracy. It can detect stronger community structure than others. The experimental performance convinces the ACSS algorithm's effectiveness for community detection.

5   Conclusions and Future Work

In order to address the issue of fewer misclassified nodes and peripheries, ACSS based on the similarity sequence is proposed for community detection. Based on the *Jaccard* metric, we use a new node metric to measure the similarity of two nodes. A sequence in descending order is first constructed based on the node similarity. Then pairs of nodes are merged until all the nodes in a network are traversed. Finally, the agglomerative process is taken based on the community similarity, resulting in the optimal division of the ACSS algorithm. ACSS is tested on four real networks and some computer-generated networks. Compared with other algorithms, ACSS can detect stronger community structure with high clustering accuracy.

Besides, it can do more detailed division for the real community structures, and detect more fine-grained communities. In addition, because it is based on similarities, the nodes with a high similarity value must be in the same community, then it will result in a higher community value.

However, in real social networks, the overlapping phenomenon may occur among communities. Thus, as future work, we will focus on detecting overlapping communities to further improve the ACSS algorithm.

Acknowledgment. This research is funded by the National Natural Science Foundation of China under grant No. 61272406 and the Fundamental Research Funds for the Central Universities, HUST: 2013TS101.

References

1. Newman, M.E.J., Girvan, M.: Finding and Evaluating Community Structure in Networks. Physical Review E 69(2) 026113 (2004)
2. Narayanan, T., Subramaniam, S.: A Newtonian Framework for Community Detection in Undirected Biological Networks. IEEE Transactions on Biomedical Circuits and Systems 8(1), 65–73 (2014)
3. Karrer, B., Levina, E., Newman, M.: Robustness of Community Structure in Networks. Physical Review E 77, 046119 (2008)
4. Van Gennip, Y., Hunter, B., Ahn, R., et al.: Community Detection Using Spectral Clustering on Sparse Geosocial data. SIAM Journal on Applied Mathematics 73(1), 67–83 (2013)
5. Newman, M.E.J.: Community Detection and Graph Partitioning. Europhysics Letters 103(2), 28003 (2013)
6. Newman, M.E.J.: Fast Algorithm for Detecting Community Structure in Networks. Physical Review E 69(6), 066133 (2004)
7. Duch, J., Arenas, A.: Community Detection in Complex Networks Using Extremal Optimization. Physical Review E 72(2), 027104 (2005)
8. Blondel, V., Guillaume, J., Lambiotte, R.: Fast Unfolding of Communities in Large Networks. J. Stat. Mech., P10008 (2008)

9. De Meo, P., Ferrara, E., Fiumara, G., Provetti, A.: Mixing Local and Global Information for Community Detection in Large Networks. Journal of Computer and System Sciences 80(1), 72–87 (2014)
10. Qi, X., Tang, W., Wu, Y., Guo, G., Fuller, E., Zhang, C.: Optimal Local Community Detection in Social Networks Based on Density Drop of Subgraphs. Pattern Recognition Letters 36, 46–53 (2014)
11. Zhang, X., Wang, L., Li, Y., Liang, W.: Extracting Local Community Structure From Local Cores. In: Xu, J., Yu, G., Zhou, S., Unland, R. (eds.) DASFAA Workshops 2011. LNCS, vol. 6637, pp. 287–298. Springer, Heidelberg (2011)
12. Nguyen, N., Dinh, T., Xuan, Y., Thai, M.T.: Adaptive Algorithms for Detecting Community Structure in Dynamic Social Networks. In: Proceedings of the IEEE INFOCOM, Shanghai, China, April 10-15, pp. 2282–2290 (2011)
13. Thang, N., Dinh, N.P., Nguyen, M.T.: Thai: An Adaptive Approximation Algorithm for Community Detection in Dynamic Scale-free Network. In: 2013 Proceedings of the IEEE INFOCOM, pp. 55–59 (2013)
14. Hafez, A.I., Al-Shammari, E.T., ella Hassanien, A., et al.: Genetic Algorithms for Multi-Objective Community Detection in Complex Networks. In: Pedrycz, W., Chen, S.-M. (eds.) Social Networks: A Framework of Computational Intelligence. SCI, vol. 526, pp. 145–171. Springer, Heidelberg (2014)
15. Yang, B., Liu, D., Liu, J., Jin, D., Ma, H.: Complex Network Clustering Algorithms. Journal of Software 20(1), 54–66 (2009)
16. Girvan, M., Newman, M.E.J.: Community Structure in Social and Biological Networks. Proceedings of the National Academy of Sciences of the United States of America 99(12), 7821–7826 (2002)
17. Yang, B., Cheung, W., Liu, J.: Community Mining from Signed Social Networks. IEEE Transactions on Knowledge and Data Engineering 19(10), 1333–1348 (2007)
18. Jin, D., Yang, B., Liu, J., Liu, D., He, D.: Ant Colony Optimization Based on Random Walk for Community Detection in Complex Networks. Journal of Software 23(3), 451–464 (2012)
19. Huang, F., Zhang, S., Zhu, X.: Discovering Network Community Based on Multi-Objective Optimization. Journal of Software 24(9), 2062–2077 (2013)
20. Lancichinetti, A., Radicchi, F., Ramasco, J., Fortunato, S.: Finding Statistically Significant Communities in Networks. PLoS ONE 6(4), e18961 (2011)
21. Pan, Y., Li, D.H., Liu, J., Liang, J.: Detecting Community Structure in Complex Networks via Node Similarity. Physica A 389, 2849–2857 (2010)
22. Hu, Y., Li, M., Zhang, P., Fan, Y., Di, Z.: Community Detection by Signaling on Complex Networks. Physical Review E 78(1), 16115 (2008)
23. Fortunato, S.: Community Detection in Graphs. Physics Reports 486(3), 75–174 (2010)
24. Leicht, E.A., Holme, P., Newman, M.E.J.: Vertex Similarity in Networks. Physical Review E 73(2), 026120 (2006)
25. Leskovec, J., Lang, K.J., Mahoney, M.: Empirical Comparison of Algorithms for Network Community Detection. In: Proceedings of the 19th International Conference on World Wide Web, pp. 631–640. ACM (2011)
26. Kannan, R., Vempala, S., Vetta, A.: On Clusterings: Good,Bad and Spectral. Journal of the ACM 51(3), 497–515 (2004)
27. Radicchi, F., Castellano, C., Cecconi, F., Loreto, V., Parisi, D.: Defining and Identifying Communities in Networks. Proceedings of the National Academy of Sciences of the United States of America 101(9), 2658–2663 (2004)

28. Zachary, W.W.: An Information Flow Model for Conflict and Fission in Small Groups. Journal of Anthropological Research 33(4), 452–473 (1977)
29. Danon, L., Diaz-Guilera, A., Duch, J., Arenas, A.: Comparing Community Structure Identification. Journal of Statistical Mechanics: Theory and Experiment 2005(9), P09008 (2005)
30. Lusseau, D.: The Emergent Properties of a Dolphin Social Network. Proceedings of the Royal Society of London 270(suppl. 2), S186–S188 (2003)
31. Newman, M.E.J.: Modularity and Community Structure in Networks. Proceedings of the National Academy of Sciences of the United States of America 103(23), 8577–8582 (2006)

A Self-learning Clustering Algorithm
Based on Clustering Coefficient

MingJie Zhong[1], ZhiJun Ding[1,*], HaiChun Sun[1], and PengWei Wang[2]

[1] The Key Laboratory of Embedded System and Service Computing,
Ministry of Education, Tongji University, Shanghai 200092, China
zmj915good@163.com, Zhijun_ding@outlook.com,
sunhaichun1985@126.com
[2] Department of Computer Science, University of Pisa, Pisa 56127, Italy
pwei.wang@gmail.com

Abstract. This paper presents a novel clustering algorithm based on clustering coefficient. It includes two steps: First, k-nearest-neighbor method and correlation convergence are employed for a preliminary clustering. Then, the results are further split and merged according to intra-class and inter-class concentration degree based on clustering coefficient. The proposed method takes correlation between each other in a cluster into account, thereby improving the weakness existed in previous methods that consider only the correlation with center or core data element. Experiments show that our algorithm performs better in clustering compact data elements as well as forming some irregular shape clusters. It is more suitable for applications with little prior knowledge, e.g. hotspots discovery.

Keywords: clustering algorithm, clustering coefficient, self-learning clustering.

1   Introduction

Clustering is an unsupervised learning method. Its objective is to group data elements into some clusters, and make data elements in the same cluster have a high similarity while those in different clusters have a low similarity. Clustering algorithms have been widely studied, and the results have been applied in mass of application domains, such as e-commerce [1], bioinformatics [2], speech recognition [3], intelligent retrieval [4,5] and text classification [6, 7, 8].

Most of the existing clustering algorithms are based on some manually set parameters like cluster's number, density or size [9]. A loop is executed until a termination condition is reached. They work well in some specific applications. But, there are some limitations. First, most of them specialize in forming some definite shapes. Influenced by the characteristics of the algorithm, they may lose some objective shapes. E.g., results of K-means [10] are some star structure clusters. Second, setting a parameter for all clusters is unsuitable. E.g., DBSCAN [11] sets global *radius* and *minPts*

---

* Corresponding author.

B. Benatallah et al. (Eds.): WISE 2014, Part I, LNCS 8786, pp. 79–94, 2014.
© Springer International Publishing Switzerland 2014

value to all clusters. In fact, to show each cluster's structure honestly, their termination condition should be self-learning. Moreover, most of these algorithms ignore the correlation among data elements in one cluster. K-means and its derivate algorithms such as K-medoids [12] clustering only consider the correlation between elements and medoids. In this paper, a novel clustering algorithm is proposed, which can learn the structure of cluster objects autonomously, perform clustering dynamically and take correlation between each other into account, thereby forming some irregular shape clusters.

The rest of this paper is organized as follows. Sect.2 is the related work in clustering algorithms. Sect.3 describes the basic concepts of this self-learning clustering algorithm. Sect.4 is the clustering algorithm implementation. Sect.5 provides experiments and analysis. Sect. 6 draws some concluding remarks and discussions.

2   Related Work

Clustering algorithms can be divided into several categories, i.e., partition-based, density-based, hierarchical, spectral, and artificial intelligence-based.

K-means [10] is a kind of partition clustering algorithms. It needs to estimate k value, starts with k initial cluster centers, and assigns each data element to its closest cluster mean. In iterations, cluster means will be recalculated and updated to a new mean according to cluster's constituent data. It will work on until its convergent function is less than a certain threshold or all the means never changed. The initial k value has a great influence on the final results of clustering.

DBSCAN [13] is a density-based clustering algorithm. It defines a ε as the radius around a data element and a *minPts* as the threshold number of data elements in a certain radius region. The density associated with a data element is the number of data elements in a region of ε. It does clustering by putting density-connected data elements into clusters.

CURE [14] and ROCK [15] are both hierarchical clustering algorithms. The CURE algorithm takes each data element as a cluster, and then merges them according to closeness between two clusters. The ROCK algorithm also takes each data element as a cluster, but it merges them based on inter connections between two clusters.

Spectral clustering, according to Shi[16] *et al.* is built on the basis of spectral graph theory. Liu *et al.* [17] try to use Spectral clustering for Chinese words clustering. It uses similar matrix and diagonal matrix to build a Laplacian matrix, and then calculates k eigenvectors corresponding k minimum eigenvalues. At last, it uses k-means to cluster elements with matrix composed by k eigenvectors.

Vesanto and Alhoniemi propose a self-organizing map (SOM) [18] algorithm, which uses artificial neural network for clustering. It can deal with data elements incrementally. And it is useful to visualize high-dimensional data elements by two steps: training and mapping. Training is to build a map with training data elements and mapping is to assign new data elements.

In the case of small and medium-scale data element set, k-means can converge and find globular clusters quickly. However k-means together with all its derivative algorithms cannot find non-globular clusters. In fact, estimating k value seems difficult because of the relatively large number of data elements. DBSCAN can find non-globular clusters. But it cannot reflect the density changing trend of the data element set very well. CURE ignores the information about the aggregate interconnectivity of data elements in two clusters, while ROCK ignores the information about the closeness of two clusters [9]. Compared with the traditional clustering algorithms, spectral clustering has an advantage of converging to global optimal solution on spatial clustering and does clustering analysis well in the case of large data element set. But the construction of similar matrix in this algorithm has a great influence on the final results of elements clustering. Moreover, there are still big challenges in dimensionality reduction of high-dimensional similarity matrix, enhancement of analysis efficiency, determination of clustering number and selection of eigenvector. For SOM, it needs a long network training time to preset the structure and data elements number of competitive layer network. When lacking of learning mode, the results are dependent on input sequence.

3   Cluster Cohesiveness

A goal of clustering algorithm is to make data elements in the same cluster have higher similarity while those in different clusters have lower similarity. To get an appropriate cluster number by learning the structures of clusters, a criterion to judge the similarity among data elements within the cluster should be needed. We learn from graph theory and introduce a concept of clustering coefficients.

Clustering coefficient [19] is used to describe the cohesiveness among vertices in a graph. Specifically, it is the cohesiveness among adjacent vertices of a vertex. Clustering coefficient is divided into local clustering coefficient and average clustering coefficient [19]. The former can be used to measure the cohesiveness near one vertex, while the latter assesses the average cohesiveness of all vertices in a graph.

Suppose C_i is defined as the local clustering coefficient of a vertex v_i. As noticed in Ref. [19], it is equal to the number of edges among vertices which are adjacent to v_i divided by the maximum possible number of edges among the vertices which are adjacent to v_i:

$$C_i = \frac{\left|\{e_{jk} : v_j, v_k \in N_i, e_{jk} \in E\}\right|}{\frac{|N_i|(|N_i|-1)}{2}} \tag{1}$$

Note that, N_i is the collection of v_i's neighbor vertices and $|N_i|$ represents the number of vertices in N_i. E is the collection of edges and e_{jk} is the edge from v_j to v_k. For undirected graph, the maximum number of edges connected to each other among $|N_i|$ number vertices is $|N_i|(|N_i|-1)/2$, so that is the denominator of C_i.

We can find that the vertex A's local clustering coefficients in Fig. 1(a) and Fig. 1(b) are equal.

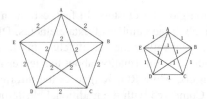

Fig. 1. Two complete graph with different edge weight

If the edge weight is a distance size between two vertices, we may find the relationship between vertices in the right graph is closer than the left graph. If the edge weight is a relationship size between two vertices, we may find the relationship between vertices in the left graph is closer than the right graph. Therefore, in the calculation of graph clustering coefficient, edge weights also need to be taken into account. After considering the edge weights, the formula of local clustering coefficient is:

$$C_i = \frac{\left|\{e_{jk} : v_j, v_k \in N_i, e_{jk} \in E\}\right|}{\frac{|N_i|(|N_i|-1)}{2}} \cdot \frac{\sum_{j \in N_i} e_{ji}}{|N_i|} \tag{2}$$

Here $\frac{\sum_{j \in N_i} e_{ji}}{|N_i|}$ is the average edge weight of neighbor vertices which are adjacent

to v_i. In Fig.2, the blue vertex in left graph has three neighbor vertices (i.e., vertex 2, 3, 4). There are three edges among these vertices, so vertex 1's local clustering coefficient is $3/c_3^2$. In the right graph, the vertex 1's local clustering coefficient can be

calculated by the formula (2), that is $3 / \left(c_3^2 \cdot \frac{s_{12} + s_{13} + s_{14}}{3} \right) = \frac{3}{s_{12} + s_{13} + s_{14}}$

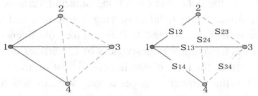

Fig. 2. Unweighted graph and weighted graph

The average clustering coefficient of the entire graph can be calculated according to local clustering coefficient of each vertex. In Ref. [19], it is defined as $\overline{C} = \frac{1}{n}\sum_{i=1}^{n} C_i$. The average clustering coefficient measures the cohesiveness of a graph as a whole.

Based on the average clustering coefficient, we introduce weighted average clustering coefficient to compute intra-class cohesiveness, defined as:

$$intra(\overline{A}) = \frac{\sum_{v_i \in V_A, e_{ij} \in E_A} C_i}{|V_A|} \tag{3}$$

Here C_i is vertex v_i's local clustering coefficient calculated by formula (2), $|V_A|$ is the number of vertices in cluster A.

When we consider the correlation between two graphs, we focus on the vertices connected to each other but located in different graphs. Assuming some vertices are connected between graph $A = (V_A, E_A)$ and $B = (V_B, E_B)$. We define inter-class clustering coefficient as:

$$inter(A',B) = \frac{\sum_{v_i \in V_{A'}, e_{ij} \in U} C_i}{|V_{A'}|}, inter(B',A) = \frac{\sum_{v_i \in V_{B'}, e_{ij} \in U} C_i}{|V_{B'}|}, \quad U = E_A \cup E_B \qquad (4)$$

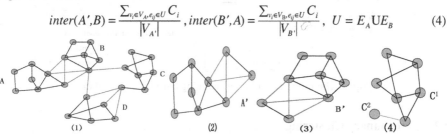

Fig. 3. (1) Graphs A, B, C and D with intersection, (2) Graph A' is composed of graph A and such vertices in graph B with intersection, (3) Graph B' is composed of graph B and such vertices in graph A with intersection, (4) Graph C can be divided into C^1 and C^2 with intersection

Graph A' consists of graph A and the vertices in graph B which are connected to graph A. Similarly, graph B' is composed of graph B and the vertices in graph A which are connected to graph B. If $inter(A',B)$ is bigger than $intra(\overline{A})$, we can join those vertices in graph B to A, and this can improve the average clustering coefficient of graph A. While both $inter(A',B) \geq intra(\overline{A})$ and $inter(B',A) \geq intra(\overline{B})$, this means merging A and B can get a new graph with a higher intra-class degree than A and B alone.

Similarly, if graph C can be split into graph C^1 and C^2, both $inter(C^1,C^2) < intra(\overline{C^1})$ and $inter(C^2,C^1) < intra(\overline{C^2})$, this means C^1 and C^2 have higher intra-class degree than C.

4   A Self-learning Clustering Algorithm

Most of the previous clustering algorithms are based on a value of presetting cluster number, which requires users have some prior knowledge. The proposed clustering algorithm starts work without a preset cluster number. We would get clusters through a self-learning process based on weighted clustering coefficient.

4.1   The Framework of A Self-learning Clustering Algorithm

The proposed algorithm includes two steps. First, sort the correlation by descending order, calculate the correlation convergence rate and build the data element network model using k-nearest-neighbor graph. Second, split and merge according to intra-class and inter-class clustering coefficient.

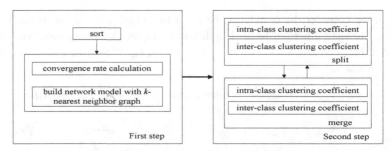

Fig. 4. The framework of a self-learning clustering algorithm

4.2   Preliminary Clustering

Given a big data element matrix whose entries list the correlation between data elements, many models can be applied to present a graph. Considering that complex correlation (in the worst, the data element matrix is a high-dimensional sparse matrix) may make model tend to a complete graph, which is not conducive to using clustering coefficient to clustering. We map data element set to vertex set and choose a common mutual k-nearest neighbor graph[9] in which each vertex has only k neighbor vertices at most. We connect vertex v_i to vertex v_j if v_j is among the k-nearest neighbors of v_i. However, since the neighborhood correlation is not symmetric, this definition leads to a directed graph. In this regard, we connect v_i and v_j only when v_i is one of the k-nearest neighbors of v_j and v_j is one of the k-nearest neighbors of v_i.

To reduce the effects of the noise data, we adopt the max-correlation approach to solve this problem. It is easy to see that vertices with smaller correlation have a higher possibility of being noise data than that with bigger correlation. Therefore, we first sort the pair vertices of matrix according to correlation by descending order. Second, get the biggest pair vertices ($l(ab)$) with their correlation w_{ab}. Check whether these two vertices have been assigned to a cluster, if neither, create a new cluster, and join vertex a and b in this cluster. If only one of the two vertices (E.g. a) has been assigned to a cluster, then we should analyze the structure of the cluster containing a to check whether vertex b can join this cluster. If not, create a new cluster and put vertex b in it. If vertices a and b are in different clusters, we should establish a link between these two clusters. It is for adding an interaction between these two clusters, and the interactive relationship is w_{ab}.

Note that, trim(a, b) is a function to check if a and b are both among k-nearest neighbors of each other. If yes, add an edge between vertex a and b. Otherwise, do nothing.

Clustering requires the resulted clusters have high cohesiveness, embodied in the correlation between two vertices and connectivity over all vertices. So in the preliminary clustering process, it needs to judge if w_{ab} satisfies the structure of a cluster. We can predict the vertices correlation threshold that can be accepted by a cluster according to the current vertices correlation and number. Therefore, it needs to get the overall correlation among vertices to estimate cluster correlation convergence rate.

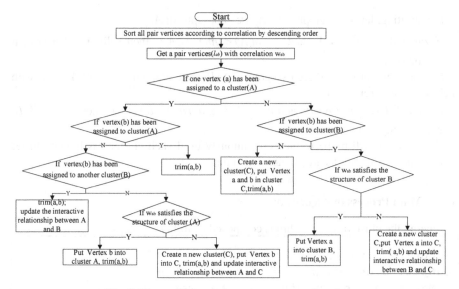

Fig. 5. The preliminary clustering process of vertices

Convergence rate, that means the rate of all the sorted correlation weight convergence by descending order. We can get this value by correlation statistic overall vertices pair and curve fitting algorithms [20].

Suppose there are n_1 vertices in a cluster, $avgn_1$ is its average cluster correlation, y'_{n_1} is the convergence rate. If $w_{ab} \leq (1 + y'_{n_1}) * avg \, n_1$, we need to add the vertex into this cluster. Otherwise, we cannot do that.

Through the above steps, we can get a result of preliminary clustering process, as shown in the Fig.6 (1).

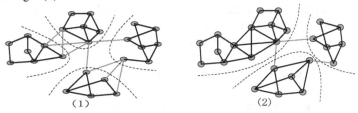

(1)   (2)

Fig. 6. (1) Possible clusters resulted from preliminary clustering process (2) New clusters resulted by splitting and merging from Fig.6 (1)

4.3   Self-learning Clustering Process

In Fig.6 (1), we can see that there are some individuals or parts of internal vertices need to be divided out. Also, some clusters need to be merged, as interconnectivity between them are more closely. There are two standards can be used to judge the problems of splitting and merging.

For splitting, let A is a graph, A_1', A_2' are subgraphs of A.

If $intra(\overline{A}) < inter(A_1', A_2')$ and $intra(\overline{A}) < inter(A_2', A_1')$, then split A; otherwise, it cannot be split.

For merging, let A and B are two graphs, $A+B$ is the graph made up of A and B with their interactive relationship.

If $inter(A', B) \leq intra(\overline{A+B})$ and $inter(B', A) \leq intra(\overline{A+B})$ then merge A and B; otherwise, they cannot be merged.

This process will be performed continuously until each cluster has higher cluster cohesiveness or iterations are greater than a threshold.

4.4   Main Process in Algorithm

In general, the clustering algorithm is composed of following steps:

```
Input: A data element matrix (P)
Output: clusters with vertices in them
1.Map data element set to vertex set (A)
2.Get pair vertices with weight from A and store them
in pairWeightList
3.Sort pairWeightList according to weight by descending
order, store them in dataList
4.Get convergence rate by correlation statistic overall
vertices pair and curve fitting algorithms
5.while(Get a pair vertices (i.e. w₁ ,w₂)!= null from
      dataList){
6.  re = 1/ correlation of (w₁,w₂)
7.  if (neither of w₁, w₂ in an existed cluster){
8.    create a new cluster, put w₁ and w₂ in it, set an
      edge from w₁ to w₂, update avg(C)}
9.  if (only one vertex(i.e. w₁) in an existed
      cluster(i.e. C){
10.     if(re<avg(C)*(1+rate(C))&& Vol(w₁)<k
        &&Vol(w₂)<k){
11.       put w₂ in cluster C, set an edge from w₁ to
          w₂, update avg(C)}
12.     else{
13.       create a new cluster, put w₂ in new cluster,
          trim (w₁,w₂)}}
14. if(one vertex(i.e. w₁) in a cluster(i.e. C), w₂ in
      another cluster(i.e. D)){
15.     if(Vol(w₁)<k && Vol(w₂)<k){
16.       set an edge from w₁ to w₂
17.       put inter(C|D,w₁|w₂|re) in interaction map
          (interMap)}}
```

```
18.}// end while
19.while(flag){
20.   for(cluster c_i: cluster set){
21.      if(intra(c_i)<inter(c_i^1,c_i^2) || intra(c_i)<inter(c_i^2,c_i^1))
22.         Split c_i, flag = false, update interMap
23.   }//end for
24.   Sort interMap by interaction descending order
25.   for(inner in : interMap){
26.   get in(c_p/c_q, interaction)
27.
if(intra(c̄_p)<inter(c'_p,c_q) && intra(c̄_q)<inter(c'_q,c_p))}
28.         value = (inter(c'_p,c_q)+inter(c'_q,c_p)
                     -intra(c̄_p)-intra(c̄_q))/2
29.         put mergeQueue(c_p/c_q,value), flag = false }
30.   }//end for
31.   Sort mergeQueue according to value by desc order
32.   for(item it : mergeQueue)
33.         merge it, update interMap
34.}//end while
```

Here, *pairWeightList is* a list and its elements are data with form as "$w_1|w_2$, *correlation*". *dataList* is a copy list of sorted *pairWeightList* by *desc*, *avg(C)* is the average correlation in cluster C, $Vol(w_l)$ is the number of vertex w_l's neighbor, k is the parameter of k-nearest neighbor graph. For *interMap*, it is a map storing clusters' interaction ship, *in(c_p/c_q, interaction)* is an key-value element of *interMap*, *mergeQueue* is a data structure storing pair clusters which needed to merge in next iteration.

For time complexity, it's necessary to pay attention to splitting process and merging process. It takes $O(n^3)$ to calculate one cluster's average clustering coefficient. For splitting, it needs one traverse to check vertices which should be split from the original cluster. Thus, it takes $O(n^4)$ to complete splitting process in one cluster. For merging, it takes $O(n^3)$. Suppose there are n vertices in total, and k clusters in result, the running time for all vertices is $\sum_{i=1}^{k} n_i^4 \leq n^4$, subject to $\sum_{i=1}^{k} n_i = n$. To reach the termination conditions of splitting and merging, a flag is set to present the iteration times. The total running time for splitting or merging is mn^4 in which m is the iteration times. Other than the splitting and merging process, the rest of sort, curve fitting and the preliminary clustering process is $O(n\log n)$, $O(n^3)$ and $O(n)$ respectively. Hence, the time complexity is $O(n^4)$. In fact, the algorithm is efficient as numbers in clusters after preliminary clustering process are always small.

5   Experiments and Analysis

In this section, we perform extensive experiments on keywords clustering, specifically, on Chinese webpage keywords. We first crawled many webpages from www.ifeng.com in blog, news, technology, finance and education sections by spider. Then we used cx-extractor [21] to extract the main content of each webpage and ANSJ [22] Chinese word analyzer to do participle. After we got Chinese keywords and correlation among them, we did clustering by k-medoids, DBSCAN and our algorithm, respectively. In this paper, the correlation of keywords is based on co-occurrence. We can find that the correlations among keywords are transitive in the same text, but in the scope of the entire webpages set, the transitivity is not established.

5.1   Words Correlation Calculation

"Term frequency–inverse document frequency" (TFIDF) is one of the most commonly used term weighting schemes in today's information retrieval systems [23]. $TF(x,d)$ is the term frequency of keyword x in webpage d, and $IDF(x,D)$ is the inverse document frequency of keyword x in webpage set D. The word's TF-IDF value is calculated as follows:

$$TFIDF(x,d,D) = TF(x,d)* IDF(x,D) \tag{5}$$

Then we define the correlation $r(x_1,x_2,d)$ between keywords x_1 and x_2 in webpage d as follows:

$$r(x_1,x_2,d) = \frac{1}{\sqrt{1 - TFIDF(x_1,d,D)^2 - TFIDF(x_2,d,D)^2}} \tag{6}$$

The correlation $r(x_1,x_2,D)$ between keywords x_1 and x_2 in webpage set D can be calculated as follows:

$$r(x_1,x_2,D) = \frac{\sum r(x_1,x_2,d_i)}{\log n} \tag{7}$$

Here d_i is a webpage contains keywords x_1 and x_2, n is the number of webpages contains keywords x_1 and x_2.

5.2   Experimental Results

Two types of experiments were conducted. First, we got a set of webpages during a certain time and did clustering by k-medoids, DBSCAN and the proposed algorithm (CBCC), respectively. Second, we did clustering by the proposed algorithm in three consecutive times. The data source is at http://pan.baidu.com/s/1b8WGm. Here the experiments results are translated from Chinese.

5.2.1   Comparison Experiment

In this experiment, we got 934574 pair keywords with correlation from 25498 web-pages from 2014.02.15-2014.03.15. We got the first 2000 pair keywords set according to correlation by desc and did clustering by k-medoids, DBSCAN and the proposed algorithm, respectively. Here are some keywords clusters got from k-medoids and the proposed algorithm.

Table 1. Some keywords clusters got from k-medoids and the proposed algorithm (CBCC)

k-medoids	villages, powers, Yunnan, population, housing, usufruct, agriculture, farmer, rural, rural land, transfer fee, foreign capital, labor, health, history, par, towns, migrants, academics, bottom line, building material, floor area, contracted land, governmental, government departments, culture, physical, phenomena, land use, science research fund, management rights, result, administration, Yangtze River Delta, collective land, area, budget
CBCC	collective land, housing, house, house price, land, usufruct
k-medoids	reason, place, fund, Central Bank, suggestion, profit rate, quantity, demonstration area, red envelope, manager, concept, currency, capital, Liaoyang city
CBCC	standard, area, risk, collectivity, interests, Central Bank, organization, government, region, place, opportunity, quantity, process, assets, suggestion, extent, policy, clause
k-medoids	thing, product, public, business, map, general public, entity, airliner, mobile phone, alipay, license tag, offline, video, account, banking industry
CBCC	resource department, offline, property, general public, alipay, Deming xu
k-medoids	chairman, representation, insurance, pensions, interests, weekly report, media, students, officials, yardstick, salary, gap, income, government, news, channels, science and technology, space, group, reporter, resource, demand, Jack ma
CBCC	net profit, income, salary, group
k-medoids	center, Jingdong, tradition, institution, Beijing, international, committee member, giant, technology, fundamental, monkeys, key point, domain
CBCC	international, area, technology, center

Here the initial number of clusters for k-medoids is 200 and the parameter k of k-nearest neighbor graph in CBCC is 4. Based on the results above, we can find that:

For k-medoids, keywords in the same cluster have a high correlation with the center keyword, but it doesn't guarantee high correlation between pairwise keywords.

For the proposed algorithm, each cluster has no center keyword, but keywords in the same cluster have a high correlation between pairwise keywords.

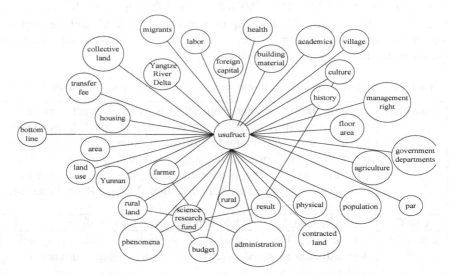

Fig. 7. The resulted graph of k-medoids with center keyword "usufruct"

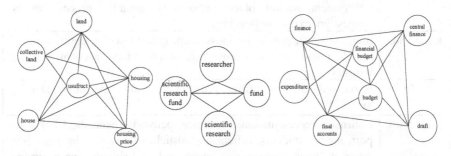

Fig. 8. Resulted graphs of "usufruct", "scientific research fund" and "budget" by CBCC

From Fig.7, we can find that "right to use" is a center keyword, and all other keywords have a relationship with it. But these other keywords have a low relationship between each other. From Fig.8, we can find that there is no center keyword in each cluster, and the keywords in the same cluster have close relationship between each other. Therefore, the former method shows the relationship with center keyword, reflecting multiaspect of center keyword, while the latter method shows the relationship with each other in cluster, focusing on such an aspect of all keywords in cluster.

Although DBSCAN is an acentric clustering algorithm, but it still ignores the correlations between each other. The result shows that the number of resulted clusters is always 2-4, while number of keywords in each cluster is either less than 10 or more than for thousand, which can be shown in Table 2.

Note that, one of the resulted clusters is the noise cluster. If the number is 2, it means that only one cluster is non-noise cluster, since the other one is a noise cluster.

The smaller cluster number make bigger words number in clusters. Although, it is hard to draw a graph of a resulted cluster containing so many keywords, it shows in sideways that DBSCAN does not take correlations of one another into account. Besides, it also shows that DBSCAN is unfit for great changeable density clustering.

Table 2. The numbers of the resulted clusters got from DBSCAN

radius / minPts	0.05	0.055	0.06	0.065	0.07	0.075	0.08	0.085
3	4	4	4	4	4	4	4	4
4	2	2	2	2	2	2	2	2
5	2	2	2	2	2	2	2	2
6	2	2	2	2	2	2	2	2

5.2.2 Theme Evolving Process

In this experiment, we did clustering by our algorithm with 5661 webpages during 2014.02.21-2014.03.02; with 18527 webpages during 2014.03.03-2014.03.13; with 583 webpages during 2014.03.14-2014.03.23. Here are some clustering results containing such key words in different time bucket.

Table 3. Some clustering results containing such key words in different time bucket by CBCC

Keywords	Time bucket	Clustering result
passenger	2014.02.21-2014.03.02	customer, taxi, passenger, vehicles, user
	2014.03.03-2014.03.13	warship, Malaysia, Beijing Airport, passport, passenger, conference, mobile phone, mainland
	2014.03.14-2014.03.23	working group, passenger, secret worry
taobao	2014.02.21-2014.03.02	statement, draft, team, message, website, engineering, plan, finance, fund, taobao, intelligent, platform, global, equipment, way, science and technology, trading platform
	2014.03.03-2014.03.13	Ali, platform, WeChat, founder, supply chain, taobao
	2014.03.14-2014.03.23	Ali, profit, team, taobao
central bank	2014.02.21-2014.03.02	income, net profit, concept, Central Bank, government, industry, place, quantity, gross margin rate, funds, ratio, extent, policy, affiliated party, product quality, official, group
	2014.03.03-2014.03.13	Central Bank, NPC member
	2014.03.14-2014.03.23	entity, customer, offline, settlement, Central Bank, QR code, People's Bank tradition, alipay

From these results, we can find the theme evolving process over time. For keyword "passenger", it appears along with "customer", "taxi", "vehicles", "user" during 2014.02.21-2014.03.02. These keywords are commonly seen in transportation scene in daily life; but it appears along with "warship", "Malaysia" when Malaysia Airlines Flight 370 lost communication at March 8th, 2014. Then during 2014.03.14-2014.03.23, it appears together with "workgroup", "secret worry" as the event of Malaysia Airlines Flight 370 has passed for one week. For word "Central Bank", it appears on financial scene during 2014.02.21-2014.03.02 while it appears with "NPC member" just because the NPC and CPPCC were held during 2014.03.03-2014.03.13. Then "Central Bank" appears together with "customer", "alipay", "QR code" just because the Central Bank released a series of policies to restrain payment of QR code scanning, and alipay's balance payment.

6   Conclusion

In this paper, we propose a novel clustering algorithm which takes the correlation among data elements into account, and calculate cluster cohesiveness with clustering coefficient. The whole clustering process has two steps. First step is preliminary clustering by k-nearest-neighbor and correlation convergence. Second step is splitting and merging the results of preliminary clustering process with intra-class and inter-class concentration degree based on clustering coefficient.

We perform extensive experiments of Chinese keywords on webpages. Our experiments results demonstrate that k-medoids tend to cluster hypersphere, in which we may find a medoid. In fact, if the clustering object is an asphere, in which no appropriate medoid exists, k-medoids will perform badly. For DBSCAN, it can discover arbitrary shapes no matter whether there is medoids in them, but it cannot reflect the dynamic changes of density. For the proposed algorithm, we do not need medoid during clustering, and can reflect such dynamic change by preliminary clustering with convergence rate. In addition, neither k-medoids nor DBSCAN takes the correlation between each other into account. K-medoids only considers the correlations with medoid and DBSCAN only considers the correlations to core data element. The proposed algorithm takes the correlation between each other into account, which make the resulted clusters have higher concentration degree. Experiments prove that, our algorithm can perform better in hotspots discovery and hotspots evolution. Otherwise, it also applies to interpersonal relationship analysis and community detection.

With respect to the algorithm efficiency, some graph theory techniques are employed in our clustering algorithm, and the time complexity is $O(n^4)$. There is a need to improve time complexity of the proposed algorithm, which is exactly our future work.

Acknowledgment. This work is partially supported by National Basic Research Program of P.R. China (973 Program) under Grant No. 2010CB328101, the National Natural Science Funds of P.R. China under Grants No. 90818023, HongKong, Macao and Taiwan Science and Technology Cooperation Program of China under Grant No.2013DFM10100.

References

1. Sarwar, B.M., Karypis, G., Konstan, J., et al.: Recommender systems for large-scale e-commerce: Scalable neighborhood formation using clustering. In: Proceedings of the Fifth International Conference on Computer and Information Technology (January 2002)
2. Roy, P.J., Stuart, J.M., Lund, J., et al.: Chromosomal clustering of muscle-expressed genes in Caenorhabditis elegans. Nature 418(6901), 975–979 (2002)
3. Momtazi, S., Sameti, H., Bahrani, M., et al.: A POS-based fuzzy word clustering algorithm for continuous speech recognition systems. In: 9th International Symposium on Signal Processing and Its Applications, ISSPA 2007, pp. 1–4. IEEE (2007)
4. Momtazi, S., Klakow, D.: A word clustering approach for language model-based sentence retrieval in question answering systems. In: Proceedings of the 18th ACM Conference on Information and Knowledge Management, pp. 1911–1914. ACM (1914)
5. Yasukawa, M., Yokoo, H.: Term Clustering based on Lengths and Co-occurrences of Terms. ADCS 2009, 126 (2009)
6. Dhillon, I.S., Mallela, S., Kumar, R.: Enhanced word clustering for hierarchical text classification. In: Proceedings of the Eighth ACM SIGKDD International Conference on Knowledge Discovery and Data Mining, pp. 191–200. ACM (2002)
7. Wu, Y.C., Yang, J.C.: A Weighted Cluster-based Chinese Text Categorization Approach: Incorporating with Word Clusters. In: 2012 IIAI International Conference on Advanced Applied Informatics (IIAIAAI), pp. 279–282. IEEE (2012)
8. Sebastiani, F.: Machine learning in automated text categorization. ACM Computing Surveys (CSUR) 34(1), 1–47 (2002)
9. Karypis, G., Han, E.H., Kumar, V.: Chameleon: Hierarchical clustering using dynamic modeling. Computer 32(8), 68–75 (1999)
10. Jain, A.K.: Data clustering: 50 years beyond K-means. Pattern Recognition Letters 31(8), 651–666 (2010)
11. Ester, M., Kriegel, H.P., Sander, J., et al.: A density-based algorithm for discovering clusters in large spatial databases with noise. In: KDD, vol. 96, pp. 226–231 (1996)
12. Park, H.S., Jun, C.H.: A simple and fast algorithm for K-medoids clustering. Expert Systems with Applications 36(2), 3336–3341 (2009)
13. Ertöz, L., Steinbach, M., Kumar, V.: Finding clusters of different sizes, shapes, and densities in noisy, high dimensional data. In: SDM (2003)
14. Guha, S., Rastogi, R., Shim, K.: CURE: an efficient clustering algorithm for large databases. ACM SIGMOD Record 27(2), 73–84 (1998)
15. Guha, S., Rastogi, R., Shim, K.: ROCK: A robust clustering algorithm for categorical attributes. In: Proceedings of the 15th International Conference on Data Engineering, pp. 512–521. IEEE (1999)
16. Shi, J., Malik, J.: Normalized cuts and image segmentation. IEEE Transactions on Pattern Analysis and Machine Intelligence 22(8), 888–905 (2000)
17. Liu, Y., Nan, W., Zheng, T.: Spectral clustering for Chinese word. In: Sixth International Conference on Fuzzy Systems and Knowledge Discovery, FSKD 2009, vol. 1, pp. 529–533. IEEE (2009)
18. Vesanto, J., Alhoniemi, E.: Clustering of the self-organizing map. IEEE Transactions on Neural Networks 11(3), 586–600 (2000)

19. Soffer, S.N., Vázquez, A.: Network clustering coefficient without degree-correlation biases. Physical Review E 71(5), 057101 (2005)
20. Guest, P.G., Guest, P.G.: Numerical methods of curve fitting. Cambridge University Press (2012)
21. https://code.google.com/p/cx-extractor/
22. https://github.com/ansjsun/ansj_seg
23. Aizawa, A.: An information-theoretic perspective of tf–idf measures. Information Processing & Management 39(1), 45–65 (2003)

Detecting Hierarchical Structure of Community Members by Link Pattern Expansion Method

Fengjiao Chen and Kan Li

Beijing Institute of Technology, Beijing, China
cfjmonkey@hotmail.com,likan@bit.edu.cn

Abstract. Community structure is an important property of complex networks, which is generally described as densely connected nodes and similar patterns of links. Hierarchy is a common property of networks. Different members have different belonging coefficients to the community, e.g. core members and boundary members, who are at different levels in the hierarchy of community. In this paper, a novel structure is presented, called hierarchical structure of members (HSM), which shows the relationships among members and multi-resolution of the community. A hierarchical link-pattern expansion method is proposed to detect HSM. First, we use the most similar link patterns to detect the seed communities which include both clique structures and star structures. Next, we define the influence between members to expand the community hierarchically. The experiment explores the hierarchical structure of members and the comparison with competitive algorithms on real-world networks demonstrates our method has stronger ability to detect communities.

Keywords: community detection, link expansion, hierarchical structure of members.

1 Introduction

Groups of similar nodes construct the community which is often considered as the functional organizations or building blocks of the complex network[4], such as departments of the company. Nodes in the community are also hierarchical, such as the leader-follower hierarchy in a department of the company. Here, we call the hierarchy in the community as *hierarchical structure of members (HSM)*. Focusing on a community, HSM divides the nodes into multiple levels according to their belonging coefficients to the community. The first level has the highest belonging coefficient where nodes are the core members. Then the community is composed of different levels of members. Varying from high level to low level, the structure of the community varies from core members to boundary members.

HSM is helpful to understand the community. From a macro perpective, it shows how the community structure varies. Take the communities in a social network as an instance, we can know who are interested in the community and predict the potential people who may be interested in. From a micro perspective,

B. Benatallah et al. (Eds.): WISE 2014, Part I, LNCS 8786, pp. 95–104, 2014.
© Springer International Publishing Switzerland 2014

it shows the relationships of members. In the collaborative filtering of recommender system, it's important to evaluate the similarities between users. Using HSM, we can not only know what the users both like but also what they both dislike or both like a little. Thus, the similarities between users can be evaluated more accurately.

There are challenges to detect the HSM. First, the seed communities can be clique structures and star structures. In coauthor network, seeds tend to be clique structures(Fig. 1a). However, in PPI network, seeds tend to be star structures(Fig. 1b). As far as we know, current algorithms cannot detect these two structures simultaneously. Second, there are correlations between members that make them stay at the same level. How to measure these correlations is a challenge.

In this paper, a hierarchical link-pattern expansion method is proposed, which overcomes these challenges. First, the most similar link patterns are used as seed communities which can be clique structures and the star structures. Second, we define the influence between members to measure the correlation. Results on the real-world networks demonstrate that the HSM is meaningful and the revealed communities is of good quality.

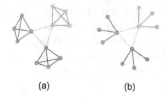

(a)   (b)

Fig. 1. Clique structure(a) and star structures(b)

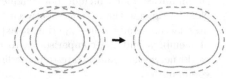

Fig. 2. Combination of two communities. Solid line indicates community and dash line indicates community neighborhood.

2   Related Work

The hierarchical structure of nodes is studied extensively in biological networks and social networks[17,5,19]. All of these prior works suggest that the hierarchy is an important feature of natural, artificial and social networks. Members in the community are also hierarchical that different levels of hierarchy indicate different belonging coefficients to the community. Searching from the core level to the boundary level can be seen as a construction process of community. Local expansion methods can describe the construction process[12]. Many other methods such as generation model[13,16], modularity[2], Markov random walk model[9] and speaker-listener model[21] are combined with the local expansion process to reach an efficient result. Lancichinetti et al.[12] varied the parameter alpha in the fitness function and explored the whole hierarchy from the entire network down to the single nodes. However, the alpha is not determined. To improve that, Havemann et al.[8] calculated all the proper values of parameter alpha which represent stable community structures, but they didn't give belonging coefficients for each level. There's lack of an algorithm to partition the members into levels by belonging coefficient.

On the other hand, seed community plays a fundamental role in the community. There are three kinds of methods to detect seed communities. The first one is clique based method, which can help to detect densely connected structures[14,8]. However, as the star structure often appeared in the real-world networks(e.g. grassland network[3]), clique based methods cannot capture such a sparse structure. The second one is based on the node's neighborhood which has been theoretically studied[7], but it cannot detect the star structure in Fig. 1 either. The last one uses single nodes or links[16,12] as seeds, which are obviously meaningless as a community.

In this paper, a hierarchical link-pattern expansion method is proposed, which can detect the hierarchical structure of members. Comparisons are conducted with three competitive methods, HLC[1], GCE[14], OSLOM[13]. HLC is based on the links, which has good performance in detecting dense communities. GCE is a local expansion method based on the clique which can detect densely connected structures fast and accurately. OSLOM is an algorithm based on the nodes which can detect significant community structures. Comparison result shows that our method has better performance on detecting communities.

3   ECMO Method

3.1   Problem Definition

To simplify the problem, here we mainly focus on the undirected, unweighted and simple network. Different from the traditional hierarchical structure of communities, the hierarchical structure of members for community C is defined as

$$Hier(C) = \{level_i\}, i = 1...m \tag{1}$$

where m is the number of levels. Each member belongs to one of the levels. Each $level_i$ is a subset of members and members in higher level have higher belonging coefficients.

In order to detect the HSM, we propose a method called ECMO. ECMO is a hierarchical link-pattern expansion method including two parts: detection of seeds and expansion of communities. Inspired by the idea in [1], we consider the community of links rather than nodes. Similarly, the corresponding neighbors of a link are defined as the links having common nodes. The degree of a link is the number of its neighbors. Besides, the neighbors of a community are defined as the links that are outside the community but neighboring the links inside the community.

3.2   Detection of Seed Communities

Seed communities are the initial or core level of the community which contains the members with the strongest belonging coefficient. Detection of seeds should handle both clique structures and star structures. There's a common property of these two structures: the intra-community links have similar topological relations, in other words, they have similar neighbors. Therefore, a link and its most similar neighbor link(s) should be in the same community. These most similar link patterns compose the seed communities. A seed community is a set of links, that is $Seed = \{e_{ij}\}$. $\forall e_{ij} \in Seed, \exists e_{pq} \in Seed$, satifies $e_{pq} = argmax\{Sim(e_{ij}, e_{p'q'})\}$ where $e_{p'q'}$ is the neighbor of e_{ij} and $Sim(\cdot, \cdot)$ is the similarity between links.

The similarity between the neighbor links is the number of their neighbors who are still neirghbors, normalized by the total number of their neighbors. The equation can be simplified as

$$Sim(e_{ik}, e_{jk}) = \frac{|n_+(i) \cap n_+(k) \cap n_+(j)|}{|n_+(i) \cup n_+(k) \cup n_+(j)|} \tag{2}$$

where the $n_+(i)$ is defined as the set of node i's neighbors and it self. Moreover, we combine the seeds who have similar node sets, which can reduce redundant seeds. The total computational complexity for detection is $O(m * k * k)$ time, where m is the number of links and k is the average degree of links ($k \ll M$).

This method can detect seed communities including both clique structures and star structures. The seed communities can get rid of single link community which is obviously not a community structure.

3.3   Expansion of Community

After revealing them, we expand each seed community to obtain the HSM. From the view of communities, all the links can be added into the community with different belonging levels. We choose the links in the neighborhood of a community with the strongest belonging coefficients to expand the community. Consistently, we measure the belonging coefficient $Bl(e_{ij})$ of a link e_{ij} to the community C as the sum of similarities with the links inside the community averaged by the link degree.

$$Bl(e_{ij}) = \frac{\sum_{e_{pq} \in C} Sim(e_{ij}, e_{pq})}{d(e_{ij})} \tag{3}$$

where e_{ij} is the neighbor of community C, $d(e_{ij})$ is the degree of e_{ij} and e_{pq} is the neighbor of e_{ij} who's in the community C. The denominator can prevent the belonging coefficient from biasing to the links with large degrees. The belonging coefficient also ranges from zero to one inclusive, which is well normalized. If the link is not the neighbor of the community, its belonging coefficient is zero. If the link is in a sole clique or star community, the belonging coefficient is one. After adding the links with the strongest belonging coefficient, candidates may be influenced and their belonging coefficients can rise higher than the added one's. This phenomena shows that some members have strong influences on each other and we should add them together. The influence of e_{ik} on e_{jk} is defined as

$$Inf(e_{ik}, e_{jk}) = Bl(e_{jk}) + \frac{Sim(e_{ik}, e_{jk})}{d(e_{jk})} - Bl(e_{ik}) \tag{4}$$

where e_{ik} is the link who are just added in the community. If the influence of e_{ik} on e_{jk} is positive, they should be at the same level. Using this step, we can construct the level in the HSM. A level in the HSM is a set of members. $level = \{e_i\}, i = 1...m$, satisfies $Inf(e_i, e_{i+1}) > 0$, where i indicates the order of joining the community and m is the size of the level.

As for detecting all the communities, many seed communities will be expanded to similar community structures (Fig. 2). Then, we combine similar communities with the measure of $\delta(S, S') = |S \cap S'| / min(|S|, |S'|)$[14], where S and S' indicate the set of nodes in each communities respectively. For each pair of communities, we combine the communities with higher similarity than threshold which can be set by the user, and a default value 0.75 is also provided which often has well result [14].

Finally, some implement skills are provided to reduce the computational complexity. At the beginning, we sort each set of nodes' neighbors with time $O(n * k * logk)$ to reduce the time of computing the link similarity from $O(k * k)$ to $O(k)$, where k is the average degree among all the links, n is the number of nodes and m is the number of links. As for seed communities, we use union-find set to detect the most similar link patterns which costs $O(m * k * k + m * logm)$ time at worst. While expanding the community, storing the candidates in the priority heap will just cost $O(logD)$ time to get the links with the strongest belonging (D is number of candicates). And the worst computational complexity for expanding a community is $O(m(klogm + m))$.

Notice that, what we group into the community are the links of the network while most community detection algorithms focus on the nodes. A simple conversion is assigning the nodes to the communities where the connected links belong. If a link is added to the community then the end node, which is not in the community, will be added to the community with the same belonging level as the link.

4   Experiments

In this section, the performance of our method is shown in three aspects. First, we show the performance of the detection of seeds. Second, we show the hierarchical structure of members in the community. Finally, comparing with competitive algorithms, we evaluate the quality of the revealed community structures. All the networks used in experiments are real-world ones which can make the results more realistic.

4.1   Seed Community

We stress the importance of the seed community and use the link-pattern algorithm to detect them. Here, the detection results are shown in three real-world networks which include both clique structures and star structures.

The first example is the coauthorship network of scientists working on network theory and experiment[18], where nodes represent scientists and unweighted links connect pairs of scientists who have coauthored at least one paper together. Fig. 3a shows the seed communities revealed by our algorithms (marked by colors on links). In this figure we can see that the nodes are densely connected and many cliques are revealed. This result is consistent with [14]. In Fig. 3b, although there's a overlapping node between the two clique, we can still detect them correctly because of the link community.

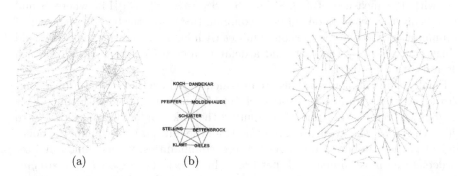

(a)   (b)

Fig. 3. Seeds of coauthor network   **Fig. 4.** Seeds of Uetz network

The second example is the protein-protein interaction network (PPI) from Saccharomyces cerevisiae screened by Uetz[20], where nodes represent proteins and unweighted edges represent the interaction between two proteins. This network is very sparse and has many star structures. Fig. 4 shows that many star structures are successfully revealed, according to the idea of the most similar link pattern.

In the third example, grassland network[3], there's also many star structures. Shown in Fig. 5, nodes represent the species (plants, herbivores, parasitoids, hyperparasitoids and hyper-hyperparasitoids are colored in red, pink, yellow, blue and purple, respectively) and links represent the food relationships between pairs of species. Although parasitoids have no interaction among themselves, they have similar links which indicate the preying on the same herbivore, that's why we group them together. Detection by the most similar link pattern can reveal meaningful seed communities.

4.2 Hierarchical Structure of Members

Detecting the hierarchical structure of members is the main goal of this paper. HSM divides nodes into levels according to their belonging coefficients. From core level to boundary level, HSM shows multi-resolution of the community. We test the algorithm on various networks, here, some classical networks are analysed.

The first one is the classical karate club network consisting of friendships between 34 members of a karate club[23]. We detect the HSM from the view of instructor (node34) in Fig. 7a. The red community in the middle presents strong

Fig. 5. Seeds of grassland network   **Fig. 6.** HSM in coauthor network

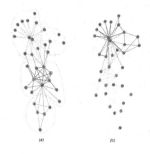

Fig. 7. HSM in karate club network   **Fig. 8.** HSM in football network

belonging that nodes are densely connected with each other. As expanding the community, we can generally see four parts (circled by dash line) with different depth of color in the network, which matches well with previous observations[4]. In addition, we can identify the relationship of them. The left-up part is weaker than the bottom part and stronger than the right part. Former researches can indicate the combination between the parts but ignore the relationship among all these small parts especially the relationship between parts in different bigger groups. From the view of president, we can have similar conclusions(Fig. 7b).

The second one is the coauthor network introduced before. We have shown its seed communities in Fig. 3 and we observe one of the seeds. Part of the network near the seed community (marked by deep red) is showed in Fig. 6. People at the seed level studied the basic theory about MagnetoEncephaloGraphy (MEG). At the next level, people extended the MEG and then people at the third level combined the phase synchronization into MEG. People at different levels express different points of research.

The third example is the American football network[6], where nodes represent teams and links represent the games between two teams. In Fig. 8, teams are divided into conferences labeled by different colors according to their games frequency. The HSM of the yellow communities is shown. We can see that games between teams in the same conferences are generally on similar colors. In addition, we can learn that the yellow conference has stronger relationship between the orange conference than the green conference. The relationship between each conference can be easily figured out through the HSM.

4.3 Community Quality

As mentioned above, community consists of densely connected nodes and similar patterns of links. So we choose the density to measure the quality of the community:

$$Density = \frac{1}{K} \sum_c \frac{2 * m_c}{n_c * (n_c - 1)} \tag{5}$$

where m_c is the number of links of the community c, n_c is the number of nodes of the community c and K is the number of communities. In our experiments, there are ten real-world networks including social networks and biological networks[1]. Three competitive community detection algorithms HLC, GCE, OSLOM are chosen. The parameters in these algorithms are default. In the HSM, members construct the community in multi-resolution. Since current methods detect one resolution of the community, in order to make comparison, we evaluate ECMO by generalizing the quality of the community at certain level of hierarchical structure. Similar with HLC, we pick up the community with highest density for comparison. The comparison results are showed in Fig. 9.

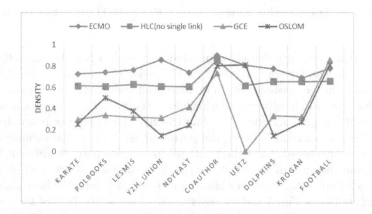

Fig. 9. Comparison of the density of community structure

ECMO performs best among most of the networks. GCE expands community by one node per iteration but stops at local optimization. Although ECMO expands by multiple nodes, it will not stop expanding until contains whole network,

---

[1] They are karate[23], coauthor[18], dolphins[15], football[6], Y2H-Union[22], NDyeast(http://www.nd.edu/~networks), Uetz[20], krogan[11], lesmis[10], polbooks (http://www.orgnet.com/) and the largest network is krogan with over 7000 links.

which help it catch better community structure. The local optimization problem is circumvented thanks to the HSM. Though HLC seems perform well on the dataset, the detecting results contains a lot of single links which are meaningless as discussed in section 2. In this case, these single links are removed and the rest of the results are assessed as shown in Fig. 9. In coauthor network, all the methods perform well because there are many clique structures that can be easily recognized. Although GCE and OSLOM performs well on the football networks but is inconstantly in different networks. On the contrast, ECMO allows the coexistence of clique and star structures, which has better performances on kinds of networks. As a conclusion, our algorithm can detect the community structure with competitive performance.

5   Conclusion

In this paper we propose a hierarchical link-pattern expansion method to detect the hierarchical structure of members. The seed communities are revealed by the most similar link patterns including both clique structures and star structures. The influence is defined to measure the correlations between members at the same levels. In the experiments, the detected HSM is meaningful in the networks where members play different roles at different levels of HSM. Finally, the comparison result with competitive algorithms on the real-world networks shows the better performance of our method.

The hierarchical structure of members is a new way to study complex networks and here we have taken steps towards detecting and understanding such a structure, but its full potential remains unexplored.

Acknowledgments. The Research was supported in part by Natural Science Foundation of China (No.60903071), National Basic Research Program of China (973 Program, No.2013CB329605), Specialized Research Fund for the Doctoral Program of Higher Education of China, and Training Program of the Major Project of BIT.

References

1. Ahn, Y.Y., Bagrow, J.P., Lehmann, S.: Link communities reveal multiscale complexity in networks. Nature 466(7307), 761–764 (2010)
2. Chen, D., Shang, M., Lv, Z., Fu, Y.: Detecting overlapping communities of weighted networks via a local algorithm. Physica A: Statistical Mechanics and its Applications 389(19), 4177–4187 (2010)
3. Dawah, H.A., Hawkins, B.A., Claridge, M.F.: Structure of the parasitoid communities of grass-feeding chalcid wasps. Journal of Animal Ecology, 708–720 (1995)
4. Fortunato, S.: Community detection in graphs. Physics Reports 486(3), 75–174 (2010)
5. Fushing, H., McAssey, M.P., Beisner, B., McCowan, B.: Ranking network of a captive rhesus macaque society: a sophisticated corporative kingdom. PloS One 6(3), e17817 (2011)

6. Girvan, M., Newman, M.E.: Community structure in social and biological networks. Proceedings of the National Academy of Sciences 99(12), 7821–7826 (2002)

7. Gleich, D.F., Seshadhri, C.: Vertex neighborhoods, low conductance cuts, and good seeds for local community methods. In: Proceedings of the 18th ACM SIGKDD International Conference on Knowledge Discovery and Data Mining, pp. 597–605. ACM (2012)

8. Havemann, F., Heinz, M., Struck, A., Gläser, J.: Identification of overlapping communities and their hierarchy by locally calculating community-changing resolution levels. Journal of Statistical Mechanics: Theory and Experiment 2011(1), P01023 (2011)

9. Jin, D., Yang, B., Baquero, C., Liu, D., He, D., Liu, J.: A markov random walk under constraint for discovering overlapping communities in complex networks. Journal of Statistical Mechanics: Theory and Experiment 2011(5), P05031 (2011)

10. Knuth, D.E., Knuth, D.E., Knuth, D.E.: The Stanford GraphBase: a platform for combinatorial computing, vol. 37. Addison-Wesley Reading (1993)

11. Krogan, N.J., Cagney, G., Yu, H., Zhong, G., Guo, X., Ignatchenko, A., Li, J., Pu, S., Datta, N., Tikuisis, A.P., et al.: Global landscape of protein complexes in the yeast saccharomyces cerevisiae. Nature 440(7084), 637–643 (2006)

12. Lancichinetti, A., Fortunato, S., Kertész, J.: Detecting the overlapping and hierarchical community structure in complex networks. New Journal of Physics 11(3), 033015 (2009)

13. Lancichinetti, A., Radicchi, F., Ramasco, J.J., Fortunato, S.: Finding statistically significant communities in networks. PloS One 6(4), e18961 (2011)

14. Lee, C., Reid, F., McDaid, A., Hurley, N.: Detecting highly overlapping community structure by greedy clique expansion. arXiv preprint arXiv:1002.1827 (2010)

15. Lusseau, D.: The emergent properties of a dolphin social network. Proceedings of the Royal Society of London. Series B: Biological Sciences 270(suppl. 2), S186–S188 (2003)

16. McDaid, A., Hurley, N.: Detecting highly overlapping communities with model-based overlapping seed expansion. In: 2010 International Conference on Advances in Social Networks Analysis and Mining (ASONAM), pp. 112–119. IEEE (2010)

17. Nagy, M., Ákos, Z., Biro, D., Vicsek, T.: Hierarchical group dynamics in pigeon flocks. Nature 464(7290), 890–893 (2010)

18. Newman, M.E.: Finding community structure in networks using the eigenvectors of matrices. Physical Review E 74(3), 036104 (2006)

19. Pumain, D.: Hierarchy in natural and social sciences, vol. 3. Springer (2006)

20. Uetz, P., Giot, L., Cagney, G., Mansfield, T.A., Judson, R.S., Knight, J.R., Lockshon, D., Narayan, V., Srinivasan, M., Pochart, P., et al.: A comprehensive analysis of protein–protein interactions in saccharomyces cerevisiae. Nature 403(6770), 623–627 (2000)

21. Xie, J., Szymanski, B.K., Liu, X.: SLPA: Uncovering overlapping communities in social networks via a speaker-listener interaction dynamic process. In: 2011 IEEE 11th International Conference on Data Mining Workshops (ICDMW), pp. 344–349. IEEE (2011)

22. Yu, H., Braun, P., Yıldırım, M.A., Lemmens, I., Venkatesan, K., Sahalie, J., Hirozane-Kishikawa, T., Gebreab, F., Li, N., Simonis, N., et al.: High-quality binary protein interaction map of the yeast interactome network. Science 322(5898), 104–110 (2008)

23. Zachary, W.: An information flow modelfor conflict and fission in small groups1. Journal of Anthropological Research 33(4), 452–473 (1977)

An Effective TF/IDF-Based Text-to-Text Semantic Similarity Measure for Text Classification

Shereen Albitar, Sébastien Fournier, and Bernard Espinasse

Aix-Marseille University, LSIS UMR CNRS 7296
Domaineuniversitaire de St. Jerome, 13397 Marseille Cedex 20, France
`first_name.last_name@lsis.org`

Abstract. The use of semantics in tasks related to information retrieval has become, in recent years, a vast field of research. Considering supervised text classification, which is the main interest of this work, semantics can be involved at different steps of text processing: during indexing step, during training step and during class prediction step. As for class prediction step, new text-to-text semantic similarity measures can replace classical similarity measures that are traditionally used by some classification methods for decision-making. In this paper we propose a new measure for assessing semantic similarity between texts based on TF/IDF with a new function that aggregates semantic similarities between concepts representing the compared text documents pair-to-pair. Experimental results demonstrate that our measure outperforms other semantic and classical measures with significant improvements.

Keywords: Classification, Semantics, Text-to-Text Semantic Similarity.

1 Introduction

Supervised text classification is currently a challenging research topic, particularly in areas such as information retrieval, recommendation, personalization, user profiles etc. Generally, supervised text classification methods use syntactical and statistical models for text document representation. This applies to the most popular text classification methods such as: Naïve Bayes Classifier (NB), Support Vector Machines (SVMs), Rocchio, and k Nearest Neighbors (kNN). These representation models ignore all semantics that reside in the original text that can help in text classification.

However, it is possible to use semantic resources to take into account meaning of the words in text representation in order to improve classification effectiveness. Thus, resulting text representation models can take into account synonyms, relations between words and also can resolve some ambiguities. Many researchers reported that using semantics in text classification improves its effectiveness in specific domains especially by deploying domain specific semantic resources [1].

There are several possibilities for involving semantics during the process of supervised texts classification. In this work, we are interested in involving semantics in class prediction step using, text-to-text semantic similarity measures. Hence, we propose a new text-to-text semantic similarity measure (TF/IDF based), called in this

B. Benatallah et al. (Eds.): WISE 2014, Part I, LNCS 8786, pp. 105–114, 2014.
© Springer International Publishing Switzerland 2014

article SemTFIDF, and we present an experimental study to evaluate it in the context of text classification. In addition, we compare it with another text-to-text semantic similarity measure proposed in the literature (IDF based) called semIDF in this article, and also with the well-known classical similarity measure Cosine that is usually deployed in the Vector Space Model. These experiments are carried out in the biomedical domain using the Ohsumed corpus and domain specific knowledge base Unified Medical Language System (UMLS®) and Rocchio with Cosine as the baseline [2].

Second section reviews state of the art methods deploying semantics in classification or other tasks related to information retrieval or data mining. Third section focuses on the use of semantics during class prediction step and presents our new measure (SemTFIDF) based on TF/IDF and suitable for supervised text classification. Fourth section presents experimental setup that we used to evaluate our new measure. Fifth section analyses the experimental results obtained with Cosine classical similarity measure and these two text-to-text similarity measures (SemIDF and SemTFIDF). Finally, we conclude and present our perspectives for future works.

2   Involving Semantics in Supervised Text Classification

Typically, most of supervised text classification techniques are based on statistical and probabilistic hypothesis in both training and classification procedures. As for text representation or indexing, the importance of a term to a document is assessed using the frequency of its occurrences in the document. So far, the intended meaning of terms and the relations among them are not treated or used in text classification. In other words, semantics and relatedness behind literally occurring words are missing in classical text classification techniques. However, last few years have seen different approaches seeking to introduce semantics during indexing, training and prediction.

Involving Semantics in Indexing. Semantics can be used during indexing for a semantic text representation. Indeed, vector-based (binary or TF/IDF) representations, used by these classical supervised classification methods, enable semantic integration or "conceptualization" that enriches document representation model using background knowledge bases [1, 3].To involve semantic features in indexing, state of the art approaches used either implicit semantics through topic modeling[4] or explicit semantics derived from structured resources and used as new features for text representation[1, 6]. Other approaches use either type in semantic kernels to support some supervised classification techniques [5].

Involving Semantics in Training. In these approaches, concepts replace words in text representation. In addition, the hierarchy and the relations among the added concepts are taken into consideration in the training step which affects the learned model, so the classification model is either the entire ontology or part(s) of its hierarchy. Both works [7, 8] used the hierarchical structure of semantic resources to involve related concepts in text representation. Authors in [8] used propagation algorithm to propagate the weights of identified concepts in patents to their superconcepts. Furthermore, authors in [9] used similar concepts in order to enriched text representation and proposed the approach Enriching vectors. Similarities among concepts are assessed using relations between concepts in the semantic resource. Both Generalization [7, 8] and Enriching vectors [9] involve semantics in the classification model implicitly.

Involving Semantics in Class Prediction. According to the literature, most research focused on enriching text representation with semantics and used classical techniques for prediction[10]. Only few works tried to involve semantics in class prediction by proposing new Text-To-Text Semantic Similarity Measures like Semantic Trees or Concept Forest in[10, 11]. Both works involve explicitly the hierarchy of ontology in text representation and training as a classification model. As for assessing the similarity between two documents, authors chose to use a relatively simple formula inspired from the classical cosine measure and reported significant improvement in classifying web documents according to Yahoo! categories.

New semantic approaches for assessing text-to-text similarities seem to be feasible using semantic similarities among concepts pair-to-pair. In fact, such approaches involve semantics in document comparison and in class prediction as well by discovering similarities between texts considering semantically similar terms in addition to lexically similar ones. According to the literature, assessing the semantic similarity between concepts of semantic resources has attracted the attention of many researchers which resulted in proposing numerous semantic similarity measures[12].

3   Text-to-Text Semantic Similarity Measures

In this section, we are interested in involving semantics in the prediction step of text classification process, particularly, through Text-To-Text Semantic Similarity Measures. In fact, some classifiers in the vector space like Rocchio use this kind of measures in class prediction as the criterion with which they choose the most similar class for a treated document. We propose a new measure for assessing semantic similarity between two Bag of Concepts (BOCs) representing two text documents (or a document and a centroïd in the case of a Rocchio classifier). First, we present some related work on text-to-text semantic similarity measures. Then, we present a new text-to-text semantic similarity measure based on a new aggregation function based of TF/IDF weighting scheme.

3.1   Related Works on Semantic Text-to-Text Similarity

In [13], authors proposed an aggregation function that assesses the semantic similarity between two groups of concepts using the mean of similarities of all combinations of pairs of concepts between these groups. Azuaje, Wang [14] proposed a similar aggregation function that takes into consideration maximum semantic similarities between each concept of g_1 and all concepts from g_2 and *vice versa*. Authors in [8] proposed a propagation algorithm to attribute weights to subsumers involving them in text representation. Furthermore, authors proposed a new text-to-text similarity measure based on these weights as well as the semantic similarity between concepts pair-to-pair. This new similarity measure is the prediction criterion that replaces classical text-to-text similarity of the vector space model like Cosine. Authors reported better clustering of patents using semantic similarities [8].

Authors in [15, 16] developed a different aggregation function (that we refer to later by SemIDF) for comparing short texts or phrases using semantic similarities and Inverse Document Frequency – Idf of text concepts. This function improved significantly text-to-text similarity on Microsoft paraphrase corpus [17] as compared to the classical Cosine similarity measure [15]. It demonstrated high accuracy when applied to automatic short answer grading [16]. The main drawback is that this approach ignores all dependencies between words in sentences.

In the context of text classification, we tested most of the similarity measures presented in this section using the tools and resources presented in section 4. According to results of our preliminary tests, only the measure proposed by Mihalcea et al. 2006 [15] demonstrated some satisfactory results.

3.2   A TF/IDF-Based Text-To-Text Semantic Similarity Measure (SemTFIDF)

We propose a new aggregation function for assessing text-to-text semantic similarity that adapts the previous measure to text classification by using TF/IDF weights instead of IDF weights. In fact, TF/IDF reflects how a feature is important to a document in a corpus. Thus, our measure takes into consideration the importance and the specificity of a feature to each of the compared documents instead of its importance to the corpus in general.

This measure is applied on indexed conceptualized documents represented as BOCs. This measure aggregates semantic similarities between concepts of the compared documents pair-to-pair. An aggregation function calculates the semantic similarity between the compared documents using their representation, and the semantic similarities between their concepts pair-to-pair that are stored in the semantic proximity matrix. This measure can be used in decision-making in order to involve semantics in class prediction of supervised text classification.

Given two text documents represented in the same feature space as BOCs and weighted using TF/IDF scheme, we propose a new measure for assessing the semantic similarity between these documents according to the following formula:

$$SemSim(T_1, T_2) = \frac{1}{2}\left(\frac{\sum_{c \in T_1} maxSim(c, T_2) * TFIDF_1(c)}{\sum_{c \in T_1} TFIDF_1(c)} + \frac{\sum_{c \in T_2} maxSim(c, T_1) * TFIDF_2(c)}{\sum_{c \in T_2} TFIDF_2(c)}\right) \quad (1)$$

where: $maxSim(c, T_2)$ is the maximum similarity between the concept (c) and all concepts representing (T_2), and $TF/IDF_1(c)$ is the weight of the concept c in document T_1.

In fact, this measure is a TF/IDF-weighted average of maximum similarities between each concept from the first document with all concepts representing the second one and vice versa. In the following, we present an experimental study in the context of supervised text classification. Next section presents the platform used in this study.

4   Experimental Setup

In order to assess the effect of Text-To-Text Semantic Similarity Measures, we use the experimental platform illustrated in Fig. 1. This platform uses Rocchio for training and prediction as the classification technique. This technique deploys TFIDF as a weighting scheme and *Cosine*, *SemIDF* and *SemTFIDF*, our contribution, as similarity measures. This section presents resources and tools used in this experimental study.

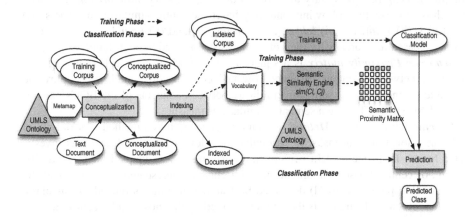

Fig. 1. Platform for supervised text classification deploying Semantic Similarity Measures

Unified Medical Language System *(UMLS®)* was developed at the National Library of Medicine (NLM) in the intent to model the language of biomedicine and health and to help computers understand the language of medicine. It organizes concepts of the various source vocabularies (like MeSH, SNOMED-CT, etc.) according to their senses grouping common concepts together. We choose to use SNOMED-CT exclusively as it provides a large nomenclature on clinical terms.

Ohsumed corpus[18] is composed of abstracts of biomedical articles of the year 1991 retrieved from the Medline database indexed using MeSH (Medical Subject Headings). The corpus is divided into Training and Test sets, so experiments are done in two phases: Training and Test. In this work, we restricted this corpus to the five most frequent classes listed in in Table 1.

Table 1. Ohsumed Corpus

Category	Training	Test
C04	972	1251
C23	976	1181
C06	588	632
C14	1192	1256
C20	502	664
Total	4230	4984

MetaMap. The major goal of MetaMap [19] developers at the NLM was to improve medical text retrieval using UMLS Metathesaurus. Indeed, MetaMap can discover links between medical text and the knowledge in the Metathesaurus.We applycomplete conceptualization using MetaMap results as described in our earlier work [3]using UCI's of concepts for conceptualization implies using concepts as features during indexing, documents are thus represented as bags of concepts (BOCs).

Semantic Similarity Engine. As shown in Fig.1, the semantic similarity Engine computes, using the vocabulary and UMLS ontology, *the semantic proximity matrix*. We chose to use the following ontology-based semantic similarity measures: (i) **cdist**[20]; (ii) **wup**[21]; (iii) **lch**[22]; (iv) **zhong**[23] and (v) **nam**[12].Our choice of ontology-based measures is for their efficiency as compared to other families.

Semantic Proximity matrix is a square matrix in which each cell represents the similarity between elements to which row and column correspond. We deploy the semantic engine to assess semantic similarity between concepts of the BOCs pair-to-pair in SNOMED-CT using a semantic similarities measure.

Rocchio Classification Method. Rocchio or centroïd-based method is widely used in Information Retrieval tasks, in particular for relevance feedback and was investigated for the first time by J.J.Rocchio[2]. Afterwards it was adapted for text classification.For centroïd-based classification, each class is represented by a vector at the center of the sphere (centroïd) delimited by training documents related to this class. The class of a new document is the one represented by the most similar centroïd.

In this work, we consider Rocchio an adequate baseline text classifier for its efficiency and simplicity in addition to its extendibility with semantic resources at both levels: text representation and similarity calculation (see section 2).Training is realized on the corpus and so five class centroïds are calculated for each of the classes. As for prediction step, the test document vector is compared to each of the centroïds learned during training. The platform uses two Text-To-Text Semantic Similarity Measures along with Cosine as a baseline to assess the similarity between the vector of the document and the vector of the centroïd.

5   Experimental Results

In these experiments, the platform executes classification five times once for each of the semantic proximity matrices and once for each aggregation function. Rocchio learns once a unique classification model as a set of centroïds. As for classification, Rocchio uses each of (*semIDF* and *semTFIDF*) in prediction using one of the five proximity matrices resulting in $5 * 2 = 10$ executions. The detailed results from these executions that are related to each semantic similarity measure (between concepts pair-to-pair) are grouped together to analyze the impact of Text-To-Text Semantic Similarity measures on the effectiveness of Rocchio.

In this work we used three evaluation measures: *Precision*, *Recall*, and F_β-*Measure*. Most classification techniques emphasize on either Precision or Recall, thus we use their harmonic mean in Fβ-Measure which is more significant [24].

In next subsections, we use as a baseline of comparison Rocchio with *Cosine* classical similarity measure applied on conceptualized Ohsumed using the CUIs of the best mapped concepts. First we present experiments using a text-to-text semantic similarity measure based on IDF proposed in the literature[15, 16] (cf. 3.1 section), and then, we present experiments using our text-to-text semantic similarity measure based on TF/IDF (cf. 3.2 section).

Results using SemIDFMeasure. Results of these experiments are detailed in Fig. 2a. We notice that using this semantic similarity measure for prediction in Rocchio did not improve its performance at MacroAveraged level. Nevertheless, local significant improvements occurred when treating documents related to (C06) that is one of the least populated classes in the training corpus. This improvement varied from (5.44%) using *wup* to (15.16%) using *lch* resulting in F1-Measure ranging between (57.65%) and (62.97%). These improvements are statistically significant according to McNemar test. Other improvements occurred as well: the first is significant using *lch* on (C04) and the second using *cdist* on (C14). Note that the class (C06) is the least populated class among the five considered classes.

Results using SemTFIDF Measure. Using our TF/IDF-based semantic similarity measure (SemTFIDF) for prediction improved the classification results of (C06). Detailed results are in Fig. 2.b. This improvement is high using all of the five semantic similarity measures ranging between (16.46%) and (18.13%) for *nam* and *wup* respectively. These improvements led to a better F1-Measure in the range [63.68%, 64.60%] as compared with results using Cosine as similarity measure on the same class (54.68%). Using all measures, except for *nam*, improved the F1-Measure of classes (C04) and (C14), these improvements are lower if compared to those on (C06). As for (C04), the improvements ranged from (2.75%) to (4.96%) using *zhong* and *lch* respectively resulting in F1-Measure in [74.65%, 76.25%]. On the other hand, improvements treating (C14) ranged from (0.18%) to (3.67%) using *wup* and *cdist* respectively resulting in F1-Measure in [73.01%, 75.55%]. Only three similarity measures *cdist*, *lch* and *zhong* increased Rocchio's Macro F1-measure.

According to previous observations, the maximum increase in F1-Measure occurred when treating the class (C06) and is of a percentage of (18.13%) using *lch* for Semantic Text-To-Test Similarity measure. In fact, this class is the least populated class in the corpus and Rocchio with Cosine obtained on the completely conceptualized corpus a relatively low value of F1-Measure for this class. These improvements at class level influenced the MacroAveraged F1-Measure with a gain ranging from (0.20%) to (2.27%) using semantic similarities *lch* and *cdist* respectively. In fact, the overall performance of Rocchio using Cosine on the conceptualized corpus is significantly different from its performance on the corpus after applying our measure according to McNemar test and using two semantic similarity measures *zhong* and *dist*.

Using *cdist*, *lch* or *zhong*, the increase in F1-Measure at class level increased the MacroAveraged F1-Measure. This approach has no impact on the weighting scheme which makes it less sensitive than others of different ranges of values retuned by these measures. Rocchio with this measure gave best results by using *cdist* as a semantic similarity measure; this resulted in a MacroAveraged F1-Measure of (65.32%) (see fig. 2.b). Note that *cdist* returns low values in the range [0, 1].

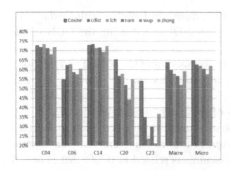 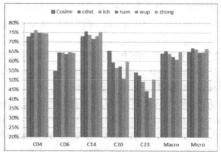

Fig. 2. a.(left)Results of applying Rocchio (Cosine) and SemIDF measure- **Fig. 2b.**(right) Results of applying Rocchio (Cosine) and SemTFIDF measure on Ohsumed using F1-measure

Discussion. Table 2 illustrates the difference between both text-to-text semantic similarity measures and Cosine. Both measures use cdist with which we obtained best classification results at macro level. SemTFIDF measure outperforms SemIDF measure and Cosine at the macro level. Moreover, it outperforms SemIDF measure for all the class. In fact, the SemTFIDF measure takes into account the TF/IDF weighting model in assessing similarities between a document and a centroid. Thus, it is essential to an aggregation function to take into account language and text statistics in assessing similarities. More precisely, first all semantic similarity measures improved Rocchio's performance for the class C06. Nevertheless, only three cases using our SemTFIDF measure improved results at MacroAveraged level. Best overall performance occurred with Rocchio and cdist similarity measure with a MacroAveraged F1-Measure of (65.32%). Both similarity measures: wup and lch, improved the performance of Rocchio at class level.

Second, we distinguish two important points for developing Semantic Text-To-Text Similarity Measures. The first point is that these measures worked with the five similarity measures and especially with *cdist*, *lch* and *zhong*. This means that they are less sensitive to differences between the ranges of the values returned by these measures.

Table 2. Comparison, using F1-measure, betweenCosineusing TFIDF, SemIDFandSemTFIDFmeasures (* forsignificantincreasesaccordingtoMcNemartest)

Category	C04	C06	C14	C20	C23	Macro	Micro
Cosine	72,65	54,68	72,88	65,20	53,96	63,87	64,81
SemTFIDF	**74,75 ***	**64,56 ***	**75,55 ***	59,31	52,45	**65,32 ***	**66,91 ***
SemIDF	71,90	62,56 *	73,46	56,74	35,07	59,95	62,74

Third, least populated classes like (C06) are challenging for classification technique as compared to other classes for which the classification model is much easier to learn. However, Text-To-Text Semantic Similarity Measures helped the classifier distinguish this class with a maximum gain reaching (18.13%) in the case of our measure using lch. Similar to our observations after applying conceptualization, the class "C06" is among the least populated classes as compared to others and so using Text-To-Text Semantic Similarity Measures might result in a better identification of this class which led to better results.

6   Conclusion

In this work, we proposed a new text-to-text semantic similarity measure based on TF/IDF and we evaluated it as a prediction criterion for supervised text classification using Rocchio. We tested this new measure and compared it with another text semantic similarity measure based on IDF proposed in the literature, along the Cosine classical similarity measure that are usually used with BOW representation model. We tested these measures in the biomedical domain on the Ohsumed corpus, using domain specific knowledge base UMLS®.

According to our experimental results, it appears relevant to use text-to-text semantic similarity measures for prediction in centroïd-based classification as it modifies the behavior of the classifier and can improve its effectiveness, particularly with the new TF/IDF based text-to-text similarity measure that we propose. However, resulting performance is dependent on the semantic similarity measure used in assessing similarities between concepts and the aggregation function used in prediction. Consequently, it necessary to develop text-to-text semantic similarity measures, those are adapted to the application context.

Finally, we assume that semantic similarities are more adequate than classical similarities like Cosine in comparing texts represented as BOCs. In other words, we recommend using semantic similarities when concepts are used as features in the vector space model.As for future work, we intend to evaluate other factors that may influence the performance of our measure. In addition, we intend to evaluate its influence on other tasks related to information retrieval such as question answering and centroïd-based clustering.

References

1. Bloehdorn, S., Hotho, A.: Boosting for text classification with semantic features. In: Mobasher, B., Nasraoui, O., Liu, B., Masand, B. (eds.) WebKDD 2004. LNCS (LNAI), vol. 3932, pp. 149–166. Springer, Heidelberg (2006)
2. Salton, G.: The SMART Retrieval System-Experiments in Automatic Document Processing 1971. Prentice-Hall, Inc. (1971)
3. Albitar, S., Fournier, S., Espinasse, B.: The Impact of Conceptualization on Text Classification. In: Wang, X.S., Cruz, I., Delis, A., Huang, G. (eds.) WISE 2012. LNCS, vol. 7651, pp. 326–339. Springer, Heidelberg (2012)
4. Blei, D.M., Ng, A.Y., Jordan, M.I.: Latent dirichlet allocation. Journal of Machine Learning Research 3, 993–1022 (2003)
5. Bloehdorn, S., Moschitti, A.: Combined syntactic and semantic Kernels for text classification. In: Amati, G., Carpineto, C., Romano, G. (eds.) ECIR 2007. LNCS, vol. 4425, pp. 307–318. Springer, Heidelberg (2007)
6. Albitar, S., Fournier, S., Espinasse, B.: Conceptualization Effects on MEDLINE Documents Classification Using Rocchio Method. In: Web Intelligence 2012, pp. 462–466 (2012)
7. Hotho, A., Staab, S., Stumme, G.: Text clustering based on background knowledge (2003)

8. Guisse, A., Khelif, K., Collard, M.: PatClust: une plateforme pour la classification séman-tique des brevets. In: Conférence d'Ingénierie des Connaissances, Hammamet, Tunisie (2009)

9. Huang, L., et al.: Learning a concept-based document similarity measure. J. Am. Soc. Inf. Sci. Technol. 63(8), 1593–1608 (2012)

10. Peng, X., Choi, B.: Document classifications based on word semantic hierarchies. In: International Conference on Artificial Intelligence and Applications (AIA 2005), pp. 362–367 (2005)

11. Wang, P., et al.: Improving Text Classification by Using Encyclopedia Knowledge. In: Proceedings of the 2007 Seventh IEEE International Conference on Data Mining 2007, pp. 332–341. IEEE Computer Society (2007)

12. Al-Mubaid, H., Nguyen, H.A.: A Cluster-Based Approach for Semantic Similarity in the Biomedical Domain. In: 28th Annual International Conference of the IEEE Engineering in Medicine and Biology Society, EMBS 2006 (2006)

13. Rada, R., et al.: Development and application of a metric on semantic nets. IEEE Transactions on Systems, Man and Cybernetics 19(1), 17–30 (1989)

14. Azuaje, F., Wang, H., Bodenreider, O.: Ontology-driven similarity approaches to supporting gene functional assessment. In: Proceedings of the ISMB 2005 SIG Meeting on Bio-Ontologies (2005)

15. Mihalcea, R., Corley, C., Strapparava, C.: Corpus-based and knowledge-based measures of text semantic similarity. In: Proceedings of the 21st National Conference on Artificial Intelligence, vol. 12006, pp. 775–780. AAAI Press, Boston

16. Mohler, M., Mihalcea, R.: Text-to-text semantic similarity for automatic short answer grading. In: Proceedings of the 12th Conference of the European Chapter of the Association for Computational Linguistics. Association for Computational Linguistics, Athens (2009)

17. Dolan, B., Quirk, C., Brockett, C.: Unsupervised construction of large paraphrase corpora: exploiting massively parallel news sources. In: Proceedings of the 20th International Conference on Computational Linguistics 2004, p. 350. Association for Computational Linguistics, Geneva (2004)

18. Hersh, W., et al.: OHSUMED: an interactive retrieval evaluation and new large test collection for research. In: 17th Annual International ACM SIGIR Conference on Research and Development in Information Retrieval. Springer-Verlag New York, Inc., Dublin (1994)

19. Aronson, A.R., Lang, F.M.: An overview of MetaMap: historical perspective and recent advances. J. Am. Med. Inform. Assoc. 17(3), 229–236 (2010)

20. Caviedes, J.E., Cimino, J.J.: Towards the development of a conceptual distance metric for the UMLS. J. of Biomedical Informatics 37(2), 77–85 (2004)

21. Wu, Z., Palmer, M.: Verbs semantics and lexical selection. In: Proceedings of the 32nd Annual Meeting on Association for Computational Linguistics 1994, pp. 133–138. Association for Computational Linguistics, Las Cruces (1994)

22. Leacock, C., Chodorow, M.: Combining Local Context and WordNet Similarity for Word Sense Identification. In: Fellbaum, C. (ed.) WordNet: An Electronic Lexical Database (Language, Speech, and Communication), pp. 265–283. The MIT Press (1998)

23. Zhong, J., Zhu, H., Li, J., Yu, Y.: Conceptual Graph Matching for Semantic Search. In: Priss, U., Corbett, D.R., Angelova, G. (eds.) ICCS 2002. LNCS (LNAI), vol. 2393, pp. 92–106. Springer, Heidelberg (2002)

24. Sebastiani, F.: Text Categorization. In: Encyclopedia of Database Technologies and Applications 2005, pp. 683–687. Idea Group (2005)

Automatically Annotating Structured Web Data Using a SVM-Based Multiclass Classifier

Daiyue Weng, Jun Hong, and David A. Bell

School of Electronics, Electrical Engineering and Computer Science,
Queen's University Belfast, Belfast BT7 1NN, UK

Abstract. In this paper, we propose a new learning approach to Web data annotation, where a support vector machine-based multiclass classifier is trained to assign labels to data items. For data record extraction, a data section re-segmentation algorithm based on visual and content features is introduced to improve the performance of Web data record extraction. We have implemented the proposed approach and tested it with a large set of Web query result pages in different domains. Our experimental results show that our proposed approach is highly effective and efficient.

Keywords: Web data annotation, Web data extraction, Web database.

1 Introduction

The volume of structured data on the Web has been increasing enormously. Such data is usually returned from the back-end databases in response to user-submitted queries, and presented in the form of data records, which are enwrapped in query result pages. Automatically extracting data from query result pages enables Web applications that integrate data from multiple sources, such as price comparison sites, and vertical search engines. Figure 1 shows a snapshot of a query result page from *abebooks.co.uk*, which has a single column containing two data records, where each record represents a book with several data items, e.g., title, price, and author etc.

The semantics of the structured data in a query result page can be easily understood by humans when the page is displayed. However, it is difficult for computers to understand such semantics as no explicit associations between data items and their labels are encoded. Some data items may even not have associated labels as their semantics is clear to humans. The problem of *Web data annotation* is to assign labels to data items, which is crucial to Web data integration systems. The existing approaches [8,13] require the extracted data items to be aligned prior to being annotated. In this case, it is the columns of data items that need to be annotated. Since data alignment and data annotation are done in two separate phases, the errors occurred in data alignment will be carried into data annotation. This further affects annotation accuracy. In this paper, we propose an approach that automatically annotates data items in data records without them being aligned first. In this case, the data items are individually

B. Benatallah et al. (Eds.): WISE 2014, Part I, LNCS 8786, pp. 115–124, 2014.
© Springer International Publishing Switzerland 2014

Fig. 1. An example query result page from the books domain

annotated. The advantages of our approach is that the annotated data items can then be easily aligned by their labels.

Our proposed approach consists of three steps. First, we use a vision-based page segmentation (VIPS) algorithm [3] to segment a query result page into a Visual Block tree, which will be introduced later. Second, we identify data records in the page based on its Visual Block tree by using the techniques proposed in [14]. A data section re-segmentation technique is applied to refine the process of data record extraction. Third, if the data items in a page are presented in tables, we use a table annotator to annotate them. Otherwise each data item is assigned a predefined label using a trained SVM-based [12] multi-class classifier.

In summary, the paper makes the following contributions: *1)* We use a SVM classifier to assign labels from a domain schema to data items, without having to rely on labels from the query result pages or other third-party sources. To the best of our knowledge, we are the first to use a SVM classifier for annotating the data items and then aligning them by their labels. *2)* Using a domain schema ensures the consistency of Web data annotation, which resolves the semantic heterogeneity across different Web databases. *3)* Our table annotator can identify tables containing data records, and then locate their corresponding headers as labels. *4)* We invent a data section re-segmentation algorithm, which improves the precision and recall of our data record extraction and works efficiently.

2   Fundamentals

2.1   Query Result Page Representation

The content of a query result page is typically organized into different regions, which contain semantically related contents, e.g., advertisements, sponsor links, query results, etc. Visual cues (e.g., lines, spaces, font sizes, background colours, etc.) can be used to distinguish regions from each other. We employ the VIPS [3] algorithm to represent a query result page as a Visual Block tree. For example, in Figure 1 the basic visual blocks for the top data record are highlighted.

2.2   Visual Blocks and Data Items

Based on the content that a leaf node contains, we categorize them as follows: *1)* those leaf nodes that contain actual data items, i.e., those nodes that contain texts representing the values of some attributes, e.g., in Figure 1, "Cracking Da Vinci's Code:...." represents the book title. *2)* those leaf nodes that contain texts representing the semantic labels of the data items that follow them, e.g., in Figure 1, "Bookseller Rating" indicates the ratings of the books. *3)* the other leaf nodes that are neither data items nor semantic labels, but that direct users to relevant information about the data record, e.g., in Figure 1, "Destination, Rates&Speeds" contains links that lead to more information for the users. *4)* those leaf nodes that contain images, e.g., the book images in Figure 1. We assign leaf nodes of types 1 and 4 with their corresponding labels, and label the others as "noise". Since the contents those leaf nodes of types 2 and 3 contain are not of interest.

3   Data Record Extraction Based on Visual Cues

We briefly review our data record extraction technique and introduce the improvements we have in this papaer. Given a query result page, our algorithm first identifies the data section from the page using two observations: *a)* the area size of a data section is usually large relative to the size of the whole page; *b)* the query terms often re-appear in the data records. We then remove noisy blocks from the data section based on statistics that characterize the importance of each block within the data section. The less important blocks are identified as noisy blocks and removed. Third, data records are extracted from the data section by continuously clustering basic visual blocks that are vertically proximate and horizontally aligned in each data record. Each cluster is then a data record. For example, in Figure 1, the visual blocks are aligned and proximate with each other, and form a data record.

Extracting data records relies on alignment and proximity has two potential problems: *1)* the false record problem, i.e., some extracted records are formed up with noisy blocks only, which remained on the top of the data section; *2)* the data record granularity problem, i.e., multiple data records form up a large data record, since they are vertically adjacent to each other. To solve these problems, we designed a new algorithm - *data section re-segmentation*, as an extension to our previous technique for data record extraction.

After the last step, we now have a set of extracted data records and treat them as *candidate data records* (CDRs). The leaf nodes horizontally form up a number of *node groups* (NGs) in the data section. We observe that: *1)* every data record is composed of several NGs; *2)* the top node group of each data record can be used as boundaries to re-segment the data section, we call these groups *top record node groups* (TRNGs). We also observe that the TRNGs of the same page are usually similar in terms of:

1. *Group shape* - a sequence of left positions of each blocks in the group;
2. *Group position* - the left position of the leftmost block in the group;
3. *Presentation style* - all the presentation styles found in the group;
4. *Content type* - the content of each block is likely to be "text", "link", "link&text", "image", "image&text", or "blank (default type)".

The similarity between two NGs g_1 and g_2 is defined as a weighted sum of the similarities on the four features between them:

$$Sim(g_1, g_2) = \quad w_1 * simGS(g_1, g_2) + w_2 * simPS(g_1, g_2)$$
$$+ w_3 * simGP(g_1, g_2) + w_4 * simCT(g_1, g_2) \qquad (1)$$

The feature weights and the similarity threshold are trained using a genetic algorithm based method [7] and the details are given in Section 5.1. Each similarity measure is defined as follows:

1. **Group Shape (simGS)**. It measures the maximum difference between the sequences of left positions of each blocks in the two groups.
2. **Group Position (simGP)**. It measures the difference between the left positions of the leftmost blocks in the two groups.
3. **Presentation Style (simPS)**. It calculates the percentage of the number of common presentation styles over the total number of presentation styles in the two groups.
4. **Content Type (simCT)**. It computes the edit distance between the content types of the two groups. We define the content type of a NG as a sequence of content types of blocks in that group.

It is obviously that if all the TRNGs are correctly identified, then all the records will be identified correctly. For example, in Figure 1, the TRNGs of the two data records are all composed of a record number (text), a book image, a book title (link text), and an icon ("Add to Basket"). They have the same group position of 10 and group shape of $(10, 28, 176, 786)$. They use the same font size (12), font weight (700), background color (transparent), font style (normal), etc. They clearly set boundaries between the two records. We use voting to identify all the TRNGs. First, the topmost NG of every candidate record is extracted, we then have a set of *candidate top record node groups* (CTRNGs). Second, we observe that the majority of the candidate records are real records. We then do similarity comparisons among the CTRNGs. If any CTRNG is similar to over 80% (*similarity ratio*) of all the CTRNGs, we will select it as a *seed* and used to identify all the other TRNGs in the data section.

The data section re-segmentation algorithm works as follows. First, the CDRs and their NGs are sorted in ascending order based on their top positions. The top NG and the rest NGs of every CDR are then put into two groups, called "Cand-TopGrs" and "TempNodeGrs" respectively. We then do similarity comparisons on each NG against the others in "CandTopGrs" to find the seed, which is then used to identify all the TRNGs in "CandTopGrs" and "TempNodeGrs". Finally, we re-segment the CDRs that the NGs in between consecutive $TRNGs[i]$ and $TRNGs[i+1]$ (including $TRNGs[i]$) are put into a group as a data record. The last boundary along with the rest of NGs in "TempNodeGrs" will form the last record of the data section.

Data record extraction benefits data annotation in two aspects: *1)* it can effectively filter out the noisy information that remain in the data section, which makes our annotation system focus on the data of interest; *2)* it significantly cuts down the amount of data that need to be labeled, which reduces the time for annotating every query result page.

4   Our Approach

4.1   Web Data Annotation Using SVMs

Web data annotation can be viewed as a multiclass classification problem, where each data item is assigned one label only from a set of labels. As shown in [11], assigning a label to a data item can be determined by a number of contextual and structural features in a query result page. We have observed that data items with different semantics can be linearly separated by these features. This motivates the use of a SVM-based approach to the annotation problem.

We observe that every Web database has a query interface which is associated with a *local query interface schema* (LIS). Every query result page in response to a specific query also has a *local result schema* (LRS). There are some attributes in both LIS and LRS. In this case, we can use the attribute names in LIS to label the corresponding attributes in LRS. However, some attributes in LRS are not in LIS since they are neither needed nor suitable for query conditions. For example, a bookseller inventory number is not needed in a user query in the book domain. We also observe that Web databases in the same domain usually share a number of attributes but they often use different labels for the attributes with the same semantics. For example, labels "save" and "discount" both refer to either a amount or percentage of reduction from the original price of a product. The first phenomenon is called *local query interface schema inadequacy* problem, and the second is referred to as the *inconsistent label* problem [8]. To solve these problems, we propose to use a domain schema from schema.org, which contains a number of attributes to represent the structured data in the domain so that data items with the same semantics will have the same label. This domain-dependent global schema is denoted by Y. We represent each data item by a binary feature vector $x = (x_1, x_2, x_3, ..., x_n)$, where x_i is the value of the corresponding feature of the data item. We use binary feature vectors because they lead to efficient algorithms. We also transform any non-binary feature into a corresponding set of binary features. We formally define Web data annotation as a multiclass classification problem as follows:

Given a training set of data items in a particular domain, $D = \{(x_1^1, y_1^1), ..., (x_{l_1}^1, y_{l_1}^1), ..., (x_1^m, y_1^m), ..., (x_{l_m}^m, y_{l_m}^m)\}$, where x_i^k is a l-dimensional binary vector

$x_i^k \in X \subseteq R^l$ and the superscript k denotes that $\boldsymbol{x}_i$ belongs to class k, $y_i^k \in Y = \{y^1, y^2, ..., y^m\}$ is a set of labels. $m(m-1)/2$ classifiers are constructed, where each is trained by learning a function $f_{ij}(\boldsymbol{x}_i) = sign(\boldsymbol{w}^{ij} \cdot \boldsymbol{x}_i + b^{ij})$ using training examples $(\boldsymbol{x}_a^i, y_a^i)$ and $(\boldsymbol{x}_b^j, y_b^j)$ from the ith and the jth classes, where $\boldsymbol{w}^{ij}$ is the weight vector, and b^{ij} is the threshold.

Given a set of unlabeled data items $D' = \{\boldsymbol{x}_1, \boldsymbol{x}_2, \boldsymbol{x}_3, ..., \boldsymbol{x}_n\}$ in a query result page of a particular domain and a set of labels Y' for the domain. We want to assign label y_j' to $\boldsymbol{x}_i$, if $f_{ij}(\boldsymbol{x}_i)$ predicts $\boldsymbol{x}_i$ belongs to class j.

If $\boldsymbol{x}_i$ is assigned multiple labels, after being classified through all the SVMs. We will find the most suitable label for $\boldsymbol{x}_i$. The standard SVM does not produce a calibrated posterior probability. We instead fit a sigmoid function to the output of each SVM to produce a posterior probability $P(y_i = 1|\boldsymbol{w} \cdot \boldsymbol{x}_i + b)$ [10]. We can then choose the label with the maximal $P(y_i = 1|\boldsymbol{w} \cdot \boldsymbol{x}_i + b)$.

4.2   Creating a Set of Labels

For each domain, we create a set of labels from a domain schema on schema.org [1]. Schema.org provides a collection of shared vocabularies that can be used to markup structured data on the Web and understood by the major search engines: Google, Microsoft, Yandex and Yahoo. For example, "Title", "Author", "Bookseller", etc. are used in book domain.

4.3   Selecting Features for SVMs

We use a set of visual, content and contextual features in the feature vector of each leaf node:

- **Visual Features.** We consider different visual features, including font size, font weight, background colour and relative position of a leaf node in its data record (e.g., top, bottom, etc.). These features are extracted from each leaf node.
- **Content and Contextual Features.** We also consider a variety of content and contextual features. These features are extracted from the content that a leaf node contains and its context. They are described below:
  - Data Type: the data type of a block is defined as a sequence ($Date, Time, Currency, Code, Number, String$). The text string of each data type has a particular pattern which is recognized by a regular expression.
  - Text: whether a block contains text content.
  - Link Text: whether a block contains a link text.
  - Image: whether a block contains an image.
  - Order: the ordering of a block in a data record when it was extracted from the record, i.e., begin or end.
  - Occurrence: whether the number of occurrences of the same text across all data records is no more than 1.
  - Attribute Name: whether a block contains one of the labels in a given label set for a domain.

Each feature can be represented as either a binary state or a set of pre-defined values $v_k, k = 1, ..., n$. In order to generate a feature vector, a feature $f_i, i = 1, ..., n$ for a block b_j is defined as a sequence of binaries $a_i = \{a_1, a_2, a_3, ..., a_k\}$:

$$a_i = \begin{cases} 1 & \text{if } f_i \text{ is true,} \\ 0 & \text{otherwise} \end{cases} \quad f_i \text{ is a binary feature} \tag{2}$$

$$a_k = \begin{cases} 1 & \text{if } f_i \text{ has a value } v_k, \\ 0 & \text{otherwise} \end{cases} \quad \begin{array}{l} f_i \text{ is a non-binary fea-} \\ \text{ture,} \\ k = 1, \ldots, n \end{array} \tag{3}$$

For example, if the data type of a block b_j is "Date", then the data type feature of b_j is encoded as $[100000]$, which corresponds to ($Date, Time, Currency, Code, Number, String$).

4.4 Learning Weight and Parameters of the SVMs

We employ the *sequential minimal optimization* (SMO)[9] algorithm to optimize *Lagrange multipliers* α_i of SVM. It also handles input vectors for linear SVM that are sparse and binary very efficiently.

The SVM performance is also dominated by C, which penalizes training examples that fall on the wrong side of the decision boundary. The training of C will be discussed in Section 5.1.

4.5 Table Annotator

Many Web databases use tables to present query results. A Web table is a two-dimensional presentation of logically related groups of data items [6]. Every table has headers, which explicitly explain the meanings of different columns, and are located at the top of the table. We use these headers as labels to annotate data items in a table. Our table annotator works as follows:

- First, we decide if the data items of a page are organized in a table by checking: *1)* if the data items of each data record are laid on a row. If so, we first find the top row, and then we cluster data items from other rows that have the same left positions with the data items in the top row. For each cluster, we want to see: *2)* whether it is actually a table column by checking if the data items of the cluster are vertically adjacent to each other.
- Second, if we do have a table, the previously identified clusters become *table data item clusters*. Since the headers are usually vertically close and left aligned to data items of the same columns. We then locate leaf nodes that are not in the data records, but are left aligned with and have the least vertical spacing from the first data items of the table data item clusters, and make their texts as the labels to annotate corresponding data item clusters.

5 Experiments

5.1 Methods and Data Sets

We prototyped our proposed approach using Visual C++ and LIBSVM [4]. To obtain the SVM models, for each domain, one training file was built that contains training examples retrieved from the extracted data records in training pages, which are randomly selected from each website in that domain. In order to get good generalization, we conduct the four-step training procedure introduced in [5] to determine parameter C. The size of the training set is gradually increased by increasing the number of training examples of each class/label until the cross validation is stabilized.

To optimise the feature weights and the similarity threshold, we randomly generate an initial population of 50 vectors, each of which contains the weights and the threshold in the interval $(0, 1)$. The population evolves from generation to generation through crossover, mutation and reproduction. Crossover produces two offspring from two parents with probability 0.9. Mutation is performed on the elements of each child vector with probability 0.01. Each element is either increased or decreased by 0.01 with equal chance. Reproduction uses *roulette wheel selection* to choose vectors proportional to their fitness values. The genetic algorithm terminates when the precision and recall converge at 90%. The fittest vector of all generations is chosen as the optimum.

We collected 200 query result pages from 20 Web databases [2] belonging to 5 domains - Books, Jobs, Movies, Music and Hotels. For each web database, 10 result pages were collected after manually submitting 10 different queries via its query interface. For data record extraction, we compare our algorithm with DEPTA [15], which is a state of the art Web data extraction system based on DOM tree analysis. For Web data annotation, we do not compare our approach with other systems for the following reasons: *a*) some Web data extraction systems do not perform Web data annotation; *b*) there is no standard testbed for Web data extraction; *c*) most Web data extraction systems and their data sets are inaccessible.

5.2 Performance Evaluations

For data record extraction, we use two common measures, i.e., *recall* and *precision* to evaluate the performance of our approach. The recall is the percentage of the number of data records that have been correctly extracted over the total number of data records on a result page. The precision is the percentage of the number of data records that have been correctly extracted over the total number of data records that have been extracted.

For Web data annotation, we use *accuracy* to evaluate our proposed method. It is defined as the percentage of the number of correctly annotated data items over the total number of data items.

5.3 Experimental Results on Data Record Extraction

The experimental results of our data record extraction technique compared with DEPTA is shown in Table 1. It depicts among the number of extracted data records (EXT), how many of them have been correctly extracted (COR) and the actual number of data records (ACT) in the data sets, whereon the precision and recall are also calculated. The performance of our approach is significantly better than DEPTA.

Table 1. Comparison results for Web data record extraction

Domain	Our Approach			DEPTA		
	COR	ACT	EXT	COR	ACT	EXT
Books	692	719	701	536	719	570
Hotel	844	881	863	802	881	824
Jobs	1331	1353	1335	250	1353	255
Movie&Music	1007	1024	1011	980	1024	1018
Total	3874	3977	3910	2568	3977	2667
Recall	97.41%			64.57%		
Precision	99.08%			96.29%		

5.4 Experimental Results on Web Data Annotation

The data items in Jobs are represented in tables, and are annotated using the table annotator. The labels assigned for the other domains are predicted by the SVMs. Table 2 shows the total (ACT) number of data items, and the label assignment accuracy (ACCY) given by the number of correct annotation (COR) and incorrect annotation (WRG) for all the data items in the 5 domains. It can be seen that our approach achieved high accuracy of around 95% in total. The accuracy is above 93% across all domains. Very good label assignment accuracy was delivered of above 98% for Jobs.

Table 2. Web data annotation accuracy of our approach

Domain	ACT	COR	WRG	ACCY
Books	7125	6761	364	94.89%
Hotel	8077	7577	500	93.81%
Jobs	6926	6838	88	98.73%
Movie&Music	7708	7227	481	93.76%
Total	29836	28403	1433	95.20%

In terms of annotation speed, the elapsed time for annotating the data items extracted from each page is very short. The average elapsed time for classifying different sizes of data sets is calculated by repeating the same experiment 1000 times and computing the mean. It is between 5 to 50 milliseconds when using SVM classification for data annotation, and 15 to 120 milliseconds for using table annotator, depending on the complexity of the page.

6   Conclusions

In this paper, we presented an automatic approach for Web data annotation. This work extends our previous work on Web data extraction, and improves the performance of the technique on both precision and recall. Our approach employs a table annotator to annotate data items that are presented in tables, and uses the SVM to annotate other data items in multi-class classification. We utilize a set of visual and content features that reside in the data items. In particular, our approach covers the situation when the query interface schema is absent. The experimental results show that our approach is effective and accurate.

References

1. http://schema.org
2. The UIUC web integration repository. Computer Science Department, University of Illinois at Urbana-Champaign (2003), http://metaquerier.cs.uiuc.edu/repository
3. Cai, D., Yu, S., Wen, J.-R., Ma, W.-Y.: Extracting content structure for web pages based on visual representation. In: Zhou, X., Zhang, Y., Orlowska, M.E. (eds.) APWeb 2003. LNCS, vol. 2642, pp. 406–417. Springer, Heidelberg (2003)
4. Chang, C.-C., Lin, C.-J.: Libsvm: A library for support vector machines. ACM Trans. Intell. Syst. Technol. 2(3), 27:1–27:27 (2011)
5. Chen, Y.-W., Lin, C.-J.: Combining svms with various feature selection strategies. In: Guyon, I., Nikravesh, M., Gunn, S., Zadeh, L.A. (eds.) Feature Extraction. STUDFUZZ, vol. 207, pp. 315–324. Springer, Heidelberg (2006)
6. Gatterbauer, W., Bohunsky, P., Herzog, M., Krüpl, B., Pollak, B.: Towards domain-independent information extraction from web tables. In: WWW 2007, pp. 71–80 (2007)
7. Goldberg, D.E.: Genetic Algorithms in Search, Optimization and Machine Learning. Addison-Wesley Longman Publishing Co., Inc. (1989)
8. Lu, C.Y., He, H., Zhao, H., Meng, W., Yu: Annotating search results from web databases. IEEE Trans. on Knowl. and Data Eng. 25(3), 514–527 (2013)
9. Platt, J.C.: Fast Training of Support Vector Machines Using Sequential Minimal Optimization. In: Advances in Kernel Methods, pp. 185–208 (1999)
10. Platt, J.C.: Probabilistic outputs for support vector machines and comparisons to regularized likelihood methods. In: Advances in Large Margin Classifiers, pp. 61–74 (1999)
11. Su, W., Wang, J., Lochovsky, F.H.: Ode: Ontology-assisted data extraction. ACM Trans. Database Syst. 34(2), 1–12 (2009)
12. Vapnik, V.N.: The nature of statistical learning theory. Springer-Verlag New York, Inc., New York (1995)
13. Wang, J., Lochovsky, F.H.: Data extraction and label assignment for web databases. WWW 2003, 187–196 (2003)
14. Weng, D., Hong, J., Bell, D.A.: Extracting data records from query result pages based on visual features. In: Fernandes, A.A.A., Gray, A.J.G., Belhajjame, K. (eds.) BNCOD 2011. LNCS, vol. 7051, pp. 140–153. Springer, Heidelberg (2011)
15. Zhai, Y., Liu, B.: Structured data extraction from the web based on partial tree alignment, 1614–1628 (2006)

Mining Discriminative Itemsets in Data Streams

Majid Seyfi, Shlomo Geva, and Richi Nayak

Data Science Discipline, Science and Engineering Faculty,
Queensland University of Technology, Brisbane, QLD, Australia
{m.seyfi,s.geva,r.nayak}@qut.edu.au

Abstract. This paper presents a single pass algorithm for mining discriminative Itemsets in data streams using a novel data structure and the tilted-time window model. Discriminative Itemsets are defined as Itemsets that are frequent in one data stream and their frequency in that stream is much higher than the rest of the streams in the dataset. In order to deal with the data structure size, we propose a pruning process that results in the compact tree structure containing discriminative Itemsets. Empirical analysis shows the sound time and space complexity of the proposed method.

Keywords: Data stream mining, discriminative Itemsets, tilted-time window model, prefix tree.

1 Introduction

A data stream is defined as a continual sequence of transactions that are transmitted with fast speed over a period of time [1]. Frequent pattern mining in single data stream has attracted high attention in the last decade [2]. These studies can be categorized according to update interval, different window models and the type of approximation [2]. In spite of the abundance of works in frequent pattern mining in single data stream, there is not much research done on multiple data streams to capture interesting trends, patterns and exceptions. Challenges raised by the combinatorial explosion of Itemsets, demanding high time and space consumption, are faced when frequent pattern mining is applied to large and   fast growing data streams [2].

An emerging research area in stream mining is discriminative pattern mining. Discriminative Itemsets are frequent in the target data stream and their frequency in that stream is much higher than the rest of the streams in the dataset. Compared to frequent Itemsets and sequential patterns, discriminative Itemsets are more directed for the purpose of comparison between data streams and carry more valuable information. They can be used for highlighting the differences between data streams and for building a classifier for behavior prediction mining.

There are several examples where the significance of mining discriminative patterns in data streams can be demonstrated. In network traffic measurements, looking for the concurrent activities of one user, that are more frequent in comparison to the rest of the group activities in the whole network, can lead to meaningful information for anomaly detection. Mining discriminative patterns can be effectively used in web

B. Benatallah et al. (Eds.): WISE 2014, Part I, LNCS 8786, pp. 125–134, 2014.
© Springer International Publishing Switzerland 2014

page document clustering as well as in personalization of search engines and news delivery services. The frequent patterns are generally frequent in all data streams and may not be useful in these applications due to not being distinctive [3]. Analyzing the clickstream data for identifying the web pages that are visited by a specific user (or a group of users) more frequently than other users (or groups), can lead to improved personalized services. In dynamic tracing of stock market fluctuation, Itemsets that occur more frequently in one stock as compared to other ones are of interest. The discriminative patterns are useful for fraud detection if a group of customers are buying certain items more frequently than the rest of the population to have monopoly in the market.

An essential issue inherent in all these applications is to find Itemsets that can distinguish the target stream from all other streams. In this paper, we define the problem of mining "discriminative Itemsets" in data streams. Let two data streams be S_1 and S_2 over the time. We find the Itemsets which are frequent in S_1 and their frequencies are higher than the same Itemsets in S_2 by threshold θ.

To the best of our knowledge, ours is the first work on mining discriminative *Itemsets* in data streams. The process of mining discriminative Itemsets has more complexities than frequent Itemsets mining due to the processing with multiple data streams. A discriminative Itemsets mining algorithm has to deal with the combinatorial explosion of Itemsets generated from multiple streams. The Apriori property defined for frequent Itemsets is not applicable here as the subset of discriminative Itemsets can be non-discriminative. Expanding the existing frequent Itemsets mining algorithms for discriminative Itemsets mining in streams will produce inefficient results.

Only a handful of work has been published in the area of discriminative pattern mining in data streams and that too for items mining only [3, 4]. The process of discriminative Itemset mining with these methods would be time and space consuming. The hierarchical counter method [3] is designed for counting frequency of all the items including infrequent ones, and this would not be acceptable in Itemset mining because of the explosion in the number of Itemset combinations. The hybrid method [3] can possibly be expanded based on an improved hash function. However, it will face major challenges such as very large number of Itemsets generation which will require a huge size of hybrid structure and a very complex hashing process. The hybrid method also shows the approximate frequencies of Itemsets and the efficiency of this method highly depends upon the hashing function and the group of Itemsets assigned to the same bucket.

In this paper, we propose the novel in-memory DISTree structure based on FP-Tree [5] and develop a single-pass algorithm for mining discriminative Itemsets in data streams using the tilted-time window model. The discriminative Itemsets are identified based on their discriminative level. The proposed method is tested on various datasets showing different characteristics such as transaction size, frequent-pattern size, length and various thresholds. Empirical analysis shows the time and space complexity with high precision. The proposed method is able to produce discriminative Itemsets with 100% recall and accuracy.

More specifically, the following contributions are made in this paper: (1) defining the problem of mining "discriminative Itemsets" in data streams; (2) introducing the novel in-memory DISTree structure holding discriminative Itemsets; and (3) developing the single pass algorithm for mining the discriminative Itemsets in data streams based on the tilted-time window model.

2   Problem Statement

Let $T = \{e_1 \dots e_i, e_{i+1} \dots e_n\}$ be a transaction holding the set of lexicographically ordered items in a given alphabet of items Σ. Let S_i and S_j be two data streams of n_i and n_j lengths respectively that consist of different length transactions. Let an Itemset I be a subset of transactions in data streams. The frequency of Itemset I in data stream S_i is denoted as $f_i(I)$. The frequency ratio of I in S_i is defined as $r_i(I) = f_i(I)/n_i$. Discriminative Itemsets can be defined as Itemsets which are relatively frequent in S_i but relatively infrequent in S_j. In other words, discriminative Itemsets have frequency ratio in the target stream higher than the discriminative level threshold as compared to the other streams. This condition can be defined as:

$$R\ (I) = r_i(I)/r_j\ (I) = (f_i\ (I)\ n_j)\ /\ (f_j\ (I)\ n_i) \geq \theta \tag{1}$$

Where $\theta > 1$ is a user-defined parameter with no upper bound and it is called as the discriminative level. The larger its value, the more discriminative the Itemset is. To deal with the situation when $f_j(I)=0$ a user specified minimum support threshold $0 < \varphi < 1/\theta$ is used. In these cases, Itemset I becomes discriminative if its frequency is greater than $\varphi\theta n_i$ i.e., $(f_i(I) \geq \varphi\theta n_i)$.

Definition1. Discriminative Itemsets: Let S_i and S_j be two data streams of current size of n_i and n_j respectively that contain varied size transactions of items in Σ, a user defined discriminative level threshold $\theta > 1$ and a minimum support threshold $\varphi \in (0, 1/\theta)$. The problem of mining discriminative Itemsets (*DI*) in S_i against S_j is defined as:

$$DI = \{I \subseteq \Sigma\ /\ (f_i(I)\ \geq\ \varphi\theta n_i)\ \&\ (n_j f_i(I)\ /\ n_i f_j(I)\ \geq\ \theta)\} \tag{2}$$

A discriminative Itemsets mining algorithm has to be memory-efficient in order to deal with the exponential number of Itemsets. The Apriori property is not valid in discriminative Itemsets, consequently not every subset of discriminative Itemset is discriminative. The discriminative Itemsets mining process must support pruning of the non-discriminative Itemsets and keeping the discriminative ones. Even in non-streaming environment, this process is time consuming. The process becomes more cumbersome for mining discriminative Itemsets from data streams.

3   The Proposed Method

The DISTree structure is proposed to process and store discriminative Itemsets existing in the current batch of transactions in data streams. Using the DISTree structure, we present a method for mining discriminative Itemsets incorporating the historical tilted-time windows model to combine the patterns of all batches in the stream data. The following data structures are used in the proposed method:

FP-Tree. We adapt the prefix tree structure as proposed in [6] with the following change. Each node in the tree is changed to contain two counters, a counter for holding frequency of Itemsets in the target data stream and the other counter for holding the frequency of Itemsets in the general data stream. An example FP-Tree structure is shown in Fig 2.

Header-Table. This tabular structure contains all the items in the data stream in an ascending order. Each item has two linked-lists pointing to the Itemsets in FP-Tree and DISTree ending with that item. These linked-lists are used during DISTree construction and pruning. An example Header-Table structure is shown in Fig 2.

DISTree. DISTree is a labeled prefix tree structure in which each node contains two counters to represent frequency of the Itemsets in two data streams. For instance, a node would contain two counters f_1 and f_2 showing the frequency of the Itemset in the sequences ending with that node in the streams S_1 and S_2 respectively. Fig 3 shows an example of DISTree. The FP-Tree structure is built according to the first incoming batch of transactions and then the combinations of each pattern in FP-Tree are generated by traversing through the linked-lists in the Header-Table structure using the FP-growth method [6] and the results are stored in the DISTree sequentially. During this process, the Itemsets that are found non-discriminative will be deleted from the DISTree structure.

Tilted-time windows. The tilted-time window structure [5] shows the historical frequencies of discriminative Itemsets. The new incoming batch of transactions, fitted in the current window frame, is processed and the newly discovered discriminative Itemsets are saved in the new time frame. The older results are merged with older timeframes according to the logarithmic structure as shown in Fig 1 to have them displayed in larger granularities. The number of discriminative and sub-discriminative Itemsets is much smaller than all the Itemsets present in the streams, and the used space for storing them in the titled-time window model is not large.

Fig. 1. Tilted-time window model

We now explains the process of building the efficient DISTree structure holding the discriminative Itemsets that exist in the target stream in comparison to the rest of streams in a stream dataset.

3.1   DISTree Construction

The prefix tree structure (FP-Tree) is created based on the incoming transactions of data stream. For each new transaction in stream S_i, either a new prefix or sub-prefix is added and the frequencies of related nodes are set; or the frequency of the Itemsets is updated for the related stream in the desired prefix if the Itemset already exists in the tree due to being present in the past transactions. Fig 2 shows the Header-Table and FP-Tree for the running example in Table 1. As shown in Fig 2, the frequency of each Itemset is saved separately for each stream in the FP-Tree with using the counters.

We first show the process of constructing DISTree, without applying any pruning that controls the size. Fig 3 shows an equivalent DISTree that will be constructed for the given datasets if a pruning process is not applied to control the size of DISTree. DISTree is constructed by generating all the combinations of Itemsets existing in FP-Tree with bottom-up traversing of Header-Table. In the process of generating all the combinations of Itemsets ending with the traversed item in Header-Table, the generated Itemsets are checked for if they are discriminative or non-discriminative.

Based on definition 1 we are interested in Itemsets which are frequent in S_i and their frequency ratio in S_i is θ times more than their frequency ratio in rest of the streams. However Itemsets that don't meet this condition cannot be pruned as they can be part of another discriminative Itemsets. We define two more categories of Itemsets namely: non-discriminative and sub-discriminative.

Definition 2. Non-Discriminative Itemsets: Let the relaxation ratio ε is defined as the error rate. The discriminative Itemset can be reported by the defined error rate if its frequency ratio in S_i is not θ/ε times less than its frequency ratio in the rest of the streams; otherwise, it is non-discriminative.

$$NDI = \{I \subseteq \textstyle\sum / (f_i(I) \leq \ \varepsilon\varphi\theta n_i) \ // \ (n_j f_i(I) \ / \ n_i f_j(I) \leq \varepsilon\theta)\} \tag{3}$$

Definition 3. Sub-Discriminative Itemsets: Let α is defined as the *sub-discriminative level*. The Itemset is sub-discriminative if its frequency ratio in S_i is not less than θ/α compared to rest of the streams in the whole history of the streams.

$$SDI = \{I \subseteq \textstyle\sum / (f_i(I) \geq \ \alpha\varphi\theta n_i) \ \& \ (n_j f_i(I) \ / \ n_i f_j(I) \geq \ \alpha\theta)\} \tag{4}$$

For the tilted-time window model, a pre-defined window size, made of batches of transactions, is defined based on time. The discriminative Itemsets are of interest; however, the sub-discriminative Itemsets should also be kept during the entire process as in the future they may become discriminative in the historical titled-time window model by merging several window frames. The user defined sub-discriminative level α (in the range of 0 and 1) controls the amount of sub-discriminative Itemsets generation. With smaller α, a large number of sub-discriminative Itemsets is produced. The non-discriminative Itemsets according to definition 2 would not be included in DIS-Tree if they are not subset of any discriminative or sub-discriminative Itemsets.

An Example: Consider the following example with two data streams of same length ($n_1=n_2=15$) as shown in Table 1. With discriminative level $\theta=3$ and minimum support threshold $\varphi=0.05$, DISTree (without pruning) can be formed using FP-Tree and Header-Table (as shown in Fig 3). For simplicity the error rate is considered as $\varepsilon=1$ i.e., no error generated. We also considered $\alpha=1$, i.e. no sub-discriminative Itemsets will be kept during the multiple windows processing.

Table 1. A very small size streams data

S/T	1	2	3	4	5	6	7	8	9	10	11	12	13	14	15
S_1	abc	abc	abc	ab	ace	bc	bc	bc	bc	b	cde	ce	ce	ce	ce
S_2	abcd	ac	ac	ac	ac	a	a	bcd	bcd	cd	cd	cd	cd	c	c

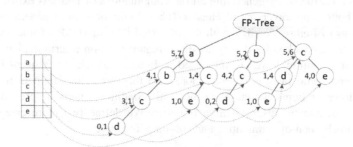

Fig. 2. The Header-Table and FP-Tree

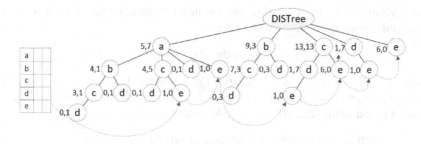

Fig. 3. The Header-Table and DISTree structure without pruning (the full tree size is only for display and is not generated)

3.2   DISTree Pruning

Construction of DISTree for fast growing data streams consumes large memory space. With each unique Itemset, a specific path made of nodes with counters needs to be added. Every possible combination needs to be generated as the discriminative Itemsets do not follow the Apriori property. It may not be feasible for holding the entire DISTree in memory for large datasets and, therefore, it becomes necessary that DISTree should be of compact size in order to allow traversing for making mining discriminative Itemsets feasible. We define a pruning process to make the DISTree in a tolerable size.

Let I_e be the Itemset ending with item e. if $f_i(I_e) \leq \varepsilon\varphi\theta n_i$ or $r_i(I_e)/r_j(I_e) < \theta/\varepsilon$ then I_e is not discriminative. The item e is tagged as non-discriminative and removed from DISTree if it is a leaf node.

In Fig 3, the node e in the Itemset $ace_{1,0}$ will be tagged as non-discriminative. This node will be deleted from the branch as it is a non-discriminative leaf node. These deletion save reasonable amounts of time and space while constructing DISTree by reducing the size of DISTree.

Fig 4 shows the final state of DISTree for the input streams given in Table 1. The significant difference in the size of (non-existent) DISTree generated without deletion process (as shown in Fig 3) and the final DISTree generated with deletion process (as shown in Fig 4) can be clearly seen.

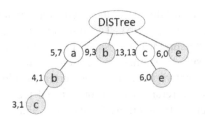

itemset	Frequency
{a b c}	(3,1)
{a b}	(4,1)
{b}	(9,3)
{c e}	(6,0)
{e}	(6,0)

Fig. 4. The final DISTree and Discriminative Itemsets

3.3   Result Phase: Generation of Discriminative Itemsets

The DISTree is generated for the current window size. The Itemsets in DISTree that satisfy definitions 1 and 3 are saved as discriminative and sub-discriminative Itemsets respectively based on the user-defined parameters θ, φ, ε and α. Fig 4 shows all the discriminative Itemsets discovered from these two data streams. This can be observed that Itemset {a b c} is discriminative; however subset {a} is found non-discriminative. This property distinguishes the discriminative mining from frequent mining.

```
1.  While not end of streams do
2.      While W_current is not full do
3.          Read the current batch of transactions B_k;
4.      End while;
5.      For all Itemsets in the current batch do
6.          Make the FP-Tree and Header-Table based on [6];
7.      End for;
8.      For all Items in Header-Table do {bottom to up}
9.          Make DISTree;
10.         Check DISTree (Dis, sub-Dis & non-Dis nodes);
11.         Delete the non-Dis Itemsets;
12.     End for;
13.     For all remained Itemsets in DISTree do
14.         Update window model by dis and sub-dis Itemsets;
15.     End for;
16.     Return the discriminative Itemsets (DI) in the
                W_current and historical tilted-time windows;
17. End while;
```

The discriminative Itemsets mining algorithm starts by initializing the DISTree root node as an empty tree and building the FP-Tree and Header-Table using FP-Growth [6] for each new incoming batch of transactions. DISTree is updated with all combinations of patterns generated using the Itemsets in FP-Tree pointed by Header-Table. In the next step, the newly added Itemsets in DISTree are checked based on Definitions 1, 2 and 3, and considered as discriminative, sub-discriminative or non-discriminative Itemsets. Any non-discriminative Itemsets that were leaf nodes are deleted.

By finishing the process for the current window size, the tilted-time window structure is updated based on logarithmic time span [5]. The discriminative and sub-discriminative Itemsets discovered in the current batch, saved in the current window frame, and the old discriminative and sub-discriminative Itemsets are merged in the tilted-time windows model. The discriminative Itemsets in the current window frame and the historical tilted-time windows are returned as an output and the process is continued for each new incoming batch of transactions for the next window frame. We explain the algorithm below detailing the steps required for mining discriminative Itemsets.

4   Results and Discussion

The time and space complexities have been reported as Execution Time and Memory Usage respectively. The method was implemented in C++. Experiments were conducted on a desktop computer with an Intel Core (TM) Duo E2640 2.8GHz CPU and 8GB main memory running 64bit Microsoft Windows 7 Enterprise. Because of space problem, we evaluate the DISTree algorithm by focusing only on the current batch of transactions and the performance of algorithm in the tilted-time window model for Historical-DISTree will be given in future publications.

Experiments are conducted with datasets generated by the IBM synthetic data generator [7]. Two input datasets, made of two data streams S_1 and S_2 with different size, average length of transactions, average length of frequent patterns and number of transactions, have been generated with *10k* numbers of unique items (as shown in Table 2). The first stream is of smaller size than the second one as it is the target stream whereas the second stream is the general stream. The *T$:I$:D$K* format is used in displaying the streams where T refers to the average length of transactions, I refer to the average size of frequent Itemsets and D shows the number of transactions in the dataset.

Table 2. Input datasets

Datasets		Description	Size (kb)
D_1	S_1	T3:I4:D16K	265 kb
	S_2	T3:I7:D116K	1,952 kb
D_2	S_1	T15:I7:D5K	101 kb
	S_2	T15:I10:D16K	332 kb

We used the same average length of transactions (T) in both streams assuming the stream data of same domain exhibit similar behavior. Two streams have different length frequent Itemsets, I, to support the generation of higher number of discriminative Itemsets. We select smaller average size of frequent Itemsets for S_1 compare to S_2 as S_2 is general trend and, in real world applications, it is expected to have larger size for frequent patterns as it includes many individual streams.

As can be seen from Figs 5 and 6 there is linear relation-ship between time (or space) and discriminative level threshold θ. By increasing θ, the time (or space) complexity decreases, as based on definition 1, the number of discovered frequent Itemsets would be lower for larger values of θ. In these charts at some point, a big fall in trends can be observed. In these points, for the defined θ values, the number of discovered discriminative Itemsets decreases dramatically as the number of infrequent items increases and the higher number of Itemsets are pruned in FP-Stream.

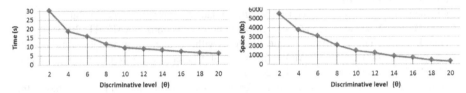

Fig. 5. Time and space complexities for D_1 ($\varphi=0.05\%$)

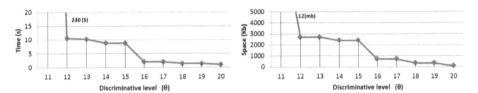

Fig. 6. Time and space complexities for D_4 ($\varphi=0.01\%$)

5   Conclusion and Future Works

In this paper, we proposed an efficient single pass algorithm for finding the discriminative Itemsets in data streams based on the tilted-time window model. Discriminative Itemsets focus on defining a stream distinctly in comparison to rest of the streams in the collection. The proposed algorithm introduces an in-memory structure, DIS-Tree, generated based on the FP-Stream method. To control the size of the data structures, we defined a pruning process for pruning the unpromising Itemsets from the prefix tree structure.

Empirical analysis shows the performance of the proposed method on the big datasets with simple complexity and the complex datasets generated in small size. The method works well in smaller discriminative level thresholds. In future, we propose more heuristics to improve the proposed method efficiency. We develop the algorithm for mining discriminative Itemsets in data streams using the sliding window model and make a classification method based on the discovered discriminative Itemsets and rules.

Acknowledgement. We would like to express our appreciation to CRC Smart Services for the partially financial funding for this research work.

References

1. Manku, G.S., Motwani, R.: Approximate Frequency Counts over Data Streams. In: International Confference on Very Large Data Bases, pp. 346–357 (2002)
2. Cheng, J., Ke, Y., Ng, W.: A survey on algorithms for mining frequent itemsets over data streams. Knowledge Information System 16(1), 1–27 (2007)
3. Lin, Z.H., et al.: Mining discriminative items in multiple data streams. World Wide Web-Internet and Web Information Systems 13(4), 497–522 (2010)
4. Seyfi, M.: Mining discriminative items in multiple data streams with hierarchical counters approach. In: Fourth International Workshop on Advanced Computational Intelligence (IWACI), pp. 172–176 (2011)
5. Giannella, C., et al.: Mining frequent patterns in data streams at multiple time granularities. Data Mining: Next Generation Challenges and Future Directions. AAAI/MIT (2003)
6. Han, J., Pei, J., Yin, Y.: Mining frequent patterns without candidate generation. In: Proceedings of the 2000 ACM SIGMOD International Conference on Management of Data, SIGMOD 2000, vol. 29(2), pp. 1–12 (2000)
7. Agrawal, R., Srikant, R.: Fast algorithms for mining association rules in large databases. In: Proceedings of the 20th International Conference on Very Large Data Bases, VLDB, pp. 487–499 (1994)

Modelling Visit Similarity Using Click-Stream Data: A Supervised Approach

Deepak Pai[1], Abhijit Sharang[2],
Meghanath Macha[1], and Shradha Agrawal[1]

[1] Adobe Research, India
{deepak.pai,mmacha,shraagra}@adobe.com
[2] Indian Institute of Technology, Kanpur, India
abhisg@iitk.ac.in

Abstract. Identifying and targeting visitors on e-commerce website with personalized content in real-time is extremely important to marketers. Although such targeting exists today, it is based on demographic attributes of the visitors. We show that dynamic visitor attributes extracted from their click-stream provide much better predictive capabilities of visitor intent. In this work, we propose a mechanism for identifying similar visitor sessions on a website based on their click-streams. Novel techniques for extracting features from visitor clicks are employed. Large margin nearest neighbour (LMNN) algorithm is used to learn a similarity metric between any two sessions. Further the sessions are classified into purchasers and non-purchasers using k-nearest neighbour (kNN) classification. Experimental results showing significant improvements over baseline algorithms based on Hidden Markov Model(HMM), support vector machine (SVM) and random forest are presented on two large real-world data sets.

Keywords: click-stream, metric learning, kNN classification, e-Commerce.

1 Introduction

Organizations are embracing online platforms to increase their reach and reduce operational cost of brick and mortar stores. Of late even software companies are moving towards online purchase and subscription models. Although web-visitors form a significant chunk of customer base for these retailers, the conversion rates are not very encouraging yet. E-commerce sites typically see conversions in the range of 1-3 % [14], which is much less compared to conversions in offline world. Hence organizations are interested in analysing online visits to understand low conversion rates and influence customers to make purchases. Marketing techniques like web-page personalization, targeted campaigns, promotional offers etc. are routinely adopted to persuade visitors. While these techniques increase the conversions, they incur significant marketing budget. Marketing to visitors who do not convert, amounts to wasted marketing dollars. Hence marketers would

B. Benatallah et al. (Eds.): WISE 2014, Part I, LNCS 8786, pp. 135–145, 2014.
© Springer International Publishing Switzerland 2014

like to know appropriate visitors, who would respond positively to a targeted campaign. We propose an approach to identify this target group for any marketing campaign, based on visitor click-streams.

Click-stream refers to the series of URLs visited by the customer. The information is available from logs generated by the servers or javascript tags embedded into the web-pages. Besides URL information, meta-data such as geographic location, purchase information, page categories, campaign information, referrer pages, etc. are also collected. This information is used to gain insight into the user intent, and subsequently predict potential leads, churn or any such desired metric. Clustering of visitor sessions by appropriate features generates market segments for targeted campaigns. Look-a-like modelling is another technique used by marketers to analyse visits. Applications of such modelling techniques include fraud detection, web page personalization, offline reporting, market segmentation, offline targeting, real-time online targeting etc.

At the heart of modelling techniques such as look-a-like modelling, clustering etc. lies the metric for identifying similarity of a visit pair. The features and their combination used for deriving such a metric has a large influence on the outcome of these techniques. Our primary contribution in the proposed work lies in building relevant features from click-stream data and then devising an appropriate metric for identifying a similarity score among visits. The contributions in this paper are multi-fold. Firstly we define the features for clustering among the multitude of features available from click-stream. Secondly we learn the feature weights using a supervised learning algorithm based on LMNN. Finally we show that the metric thus learnt is actually useful, by classifying the data using a kNN classifier. The superiority of the feature engineering and classification scheme is shown by comparing it with state-of-the-art algorithms like HMM, SVM and random forests.

The rest of the paper is organized as follow. We first describe related work in the areas of click-stream analytics. Next we define our problem in formal terms and our proposed solution approach. In the subsequent sections, we describe our experimental set-up, results and discuss the relevance of the results. Finally we conclude and discuss possible future enhancements.

2   Related Work

Classification of visitors based on click-streams is reasonably well explored area. Markov chains and Hidden Markov models (HMMs) are commonly applied to model click data. Scott and Hann [15] classify visitor sessions into one of predefined categories, namely decisive shoppers, deliberators, and never buyers, with a nested HMM. The higher level HMM models visits across sessions, while the nested lower level HMM captures clicks in a given session. Montgomery et. al. [11] have modelled browsing behaviour as dynamic multinomial probit models. Further they propose to capture previous clicks into higher order Markov models. HMMs are also used by Ypma et. al. [20] to categorize web-pages. Further they clustered users based on the observed click-streams. Sismeiro and Bucklin [3] propose a Tobit model capturing browser behaviour. Empirical results showing variation in browsing intent with depth of visit and number of repeat visits are presented.

Recent work has explored visitor clustering from their on-site behaviour. Li, Tian, and Xing [7] compute similarity of visitor sessions based on web-page attributes using Euclidean distance measure. The similarity is further used to cluster visits using k-medoids clustering algorithm. Petridou et. al. [13] cluster visitors considering the visit behaviour and time of viewing. Moe et. al. [10] categorize users into directed buyers, hedonic browsers, search deliberators, and knowledge seekers. Visits are clustered into one of these groups based on the browsing characteristics across various web-page categories.

Prior work has concentrated on modelling click-streams as first or second order Markov chains [11]. Intuitively it feels more likely that visitor choices are dependent on their entire history of clicks in the current as well as certain previous sessions. However higher order Markov models are more complicated to model, especially in BigData scenario. HMMs address this problem to a certain extent, but prior work of modelling click-streams with HMMs seem limited. Moreover, HMMs show significant bias towards the majority class, which limits their application in case of data with class imbalance, such as ours. Additionally, capturing metrics like visit counts, repeat visits, and visit duration into HMM observation sequence is non-trivial and noticeably previous work on HMM and Markov chains have not captured this information. Moreover,our experiments showed that the performance of HMMs was poor and comparable only to random oracle. We observe from our approach that features are significant indicators of visitor intent. The available clustering techniques consider such metrics, however each dimension is given equal weight. Mahalanobis distance [8] better captures the distance between two points, in a transformed dimension. Blitzer, Weinberger, and Saul [2] propose a method for learning Mahalanobis distance from labelled data points as a supervised learning algorithm. The algorithm is optimized to improve kNN classification results. Such an algorithm has been effectively used in literature [9] for variety of applications. To the best of our knowledge, there has been no previous work in exploring LMNN algorithm with Mahalanobis distance for classifying or clustering visitor sessions. In this work we present our results of such an exploration and show improvements over baseline algorithms, empirically on two large real-world data-sets.

3 Problem Definition

We formally define our problem in this section. We start with an intuition to the proposed solution.

We model the problem at a session level rather than a visitor level, since different sessions of the same user are not necessarily the same. Online purchase cycle of users interested in making a purchase could be broken down into logical stages. Such techniques have been used effectively by [10,16]. Initial user sessions are geared towards deliberating on the product of purchase since the user is more

interested in looking at various available options to choose from. Hence the visits
are concentrated on informative pages, which provide data to make a decision.
At the end of this phase the user has a narrowed down a list of items, among
which, he would make a purchase. Later stages involve repeated visits to items
in the interest set, ending with a purchase. However, most visitors on the website
arrive only to browse and update their knowledge about certain products without
having an intent of purchasing. Moreover, in case of e-commerce websites, visitor
interest changes rapidly. A visitor interested in purchasing a gaming console
today, might be interested in purchasing books, few months later. Combining
information across visitor sessions, therefore results in a heterogeneous mixture
of interests and does not yield good results. Intuitively identifying visitors in
deliberation phase and influencing them would be more useful than targeting
users with non-purchase intent or users who have already made a decision. Hence
our proposed work concentrates on modelling visits at a session level rather than
visitor level.

Let $\mathbb{S} = \{S_1, S_2, ..., S_n\}$ be the set of user sessions or visits.

Let $S_i = \{C_i^1, C_i^2, ..., C_i^m\}$ be the click-stream with C_i^k representing k^{th} click of
i^{th} session. C_i^k is a vector of features for every click. The feature vector is formed
from the information derived from meta-data corresponding to the clicks.

Let $C_i^k = \{f_{ik}^1, f_{ik}^2, ..., f_{ik}^p\}$ be the feature values corresponding to k^{th} click of
i^{th} session. Let p_i be the binary label associated with session S_i with 1 representing a purchase.

We want to compute a similarity metric Sim_{ij} for any given pair of sessions
S_i and S_j, based on $\mathbb{F}$. Hypothetically similar sessions will share the same label
as compared to dissimilar sessions.

Given the similarity matrix $Sim = (sim_{ij})_{i,j=1..n}$, we need to label points
in the test-set, based on the labels of training-set in their neighbourhood. Let
$S_{i,test}$ be any point in the test-set, that needs to be labelled.

Let $S_{i,train} = \{S_{i,train}^1, S_{i,train}^2, ..., S_{i,train}^k\}$ be the data points in the training-
set, which are in the neighbourhood of $S_{i,test}$. The label of $S_{i,test}$ is decided by
majority voting of elements of $S_{i,train}$.

4   Approach

Our approach broadly consists of categorising the URLs, sessionizing URLs into
user sessions, selecting features that best capture the visitor behaviour on the
website in a given session, learning a metric to find similar sessions and finally
assigning labels to sessions based on the learnt metric. We discuss the details in
the subsequent sections.

4.1   Categorisation

A reasonably large website consists of ten or hundred thousand URLs and thou-
sands of unique URLs. To consider each of these URLs as a separate dimension
would lead to very sparse matrix on which learning algorithms do not perform

well. Moreover scaling the algorithms to such large feature dimensions is not trivial. To deal with this problem [15,11,10] categorised the URLs into pre-defined categories. Following a similar approach we initially categorised the URLs into pre-defined categories.

Let $U = \{u_1, u_2, ..., u_n\}$ be the unique URLs and $Cat = \{cat_1, cat_2, ..., cat_p\}$ be the pre-defined categories. We define a mapping function $\mathbb{M}$, which maps U to Cat.

$\mathbb{M}(u_i) = cat_j \ \forall \ u_i \in U \mid cat_j \in Cat$

Henceforth our analysis would be on the transformed space Cat.

4.2 Sessionization

From the raw click-stream data, we need to form logical sessions corresponding to user visit. The sessionization is done based on thirty minutes of inactivity. For every user, from the start of his session, till the point of inaction for thirty minutes, all URLs are grouped to form a single visitor session.

4.3 Feature Selection

We first compute the latent feature–time spent on the category. This is computed by considering the difference in time-stamps of subsequent clicks. After analysing all available features / meta-data for each click, visit count and the visit duration of the category are found to be the most indicative features. The categories defined in the previous section $Cat(Metric, S_i)$ form our features, where $Metric$ is either visit count or visit duration. First we compute the raw feature values for each session from the click-stream. Visit count values for a category cat_j in any given session S_i is the frequency of visits to the category in the session.

$$Cat(VisitCount, S_i) = \{cat_i^1, cat_i^2, ..., cat_i^p\} \text{ where } cat_i^j = Frequency(cat_j) \text{ in } S_i$$

$$cat_i^j = \sum_{j=1}^{p} v_i^j \mid v_i^j = \begin{cases} 1 & \text{if } \mathbb{M}(f_{ik}^l) = cat_j \\ 0 & \text{otherwise} \end{cases} \forall \ C_i \in S_i \text{ and } f_{ik}^l = URL.$$

Similarly, we compute raw feature values for the metric, visit duration as summation of visit durations to the categories over all clicks in that session and normalize the values by computing the term frequency-inverse document frequency (tf-idf) score from the click-stream. tf-idf [6,1,19] is widely used in text processing, where individual words or n-grams form the feature vector. In our scenario, categories, sessions, and set of all sessions are analogous to words / n-grams, documents, and document corpus respectively. The need for normalization is intuitive. People on an average might spend more time on product-1 as compared to product-2 or an individual user-1 might take more time looking through product reviews than user-2. Normalizing the features will remove such biases.

4.4   Metric Learning

As the next step, we learn weights for each feature in our feature vector $\mathbb{F}$. The weights are then used to learn a metric of distance between any two data points. One of the most common metrics is the **Euclidean metric** which assigns equal weights to all features and is hence not optimal. This problem is tackled by taking the Euclidean distance in a transformed space where the variance of the dimensions in the feature vector is taken into account. The transformed space can be achieved by introducing a matrix L into the distance function.

$$\mathbb{D}(\boldsymbol{x_i}, \boldsymbol{x_j}) = (\boldsymbol{x_i} - \boldsymbol{x_j})^T L^T L (\boldsymbol{x_i} - \boldsymbol{x_j})$$

The above equation transforms to

$$\mathbb{D}(\boldsymbol{x_i}, \boldsymbol{x_j}) = (L\boldsymbol{x_i} - L\boldsymbol{x_j})^T (L\boldsymbol{x_i} - L\boldsymbol{x_j})$$

which is the Euclidean distance in a transformed space. The metric is called Mahalanobis metric. Blitzer, Weinberger, and Saul [2] propose learning a Mahalanobis distance metric for kNN classification from labelled examples. In the subsequent sections we briefly discuss about the algorithm and the proposed improvements.

Large Margin Nearest Neighbour. In this section we briefly describe the LMNN approach. The algorithm is a measure to improve the accuracy of the kNN classification algorithm. The goal of the algorithm is to ensure that all k neighbours for a test point belong to the same class. This is achieved by learning a positive semi-definite matrix M, which transforms the original vector space to transformed space. The matrix $M = L^T L$ is arrived at by solving a convex optimisation problem with additional constraints. Thus the problem is reformulated to be semi-definitive program(SDP) [17]. The distance function is hence formulated as

$$\mathbb{D}(\boldsymbol{x_i}, \boldsymbol{x_j}) = (\boldsymbol{x_i} - \boldsymbol{x_j})^T M (\boldsymbol{x_i} - \boldsymbol{x_j})$$

Thus the SDP is formulated to minimize the following cost function:

$$\sum_{ij} \eta_{ij} \mathbb{D}(\boldsymbol{x_i}, \boldsymbol{x_j}) + c \sum_{i,j,k} \eta_{ij} (1 - p_{ik}) [1 + \mathbb{D}(\boldsymbol{x_i}, \boldsymbol{x_j}) - \mathbb{D}(\boldsymbol{x_i}, \boldsymbol{x_k})], \qquad (1)$$

where $p_{ij} \in \{0, 1\}$ is used to indicate whether labels p_i and p_j match or not. $\eta_{ij} \in \{0, 1\}$ is used to indicate whether $\boldsymbol{x_i}$ and $\boldsymbol{x_j}$ are target neighbours or not. For every data point x_i, set of k target neighbours are initially identified. The target set, η_{ij} remains unchanged during learning. c is a positive constant generally set by cross validation. c is a positive constant

Additional constraints are

$$\mathbb{D}(\boldsymbol{x_i}, \boldsymbol{x_l}) - \mathbb{D}(\boldsymbol{x_i}, \boldsymbol{x_j}) \geq \mathbb{D}(\boldsymbol{x_i}, \boldsymbol{x_j}) - \mathbb{D}(\boldsymbol{x_i}, \boldsymbol{x_l})$$
$$1 + \mathbb{D}(\boldsymbol{x_i}, \boldsymbol{x_j}) - \mathbb{D}(\boldsymbol{x_i}, \boldsymbol{x_l}) \geq 0$$
$$M \succeq 0$$

The first term in eq. (1) penalises large distances between each data point and its target neighbours, while the second term penalizes small distances between each data point and all other data points with different labels. The final constraint restricting M to be positive semi-definitive.

4.5 kNN Classification

Finally, we apply a straight forward kNN classifier with majority voting to label the test data.

5 Experimental Set-Up

In this section we describe our experimental set-up, data-set used and the results of our explorations. We evaluate our approach on two large real-world datasets: the first from a software company that sells products and subscriptions online, and the second from an e-Commerce retailer specializing in the sale of hiking and camping products. Also, we show that the nature of our data-sets is quite generic and hence the proposed algorithms could be readily applied to any other data-set.

5.1 Data Description

Our dataset consisted of a single day worth of click-streams for each company.

Data-Set I. The first data-set had $19,737,907$ sessions. The number of purchases were 1444 giving us a conversion rate of 0.007%. The average click length was 2.7. Average time spent by users in a session was 186 seconds.

Data-Set II. The second data-set had $211,577$ sessions. The number of purchases were 3459 giving us a conversion rate of 1.6%. The average click length was 11. Average time spent by users in a session was 298 seconds.

5.2 Categories

As described in the earlier section, we categorize URLs into appropriate categories. The categories are predefined based on their relevance to the data-sets. In case of our first data-set the categories used were Cat = {Home, Product, Account, Info, Trial, ShopCart, Order, Help}. Categories Cat = {Home, Product, Account, Category, Info, ShopCart, Search, Order}, were formed in case of the second data-set.

6 Results and Discussion

In this section we present the results of our analysis. Further we discuss the results and ways this information could be used by marketers to perform online or offline targeting of customers. We compare the results of our algorithm, against results from HMM, SVM and random forest, that are used as the baseline algorithms.

6.1 Baseline Algorithms

We compare our results against multiple baseline algorithms. Below we briefly describe the set-up of each algorithm.

HMM. We build HMM model of the click data on the lines of [11]. The webpages are categorized as described earlier. The categories form the set of observations. We model the HMM to have two hidden states indicative of purchasers and non-purchasers. Initial state probabilities, state transition probabilities and emission probabilities are computed as prior from the training data. Viterbi and BaumWelsh algorithms are used for training purposes and compute the HMM parameters. The model thus built is used to predict the visitor state from the observation sequences in test data.

SVM. We train a SVM classifier with linear, polynomial, radial and sigmoid kernels. Visitor information such as country, referrer search engine, demographic region, number of clicks and total time spent on the website are used as features.

Random Forest. Here, we use the same features as in SVM to train a random forest classifier. The model thus built is used to predict labels for the test data.

6.2 Validation

We are interested in classifying user sessions, as early as possible in the session. We split the click-streams vertically into 25, 50, and 75 % of their total length. As we are interested in identifying visitors who are likely to make a purchase, these visitors form our positive class. All metrics that are reported are for positive (minority) class. We divide our data into training and test data with a 80-20 split. LMNN algorithm is trained to arrive at the covariance matrix $\mathbb{M}$. The Mahalanobis distance between any two sessions Sim_{ij}, which forms our distance metric is computed as described earlier. We compute the similarity metric for sessions in the test data to find k nearest neighbours. We experimented with values of k ranging from 5 to 15 and selected 13 as an appropriate value of k based on empirical evidence. Labels for each point in the test-set are then computed based on majority voting of training data points in the neighbourhood. We looked at the variation in precision and recall over different values of k for

25% split of the data and observed that the best results are obtained for $k = 13$. While best precision values are observed at $k = 13$, recall values remain constant. The precision recall values for $k = 13$ and 25% split of click-stream are shown in Table 1. Intuitively kNN results in best classification when only the most relevant neighbours are used for classification, with any increase or decrease in the number of neighbours resulting in data points from other classes dominating the classification, especially with large class imbalance. We combine the precision and recall values into F1 score, to better compare with the performance of baseline algorithms. F1 values of the proposed algorithm in comparison with the baseline algorithms are also shown in the same table. Similarly P-R values and F1 scores for other data splits (50 and 75) are shown in Tables 2- 3 respectively. Value of w_1 and w_0 are set by cross validation.

6.3   Discussion

We observe that our algorithm performs significantly better than the baseline algorithms. Results show higher precision, recall, and F1 values for our algorithm as compared to the baseline algorithms. SVM classifier always predicted the majority class resulting in precision, recall and F1 score being all 0. Hence we have not included the results of SVM in the result tables. HMM and random forests are seen to perform extremely poorly. This is primarily due to the huge class imbalance in the data. HMM for instance has a near cent percent start probability of visitor being a non-purchaser. The transition model built is diagonal heavy indicating that the visitors are most likely to remain in the category that they are in. Eventually the model predicts the visitor being a non-purchaser most of the time. A similar behaviour is observed in case of random forest as well. Not surprisingly the precision and recall values are very low for both baseline algorithms. Variations of LMNN with Euclidian distance and Mahalanobis distance show comparable performance. The results are consistent across various splits of the click-streams. Moreover it is observed that the precision and recall improve as we observe more data i.e. as we move from 25 to 50% or 50 to 75% data split. Comparable results are obtained across data-sets from significantly different domains, indicating the generality of the proposed approach. We observe that visitors not having a purchase intent spend more time on help related pages like blogs and forums. This is explained by the fact that these visitors have already purchased the software and are currently more interested in getting it working.Not surprisingly, visitors having a purchase intent spend time on shopping cart related pages, adding / removing items and modifying quantity of items to purchase. Visits resulting in a purchase involve lesser number of product views but larger amount of time spent on product pages as compared to visits that do not result in a purchase. This is consistent with our hypothesis that visitors in the deliberation phase look at variety of products before making a purchase decision. The actual purchase happens in the later sessions which are characteristic of decisive buyers, who look at few products that they are most interested in and spend large amount of time gathering details of these products. Visits and time spent on home page confirms this behaviour, with buyers spending much lesser time on home page. Initial visitor sessions are seen to be starting from home page, whereas decisive buyers are more likely to land on the product page they are interested in.

7   Conclusion and Future work

An algorithm to classify online visitor sessions based on their click-streams is presented. The classification is based on kNN algorithm based on a novel similarity metric. The similarity metric is computed as a Mahalanobis distance between two visitor sessions. The feature weights or covariance matrix is learnt from LMNN algorithm. Our feature engineering in combination with the classification scheme outperform baseline algorithms, based on HMM, SVM and random forest, as shown from empirical evidence with two large real-world data-sets. The consistency of the algorithm over varied data-sets, shows the generality of our approach. Although the current work focuses on identifying potential purchases, the similarity metric is equally applicable to other applications like churn prediction, look alike modelling, visitor segmentation, etc. that will be explored in future work.

Table 1. Precision Recall values for k=13 and 25 % split

	Data-set I			Data-set II		
Algorithm	Precision	Recall	F1	Precision	Recall	F1
HMM	0.026	0.264	0.047	0.026	0.27	0.047
Random Forest	0.003	0.007	0.004	0.083	0.027	0.041
LMNN-Euclidean	0.804	0.153	0.256	0.785	0.342	0.477
LMNN-Mahalanobis	0.807	0.156	0.261	0.788	0.342	0.477

Table 2. Precision Recall values for k=13 and 50 % split

	Data-set I			Data-set II		
Algorithm	Precision	Recall	F1	Precision	Recall	F1
HMM	0.027	0.26	0.049	0.026	0.258	0.047
Random Forest	0.025	0.029	0.027	0.138	0.023	0.039
LMNN-Euclidean	0.795	0.461	0.584	0.857	0.490	0.624
LMNN-Mahalanobis	0.801	0.451	0.577	0.868	0.489	0.625

Table 3. Precision Recall values for k=13 and 75 % split

	Data-set I			Data-set II		
Algorithm	Precision	Recall	F1	Precision	Recall	F1
HMM	0.023	0.184	0.041	0.024	0.197	0.043
Random Forest	0.043	0.009	0.015	0.141	0.016	0.029
LMNN-Euclidean	0.795	0.617	0.695	0.860	0.609	0.713
LMNN-Mahalanobis	0.802	0.661	0.724	0.860	0.617	0.718

References

1. Aizawa, A.: An information-theoretic perspective of tf–idf measures. Information Processing & Management 39(1), 45–65 (2003)
2. Blitzer, J., Weinberger, K.Q., Saul, L.K.: Distance metric learning for large margin nearest neighbor classification. In: Advances in Neural Information Processing Systems, pp. 1473–1480 (2005)
3. Bucklin, R.E., Sismeiro, C.: A model of web site browsing behavior estimated on clickstream data. Journal of Marketing Research, 249–267 (2003)
4. Cadez, I., Heckerman, D., Meek, C., Smyth, P., White, S.: Visualization of navigation patterns on a web site using model-based clustering. In: Proceedings of the Sixth ACM SIGKDD International Conference on Knowledge Discovery and Data Mining, pp. 280–284. ACM (2000)
5. Faloutsos, M., Faloutsos, P., Faloutsos, C.: On power-law relationships of the internet topology. In: ACM SIGCOMM Computer Communication Review, vol. 29, pp. 251–262. ACM (1999)
6. Joachims, T.: A probabilistic analysis of the rocchio algorithm with tfidf for text categorization. Technical report, DTIC Document (1996)
7. Li, J., Tian, H., Xing, D.: Clustering user session data for web applications test. Journal of Computational Information Systems 7(9), 3174–3181 (2011)
8. Mahalanobis, P.C.: On the generalized distance in statistics. Proceedings of the National Institute of Sciences (Calcutta) 2, 49–55 (1936)
9. Mensink, T., Verbeek, J., Perronnin, F., Csurka, G.: Metric learning for large scale image classification: Generalizing to new classes at near-zero cost. In: Fitzgibbon, A., Lazebnik, S., Perona, P., Sato, Y., Schmid, C. (eds.) ECCV 2012, Part II. LNCS, vol. 7573, pp. 488–501. Springer, Heidelberg (2012)
10. Moe, W.W.: Buying, searching, or browsing: Differentiating between online shoppers using in-store navigational clickstream. Journal of Consumer Psychology 13(1), 29–39 (2003)
11. Montgomery, A.L., Li, S., Srinivasan, K., Liechty, J.C.: Modeling online browsing and path analysis using clickstream data. Marketing Science (2004)
12. Newman, M.E.: Power laws, pareto distributions and zipf's law. Contemporary Physics 46(5), 323–351 (2005)
13. Petridou, S.G., Koutsonikola, V.A., Vakali, A.I., Papadimitriou, G.I.: Time-aware web users' clustering. IEEE Transactions on Knowledge and Data Engineering 20(5), 653–667 (2008)
14. Poggi, N., Carrera, D., Gavalda, R., Ayguadé, E., Torres, J.: A methodology for the evaluation of high response time on e-commerce users and sales
15. Scott, S.L., Hann, I.-H.: A nested hidden markov model for internet browsing behavior (2006)
16. Sismeiro, C., Bucklin, R.E.: Modeling purchase behavior at an e-commerce web site: a task-completion approach. Journal of Marketing Research (2004)
17. Vandenberghe, L., Boyd, S.: Semidefinite programming. SIAM Review (1996)
18. Weinberger, K.Q., Saul, L.K.: Distance metric learning for large margin nearest neighbor classification. J. Mach. Learn. Res. 10, 207–244 (2009)
19. Wu, H.C., Luk, R.W.P., Wong, K.F., Kwok, K.L.: Interpreting tf-idf term weights as making relevance decisions. ACM Transactions on Information Systems (TOIS) 26(3), 13 (2008)
20. Ypma, A., Ypma, E., Heskes, T.: Categorization of web pages and user clustering with mixtures of hidden markov models (2002)

BOSTER: An Efficient Algorithm for Mining Frequent Unordered Induced Subtrees

Israt J. Chowdhury and Richi Nayak

School of Electrical Engineering and Computer Science, Science and Engineering Faculty,
Queensland University of Technology, Brisbane, Australia
{israt.chowdhury,r.nayak}@qut.edu.au

Abstract. Extracting frequent subtrees from the tree structured data has important applications in Web mining. In this paper, we introduce a novel canonical form for rooted labelled unordered trees called the balanced-optimal-search canonical form (BOCF) that can handle the isomorphism problem efficiently. Using BOCF, we define a tree structure guided scheme based enumeration approach that systematically enumerates only the valid subtrees. Finally, we present the balanced optimal search tree miner (BOSTER) algorithm based on BOCF and the proposed enumeration approach, for finding frequent induced subtrees from a database of labelled rooted unordered trees. Experiments on the real datasets compare the efficiency of BOSTER over the two state-of-the-art algorithms for mining induced unordered subtrees, HybridTreeMiner and UNI3. The results are encouraging.

Keywords: Web mining, frequent subtrees, labelled rooted unordered trees, induced subtrees, canonical form, enumeration approach.

1   Introduction

In order to improve the Web-based applications, finding frequent patterns is a common task in Web usage mining that discovers useful information from the Web data. The web usage data, the sequences of accesses pursued by users, can be easily represented as trees [1]. The frequent subtree mining task can be used in distinguishing various users according to their common browsing behavior [2].

In this paper we study the problem of finding frequent subtrees from the database of unordered trees.

Unordered trees have shown the capability of identifying interesting relations due to not being constrained by sibling order (i.e. no fixed left-to-right order among sibling nodes) [3]. However, this distinct property makes the process of mining frequent unordered subtrees more challenging in comparison to ordered trees. Exponential candidate generation with redundancy is the main problem in mining frequent unordered subtrees. It is critical to determine a "good" growth strategy as there can be many possible ways to extend a candidate subtree due to not having sibling order constraint. Moreover, high computation and memory expense are always an issue for mining tree data. Many algorithms have been proposed to overcome these challenges

B. Benatallah et al. (Eds.): WISE 2014, Part I, LNCS 8786, pp. 146–155, 2014.
© Springer International Publishing Switzerland 2014

where they use a canonical form, and extend the candidates only that conform to the canonical form. Several canonical representations based on sorted pre-order string [4], depth-first traversal [5-7] and breadth-first traversal [8] have been proposed. These canonical forms need an additional isomorphism test for avoiding redundancy problem. Besides, the existing algorithms use extension and join operations for candidate enumeration [8, 9] , which produce a large number of candidates including invalid subtrees. Authors in [10] have developed an enumeration approach using underlying tree structure information that generates only valid subtrees, but, the method suffers from extensive memory usage.

We have previously proposed an optimal tree traversal algorithm for traversing a rooted unordered tree [11] and finding similarity amongst tree data. In this paper, we extend this traversing algorithm by introducing a new heuristic that leads towards a new definition of canonical form for representing unordered trees, called the balanced-optimal canonical form (BOCF). The BOCF can alleviate redundancy problem as it is able to represent unordered trees uniquely even in the presence of isomorphism. Using BOCF, we specify an optimal enumeration approach to systematically enumerate all frequent subtrees based on underlying tree structure information. This enumeration approach is efficient as it restricts the search, by only generating the unambiguous and valid subtrees using the underlying tree structure information. Finally, the balanced optimal search tree miner (BOSTER) algorithm is proposed for mining frequent induced unordered subtrees from a database of labelled rooted unordered trees. Empirical analysis carried out using a real data has shown the effectiveness of BOSTER over the two state-of-the-art algorithms, HybridTreeMiner [8] and UNI3[10].

2   Preliminaries

Let $T = (V, E, L)$ be a *rooted labeled unordered* tree, where $V = \{v_0, v_1, v_2, ..., v_n\}$ denotes the set of nodes with v_0 as *root* node, $E = \{(v_i, v_j)| v_i, v_j \in V\} = \{e_1, e_2, ..., e_{n-1}\}$ denotes the set of edges and L denotes the set of labels. The label is given by a function $\Phi: V \rightarrow L$ which maps nodes with unique labels. An unordered tree has no ordering relationship among the nodes except ancestor-descendent or parent-child. The ancestor-descendent relationship between two nodes is denoted by $v_i \prec v_j$, i.e. v_i is ancestor of v_j, the '$\prec$' symbol represents 'precedes'. The level of a node v_i in a tree T is denoted as $Lv(T, v_i)$ and the height of a tree T is denoted by $H(T)$.

Definition 1 (*Induced Subtrees*): A tree $T'(V', L', E')$ is an unordered *induced subtree* of a tree $T (V, L, E)$ iff: (1) $V' \subseteq V$, (2) $E' \subseteq E$, (3) $L' \subseteq L$ and the labelling of V' in T is preserved in T' (4) $\forall v_i' \in V'$, $\forall v_i \in V$ and v_i' is not the root node, then parent of $v_i' =$ parent of v_i, and (5) no left-to-right ordering among the siblings in T is preserved among the corresponding nodes in T'.

Definition 2 (*Equivalent Node*): If two nodes v_i and v_j of a tree T, have the same label originated from the same labelled parent node (parent of $v_i =$ parent of v_j) and have the same labelled child nodes then they are called *equivalent nodes*, denoted by $v_i \cong v_j$.

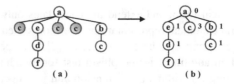

Fig. 1. The highlighted nodes are the equivalent nodes (a) and the numerical values are the weights of the respective nodes (b), for simplicity only label is used to represent a node

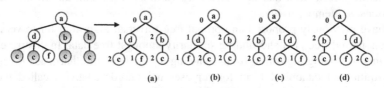

Fig. 2. Four rooted ordered trees obtained from the same rooted unordered tree

Definition 3 (*Weight of Node*): *Weight* of a node v_i ($v_i \neq v_0$) is defined as the total number of its equivalent node and denoted by w_i (fig 1).

According to the properties of unordered trees we have Lemma 1.

Lemma 1. *Weight of the root node v_0 is always zero, $w_0 = 0$. For each node $v_i \in V$ ($v_i \neq v_0$), the weight w_i ($w_i \neq w_0$) should always have a minimum value of one.*

PROOF:

1. *According to the tree structure schema no equivalent node of a root node is possible as its ancestors are undefined. Hence, the weight of the root is always zero.*
2. *Each node v_i ($v_i \neq v_0$) of tree T should have at least one equivalent node, otherwise v_i doesn't belong to that tree. Hence, the minimum weight of the node is one, $w_i = 1$. For node v_i, $w_i > 1$ if the node has more than one equivalent node.*

Definition 4 (*Mining Unordered Induced Subtree*): Let T_{db} denotes a database where each transaction is a labelled rooted unordered tree. The task of frequent induced subtree mining from T_{db} is finding all induced subtrees that have minimum support s.

Definition 5 (*Support*): Support s of a tree T' in database T_{db} is defined as the number of trees, T that has at least one occurrence of T' as an induced subtree in its structure.

3   Optimal Canonical Form

A canonical form (CF) of a tree is a representative form that can consistently represent many equivalent variations of that tree into one standard [7, 12]. The canonical forms for ordered and unordered subtrees are different. Due to having no sibling order, several ordered variations are possible from an unordered tree.

Definition 6 (*Equivalent ordered trees*): Two distinct ordered trees T_1 and T_2 are equivalent to each other if they represent same unordered tree T, denoted by $T_1 \cong T_2$.

An example of equivalent ordered trees is given in fig 2, where four ordered trees can be derived from an unordered tree. We propose to represent these ordered variations by a single canonical form following an optimal traversal so that the same unordered tree is derived from each of them.

3.1   Balanced Optimal Canonical Form (BOCF)

We have earlier developed an optimal tree search traversal algorithm [11] by reducing the traversing problem to an optimization problem called "simple assemble line balancing" [13]. Unlike existing traversal algorithms [12], our algorithm [11] works based on optimization instead of fixing left-to-right order among siblings. We propose heuristics that are applied recursively for setting the rules of traversing the whole tree. Heuristic 1 identifies a potential node during the traversal process. Heuristics 2 and 3 select the best node if multiple nodes are identified as candidates for traversal. Induction of heuristics will result in the optimal traversal balanced.

Heuristic 1. *After traversing the root node, the enumeration of available nodes satisfying the ancestral relationship ($v_i \prec v_j$) will be prioritized based on their weights.*

Heuristic 2. *If there exist two or more nodes with maximum weight, the node with maximum number of children will get priority for traversing next.*

Heuristic 3. *In case of existence of multiple nodes with equal weight and children count, the minimum lexicographical order will be used to prioritize their traversing.*

Consider the example tree in fig 1, following this traversal scheme, root node v_a will be traversed first. Next eligible nodes for traversing will be v_e, v_c, v_b as their parent node has been traversed. Node v_c will be chosen following heuristic 1. Heuristic 2 will need to be applied to choose between v_e and v_b, v_e will be traversed accordingly. Node v_b will be traversed next using heuristic 3, as the other two heuristics fail to prioritize the order between v_b and v_d. The final sequence for traversing the whole tree will be v_a, v_c, v_e, v_b, v_d, v_c, v_f, that is not restricted by depth-first or breadth-first order.

We propose a balanced-optimal canonical form for a tree represented in the optimal order obtained by this traversal. BOCF is a string representation of a tree along with four unique symbols, +1, -1, +2 and -2, that are used to represent the breadthwise movement from sibling to sibling and the depth-wise movement from a child to its parent. We use +1 and -1 for forward and backward travel towards depth, and +2 and -2 for forward and backward travel towards breadth respectively. We assume that none of these symbols are included in the alphabet of node labels.

Definition 7 (*BOCF String Representation of Unordered Tree*): The BOCF string representation of the rooted unordered tree is achieved by a guided record of sibling nodes. When a new node appears under its parent node, only the breadthwise movement from the existing rightmost sibling node is permitted.

Consider the trees in fig 2.The optimal order of the equivalent trees in fig 2 is: v_a, v_b, v_c, v_d, v_c, v_f. Using definition 7, the unique BOCF string representation of these four trees is: $0v_a$, +1, $2v_b$, +1, $2v_c$, -1, +2, $1v_d$, +1, $2v_c$, -2, $1v_f$. It should be noted that all equivalent ordered trees is represented by a unique standard form. It indicates that they all are originated from the same unordered tree. This greatly benefits unordered tree mining. The optimal traversal poses a total order on all variants of an unordered tree which guarantees the uniqueness of BOCF for a labelled rooted *unordered* tree.

3.2   Dealing with the Isomorphism and Automorphism Problem

A main challenge in defining a canonical form for unordered trees is faced when two trees are found isomorphic. If a bijective mapping exists between the set of nodes of two trees T_1 and T_2, which preserves and reflects the tree structures, then these trees are called isomorphic to each other, denoted as $T_1 \cong T_2$. The term automorphism corresponds to isomorphism of a tree to itself. It is necessary to identify which of the ordered subtrees forms an automorphism group of an unordered subtree. During candidate generation, each subtree encoding should uniquely map to a single subtree only. Existing research addresses this problem by choosing one of the trees from the automorphism group as the representative of the group, and then all other isomorphic subtrees are ordered according to the representative of the automorphism group during candidate generation [7, 8]. This ensures that, for a particular unordered subtree, its occurrences are correctly counted so that the frequency can be easily determined. However, a checking is always required to find the presence of isomorphism in a tree. This causes an additional memory and time consumption for keeping the record of the representative tree and for doing isomorphism testing.

As shown earlier, the proposed BOCF encodes an unordered tree (including all of its ordered variants which are actually isomorphic to each other) uniquely. In other words, BOCF provides a unique representation to all isomorphic trees. This ensures that trees encoded with BOCF representation will be correctly grouped and counted. Unlike other canonical forms, BOCF does not require a record of representative trees or, an extra checking during candidate generation for dealing the isomorphism problem. Moreover, BOCF can naturally handle the automorphism problem. For applying the optimal traversal, the trees need to be pre-processed so that a concise tree representation can be derived by combining equivalent nodes. Consequently the weight of each node under its parent node is calculated. We conjecture that the equivalent nodes (i.e. same labelled sibling nodes having the same child) should not be treated as distinct nodes. The order between them is not important, but, only the occurrences are important. This process allows us to avoid the isomorphism of a tree to itself, i.e. solving the automorphism problem. Consider the following example in fig 3(a) where the dotted area is showing a case of automorphism problem for the considered tree. However, the BOCF representation is derived based on the weighted tree as shown in figure 3(b) where automorphism can no longer exist.

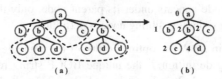

(a)   (b)

Fig. 3. Automorphism problem

4   Mining Frequent Labelled Unordered Induced Subtrees

We define an enumeration tree that lists all induced unordered subtrees in T_{db} according to their BOCF strings. We used the right-path extension and join operations for growing the enumeration tree. Previous research has shown that the right-path extension produces a complete and non-redundant candidate generation [7, 8, 14]. The use of extension alone for growing enumeration tree can be inefficient because the number of potential growth may be very large, especially when the cardinality of the alphabet for node labels is large. This shortcoming necessitates of using a join operation [7, 8]. However, a join operation often generates invalid subtrees. We propose using a tree-structure guided schema for enumeration which allows the generation of valid subtrees only. In the proposed tree structure guided enumeration approach, the underlying level and fan-out information of nodes are utilized during candidate generation.

Operations on the Enumeration Tree: The basis of our enumeration tree is as follows. An unordered N-tree (i.e. a tree with N number of nodes) BOCF is formed from the unordered $(N+1)$-tree BOCF by removing the right-most path (i.e. the right-most node along with its edge) at the bottom level.

For growing the enumeration tree we define extension and join operations using the BOCF string and the tree-structure guided schema.

Definition 8 (*BOCF-extension*): For a node v_i (fan-out $\neq 0$) of the BOCF T_1, extension is possible to apply using every frequent label v_j having level $Lv(T_1, v_i)$-1. This extension operation will result in a new BOCF T_2 in the enumeration tree where v_j will be the child of v_i. If T_1 is a N-tree BOCF, then the resultant new BOCF T_2 will be a $(N+1)$-tree with height $H(T_1)$ +1. Further extension is possible from this new right-most node v_j.

Before giving the definition of BOCF-join operation, we define equivalent groups.

Definition 9 (*Equivalent group*): If two N-node trees T_1 and T_2 have height $H(T_1) = H(T_2)$ and share first N-1 node (along with labels and weights) in common, they are considered as equivalent group, denoted by $T_1 \cong T_2$.

Definition 10 (*BOCF-join*): Join operation is a guided extension between two BOCFs, T_1 and T_2, from an equivalent group, $T_1 \cong T_2$. Assume, v_i and v_j are the corresponding right-most node of T_1 and T_2, where $w_i > w_j$ or, $w_i = w_j$ with v_i lexicographically sorts lower than v_j. By joining v_j in T_1 at the position of $Lv(T_1, v_i)$-1 will result in a new $(N+1)$ node BOCF, denoted by $T_1 \odot T_2$, of the same height as tree T_1.

Growth Rules: Candidate trees can have a large number of potential nodes to get a right-path extension. In order to restrict this growth, heuristics can be employed using BOCF definition. This will result in reduction of the number of candidates generated as well as in the reduction of the number of isomorphic subtrees. These rules support the basic formation principle of the enumeration tree, i.e. keeping the N-tree BOCF unchanged with the newly generated $(N+1)$-tree BOCF.

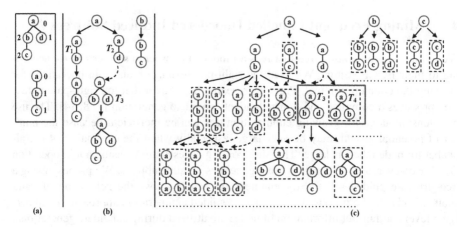

(a)   (b)   (c)

Fig. 4. Comparison between the proposed and existing enumeration techniques

Rule1: *Among all the nodes at the bottom level, the node with the maximum weight will be chosen for BOCF-extension.*

Rule2: *If there are more than two maximum weighted nodes then the node with maximum children will be chosen for BOCF-extension.*

Rule3: *If more than two maximum weighted nodes with the same number of children exist then the node that sorts lexicographically lower will be chosen for BOCF-extension.*

Consider an example database in fig 4a. We compare our enumeration tree (fig 4b) with the enumeration tree (fig 4c) generated by following the HybridTreeMiner method [8] (abbreviated as HBT here). HBT also uses the right-path extension and join operations for growing the enumeration tree, but, these are defined using a different canonical form (BFCF) [7], whereas we use BOCF and the tree-structure guided schema for growing the enumeration tree. The dotted rectangles in (fig 4c) are showing the generation of invalid subtrees in HBT. We did not show the full enumeration tree for HBT. If we continue it will grow in a much bigger size, resulting in much higher numbers of invalid subtrees. But, for our method, fig 4b is the complete enumeration tree of the considered database.

It can be clearly seen that our enumeration tree generates much less candidates in comparison to HBT enumeration tree because of producing only valid subtrees. Generation of several invalid subtrees causes extra memory space and, then, pruning of these subtrees causes additional computational cost for HBT. Moreover, our enumeration approach is more robust to the isomorphism problem. In fig 4c the enumeration tree produces two candidate trees T_3 and T_4, which are isomorphic. For counting the exact support these two should consider as same candidate. In that case an extra checking method is needed to count isomorphic trees; but our enumeration approach avoids growing any isomorphic tree. For example, in fig 4b; only tree T_3 exists, tree T_4 can't be generated. According to BOCF-join, join is supported only from T_1,"$0v_a +1$ $2v_b$" to T_2, "$0v_a +1$ 1 v_d" as $w_b > w_d$.

BOSTER Algorithm	**Grow_Enum** (C_k, level, weight, fan-out)

Input: a database T_{db} consisting of labelled rooted unordered trees represented as BOCF strings, a dictionary containing level and fan-out information of each node, a user defined minimum support (*min_sup*).
 Output: All frequent induced subtrees.
	1. **for all** $f \in C_k$ **do**
	 2. Select the right-most node of C_k using *Growth rules*;
	 3. Generate candidate C_{k+1} by adding f; //using *BOCF-extension*;
	 4. **if** support (C_{k+1}) $\geq$ *min_sup* **then**
	 5. *Result* $\leftarrow$ *Result* $\cup$ C_{k+1};
	 6. **end if**

1. *Result* $\leftarrow \emptyset$;
 2. $F1 \leftarrow$ the set of all frequent nodes;
 3.   **for all** $t_k \in F1$ **do**
 4.   **if** *fan-out*(t_k) = 0
 5.   **continue**
 6.   **end if**
 7.   *Grow_Enum* (t_k, level, weight, fan-out);
 8.   **end for**
 9. **return** *Result*;
	7. *Grow_Enum* (C_{k+1}, level, weight, fan-out);
	 8. **end for**
	 9. **for all** $C_k{'}$ such that $C_k \cong C_k{'}$ **do**
	 10. $C_{k+1} \leftarrow C_k \odot C_k{'}$; //using *BOCF-join*;
	 11. **if** support (C_{k+1}) $\geq$ *min_sup* **then**
	 12. *Result* $\leftarrow$ *Result* $\cup$ C_{k+1};
	 13. **end if**
	 14. **Grow_Enum** (C_{k+1}, level, weight, fan-out);
	 15. **end for**

Fig. 5. High level pseudo code of BOSTER algorithm

Fig 5 lists the BOSTER algorithm. The process of frequent subtree mining is initiated by scanning the tree database, T_{db}, where trees are stored as BOCF strings along with weight, level and fan-out information of each node. The *Grow_Enum* method is called recursively for growing the candidates. The frequency of every resultant candidate tree is computed according to the method used in [7, 8]. This is basically an apriori based frequency counting which gives us the exact frequent subtree list. In order to improve computational efficiency, we stop counting of a subtree as soon as the tree count reaches the minimum support value.

5   Experimental Evaluation

We have performed experiments to evaluate the efficiency of the proposed algorithm on real application data. All experiments have been conducted on a 2.8GHz Intel Core i7 PC with 8GB main memory and running the UNIX operating system. Two state-of-the-art unordered tree mining algorithms, HBT [8] and UNI3 [10] are used for benchmarking. We recorded the run time and memory usage of each algorithm and compared their performances.

In line with other research and to show scalability, three variations of the real weblog data, CSLOGS [2, 14], are used. (1) CSLOG1 - data generated from the first week web log usage consisting of 8,074 trees. (2) CSLOG12 - data generated from the first two weeks usage consisting of 13,934 trees. (3) CSLOGS - the entire data covering all weeks consisting of 59,691 trees, 716,263 nodes and 13,209 unique node labels.

Fig 6(a, b, c) and fig 7(a, b, c) compare the runtime and memory comparison of BOSTER against HBT and UNI3 respectively. For both runtime and memory comparison, BOSTER significantly outperforms HBT in all cases. However, UNI3 gave better memory consumption than BOSTER over CSLOG1 and CSLOG12. On the entire set of CSLOGS, BOSTER started to outperform UNI3 for support value less than 100. After this support value, UNI3 could not perform due to extensive memory usage (fig 7c). We allocated about 15GB memory to run UNI3, but, it still failed to execute results. UNI3 includes a large number of extra data structure to hold intermittent information for the mining process. These additional structures cause the out of memory problem when mining the large data with small support values. Moreover, both HBT and UNI3 keep record of representative trees for performing an isomorphism test that causes additional time and memory expense, but BOSTER can avoid this extra cost using BOCF string representation.

In real-life applications, memory usage can have a significant impact on the application's usability from the perspective of performance, interactivity, etc. BOSTER is able to consume less memory with yielding efficient time complexity, in comparison to the benchmarked algorithms, even in the presence of large data.

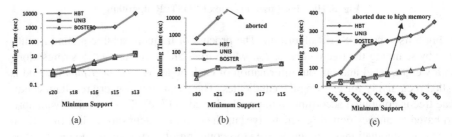

Fig. 6. Runtime comparison over CSLOG1 (a), CSLOG12 (b), full CSLOGS (c)

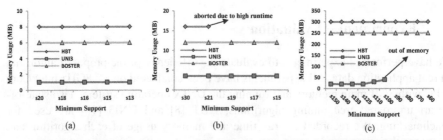

Fig. 7. Memory comparison over CSLOG1 (a), CSLOG12 (b), full CSLOGS (c)

6   Conclusion

In this paper, we presented a novel canonical form, and developed a new method of finding frequent induced subtrees from the dataset of labelled rooted unordered trees. We empirically evaluated the efficiency of the proposed algorithm, BOSTER, against the well-known algorithms in the literature, over real life datasets.

In future we will extend the proposed algorithm to find condensed representations like frequent closed patterns and we also will explore the scope for extending our canonical form to represent free trees in order to mine frequent patterns from them.

References

1. Pei, J., Han, J., Mortazavi-asl, B., Zhu, H.: Mining Access Patterns Efficiently from Web Logs. In: Terano, T., Liu, H., Chen, A.L.P. (eds.) PAKDD 2000. LNCS, vol. 1805, pp. 396–407. Springer, Heidelberg (2000)
2. Zaki, M.J., Aggarwal, C.C.: XRules: An Effective Structural Classifier for XML Data. In: Proceedings of the 9th ACM SIGKDD International Conference on Knowledge Discovery and Data Mining, pp. 316–325. ACM, Washington, D. C. (2003)
3. Wang, Y., DeWitt, D.J., Cai, J.-Y.: X-Diff: An Effective Change Detection Algorithm for XML Documents. In: Proceedings of the 19th International Conference on Data Engineering, pp. 519–530. IEEE, Vienna (2003)
4. Luccio, F., Enriquez, A.M., Rieumont, P.O., Pagli, L.: Exact Rooted Subtree Matching in Sublinear Time. Universita Di Pisa Technical Report TR-01 (2001)
5. Asai, T., Arimura, H., Uno, T., Nakano, S.-I.: Discovering Frequent Substructures in Large Unordered Trees. Springer, Heidelberg (2003)
6. Nijssenm, S., Kok, J.N.: Efficient Discovery of Frequent Unordered Trees. In: First International Workshop on Mining Graphs, Trees and Sequences. Springer, Heidelberg (2003)
7. Chi, Y., Yang, Y., Muntz, R.R.: Canonical Forms for Labelled Trees and Their Applications in Frequent Subtree Mining. Knowledge and Information System 8(2), 203–234 (2005)
8. Chi, Y., Yang, Y., Muntz, R.R.: HybridTreeMiner: An Efficient Algorithm for Mining Frequent Rooted Trees and Free Trees Using Canonical Forms. In: Proceedings of the 16th International Conference on Scientific and Statistical Database Management, pp. 11–20. IEEE, Santorini (2004)
9. Chehreghani, M.H.: Efficiently Mining Unordered Trees. In: Proceedings of the 11th IEEE International Conference on Data Mining, Vancouver, BC, pp. 111–120 (2011)
10. Hadzic, F., Tan, H., Dillon, T.S.: UNI3 - Efficient Algorithm for Mining Unordered Induced Subtrees Using TMG Candidate Generation. In: Proceedings of the 1st IEEE Symposium on Computational Intelligence and Data Mining, Honolulu, Hawaii, pp. 568–575 (2007)
11. Chowdhury, I.J., Nayak, R.: A Novel Method for Finding Similarities between Unordered Trees Using Matrix Data Model. In: Lin, X., Manolopoulos, Y., Srivastava, D., Huang, G. (eds.) WISE 2013, Part I. LNCS, vol. 8180, pp. 421–430. Springer, Heidelberg (2013)
12. Valiente. Algorithms on Trees and Graphs. Springer, Heidelberg (2002)
13. Scholl, A.: Balancing and Sequencing of Assembly Lines. Physica-Verlag, Heidelberg (1999)
14. Zaki, M.J.: Efficiently Mining Frequent Trees in A Forest: Algorithms and Applications. IEEE Transactions on Knowledge and Data Engineering 17(8), 1021–1035 (2005)

Phrase Queries with Inverted + Direct Indexes

Kiril Panev and Klaus Berberich

Max Planck Institute for Informatics, Saarbrücken, Germany
{kiril,kberberi}@mpi-inf.mpg.de

Abstract. Phrase queries play an important role in web search and
other applications. Traditionally, phrase queries have been processed using a positional inverted index, potentially augmented by selected multi-
word sequences (e.g., n-grams or frequent noun phrases). In this work,
instead of augmenting the inverted index, we take a radically different
approach and leverage the *direct index*, which provides efficient access to
compact representations of documents. Modern retrieval systems maintain such a direct index, for instance, to generate snippets or compute
proximity features. We present extensions of the established term-at-a-
time and document-at-a-time query-processing methods that make effective combined use of the inverted index and the direct index. Our
experiments on two real-world document collections using diverse query
workloads demonstrate that our methods improve response time substantially without requiring additional index space.

Keywords: information retrieval, search engines, efficiency.

1 Introduction

Identifying documents that literally contain a specific phrase (e.g., "sgt. pepper's
lonely hearts club band" or "lucy in the sky with diamonds") is an important task
in web search and other applications such as entity-oriented search, information
extraction, and plagiarism detection. In web search, for instance, such phrase
queries account for up to 8.3% of queries issued by users, as reported by Williams
et al. [24]. When not issued explicitly by users, phrase queries can still be issued
implicitly by query-segmentation methods [12] or other applications that use
search as a service.

Phrase queries are usually processed using a positional inverted index, which
keeps for every word from the document collection a posting list recording in
which documents and at which offsets therein the word occurs [25]. To process
a given phrase query, the posting lists corresponding to the words in the query
need to be intersected. This can be done sequentially, processing one word at a
time, or in parallel, processing all words at once. Either way, it is typically an
expensive operation in practice, since phrase queries tend to contain stopwords.
Posting lists corresponding to stopwords, or other frequent words, are very long,
resulting in costly data transfer and poor memory locality.

To improve the performance of phrase queries, several methods have been
proposed that augment the inverted index by also indexing selected multi-word

B. Benatallah et al. (Eds.): WISE 2014, Part I, LNCS 8786, pp. 156–169, 2014.
© Springer International Publishing Switzerland 2014

sequences in addition to individual words. Those multi-word sequences can be selected, for instance, based on their frequency in a query workload or the document collection, whether they contain a stopword, or other syntactic criteria [21, 24]. While this typically improves performance, since phrase queries can now be processed by intersecting fewer shorter posting lists, it comes at the cost of additional index space and a potential need to select other multi-word sequences as the document collection and the query workload evolve.

One may argue that phrase query processing is just substring matching – a problem investigated intensively by the string processing community. Their solutions such as suffix arrays [16] and permuterm indexes [9]), while increasingly implemented at large scale, are not readily available in today's retrieval systems.

In this work we take a radically different approach. Rather than augmenting the inverted index, we leverage the *direct index* that modern retrieval systems [3, 18] maintain to generate snippets or compute proximity features. The direct index keeps for every document in the collection a compact representation of its contents – typically a list of integer term identifiers. We develop extensions of the established term-at-a-time (TAAT) and document-at-a-time (DAAT) query-processing methods that make effective combined use of the inverted index and the direct index. Our extensions build on a common cost model that captures the trade-off between expensive random accesses (e.g., used to locate a document in the direct index) and inexpensive sequential accesses (e.g., used to read a run of consecutive postings). Taking this trade-off into account, our extended term-at-a-time query processing regularly checks whether verifying the current set of candidates using the direct index is less expensive than reading the remaining posting lists from the inverted index. Likewise, our extended document-at-a-time query processing identifies a set of words upfront, so that intersecting the corresponding posting lists results in a candidate set small enough to be verified using the direct index.

Our extensions are efficient in practice, as demonstrated by our experimental evaluation. Using a corpus from The New York Times and ClueWeb09-B as document collections and query workloads capturing different use cases, we see that our extensions improve response time by a factor of 1.25–2.5 for short queries and 6–80 for long queries. Under the assumption that the direct index is already in place, which is a reasonable one for modern retrieval systems, this improvement comes with no additional index space.

Contributions made in this paper thus include: (i) a novel approach to process phrase queries leveraging the direct index available in modern retrieval systems and requiring no additional index space; (ii) two extensions of established query-processing methods that make effective combined use of the inverted index and the direct index; (iii) comprehensive experiments on two real-world document collections demonstrating the practical viability of our approach.

Organization. In Section 2 we present the technical background of our work. Section 3 then introduces our cost model and our extended query-processing methods. Our experimental evaluation is subject to Section 4. We give an overview of related work in Section 5. Our conclusion and possible future work are presented in Section 6.

Dictionary   **Inverted Index**   **Direct Index**

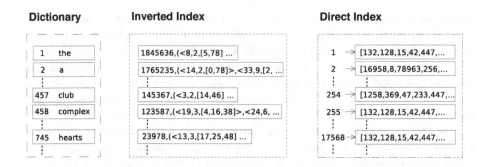

Fig. 1. Prototypic architecture of a modern retrieval system

2 Background

In this section, we introduce our notation and provide some technical background on the approach put forward in this work.

We let $\mathcal{D}$ denote the document collection. Documents therein are sequences of words from a vocabulary $\mathcal{V}$. For a word $v \in \mathcal{V}$ we let $df(v)$ denote its document frequency and $cf(v)$ denote its collection frequency. Likewise, $tf(v,d)$ refers to the term frequency of v in a document $d \in \mathcal{D}$. Phrase queries are also sequences of words from $\mathcal{V}$.

Given a phrase query $\mathbf{q} = \langle q_1, \ldots, q_{|\mathbf{q}|} \rangle$, our objective is to identify the set of all offsets in documents where the given word sequence occurs. Put differently, we only perform a phrase match – aspects of ranking documents based on relevance or proximity are orthogonal to our work.

Figure 1 depicts the prototypic architecture of a modern retrieval system [3, 18], as we assume it in this work, consisting of the following building blocks:

– a *dictionary* that maps words from the vocabulary $\mathcal{V}$ to integer term identifiers, typically implemented using a hash map, a front-coded list, or a B$^+$-tree. By assigning term identifiers in descending order of collection frequency, one can improve the compressibility of the other index structures. Besides term identifiers, the dictionary contains per-word collection statistics such as document frequencies and collection frequencies.
– an *inverted index* that maps term identifiers to posting lists containing information about the occurrences of the corresponding word in the document collection. In a positional inverted index, as used in this work, a posting

$$\langle d, tf(v,d), [o_1, \ldots, o_{tf(v,d)}] \rangle$$

consists of a document identifier d, the term frequency, and an array of offsets where the word v occurs. Note that the term frequency corresponds to the length of the array of offsets and is thus typically represented implicitly. The inverted index is the most important index structure used to match and score documents in response to a query.

- a *direct index* that maps document identifiers to a compact representation of the corresponding document's content. One possible representation is to use arrays of term identifiers, with a special term identifier to mark sentence boundaries. The direct index can be used, for instance, to generate result snippets and compute proximity features.

The concrete implementation of these building blocks depends on whether the index structures are kept in main memory or on secondary storage and whether they reside on a single machine or are spread over a cluster of machines.

Two established methods exist to process phrase queries [5] and other queries that can not profit from optimizations for ranked retrieval such as WAND [4]. Term-at-a-time (TAAT) query processing considers words from the query sequentially in ascending order of their document frequency. While doing so, it maintains a set of candidate documents that can still match the query. When considering a word from the query, the corresponding posting list is intersected with the current set of candidate documents. In the case of phrase queries, for each candidate document the set of offsets is maintained at which the query phrase can still occur. Over time, the set of candidate documents shrinks and query processing can be terminated early once no candidate documents remain. Document-at-a-time query processing (DAAT) considers words from the query in parallel, intersecting their corresponding posting lists. When the same document is seen in all posting lists, meaning that it contains all words from the query, offsets need to be inspected to see whether it contains the query phrase. Using skipping [17] the intersecting of posting lists can be greatly speeded up.

3   Phrase Queries with Inverted + Direct Indexes

In this section, we describe our cost model, characterizing the cost of using the index structures introduced above. Building on it, we develop extensions of established query-processing methods that make combined use of the inverted index and the direct index.

3.1   Cost Model

The cost of random accesses (RA) and sequential accesses (SA) differs substantially across different levels of the memory hierarchy and different storage systems. Thus, as recently reported by Wang et al. [23], on modern magnetic hard disk drives (HDDs) random accesses are 100–130x more expensive than sequential accesses, on modern solid state drives (SSDs) there is still a 2–3x difference, and even in main memory locality of reference matters.

Our cost model abstracts from hardware and implementation details (e.g., block size, cache size, and compression) as well as concrete data structures employed. Its objective is to capture the relative trade-off between random accesses and sequential accesses when using the dictionary, the inverted index, and the direct index. We let c_R denote the cost of a random access and c_S denote the cost of a sequential access. Our methods developed below are parameterized by the cost ratio c_R/c_S and seek to optimize the total cost

$$c_R \cdot \#R + c_S \cdot \#S$$

with $\#R$ ($\#S$) denoting the total number of random (sequential) accesses.

We assume the following costs for the index structures described in Section 2. Looking up the term identifier and associated collection statistics in the *dictionary* costs a single random access per word v from the query. Reading the entire posting list associated with word v from the *inverted index* incurs a single random access, to locate the posting list, and $df(v)$ sequential accesses to read all postings therein. On a HDD, for instance, assuming that the posting list is stored contiguously, this corresponds to seeking to its start followed by sequentially reading its (possibly compressed) contents. Finally, retrieving the compact representation of a document d from the *direct index* amounts to a single random access. We hence disregard (differences in) document lengths and assume that compact document representations are small enough to be fetched in a single operation.

3.2   Term-at-a-Time

Term-at-a-time query processing considers words from the query sequentially in ascending order of their document frequency. Our example phrase query "lucy in the sky with diamonds" from the introduction would thus be processed by considering its words in the order $\langle$ lucy, diamonds, sky, with, in, the $\rangle$. After having processed the first two words lucy and diamonds, the set of candidates consists of only those documents that contain both words at exactly the right distance. If this set is small enough, instead of reading the remaining posting lists, it can be beneficial to switch over to the direct index and verify which candidates contain the not-yet-considered words at the right offsets.

Let $\langle v_1, v_2, \ldots, v_n \rangle$ denote the sequence of distinct words from the phrase query q in ascending order of their document frequency (i.e., $df(v_i) \leq df(v_{i+1})$). After the first $k > 1$ words have been processed, we can compare the cost of continuing to use the inverted index against verifying the set of candidates using the direct index. More precisely, according to our cost model, the cost of continuing to use the inverted index is

$$c_R \cdot (n - k) + c_S \cdot \sum_{i=k+1}^{n} df(v_i) \,,$$

for accessing and reading the remaining $(n-k)$ posting lists. The cost of verifying all documents from the current candidate set $\mathcal{C}_k$ is

$$c_R \cdot |\mathcal{C}_k| \,.$$

Our extension of TAAT determines the two costs whenever a posting list has been completely read and switches over to verifying the remaining candidate documents using the direct index once this is beneficial.

3.3   Document-at-a-Time

Document-at-a-time query processing considers words from the query in parallel and does not (need to) maintain a set of candidate documents. This has two ramifications for our extension. First, we need to decide upfront which words from the query to process with the inverted index. Second, to do so, we have to rely on an estimate of the number of candidate documents left after processing a set of words from the query using the inverted index.

To make this more tangible, consider again our example phrase query "lucy in the sky with diamonds". We could, for instance, use the inverted index to process the words in $\{\,\text{lucy}, \text{diamonds}\,\}$ and verify the remaining candidate documents using the direct index. The cost for the former can be determined exactly. The cost for the latter depends on how many documents contain the two words at the right distance. Upfront, however, this number can only be estimated.

Let $\langle\, v_1, v_2, \ldots, v_n \,\rangle$ denote the sequence of distinct words from the phrase query $\mathbf{q}$, again in ascending order of their document frequency. Finding the cost-optimal subset I of words from the query to be processed using the inverted index can be cast into the following optimization problem

$$\operatorname*{arg\,min}_{I \subseteq \langle\, v_1, v_2, \ldots, v_n \,\rangle} \quad c_R \cdot |I| \;+\; c_S \cdot \sum_{v \in I} df(v) \;+\; c_R \cdot |\hat{\mathcal{C}}_I| \quad \text{s.t.} \quad I \neq \emptyset \,.$$

The first two summands capture the cost of processing the words from I using the inverted index; the last summand captures the estimated cost of verifying candidate documents using the direct index. We demand that at least one word is processed using the inverted index (i.e., $I \neq \emptyset$). Otherwise, all documents from the collection would have to be matched against the query phrase using the direct index, which is possible in principle but unlikely a preferable option. In the above formula, $|\hat{\mathcal{C}}_I|$ denotes an estimate of how many candidate documents are left after words from I have been processed using the inverted index.

Assuming that no additional collection statistics (e.g., about n-grams) are available, we estimate this cardinality as

$$|\hat{\mathcal{C}}_I| = |\mathcal{D}| \cdot \prod_{v \in I} \frac{df(v)}{|\mathcal{D}|} \,,$$

thus making an independence assumption about word occurrences in documents. More elaborate cardinality estimations, taking into account the order of words and their dependencies, would require additional statistics (e.g., about n-grams) consuming additional space.

It turns out that an optimal solution to our optimization problem can be determined efficiently. To this end, observe that $\{v_1, \ldots, v_k\}$ has minimal cost among all k-subsets of words from the query. This follows from a simple exchange argument: We can replace words in any k-subset by words from $\{v_1, \ldots, v_k\}$ having at most the same document frequency without increasing the overall cost. It is thus sufficient to determine the $\{v_1, \ldots, v_k\}$, consisting of the k rarest words from the query, that minimizes the overall cost. This can be done in time $\mathcal{O}(|\mathbf{q}| \log |\mathbf{q}|)$

– dominated by the cost of sorting words from the query.

Both our extensions ignore the cost for looking up in the dictionary information about the words from the query. This is a fixed cost which all query-processing methods have to pay. Further, when verifying a candidate document using the direct index, our extensions do not need to employ a string matching algorithm such as Knuth-Morris-Pratt [15], since they already know candidate offsets where the phrase can still occur.

4   Experimental Evaluation

We now present experiments evaluating the approach put forward in this work.

4.1   Setup and Datasets

Document Collections. We use two publicly available real-world document collections for our experiments: (i) The New York Times Annotated Corpus (NYT) [2], which consists of more than 1.8 million newspapers articles from the period 1987-2007 and (ii) ClueWeb09-B (CW) [1] as a collection of about 50 million English web documents crawled in 2009. Table 1 reports detailed characteristics of the two document collections.

Table 1. Document collection characteristics

	NYT	CW
# Documents	1,831,109	50,221,915
# Term occurrences	1,113,542,501	23,208,133,648
# Terms	1,833,817	51,322,342
# Sentences	49,622,213	1,112,364,572
Size of inverted index (GBytes)	2.98	60.21
Size of direct index (GBytes)	2.22	48.65

Workloads. To cover diverse use cases, we use four different workloads per document collection: (i) a query log from the MSN search engine released in 2006 (MSN), (ii) a subset of the aforementioned query log that only contains explicit phrase queries (i.e., those put into quotes by users) (MSNP), (iii) entity labels as captured in the `rdfs:label` relation of the YAGO2 knowledge base [14] (YAGO), (iv) a randomly selected subset of one million sentences from the respective document collection (NYTS/CWS). While the first two workloads capture web search as a use case, the YAGO workload mimics entity-oriented search, assuming that entities are retrieved based on their known labels. The NYTS/CWS workloads capture plagiarism detection as a use case, assuming that exact replicas of text fragments need to be retrieved. Queries in the first three workloads are short consisting of about three words on average; queries in the sentence-based workloads are –by design– longer, consisting of about twenty words on average. For each workload, we consider a randomly selected subset of 7,500 distinct queries. Table 2 reports more detailed characteristics of the considered workloads.

Table 2. Workload characteristics

	# Queries	# Distinct Queries	ø Query Length
YAGO	5,720,063	4,599,745	2.45
MSN	10,428,651	5,627,838	3.58
MSNP	131,857	105,825	3.37
NYTS	1,000,000	970,051	21.66
CWS	1,000,000	929,607	20.55

Methods under comparison are the following: (i) TaaT-I and (ii) DaaT-I as the established query-processing methods that solely rely on the inverted index, (iii) TaaT-I+D and (iv) DaaT-I+D as our extensions that also use the direct index. The latter two are parameterized by the cost ratio c_R/c_S, which we choose from $\{10, 100, 1000, 10{,}000\}$ and report with the method name. Thus, when we refer to TaaT-I+D(100), the cost ratio is set as $c_R/c_S = 100$.

Implementation. All our methods have been implemented in Java. Experiments are conducted on a a server-class machine having 512 GB of main memory, four Intel Xeon E7-4870 10-core CPUs, running Debian GNU/Linux as an operating system, and using Oracle JVM 1.7.0_25. The direct index keeps for every document from the collection an array of integer term identifiers. The inverted index keeps for every word a posting list organized as three aligned arrays containing document identifiers (in ascending order), term frequencies, and offsets. With this implementation, skipping in posting lists is implemented using galloping search. When determining response times of a method for a specific query, we execute the query once to warm caches, and determine the average response time based on three subsequent executions. Here, response time is the time that passes between receiving the query and returning the set of offsets where the phrase occurs in the document collection. To shield us from distortion due to garbage-collection pauses, we use the concurrent mark-sweep garbage collector.

4.2   Experimental Results

We compare TaaT-I+D and DaaT-I+D against their respective baseline TaaT-I and DaaT-I in terms of query response times. To determine the response time of a method for a specific query, we execute it once to warm caches, and determine the average response time based on three subsequent executions. Here, all 7,500 queries from each workload are taken into account, including those that do not return results. Table 3 reports average response times in milliseconds. Improvements of our extensions over their respective baseline reported therein are statistically significant ($p < 0.01$), as measured by a paired Student's t-test.

We observe that TAAT-I+D and DAAT-I+D perform significantly better than their respective baseline. For $c_R/c_S = 1,000$, as a concrete setting, our extensions improve response time by 1.25–2.5× for the workloads consisting of relatively short queries. For the verbose workloads consisting of sentences from the respective document collection, our extensions improve response time by 6–80×. Across both document collections we see slightly smaller improvements for DAAT-I+D. This is expected given that DAAT-I is the stronger among the baselines and DAAT-I+D has to decide upfront which words to process with which index. The performance of DAAT-I+D thus depends critically on the accuracy of cardinality estimation.

To see when our extensions are particularly effective, consider the phrase query "the great library of alexandria" from the YAGO workload. On the NYT document collection, the response time for this query drops from 444.7 ms with TAAT-I to only 5.3 ms with TAAT-I+D(1,000) and from 12.3 ms with DAAT-I to 1.3 ms with DAAT-I+D(1,000). Both TAAT-I and DAAT-I read the entire posting lists for all words from the query, resulting in a total of 3,743,120 sequential accesses. With TAAT-I+D(1,000), though, after processing the posting lists for alexandria and library, thereby making only 43,148 sequential accesses, we are left with only 65 candidates, which can quickly be verified using the forward index. Based on its cardinality estimate, DAAT-I+D(1,000) also selects the set { alexandria, library }, resulting in the same cost. However, it is difficult for our extensions to improve response times of short queries whose words have similar document frequencies. In these cases, TAAT-I+D(1,000) and DAAT-I+D(1,000) fall back onto using only the inverted index, since it is not beneficial, according to our cost model, to switch over to the forward index. For the phrase query "sky eye" on the NYT document collection, we have document frequencies $df(\text{sky}) = 35{,}798$ and $df(\text{eye}) = 80{,}830$. After processing the term sky, only 35,798 candidates are left, so that according to our cost model it is better to process the term eye using the inverted index. This results in 80,830 sequential accesses, which is substantially less expensive than performing the 35,798 random accesses required to verify candidates.

We also investigate where our gains come from and whether our extensions perform particularly well for shorter or longer queries. Figure 2 and Figure 3 show average response times by query length for NYT and CW as observed on the different workloads. Comparing the two baselines, we observe that query response time tends to increases with query length for both, but stagnates at some point for TAAT-I. This is natural since TAAT-I can often terminate query processing early once no candidates are left, whereas DAAT-I has to continue reading all posting lists. On the sentence-based workloads, in which every query

Table 3. Query response time (in ms)

	NYT				CW			
	YAGO	MSN	MSNP	NYTS	YAGO	MSN	MSNP	CWS
TaaT-I	14.13	10.94	21.87	810.01	1,767.02	787.84	2,428.85	16,962.03
TaaT-I+D(10)	4.64	4.47	9.26	12.95	1,498.60	1,948.41	3,649.31	4,067.63
TaaT-I+D(100)	4.51	4.11	8.56	10.76	860.68	915.58	2,048.24	3,161.38
TaaT-I+D(1,000)	5.76	4.77	9.89	10.10	676.23	497.94	1,557.80	1,919.78
TaaT-I+D(10,000)	7.32	5.73	11.95	11.47	699.19	480.97	1,389.16	1,500.35
DaaT-I	2.85	3.22	6.62	38.47	55.52	72.92	138.20	446.78
DaaT-I+D(10)	1.24	1.03	2.05	2.93	194.07	268.93	470.53	297.02
DaaT-I+D(100)	1.43	0.97	2.38	3.60	51.33	57.23	95.48	68.52
DaaT-I+D(1,000)	1.82	1.25	3.24	4.77	40.22	35.24	79.16	71.17
DaaT-I+D(10,000)	2.13	1.60	4.03	6.56	44.55	37.08	94.01	87.71

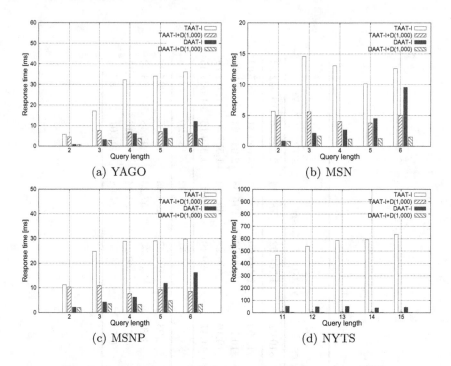

Fig. 2. Query response time by query length (in ms) on NYT

has by design at least one result, query response times of TAAT-I keep increasing with query length. On both document collections and all workloads, our extensions improve query response time consistently across different query lengths. Only when comparing DAAT-I and DAAT-I+D we observe a widening gap. As explained before, query response times increase with query length for DAAT-I, whereas our extension DAAT-I+D manages to keep them relatively constant.

Summary

Our extensions TAAT-I+D and DAAT-I+D significantly improve query response time across different query lengths, as demonstrated by our experiments on two document collections and four different workloads. It is worth emphasizing once more that our extensions solely rely on the inverted index and direct index, which are available in modern retrieval system anyway, so that their improvement comes with no additional index space.

5   Related Work

The use of phrases dates back to the early days of Information Retrieval. Since indexing all possible phrases is prohibitively expensive, researchers initially focused on identifying salient phrases in documents to index. Salton et al. [19],

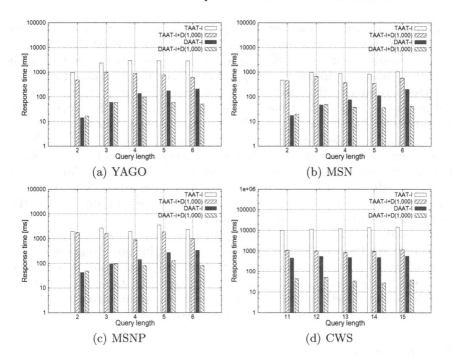

Fig. 3. Query response time by query length (in ms) on CW

Fagan [8], and Gutwin et al. [11] develop methods to this end based on statistical and syntactical analysis. With those methods, one could thus look up documents that contain one of the selected salient phrases.

To support arbitrary phrase queries, more recent research has looked into systematically augmenting the inverted index by selected multi-word sequences in addition to individual words. Williams et al. [24] develop the *nextword index*, which indexes bigrams containing a stopword in addition to individual words. They also discuss how to make effective combined use of the nextword index, a partial phrase index, which caches the phrases most often occurring in the query workload, and a standard inverted index. Chang and Po [6] describe an approach to select variable-length multi-word sequences taking into account part-of-speech information, frequency in the document collection, and frequency in a query workload. Transier and Sanders [21] present a method that works "out of the box" without any knowledge about the query workload. Their approach selects bigrams that are beneficial to keep because intersecting the posting lists of their constituting words is expensive. One shortcoming of all of these approaches is that they require additional index space. Our method, in contrast, leverages the direct index, which is anyway available in modern retrieval systems. Interestingly, Transier and Sanders [21] describe a somewhat similar approach that uses a non-positional inverted index in combination with the direct direct index. Issuing a conjunctive Boolean query against the inverted index they identify candidate documents. Those are then fetched from the direct direct index and matched against the phrase query using the Knuth-Morris-Pratt algorithm [15]. Our approach, in contrast, foregoes reading all posting lists and verifies candidates by probing specific offsets as opposed to performing a phrase match.

Other related yet orthogonal work has looked into how positional inverted indexes can be improved for specific use cases. Shan et al [20] propose a *flat position index* for phrase querying. It views the whole document collection as a single word sequence, so that posting lists consist of offsets only but do not contain any document identifiers. Only when an occurrence of the query phrase is found, the corresponding document identifier is reconstructed from a separate index. He and Suel [13] develop a positional inverted index for versioned document collections (e.g., web archives), exploiting the typically large overlap between consecutive versions of the same document.

It is worth mentioning that, at its core, phrase querying is a substring matching problem, which has been studied intensively by the string processing community. State-of-the-art solutions such as suffix arrays [16] and permuterm indexes [9] were originally designed to work in main memory. Scaling such data structures to large-scale document collections as we consider them is an exciting ongoing direction of research with good progress [7, 10, 22] in the recent past.

6 Conclusion

We have proposed extensions of the established term-at-a-time and document-at-a-time query processing methods that make combined use of the inverted index and the direct index. Given that modern retrieval systems keep a direct index for snippet generation and computing proximity features, our extensions do not require additional index space. Our experiments show that these extensions outperform their respective baseline when the cost ratio parameter is set appropriately. We observed substantial improvements in response time in particular for verbose phrase queries consisting of many words. We believe that this makes our approach relevant for applications, such as entity-oriented search or plagiarism detection, that rely on machine-generated verbose phrase queries.

As part of our future research, we plan to (i) investigate more elaborate cardinality estimation techniques and their effect on query-processing performance, (ii) refine our cost model to take into account the effect of skipping and other optimizations, (iii) adapt our approach to also handle more expressive queries (e.g., including wildcards and/or proximity constraints).

References

[1] The ClueWeb09 Dataset, http://lemurproject.org/clueweb09/
[2] The New York Times Annotated Corpus, http://corpus.nytimes.com/
[3] Brin, S., Page, L.: The anatomy of a large-scale hypertextual web search engine. Computer Networks 30(1-7), 107–117 (1998)
[4] Broder, A.Z., Carmel, D., Herscovici, M., Soffer, A., Zien, J.Y.: Efficient query evaluation using a two-level retrieval process. In: CIKM 2003 (2003)

[5] Büttcher, S., Clarke, C., Cormack, G.V.: Information Retrieval: Implementing and Evaluating Search Engines. MIT Press (2010)

[6] Chang, M., Poon, C.K.: Efficient phrase querying with common phrase index. Inf. Process. Manage. 44(2), 756–769 (2008)

[7] Culpepper, J.S., Petri, M., Scholer, F.: Efficient in-memory top-k document retrieval. In: SIGIR 2012 (2012)

[8] Fagan, J.: Automatic phrase indexing for document retrieval. In: SIGIR 1987 (1987)

[9] Ferragina, P., Venturini, R.: The compressed permuterm index. ACM Transactions on Algorithms 7(1) (2010)

[10] Gog, S., Moffat, A., Culpepper, J.S., Turpin, A., Wirth, A.: Large-scale pattern search using reduced-space on-disk suffix arrays. CoRR abs/1303.6481 (2013)

[11] Gutwin, C., Paynter, G., Witten, I., Nevill-Manning, C., Frank, E.: Improving browsing in digital libraries with keyphrase indexes. Decision Support Systems 27(1-2), 81–104 (1999)

[12] Hagen, M., Potthast, M., Beyer, A., Stein, B.: Towards optimum query segmentation: in doubt without. In: CIKM 2012 (2012)

[13] He, J., Suel, T.: Optimizing positional index structures for versioned document collections. In: SIGIR 2012 (2012)

[14] Hoffart, J., Suchanek, F.M., Berberich, K., Weikum, G.: Yago2: A spatially and temporally enhanced knowledge base from wikipedia. Artif. Intell. 194, 28–61 (2013)

[15] Knuth, D., Morris, J. J., Pratt, V.: Fast pattern matching in strings. SIAM Journal on Computing 6(2), 323–350 (1977)

[16] Manber, U., Myers, E.W.: Suffix arrays: A new method for on-line string searches. SIAM J. Comput. 22(5), 935–948 (1993)

[17] Moffat, A., Zobel, J.: Self-indexing inverted files for fast text retrieval. ACM Trans. Inf. Syst. 14(4), 349–379 (1996)

[18] Ounis, I., Amati, G., Plachouras, V., He, B., Macdonald, C., Johnson, D.: Terrier information retrieval platform. In: Losada, D.E., Fernández-Luna, J.M. (eds.) ECIR 2005. LNCS, vol. 3408, pp. 517–519. Springer, Heidelberg (2005)

[19] Salton, G., Yang, C.S., Yu, C.T.: A theory of term importance in automatic text analysis. Journal of the American Society for Information Science 26(1), 33–44 (1975)

[20] Shan, D., Zhao, W.X., He, J., Yan, R., Yan, H., Li, X.: Efficient phrase querying with flat position index. In: CIKM 2011 (2011)

[21] Transier, F., Sanders, P.: Out of the box phrase indexing. In: Amir, A., Turpin, A., Moffat, A. (eds.) SPIRE 2008. LNCS, vol. 5280, pp. 200–211. Springer, Heidelberg (2008)

[22] Vigna, S.: Quasi-succinct indices. In: WSDM 2013 (2013)

[23] Wang, J., Lo, E., Yiu, M.L., Tong, J., Wang, G., Liu, X.: The impact of solid state drive on search engine cache management. In: SIGIR 2013 (2013)

[24] Williams, H.E., Zobel, J., Bahle, D.: Fast phrase querying with combined indexes. ACM Trans. Inf. Syst. 22(4), 573–594 (2004)

[25] Zobel, J., Moffat, A.: Inverted files for text search engines. ACM Comput. Surv. 38(2) (2006)

Ranking Based Activity Trajectory Search

Wei Chen[1], Lei Zhao[1,2], Xu Jiajie[1,2], Kai Zheng[3], and Xiaofang Zhou[1,3]

[1] School of Computer Science and Technology, Soochow University, China
[2] Jiangsu Provincial Key Laboratory for Computer Information Processing Technology, Soochow University, China
[3] School of ITEE, The University of Queensland, Brisbane, Australia
wchzhg@gmail.com, {zhaol,xujj}@suda.edu.cn,
{kevinz,zxf}@itee.uq.edu.au

Abstract. With the proliferation of the GPS-enabled devices and mobile techniques, there has been a lot of work on trajectory search in the last decade. Previous trajectory search has focused on spatio-temporal features and text descriptions. Different from them, we study a novel problem of searching trajectories with activities and corresponding ranking information. Given a query q, which is attached with a set of activities and a threshold of distance, the results of ranking based activity trajectory search (RTS) are k trajectories such that the given activities are performed with the highest ranking within the threshold of distance. In addition, we also extend the query with an order, i.e., order-sensitive ranking based activity trajectory search (ORTS), which takes both the order of activities in a query q and the order of trajectories into account. It is challenging to answer RTS and ORTS efficiently due to the structural complexity of trajectory data with ranking information. In this paper, a hybrid index AC-tree and its optimized variant RAC-tree are proposed to achieve higher efficiency. Extensive experiments verify the high efficiency and scalability of the proposed algorithms.

Keywords: Trajectory Search, Ranking, Activity Trajectory.

1 Introduction

As the ubiquitousness of devices with GPS and the rapid development of wireless sensor technology, more and more people log their locations nowadays. In addition, the share of trajectories become available on more and more web sites, such as Twitter, Four-square, Facebook and Bikely, which leads to an incredible increasing of trajectory number. Trajectory search has become a popular concern for industry and academic community in the last decade. Besides, many trajectories based applications are also becoming increasingly popular, such as trip planning and trajectory recommendation.

In recent work on trajectory search, a trajectory is usually modeled as a sequence of geo-locations with keywords information. These work focuses on minimizing the distance while all keywords of a query are covered. However, this is not always reasonable. Assuming that a user wants to do some shopping and

B. Benatallah et al. (Eds.): WISE 2014, Part I, LNCS 8786, pp. 170–185, 2014.
© Springer International Publishing Switzerland 2014

have a haircut after work, conventional query will return k trajectories in which the user can fulfill his plan. However the barbershops and shopping malls in these trajectories may have a bad reputation. Having observed the weakness of previous studies, we propose a new query of searching activity trajectories with ranking information.

Consider the example demonstrated in Fig. 1. A tourist wants to conduct activities (a, c, d) on her/his trip and the total distance of trip cannot exceed 20. If the ranking information is beyond consideration, τ_1 will be the best result since the trajectory $(q, p_{1,2}, p_{1,3}, p_{1,4})$ has the minimum distance. However, τ_2 is a better choice in a sense because the rankings of activities (a, c, d) in τ_2 are higher than that in τ_1.

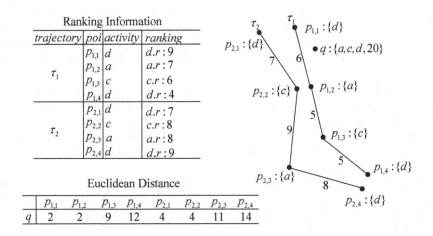

Ranking Information

trajectory	poi	activity	ranking
τ_1	$p_{1,1}$	d	$d.r:9$
	$p_{1,2}$	a	$a.r:7$
	$p_{1,3}$	c	$c.r:6$
	$p_{1,4}$	d	$d.r:4$
τ_2	$p_{2,1}$	d	$d.r:7$
	$p_{2,2}$	c	$c.r:8$
	$p_{2,3}$	a	$a.r:8$
	$p_{2,4}$	d	$d.r:9$

Euclidean Distance

	$p_{1,1}$	$p_{1,2}$	$p_{1,3}$	$p_{1,4}$	$p_{2,1}$	$p_{2,2}$	$p_{2,3}$	$p_{2,4}$
q	2	2	9	12	4	4	11	14

Fig. 1. An Example of Trajectory Matching

The query proposed in this paper is more challenging compared with the previous work. On one hand, indexing trajectory data becomes more difficult since extra ranking information should be taken into account. On the other hand, computing is becoming increasingly complex since more factors should be taken into consideration. Obviously, existing methods cannot be applied to this work directly.

In order to tackle the problem efficiently, we propose new indexes and algorithms in this paper. Firstly, a new hybrid index called AC-tree is proposed. We prune the search space fast and obtain a candidate set by retrieving AC-tree. Secondly, new algorithms are proposed to compute the maximum ranking within a threshold of distance for each candidate trajectory. Thirdly, optimized index and algorithms are proposed to make computing faster. To sum up, the main contribution of this work is to take ranking information into account while recommending trajectories. It makes the results more reasonable.

The rest of this paper is organized as follows. In Section 2, we briefly view existing work related to trajectory search. Section 3 presents the problem statement and necessary notations in this work. We introduce the index tree in Section 4 and algorithms in Section 5, which is followed by the optimization of index and algorithms in Section 6. In Section 7 we report the experimental results. This paper is concluded in Section 8.

2   Related Work

In the past decade, trajectory search has received significant attentions. Existing work [7,6,4,5] focus on searching trajectories in spatial-temporal domain without any other important features, such as, textual descriptions and activity information. [3] tackles the problem of searching trajectories by locations in spatial domain, in which multiple locations are used as the query and Euclidean distance is the only restriction. Chen et al. [5] investigate the problem of discovering popular routes from historical trajectories by given a location to a destination, in which a transfer network is established to derive the transfer probability for transfer nodes.

In existing work, [21,15,20,10,8] study the problem of predicting destinations based on trajectory data. [8] handle the problem of destination prediction by sub-trajectory synthesis, where historical trajectories are decomposed into sub-trajectories to avoid the data sparsity problem. A grid graph is proposed to represent the trajectories and a Bayesian inference framework is proposed to compute the probability in this work.

Meanwhile, with the prevalence of spatial web objects on the Internet, a lot of work appears in spatial keyword search. Existing work [14,16,18,19,1,13,22] address the problem of spatial keywords search in spatial-temporal and textual domain. Especially, [1,13] consider the fusion of keywords and trajectories. In [1], keywords are associated with the whole trajectory rather than each individual point. In addition, privacy protection is another important issue while searching trajectories [12,11,9,24,23,8] due to the increasing importance of protecting users from information leak.

[13] is the most similar work to our query, where each point in a trajectory is attached with keywords. Multiple locations with keywords are used as the query to search for trajectories that has the minimum match distance with respect to the query. A novel grid index is developed to organize trajectory data and new algorithms are also proposed to the tackle the problem efficiently. Despite the great contributions made by [13], it does not take the ranking information into account during processing. Hence, both the index structure and the algorithms of [13] are not suitable to our problem.

In order to address spatial keywords search efficiently, some hybrid index structures are proposed. In [17], a new index called IR^2-tree is proposed to prune search space in spatial and textual domain. Cong et al. [2] propose an IR-tree, which is an integration of R-tree and inverted files. By traversing IR-tree, the search space can be pruned fast with location and textual information. Zheng et al. [13] propose a novel Grid index called GAT, which utilizes both spatial information and query keywords to reduce search space.

In spite of the significant contributions of the aforementioned work, none of them take ranking information into consideration, which is an important new feature of trajectories. Meanwhile, the hybrid indexes and the corresponding algorithms are also not suitable to our search. As a result, we propose novel indexes and algorithms in this paper.

3 Problem Statement

In this section, we present all the definitions used throughout the paper. Before that, the notations used in this paper are summarized in Table 1.

Table 1. Definitions of notations

Notation	Definition
α	Activity
$\alpha.r$	Ranking of α and the value is in the range of [1-10]
τ	Trajectory
$\mathcal{D}$	Set of τ
$\hat{d}$	The threshold of distance
$p.\varphi$	The set of activities of POI p
ω	Trajectory matching in the form of (τ, q)
ω^o	Order-sensitive trajectory matching
$r(\omega)$	Ranking of a trajectory matching ω
$d(\omega)$	Distance of a trajectory matching ω

Definition 1. *Trajectory.* *Let $p = (x, y, \varphi, r)$ be a POI where x is the longitude, y is the latitude, φ denotes the activities that can be performed in the location (x, y), and r is a set of ranking information of φ. A trajectory is a sequence of POIs, denoted as $\tau = (p_1, p_2, \ldots, p_n)$.*

Definition 2. *Sub-trajectory.* *Given a trajectory $\tau = (p_1, p_2, \ldots, p_n)$, a sub-trajectory of τ is $\tau_s^e = (p_s, p_{s+1}, \ldots, p_e)$, where $1 \leq s \leq e \leq n$, denoted as $\tau_s^e \subseteq \tau$.*

Definition 3. *Trajectory matching.* *Let $q = (x, y, \varphi)$ be a query, where (x, y) is the start point of a trip and φ is a set of activities needed to be performed. Given a trajectory τ, a tuple $\omega = (\tau, q)$ is a trajectory matching if there exists a sub-trajectory $\tau_s^e \subseteq \tau$, such that $q.\varphi \subseteq \bigcup_{p_i \in \tau_s^e} p_i.\varphi$.*

Definition 4. *Distance between trajectory and query.* *Given a trajectory $\tau = (p_1, p_2, \ldots, p_n)$ and a query q, the distance between τ and q is*

$$d(\tau, q) = dis(q, p_1) + \sum_{i=1}^{n-1} dis(p_i, p_{i+1}), \qquad (1)$$

where dis is to get the Euclidean distance between two locations.

Definition 5. *Ranking of trajectory matching.* *Given a trajectory matching* $\omega = (\tau, q)$ *and a threshold of distance* d, *suppose* τ *has* n *sub-trajectories* τ_1, τ_2, ..., τ_n, *such that each* $\omega_i = (\tau_i, q)(1 \leqslant i \leqslant n)$ *is a trajectory matching, then*

$$r(\omega, \widehat{d}) = \max_{d(\omega_i) \leqslant \widehat{d}} (\sum_{\alpha \in q.\varphi} \max_{p_j \in \tau_i} (p_j.\alpha.r)), \qquad (2)$$

is the ranking of the trajectory matching ω.

In Fig. 1, $r((\tau_1, q), \widehat{d}) = r(((p_{1,1}, p_{1,2}, p_{1,3}), q), 20) = 22$. Although the sub-trajectory $(p_{1,2}, p_{1,3}, p_{1,4})$ also covers the activity set of q, its ranking is less than 22. Meanwhile, $r((\tau_2, q), \widehat{d}) = r(((p_{2,1}, p_{2,2}, p_{2,3}), q), 20) = 23$. Although the ranking of sub-trajectory $(p_{1,2}, p_{1,3}, p_{1,4})$ is higher, its distance is larger than 20.

Given a set of trajectories $\mathcal{D}$ and a query q, let $\mathcal{C}(\mathcal{C} \subseteq \mathcal{D})$ be a candidate which means for any trajectory $\tau \in \mathcal{C}$, it follows that (τ, q) is a trajectory matching. Given a positive integer k and a threshold of distance $\widehat{d}$, the *Ranking Based Activity Trajectory Search(RTS)* returns a set $\mathcal{R}(\mathcal{R} \subseteq \mathcal{C}, |\mathcal{R}| = k)$, such that $\forall \tau \in \mathcal{R}$ and $\forall \tau' \in \mathcal{C} - \mathcal{R}$ it follows that $r((\tau, q), \widehat{d}) \geq r((\tau', q), \widehat{d})$.

4   AC-Tree

Due to the uneven distribution of POIs in practice, the search regions are partitioned into different grid cells. If the number of POIs attached to a cell exceeds a threshold θ, the cell should be partitioned into four components. An AC-tree is proposed to organize all cells. Each cell corresponds to a leaf node of the AC-tree.

Non-Leaf Nodes. Each non-leaf node of the AC-tree is used to store the coordinates of four vertices of the cell and the pointers to their child nodes. If one cell is divided and the number of POIs of its i-th is 0, then the pointer to i-th child is null.

Leaf Nodes. Each leaf node stores the activities of some POIs which belong to some certain trajectories and fall in the range of the cell corresponding to the leaf node. In Fig. 2(a), trajectory τ_1 contains c and d, τ_3 contains a and c in cell 52. As a result, leaf node 52 should maintain a list to store this information.

Fig. 2(c) shows an example of an AC-tree for which the threshold $\theta = 2$. For the sake of convenience, the cell_id of the i-th sub-cell of current cell is defined as $4 \times current\_cell\_id + i$, $dis(p, q)$ is used to denotes the minimum distance between query q and any location (x, y) in cell p.

5   Algorithms of RTS and ORTS

To the best of our knowledge, there is no previous work on trajectory search considering activities, spatial and ranking information simultaneously. Given a query q and a threshold of distance $\widehat{d}$, the method of RTS consists of two steps.

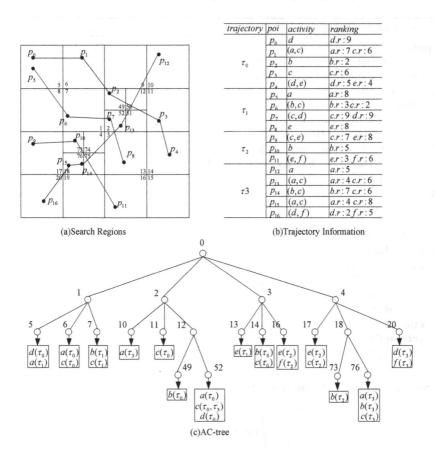

trajectory	poi	activity	ranking
τ_0	p_0	d	$d.r:9$
	p_1	(a,c)	$a.r:7\,c.r:6$
	p_2	b	$b.r:2$
	p_3	c	$c.r:6$
τ_1	p_4	(d,e)	$d.r:5\,e.r:4$
	p_5	a	$a.r:8$
	p_6	(b,c)	$b.r:3\,c.r:2$
	p_7	(c,d)	$c.r:9\,d.r:9$
	p_8	e	$e.r:8$
τ_2	p_9	(c,e)	$c.r:7\,e.r:8$
	p_{10}	b	$b.r:5$
	p_{11}	(e,f)	$e.r:3\,f.r:6$
$\tau3$	p_{12}	a	$a.r:5$
	p_{13}	(a,c)	$a.r:4\,c.r:6$
	p_{14}	(b,c)	$b.r:7\,c.r:6$
	p_{15}	(a,c)	$a.r:4\,c.r:8$
	p_{16}	(d,f)	$d.r:2\,f.r:5$

(a)Search Regions   (b)Trajectory Information

(c)AC-tree

Fig. 2. Details of AC-tree

Firstly, traversing the AC-tree to get a candidate set $\mathcal{C}$, in which all trajectories match query q. Secondly, computing the ranking of trajectory matching for each candidate trajectory. Baseline algorithm is proposed to prune the search space and get a candidate set, the detail information is presented as follows.

5.1   Traversing Index Tree

In this section, a new algorithm is developed to prune search space and get a candidate set $\mathcal{C}$. We commence this part by traversing AC-tree beginning at the root node with breadth-first strategy. The details are illustrated in algorithm 1.

Given a query q and a threshold $\widehat{d}$, assuming $q.\varphi = \{\alpha_1, \alpha_2, \ldots, \alpha_m\}$. The retrieval of candidate trajectory consists of two steps.

Algorithm 1. Traversing Index Tree

Input: query q, $\widehat{d}$, AC-tree tr
Output: candidate set $\mathcal{C}$
 1: Inserting the root node of tr into l
 2: $\mathcal{M}[*,*] \leftarrow 0$
 3: **while** $l \neq \phi$ **do**
 4:     $p \leftarrow$ the first entry of l
 5:     **if** $dis(p,q) \leq \widehat{d}$ **then**
 6:         **if** p is a $non-leaf$ $node$ **then**
 7:             Insert all child nodes of p into l;
 8:         **else**
 9:             Update matrix $\mathcal{M}$;
10:         **end if**
11:     **end if**
12:     Remove the first entry of l;
13: **end while**
14: **for** $j \leftarrow 1$ to $|\mathcal{D}|$ **do**
15:     **if** all $\mathcal{M}[i,j] = 1$ $(i \in [1, |q.\varphi|])$ **then**
16:         put trajectory τ_j into $\mathcal{C}$;
17:     **end if**
18: **end for**
19: **return** $\mathcal{C}$;

Step 1: Compute the matrix $\mathcal{M}$:

$$
\mathcal{M} = \begin{pmatrix}
p_{1,1} & p_{1,2} & \cdots & p_{1,n} \\
p_{2,1} & p_{2,2} & \cdots & p_{2,n} \\
\vdots & \vdots & & \vdots \\
p_{m,1} & p_{m,2} & & p_{m,n}
\end{pmatrix}
$$

where
$$
p_{i,j} = \begin{cases} 1, & if\ trajectory\ j\ contains\ query\ activity\ \alpha_i \\ 0, & otherwise \end{cases}
$$

In step 1, the algorithm maintains a FIFO queue l to store the nodes that should be visited. At the beginning, inserting the root node of AC-tree into l and matrix $\mathcal{M}$ is initialized to be zero. For each loop, if the minimum distance between a node p and query q is smaller than $\widehat{d}$, we insert the child nodes of p into list l, or update matrix $\mathcal{M}$ based on the list attached to p. In addition, if the distance is larger than $\widehat{d}$, there is no need to retrieve this region.

Step 2: Line 14 to 19 get a candidate set by processing matrix $\mathcal{M}$. For each trajectory τ in $\mathcal{D}$, it will be a candidate if τ contains all activities of $q.\varphi$.

5.2   Computing the Ranking of Trajectory Matching

For each trajectory τ in $\mathcal{C}$, a straightforward way to compute $r(\omega, \widehat{d})$ is to find out all possible sub-trajectories that match query q and compute $r(\omega_i, \widehat{d})$, then the algorithm returns the maximum $r(\omega_i, \widehat{d})$ as the result. However, the computation complexity of this method is too high. In the next paper, a more efficient algorithm is proposed to resolve the problem.

Given a query q, a threshold $\widehat{d}$ and a candidate trajectory $\tau = (p_1, p_2, \ldots, p_n)$, the details of computing $r(\omega, \widehat{d})$ is illustrated in algorithm 2 which has two steps.

Algorithm 2. Computing the Ranking of Trajectory Matching

Input: query q, trajectory τ
Output: $r(\omega, \widehat{d})$
1: create a link list l;
2: $vec \leftarrow 0, r(\omega, \widehat{d}) \leftarrow 0, s \leftarrow 1, e \leftarrow 1$;
3: **while** $e \leq l.length$ **do**
4:     **if** $d(\tau_s^e, q) \leq \widehat{d}$ **then**
5:         **for** $each\ \alpha \in p_e.\varphi \wedge q.\varphi$ **do**
6:             $vec[\alpha] + +$;
7:         **end for**
8:     **else**
9:         **if** $all\ vec[\alpha] > 0(\alpha \in q.\varphi)$ **then**
10:             compute $r((\tau_s^{e-1}, q), \widehat{d})$ based on Eq.(2);
11:             **if** $r((\tau_s^{e-1}, q), \widehat{d}) > r(\omega, \widehat{d})$ **then**
12:                 $r(\omega, \widehat{d}) \leftarrow r((\tau_s^{e-1}, q), \widehat{d})$;
13:             **end if**
14:         **end if**
15:         **for** $each\ \alpha \in p_s.\varphi \wedge q.\varphi$ **do**
16:             $vec[\alpha] - -$;
17:         **end for**
18:         $s \leftarrow s + 1$;
19:         continue;
20:     **end if**
21:     $e \leftarrow e + 1$;
22: **end while**
23: **return** $r(\omega, \widehat{d})$;

Step 1: For each p_i in trajectory τ, p_i is inserted into a link list l if $p_i.\varphi \wedge q.\varphi \neq \phi$ and $dis(q, p_i) \leq \widehat{d}$. Compared with using the whole trajectory τ, computation cost is reduced with l since less POIs are taken into account.

Step 2: In this section, a vector vec is used to keep tracking the number of occurrences of activity $\alpha(\alpha \in q.\varphi)$ in current sub-trajectory. For each loop, the algorithm updates vec or computes $r((\tau_s^{e-1}, q), \widehat{d})$ if necessary according to the distance between τ_s^e and q.

5.3   Computing the Ranking in Order-Sensitive Situation

In many real applications, users may want to perform their activities with an order. For instance, a worker plans to go shopping after having a haircut, i.e., the order of activities is *haircut* → *shopping*. Obviously, RTS is not applicable in this scenario.

Definition 6. *Order-Sensitive Trajectory Matching.* *Given a trajectory matching* $\omega = (\tau, q)$ *and a threshold of distance* $\widehat{d}$, $\omega = (\tau, q)$ *is called an order-sensitive trajectory matching, denoted as* $\omega^o = (\tau, q)^o$, *on condition that existing* $\omega_i = (\tau_i, q)$ *and for any pair of query activities*(q_i, q_j, $i < j$) *of* $q.\varphi$ *there exists* p_m *and* $p_n(m \leq n)$ *of* τ_i, *such that* $q_i \in p_m.\varphi$, $q_j \in p_n.\varphi$. *The ranking of* $\omega^o = (\tau, q)^o$ *is defined as:*

$$r(\omega^o, \widehat{d}) = \max_{d(\omega_i^o) \leqslant \widehat{d}} (\sum_{\alpha \in q.\varphi} \max_{p_j \in \tau_i}(p_j.\alpha.r)),$$

where $d(\omega_i^o) = d(\omega_i)$.

The *Order-sensitive Ranking Based Activity Trajectory Search(ORTS)* returns k distinct trajectories which have the maximum $r(\omega^o, \widehat{d})$.

Consider the example in Fig. 1, assuming the orders of τ_1 and τ_2 are $p_{1,1} \to p_{1,2} \to p_{1,3} \to p_{1,4}$ and $p_{2,1} \to p_{2,2} \to p_{2,3} \to p_{2,4}$. If the order of $q.\varphi$ is $a \to c \to d$, the sub-trajectory $(p_{1,2}, p_{1,3}, p_{1,4})$ is the only result. This is because the order of performing (a, c, d) in sub-trajectory $(p_{1,1}, p_{1,2}, p_{1,3})$ and $(p_{2,1}, p_{2,2}, p_{2,3})$ do not keep the order of τ_1 and τ_2.

The same with RTS, ORTS consists of two steps of traversing AC-tree and computing the ranking of trajectory matching. The algorithm 1 is still adopted in this part to prune search space. However, computing the ranking of trajectory matching in order-sensitive case is more challenging, as it needs to make the order of performing activities in a trajectory consistent with the query and try to maximize the ranking within $\widehat{d}$. Given a candidate trajectory, a naive method of ORTS is to enumerate all possible sub-trajectory matches and find the maximum $r(\omega_i^o, \widehat{d})$. Clearly, this is not efficient. A new algorithm is illustrated in the rest of this paper.

Given a trajectory $\tau = (p_1, p_2, \ldots, p_n)$ and a query q, let $q.\varphi = (\alpha_1, \alpha_2, \ldots, \alpha_m)$. We define an $m \times n$ matrix $\mathcal{R}$ such that its element $\mathcal{R}[i, j](1 \leq i \leq m, 1 \leq j \leq n)$ denotes the maximum ranking between the sub-query $q_1^i.\varphi = (\alpha_1, \alpha_2, \ldots, \alpha_i)$ and the sub-trajectory $\tau_1^j = (p_1, p_2, \ldots, p_j)$. The element of $\mathcal{R}$ is given as follows:

$$\mathcal{R}[i, j] = \max_{1 \leq k \leq j} \{\mathcal{R}[i - 1, k] + mr(\alpha_i, \tau_k^j)\} \qquad (3)$$

where $mr(\alpha_i, \tau_k^j)$ is the maximum ranking of α_i in sub-trajectory τ_k^j.

Computing the ranking of trajectory matching in order-sensitive is illustrated in algorithm 3, which consists of two steps. Firstly, creating a link list l. Secondly, computing $r(\omega^o, \widehat{d})$. Different from algorithm 2, this algorithm updates matrix $\mathcal{R}(*, *)$ according to Eq.(3).

6   Optimization

As described in Section 5, there is no need to search regions which are beyond $\widehat{d}$. However, it needs to compute the ranking of trajectory matching for each

---

Algorithm 3. Computing the Ranking in Order-sensitive Case

Input: query q,trajectory τ
Output: $r(\omega^o, \widehat{d})$
1: create a link list l;
2: $\mathcal{R}(*,*) \leftarrow 0$, $vec \leftarrow 0$, $s \leftarrow 1$, $e \leftarrow 1$, $r(\omega^o, \widehat{d}) \leftarrow 0$;
3: **while** $e \leq l.length$ **do**
4:　　**if** $d(\omega_s^e, q) \leq \widehat{d}$ **then**
5:　　　　**for** each $\alpha \in p_e.\varphi \wedge q.\varphi$ **do**
6:　　　　　　$vec[\alpha] + +$;
7:　　　　**end for**
8:　　**else**
9:　　　　**if** all $vec[\alpha] > 0 (\alpha \in q.\varphi)$ **then**
10:　　　　　　$\mathcal{R}(*,*) \leftarrow 0$;
11:　　　　　　Update each element $\mathcal{R}[i,j]$ based on Eq.(3)
12:　　　　　　**if** $\mathcal{R}[|q.\varphi|, |\tau|] > r(\omega^o, \widehat{d})$ **then**
13:　　　　　　　　$r(\omega^o, \widehat{d}) \leftarrow \mathcal{R}[|q.\varphi|, |\tau|]$;
14:　　　　　　**end if**
15:　　　　**end if**
16:　　　　**for** each $\alpha \in p_s.\varphi \wedge q.\varphi$ **do**
17:　　　　　　$vec[\alpha] - -$;
18:　　　　**end for**
19:　　　　$s \leftarrow s + 1$;
20:　　　　continue;
21:　　**end if**
22:　　$e \leftarrow e + 1$;
23: **end while**
24: **return** $r(\omega^o, \widehat{d})$;

---

trajectory in $\mathcal{C}$. It is costly especially when the cardinality of $\mathcal{C}$ is large. In order to improve the efficiency of query, we optimize the AC-tree and the algorithm of computing the ranking of trajectory matching. Two components of the optimization are illustrated as follows.

Componet 1: In this section, a RAC-tree is proposed, which is the variant of AC-tree. As depicted in Fig. 3, each non-leaf node of RAC-tree contains all activities that can be fulfilled in its child nodes. Shown in Fig.2(a), the activities that can be performed in child cells of 3 are (b, c, e, f), then this information is inserted into node 3.

The leaf nodes of RAC-tree not only contain activity information but also ranking information. Seen from Fig. 3, the entry of a list attached to a leaf node is a tuple in the form of $\alpha(\tau_i : \alpha.r)$, where τ_i is the trajectory that contains α and $\alpha.r$ represents the maximum ranking of α in current leaf node in trajectory τ_i. Besides, the definition of each element of matrix $\mathcal{M}$ is changed:

$$p_{i,j} = \begin{cases} \alpha_i.r, & \text{if trajectory } j \text{ contains query activity } \alpha_i \\ 0, & \text{otherwise} \end{cases}$$

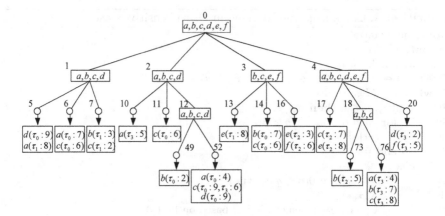

Fig. 3. RAC-tree

where $\alpha_i.r$ is the maximum ranking of query activity α_i in trajectory j within $\widehat{d}$. As shown in Fig 2(a) and Fig 2(b), cell 76 contains activities a, b and c, and the maximum rankings of them are: (4, 7, 8), so the entry attached to the leaf node 76 is $(a(\tau_3 : 4), b(\tau_3 : 7), c(\tau_3 : 8))$.

Compared with using AC-tree, the query is more efficient with the RAC-tree. On one hand, the search regions are pruned with spatial and activity information, as there is no need to consider the tree node p which contains no activity of $q.\varphi$ even if $dis(q, p) \leq \widehat{d}$. On the other hand, the candidate set $\mathcal{C}$ can be pruned fast by the RAC-tree. Given a candidate trajectory τ_j, if τ_j matches query q, it follows that $r((\tau_j, q), \widehat{d}) \leq \sum_{i=1}^{|q.\varphi|} p_{i,j}$ according to the definition of matrix $\mathcal{M}$. As a result, if $\sum_{i=1}^{|q.\varphi|} p_{i,j} < res[k]$, τ_j can be pruned from $\mathcal{C}$, where $res[k]$ denotes the current kth maximum result.

Component 2: In this section, algorithm 2 and 3 are optimized. Different from taking all $p_i(p_i.\varphi \land q.\varphi \neq \phi)$ of l into consideration, a new list $l = \{(p_1, p_1.r), (p_2, p_2.r), \ldots, (p_n, p_n.r)\}$ is created, where p_i is the POI such that $p_i.\varphi \land q.\varphi \neq \phi$ and $dis(p_i, q) \leq \widehat{d}$, and $p_i.r = \sum_{j=1}^{|q.\varphi|} \alpha_j.r$, where $\alpha_j.r$ denotes the maximum ranking of query activity α_j from p_i to the last node of l.

Given a candidate trajectory τ and its corresponding l, we have $p_s.r \geq p_{s'}.r(s' > s)$. As a result, if there exists $(p_s, p_s.r) \in l$ such that $p_s.r < res[k]$ or $p_s.r < r((\tau, q), \widehat{d})$, then there is no need to compute any $r((\tau_{s'}^{e'}, q), \widehat{d})$ since $r((\tau_{s'}^{e'}, q), \widehat{d}) \leq p_{s'}.r$.

7 Experimental Study

In this section, we conduct extensive experiments on half-real datasets to demonstrate the performance of proposed indexes and algorithms. The settings of experiment are presented in table 2.

Table 2. Settings of Experiment

Parameter	Range	Default Value
$|\mathcal{D}|$	10k-50k	50k
$|q.\varphi|$	3-7	5
results k	5-25	10
activity	30-50	40
$\widehat{d}$	4km-8km	6km

We study the query time of different algorithms. 1) Using hybrid index AC-tree to organize trajectory data and unoptimized algorithm to compute the ranking of trajectory matching, denoted as B-RTS and B-ORTS for RTS and ORTS respectively. 2) RAC-tree and upoptimized algorithm based method, denoted as R-RTS and R-ORTS. 3) We use C-RTS and C-ORTS to denote organizing trajectory data with RAC-tree and using optimized algorithm to compute the ranking. Especially, we set $\theta = 200$ in this study, which means that if the number of POIs in each grid cell exceeds 200, it should be partitioned into four components.

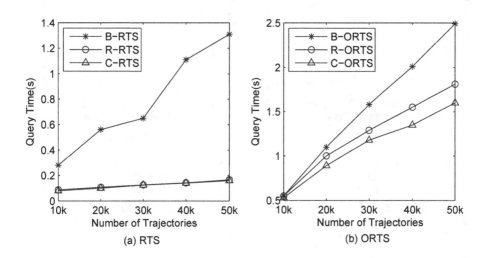

Fig. 4. Effectiveness of $|\mathcal{D}|$

Effectiveness of $|\mathcal{D}|$. First of all, we study the scalability of the three approaches by comparing the time cost of them. In Fig. 4(a) and 4(b), the cardinality of $\mathcal{D}$ varies from 10k to 50k. With no surprise, it needs more time in query with the increasing of the number of trajectories. Besides, from Fig. 4 we know that using RAC-tree to organize trajectory data outperforms AC-tree. The optimization of computing the ranking of trajectory matching is not sensitive in disorder case. However, this optimization is sensitive in order-sensitive case.

Effectiveness of Number of Activities. We also study the performance of proposed indexes and algorithms while varying the number of activities from 30 to 50. Seen from Fig. 5(a) and 5(b), with the increasing of the number of activities the query time becomes less, which is expected. For each trajectory, the number of different kinds of activities contained by which becomes smaller with the increasing of the number of activities. As a result, a trajectory is less likely to be a result and the cardinality of candidate set also becomes smaller, which results in the decreasing of query time.

Effectiveness of $\widehat{d}$. Another important concern of query is the threshold of distance. Shown in Fig.5(c) and 5(d), we set $\widehat{d}$ from 4km to 8km. With no surprise, the query spends more time with the increasing of $\widehat{d}$, even though it is

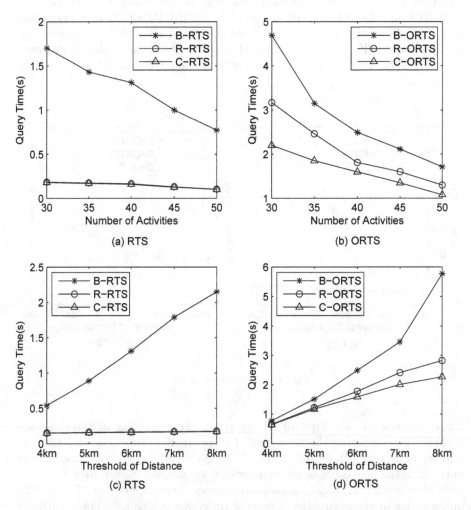

Fig. 5. Effectiveness of $\widehat{d}$ and the number of activities

not obvious for B-RTS and R-RTS. A trajectory is more likely to be a candidate and the number of candidate trajectories becomes lager, with a greater $\hat{d}$.

Effectiveness of k. Fig.6(a) and 6(b) show the query time while varying k from 5 to 25. Different from R-RTS, C-RTS, R-ORTS and C-ORTS the time cost of B-RTS and B-ORTS almost has no significant change for different k. This is because it needs to compute $r(\omega, \hat{d})$ for each trajectory in candidate set $\mathcal{C}$, which has the same cardinality for different k. For R-RTS, C-RTS, R-ORTS and C-ORTS the current kth maximum result tends to be smaller as k increases, which means that more candidate trajectories should be taken into account. Hence, the query time of them monotonously increases with the increasing of k.

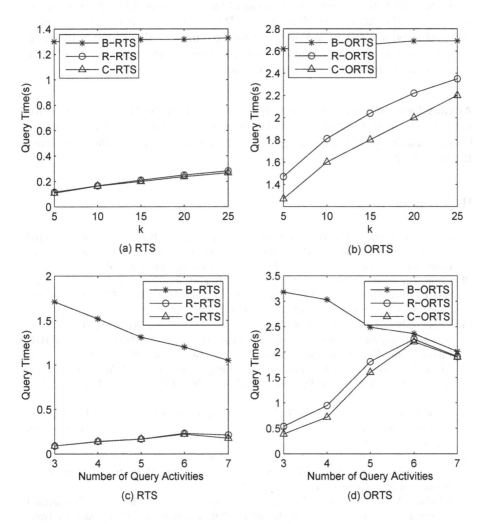

Fig. 6. Effectiveness of k and $q.\varphi$

Effectiveness of $|q.\varphi|$. Finally, we investigate the performance while changing the number of query activities. As depicted in Fig. 6(c) and 6(d), query time of B-RTS and B-ORTS monotonously decreases with the increasing of $q.\varphi$. This is because a trajectory is less likely to be a candidate and the size of the candidate set becomes smaller as $q.\varphi$ increases.

8   Conclusions

This paper studies a novel problem of searching trajectories with activities, spatial and ranking information. In order to tackle the problem efficiently, we propose an AC-tree to organize trajectory data and develop novel algorithms, named RTS and ORTS, to compute the ranking of trajectory matching. As a progress, we optimize the AC-tree by developing a RAC-tree to prune the search space, and optimized algorithms are also proposed to improve the efficiency of query. Experimental results show that the optimization of index structure and algorithms achieves high efficiency and scalability.

Acknowledgements. This work was supported by NSFC grant 61073061, 61303019, 61003044 and 61232006, Doctoral Fund of Ministry of Education of China 20133201120012, and Jiangsu Provincial Department of Education grant 12KJB520017.

References

1. Shang, S., Ding, R.G., Yuan, B., Xie, K.X., Zheng, K., Kalnis, P.: User Oriented Trajectory Search for Trip Recommendation. In: Proceedings of the 15th International Conference on Extending Database Technology, pp. 156–167 (2012)
2. Cong, G., Jensen, C.S., Wu, D.M.: Efficient Retrieval of The Top-k Most Relevant Spatial Web Objects. In: Proceedings of the 2009 ACM SIGMOD International Conference on Management of Data, pp. 337–348 (2009)
3. Chen, Z.B., Shen, H.T., Zhou, X.F., Zheng, Y., Xie, X.: Searching Trajectories by Locations-An Efficiency Study. In: Proceedings of the 2010 ACM SIGMOD International Conference on Management of Data, pp. 255–266 (2010)
4. Reza, S., Davood, R.: On Efficiently Searching Trajectories and Archival Data for Historical Similarities. In: Proceedings of the 2008 VLDB Endowent, pp. 896–908 (2008)
5. Chen, Z.B., Shen, H.T., Zhou, X.F.: Discovering Popular Routes from Trajectories. In: Proceedings of the 2011 ICDE International Conference on Data Engineering, pp. 900–911 (2011)
6. Chen, L., Ozsu, M.T., Oria, V.: Robust and Fast Similarity Search for Moving Object Trajectories. In: Proceedings of the 2005 ACM SIGMOD International Conference on Management of Data, pp. 491–502 (2005)
7. Vlachos, M., Kollios, G., Gunopulos, D.: Discovering Similar Multidimensional trajectories. In: Proceedings of the 2002 ICDE International Conference on Data Engineering, pp. 673–684 (2002)

8. Xue, A.Y., Zhang, R., Zheng, Y., Xie, X., Huang, J., Xu, Z.H.: Destination Prediction by Sub-Trajectory Synthesis and Privacy Protection Against Such Prediction. In: Proceedings of the 2013 ICDE International Conference on Data Engineering, pp. 254–265 (2013)

9. Nergiz, M.E., Atzoir, M., Sayqin, Y.: Towards Trajectory Anonymization: a Generalization-based approach. In: Proceedings of the 2008 Internatioal Workshop on Security and Privacy in GIS and LBS, pp. 52–61 (2008)

10. Jeung, H.Y., Liu, Q., Shen, H.T., Zhou, X.F.: A Hybrid Prediction Model for Moving Objects. In: Proceedings of the 2008 ICDE International Conference on Data Engineering, pp. 70–79 (2008)

11. Terrovitis, M., Manoulis, N.: Privacy Preservation in the Publication of Trajectories. In: Proceedings of the 2008 MDM International Conference on Mobile Data Management, pp. 65–72 (2008)

12. Gidofalvi, G., Huang, X.G., Perdersen, T.B.: Privacy-Preserving Data Mining on Moving Object Trajectories. In: Proceedings of the 2007 MDM International Conference on Mobile Data Management, pp. 60–68 (2007)

13. Zheng, K., Shang, S., Yuan, N.J., Yang, Y.: Towards Efficient Search for Activity Trajectories. In: Proceedings of the 2013 ICDE International Conference on Data Engineering, pp. 230–241 (2013)

14. Zhou, Y.H., Xie, X., Wang, C., Gong, Y.C., Ma, W.Y.: Hybrid Index Structures for Location-based Web Search. In: Proceedings of the 14th ACM International Conference on Informaiton and Knowledge Management, pp. 155–162 (2005)

15. Tao, Y.F., Faloutsos, C., Papadis, D., Liu, B.: Prediction and indexing of moving objects with unknown motion patterns. In: Proceedings of the 2004 ACM SIGMOD International Conference on Management of Data, pp. 611–622 (2004)

16. Chen, Y.Y., Suel, T., Markowetz, A.: Efficient Query Processingin Geographic Web Search Engines. In: Proceedings of the 2006 ACM SIGMOD International Conference on Management of Data, pp. 277–288 (2006)

17. Felipe, I.D., Hristidis, V., Rishe, N.: Keyword Search on Spatial Databases. In: Proceedings of the 2008 ICDE International Conference on Data Engineering, pp. 656–665 (2008)

18. Hariharan, R., Hore, B., Li, C., Mehrotra, S.: Processing Spatial-Keyword (SK) Queries in Geographic Information Retrieval (GIR) Systems. In: Proceedings of the 2007 SSDBM international conference on Scientific and Statical Database Management, p. 16 (2007)

19. Cao, X., Cong, G., Jensen, C.S., Ooi, B.C.: Collective Spatial Keyword Querying. In: Proceedings of the 2011 ACM SIGMOD International Conference on Management of Data, pp. 373–384 (2011)

20. Simmons, R., Browning, B., Zhang, Y., Sadekar, V.: Learning to Predict Driver Route and Destination Intent. In: ITSC, pp. 127–132 (2006)

21. Patterson, D.J., Liao, L., Fox, D., Kautz, H.: Inferring High-Level Behavior from Low-Level Sensors. In: Dey, A.K., Schmidt, A., McCarthy, J.F. (eds.) UbiComp 2003. LNCS, vol. 2864, pp. 73–89. Springer, Heidelberg (2003)

22. Long, C., Wong, R., Wang, K., Fu, A.: Collective Spatial Keyword Queries: A Distance Owner-Driven Approach. In: SIGMOD, pp. 689–700 (2013)

23. Hashem, T., Kulik, L., Zhang, R.: Privacy Preserving Group Nearest Neighbor Queries. In: Proceedings of the 13th International Conference on Extending Database Technology, pp. 489–500 (2010)

24. Abul, O., Bonchi, F., Nanni, M.: Never Walk Alone: Uncertainty for Anonymity in Moving Objects Databases. In: Proceedings of the 2008 ICDE International Conference on Data Engineering, pp. 376–385 (2008)

Topical Pattern Based Document Modelling and Relevance Ranking

Yang Gao, Yue Xu, and Yuefeng Li

Science and Engineering,
Queensland University of Technology, Brisbane, Australia
{y21.gao,yue.xu,y2.li}@qut.edu.au

Abstract. For traditional information filtering (IF) models, it is often assumed that the documents in one collection are only related to one topic. However, in reality users' interests can be diverse and the documents in the collection often involve multiple topics. Topic modelling was proposed to generate statistical models to represent multiple topics in a collection of documents, but in a topic model, topics are represented by distributions over words which are limited to distinctively represent the semantics of topics. Patterns are always thought to be more discriminative than single terms and are able to reveal the inner relations between words. This paper proposes a novel information filtering model, Significant matched Pattern-based Topic Model (SPBTM). The SPBTM represents user information needs in terms of multiple topics and each topic is represented by patterns. More importantly, the patterns are organized into groups based on their statistical and taxonomic features, from which the more representative patterns, called Significant Matched Patterns, can be identified and used to estimate the document relevance. Experiments on benchmark data sets demonstrate that the SPBTM significantly outperforms the state-of-the-art models.

Keywords: Topic model, information filtering, significant matched pattern, relevance ranking.

1 Introduction

In information filtering (IF) models, relevance features are discovered from a training collection of documents and used to represent the user's information needs of the collection. Term-based approaches, such as Rocchio, BM25, etc [2,9], are popularly used to generate term-based features due to their efficient computational performance as well as mature theories. But the term-based document representation suffers from the problems of polysemy and synonymy. To overcome the limitations of term-based approaches, pattern mining based techniques have been used to utilise patterns to represent users' interest and achieved some improvements in effectiveness [5] since patterns carry more semantic meaning than terms. Also, some data mining techniques have been developed to remove redundant and noisy patterns for improving the quality of the discovered patterns, such as maximal patterns, closed patterns, master patterns, etc [1,17,19],

B. Benatallah et al. (Eds.): WISE 2014, Part I, LNCS 8786, pp. 186–201, 2014.
© Springer International Publishing Switzerland 2014

some of which have been used for representing user information needs in IF systems [21]. All these data mining and text mining techniques hold the assumption that the user's interest is only related to a single topic. However, in reality this is not necessarily the case. For example, one news article talking about a "car" is possibly related to policy, market, etc. At any time, new topics may be introduced in the document stream, which means the user's interest can be diverse and changeable. In this paper, we propose to model users' interest in multiple topics rather than a single topic, which reflects the dynamic nature of user information needs.

Topic modelling [3, 13, 15] has become one of the most popular probabilistic text modelling techniques and has been quickly been accepted by many communities. The most inspiring contribution of topic modelling is that it automatically classifies documents in a collection by a number of topics and represents every document with multiple topics and their corresponding distribution. Latent Dirichlet Allocation (LDA) [3] is the most effective topic modelling method. It is reasonable to expect that applying LDA to IF could create a breakthrough for current IF models. However, there are two problems in directly applying LDA to IF. The first problem is that the topic distribution itself is insufficient to represent documents due to its limited number of dimensions (i.e. a prespecified number of topics). The second problem is that the word-based topic representation (i.e. each topic in a LDA model is represented by a set of words) is limited to distinctively represent documents which have different semantic content since many words in the topic representation are not often representative. Our previous work [7] incorporated data mining into topic modelling and generated pattern-based topic representation, which discovers the associations of words inner topics and alleviates the problem of semantic ambiguity of the topic representations in LDA model. However, the pattern-based topic representation can only represent the collection rather than modelling individual documents. How to utilize the pattern-based topic modelling for document representation is still an open question.

In this paper, we propose a new model, called Significant matched Pattern-based Topic Model (SPBTM), in which two parts are involved, user interest modelling (also called "document modelling" since the user interest is generated based on a collection of documents) and document relevance ranking. The user interest model is represented in terms of multiple topics and each topic is represented by patterns. More importantly, the patterns are organized into groups, called equivalence classes, based on their statistical and taxonomic features. With the structured representation, the set of more representative patterns can be identified to represent the user's information needs. Based on the user's interest model, significant matched patterns are selected to determine the relevance of a new coming document.

The remainder of this paper is organized as follows. Section 2 provides a brief background of work LDA. Section 3 and 4 presents the details of our proposed model. Then, we describe data sets, baseline models and empirical results in

Section 5. Section 6 reports related discussions, followed by related work. At last, Section 8 concludes the whole work and presents the future work.

2   Background

Latent Dirichlet Allocation (LDA) [3] is a typical statistical topic modelling technique and the most common topic modelling tool currently in use. It can discover the hidden topics in collections of documents using the words that appear in the documents. Let $D = \{d_1, d_2, \cdots, d_M\}$ be a collection of documents. The total number of documents in the collection is M. The idea behind LDA is that every document is considered to contain multiple topics and each topic can be defined as a distribution over a fixed vocabulary of words that appear in the documents. In LDA model, Gibbs sampling method is a very effective strategy for hidden parameters estimation [11] that is used in this paper. The resulting representations of the LDA model are at two levels, document level and collection level. At document level, each document d_i is represented by topic distribution $\theta_{d_i} = (\vartheta_{d_i,1}, \vartheta_{d_i,2}, \cdots, \vartheta_{d_i,V})$. At collection level, D is represented by a set of topics each of which is represented by a probability distribution over words, ϕ_j for topic j. Apart from these two levels of representations, the LDA model also generates word-topic assignments, that is, the word occurrence is considered related to the topics by LDA. Take a simple example and let $D = \{d_1, d_2, d_3, d_4\}$ be a small collection of four documents with 12 words appearing in the documents. Assuming the documents in D involve 3 topics, Z_1, Z_2 and Z_3. Table 1 illustrates the topic distribution over documents and word-topic assignments in this small collection. From the outcomes of the LDA model, the topic distribution over the whole collection D can be calculated, $\theta_D = (\vartheta_{D,1}, \vartheta_{D,2}, \cdots, \vartheta_{D,V})$, where $\vartheta_{D,j}$ indicates the importance degree of the topic Z_j in the collection D.

Table 1. Example results of LDA: word-topic assignments

Topic		Z_1		Z_2		Z_3
d	$\vartheta_{d,1}$	Words	$\vartheta_{d,2}$	Words	$\vartheta_{d,3}$	Words
d_1	0.6	w_1, w_2, w_3, w_2, w_1	0.2	w_1, w_9, w_8	0.2	w_7, w_{10}, w_{10}
d_2	0.2	w_2, w_4, w_4	0.5	w_7, w_8, w_1, w_8, w_8	0.3	w_1, w_{11}, w_{12}
d_3	0.3	w_2, w_1, w_7, w_5	0.3	w_7, w_3, w_3, w_2	0.4	w_4, w_7, w_{10}, w_{11}
d_4	0.3	w_2, w_7, w_6	0.4	w_9, w_8, w_1	0.3	w_1, w_{11}, w_{10}

3   Pattern Enhanced LDA

Pattern-based representations are considered more meaningful and more accurate to represent topics than word-based representations. Moreover, pattern-based representations contain structural information which can reveal the association between words. In order to discover semantically meaningful patterns to represent topics and documents, two steps are proposed: firstly, construct a new transactional dataset from the LDA model results of the document collection D; secondly, generate pattern-based representations from the transactional dataset to represent user needs of the collection D.

3.1   Construct Transactional Dataset

Let R_{d_i,Z_j} represent the word-topic assignment to topic Z_j in document d_i. R_{d_i,Z_j} is a sequence of words assigned to topic Z_j. For the example illustrated in Table 1, for topic Z_1 in document d_1, $R_{d_1,Z_1} = \langle w_1, w_2, w_3, w_2, w_1 \rangle$. We construct a set of words from each word-topic assignment R_{d_i,Z_j} instead of using the sequence of words in R_{d_i,Z_j}, because for pattern mining, the frequency of a word within a transaction is insignificant. Let I_{ij} be a set of words which occur in R_{d_i,Z_j}, $I_{ij} = \{w|w \in R_{d_i,Z_j}\}$, i.e. I_{ij} contains the words which are in document d_i and assigned to topic Z_j by LDA. I_{ij}, called a *topical document transaction*, is a set of words without any duplicates. From all the word-topic assignments R_{d_i,Z_j} to Z_j, we can construct a transactional dataset Γ_j. Let $D = \{d_1, \cdots, d_M\}$ be the original document collection, the transactional dataset Γ_j for topic Z_j is defined as $\Gamma_j = \{I_{1j}, I_{2j}, \cdots, I_{Mj}\}$. For the topics in D, we can construct V transactional datasets $(\Gamma_1, \Gamma_2, \cdots, \Gamma_V)$. An example of transactional datasets is illustrated in Table 2, which is generated from the example in Table 1.

3.2   Generate Pattern Enhanced Representation

The basic idea of the proposed pattern-based method is to use frequent patterns generated from each transactional dataset Γ_j to represent Z_j. In the two-stage topic model [7], frequent patterns are generated in this step. For a given minimal support threshold σ, an itemset X in Γ_j is frequent if $supp(X) >= \sigma$, where $supp(X)$ is the support of X which is the number of transactions in Γ_j that contain X. The frequency of the itemset X is defined $\dfrac{supp(X)}{|\Gamma_j|}$. Topic Z_j can be represented by a set of all frequent patterns, denoted as $\mathbf{X}_{Z_i} = \{X_{i1}, X_{i2}, \cdots, X_{im_i}\}$, where m_i is the total number of patterns in $\mathbf{X}_{Z_i}$ and V is the total number of topics. Take Γ_2 as an example, which is the transactional dataset for Z_2. For a minimal support threshold $\sigma = 2$, all frequent patterns generated from Γ_2 are given in Table 3 ("itemset" and "pattern" are interchangeable in this paper).

Table 2. Transactional datasets generated from Table 1 (topical document transaction(TDT))

T	TDT	TDT	TDT
1	$\{w_1, w_2, w_3\}$	$\{w_1, w_8, w_9\}$	$\{w_7, w_{10}\}$
2	$\{w_2, w_4\}$	$\{w_1, w_7, w_8\}$	$\{w_1, w_{11}, w_{12}\}$
3	$\{w_1, w_2, w_5, w_7\}$	$\{w_2, w_3, w_7\}$	$\{w_4, w_7, w_{10}, w_{11}\}$
4	$\{w_2, w_6, w_7\}$	$\{w_1, w_8, w_9\}$	$\{w_1, w_{11}, w_{10}\}$
	Γ_1	Γ_2	Γ_3

Table 3. Frequent patterns for Z_2, $\sigma = 2$

Patterns	supp
$\{w_1\}, \{w_8\}, \{w_1, w_8\}$	3
$\{w_9\}, \{w_7\}\{w_8, w_9\}, \{w_1, w_9\},$ $\{w_1, w_8, w_9\}$	2

4   Information Filtering Model Based on Pattern Enhanced LDA

The representations generated by the pattern enhanced LDA model, discussed in Section 3, carry more concrete and identifiable meaning than the word-based representations generated using the original LDA model. However, the number of patterns in some of the topics can be huge and many of the patterns are not discriminative enough to represent specific topics. As a result, documents cannot be accurately represented by these topic representations. That means, these pattern-based topic representations which represent user interests may not be sufficient or accurate enough to be directly used to determine the relevance of new documents to the user interests. In this section, one novel IF model, Significant matched Pattern-based Topic Model (SPBTM), is proposed based on the pattern enhanced topic representations. The proposed model consists of topics distribution describing topic preferences of documents or a document collection and structured pattern-based topic representations representing the semantic meaning of topics in documents. Moreover, the proposed model estimates the relevance of incoming documents based on Significant Matched Patterns, which are the more relevant and representative patterns, as proposed in this paper. The details are described in the following subsections.

4.1   Equivalence Class

Normally, the number of frequent patterns is considerably large and many of them are not necessarily useful. Several concise patterns have been proposed to represent useful patterns generated from a large dataset instead of frequent patterns, such as maximal patterns [1] and closed patterns. The number of these concise patterns is significantly smaller than the number of frequent patterns for a dataset. In particular, the closed pattern has drawn great attention due to its attractive features [17, 19].

Definition 1. *Closed Itemset*: for a transactional dataset, an itemset X is a closed itemset if there exists no itemset X' such that (1) $X \subset X'$, (2) $supp(X) = supp(X')$.

Definition 2. *Generator*: for a transactional dataset Γ, let X be a closed itemset and $T(X)$ consists of all transactions in Γ that contain X, then an itemset g is said to be a generator of X iff $g \subset X, T(g) = T(X)$ and $supp(X) = supp(g)$. A generator g of X is said a minimal generator of X if $\nexists g' \subset g$ and g' is a generator of X.

Definition 3. *Equivalence Class*: for a transactional dataset Γ, let X be a closed itemset and $G(X)$ consist of all generators of X, then the equivalence class of X in Γ, denoted as $EC(X)$, is defined as $EC(X) = G(X) \cup \{X\}$.

Let EC_1 and EC_2 be two different equivalence classes of the same transactional dataset. Then $EC_1 \cap EC_2 = \emptyset$, which means that the equivalence classes are exclusive of each other.

All the patterns in an equivalence class have the same frequency. The frequency of a pattern indicates the statistical significance of the pattern. The frequency of the patterns in an equivalence class is used to represent the statistical significance of the equivalence class. Table 4 shows the three equivalence classes within the patterns for topic Z_2 in Table 3, where f indicates the statistical significance of each class.

Table 4. The equivalence classes in Z_2

$EC_{21}\ (f_{21} = 0.75)$	$\{w_1, w_8\}, \{w_1\}, \{w_8\}$
$EC_{22}\ (f_{22} = 0.5)$	$\{w_1, w_8, w_9\}, \{w_1, w_9\}, \{w_8, w_9\}, \{w_9\}$
$EC_{23}\ (f_{23} = 0.5)$	$\{w_7\}$

There are two parts in the proposed model SPBTM: the training part to generate user information needs from a collection of training documents (i.e. user interest modelling or document modelling) and the filtering part to determine the relevance of incoming documents based on the user's interests (i.e. document relevance ranking).

4.2 Topic-Based User Interest Modelling

For a collection of documents D, the user's interests can be represented by the patterns in the topics of D. As mentioned in Section 3, θ_D represents the topic distribution of D and can be used to represent the user's topic interest distribution, $\theta_D = (\vartheta_{D,1}, \vartheta_{D,2}, \cdots, \vartheta_{D,V})$, and V is the number of topics. In this paper, the topic distribution in collection D is defined as the average of the topic distributions of the documents in D, i.e. $\vartheta_{D,j} = \frac{1}{M} \sum_{i=1}^{M} \theta_{d_i,j}$. The probability distribution of topics in θ_D represents the degree of interest that the user has in these topics.

By using the methods described in Section 3, for a document collection D and V pre-specified latent topics, from the results of LDA to D, V transactional datasets, $\Gamma_1, \cdots, \Gamma_V$ can be generated from which the pattern-based topic representations for the collection, $U = \{\mathbf{X}_{Z_1}, \mathbf{X}_{Z_2}, \cdots, \mathbf{X}_{Z_V}\}$, can be generated, where each $\mathbf{X}_{Z_i} = \{X_{i1}, X_{i2}, \cdots, X_{im_i}\}$ is a set of frequent patterns generated from transactional dataset Γ_i. U is considered the user interest model, and the patterns in each $\mathbf{X}_{Z_i}$ represent what the user is interested in terms of topic Z_i.

Frequent patterns can be well organized into groups based on their statistics and coverage. As discussed in Section 4.1, equivalence class is a useful structure which collects the frequent patterns with the same frequency in one group. The statistical significance of the patterns in one equivalence class is the same. This distinctive feature of equivalence classes can make the patterns more effectively used in document filtering. In this paper, we propose to use equivalence classes to represent topics instead of using frequent patterns or closed patterns.

Assume that there are n_i frequent closed patterns in $\mathbf{X}_{Z_i}$, which are $c_{i1}, \cdots, c_{in_i}$, and that $\mathbf{X}_{Z_i}$ can be partitioned into n_i equivalence classes, $EC(c_{i1}), \cdots,$

$EC(c_{in_i})$. For simplicity, the equivalence classes are denoted as $EC_{i1}, \cdots, EC_{in_i}$ for $\mathbf{X}_{Z_i}$, or simply for topic Z_i. Let $\mathbb{E}(Z_i)$ denote the set of equivalence classes for topic Z_i, i.e. $\mathbb{E}(Z_i) = \{EC_{i1}, \cdots, EC_{in_i}\}$. In the model SPBTM, the equivalence classes $\mathbb{E}(Z_i)$ are used to represent user interests which are denoted as $\mathbb{U}_E = \{\mathbb{E}(Z_1), \cdots, \mathbb{E}(Z_V)\}$.

4.3   Topic-Based Document Relevance Ranking

In terms of the statistical significance, all the patterns in one equivalence class are the same. The differences among them are their size. If a longer pattern and a shorter pattern from the same equivalence class appear in a document simultaneously, the shorter one becomes insignificant since it is covered by the longer one and it has the same statistical significance as the longer one.

In the filtering stage, document relevance is estimated to filter out irrelevant documents based on the user's information needs. For a new incoming document d, the basic way to determine the relevance of d to the user interests is firstly to identify significant patterns in d which match some patterns in the topic-based user interest model and then estimate the relevance of d based on the user's topic interest distributions and the significance of the matched patterns.

The significance of one pattern is determined not only by its statistical significance, but also by its size since the size of the pattern indicates the specificity level. Among a set of patterns, usually a pattern taxonomy exists. For example, Fig. 1 depicts the taxonomy constructed for $\mathbf{X}_{Z_2}$ in Table 3. This tree-like structure demonstrates the subsumption relationship between the discovered patterns in Z_2. The longest pattern in a pattern taxonomy, such as $\{w_1, w_8, w_9\}$ in Fig. 1, is the most specific pattern that describes a user's interests since longer pattern has more specific meanings, while single words, such as w_1 in Fig. 1, are the most general patterns which are less capable of discriminating the meaning of the topic from other topics as compared to longer patterns such as $\{w_1, w_8, w_9\}$. The pattern taxonomy presents different specificities of patterns according to the level in the taxonomy and thus the size of the pattern. Therefore, we define the pattern specificity below.

Definition 4. *Pattern specificity:* The specificity of a pattern X is defined as a power function of the pattern length with the exponent less than 1, denoted as $spe(X)$, $spe(X) = a|X|^m$, where a and m are constant real numbers and $0 < m < 1$, $|X|$ is the length of X, i.e. the number of words in X.

Definition 5. *Topic Significance*: Let d be a document, Z_j be a topic in the user interest model, PA_{jk}^d be a set of matched patterns for topic Z_j in document d, $k = 1, \cdots, n_j$, $f_{j1}, \cdots, f_{jn_j}$ be the corresponding supports of the matched patterns, then the topic significance of Z_j to d is defined as:

$$sig(Z_j, d) = \sum_{k=1}^{n_j} spe\left(PA_{jk}^d\right) \times f_{jk} = \sum_{k=1}^{n_j} a|PA_{jk}^d|^m \times f_{jk} \tag{1}$$

where m is the scale of pattern specificity (we set $m = 0.5$), and a is a constant real number (in this paper, we set $a = 1$).

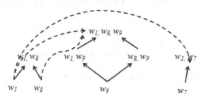

Fig. 1. Pattern Taxonomy in Z_2

In the SPBTM model, the topic significance is determined by significant matched pattern which is defined below.

Definition 6. *Significant Matched Patterns (SMPatterns)*: Let d be a document, Z_j be a topic in the user interest model, $EC_{j1}, \cdots, EC_{jn_j}$ be the pattern equivalence classes of Z_j, then a pattern X in d is considered a *matched pattern* to equivalence class to equivalence class EC_{jk}, if $X \in EC_{jk}$. Let c_{jk} be the closed pattern in EC_{jk}, a matched pattern X to EC_{jk} is considered a *significant matched pattern* to EC_{jk} if $\eta_X = \dfrac{|X|}{|c_{jk}|} \geq \varepsilon$, where $\varepsilon \in [0,1]$ is the threshold for determining the significant pattern, the higher the η_X, the more significant the significant pattern is.

The set of all SMPatterns, denoted as SM_{jk}^d, to equivalence class EC_{jk} are those matched patterns which are significantly close to the closed pattern and only a proportion (controlled by ε) of all the matched patterns in EC_{jk} are selected. Therefore, the SMPatterns SM_{jk}^d, where $k = 1, \cdots, n_j$ are considered the significant patterns in d which can represent the relevant topic Z_j.

For an incoming document d, we propose to estimate the relevance of d to the user interest based on the topic significance and topic distribution. The document relevance is estimated using the following equation:

$$Rank(d) = \sum_{j=1}^{V} sig(Z_j, d) \times \vartheta_{D,j} \qquad (2)$$

For the SPBTM, the patterns PA_{jk}^d in the topic significance $sig(Z_j, d)$ are SMPatterns in $\mathbb{U}_E$. And the specificity is calculated by the closed pattern c_{jk} in E_{jk} and η_X which represents the degree of the significance of the matched pattern X in the specific equivalence class. By incorporating Equation (1) into Equation (2), the relevance ranking of d, denoted as $Rank_E(d)$, is estimated by the following equation:

$$Rank_E(d) = \sum_{j=1}^{V} \sum_{k=1}^{n_j} \sum_{X \in SM_{jk}^d} \eta_X |X|^{0.5} \times \delta(X, d) \times f_{jk} \times \vartheta_{D,j} \qquad (3)$$

where V is the total number of topics, SM_{jk}^d is the set of significant matched patterns to equivalence class $EC_{jk}, k = 1, \cdots, n_j$ and $f_{j1}, \cdots, f_{jn_j}$ is the corresponding statistical significance of the equivalence classes, $\vartheta_{D,j}$ is the topic distribution, and

$$\delta(X, d) = \begin{cases} 1 & \text{if } X \in d \\ 0 & \text{otherwise} \end{cases} \tag{4}$$

The higher the $Rank(d)$, the more likely the document is relevant to the user's interest.

5   Evaluation

Two hypotheses are designed for verifying the IF model proposed in this paper. The first hypothesis is that, user information needs involve multiple topics, then document modelling by taking multiple topics into consideration can generate more accurate user interest models. The second hypothesis is that the proposed SMPatterns are more effective in determining relevant documents than other patterns. To verify the hypotheses, experiments and evaluation have been conducted. This section discusses the experiments and evaluation in terms of data collection, baseline models, measures and results. The results show that the proposed topic-based model significantly outperforms the state-of-the-art models in terms of effectiveness.

5.1   Data and Measures

The Reuters Corpus Volume 1 (RCV1) dataset was collected by Reuter's journals between August 20, 1996, and August 19, 1997, incorporating a total of 806,791 documents that cover a variety of topics and a large amount of information. 100 collections of documents were developed for the TREC filtering track. Each collection is divided into a training set and a testing set. According to Buckley and others [4], the 100 collections are stable and sufficient enough for high quality experiments. In the TREC track, a collection is also referred to as a 'topic'. In this paper, to differentiate from the term 'topic' in the LDA model, the term 'collection' is used to refer to a collection of documents in the TREC dataset. The first 50 collections were composed by human assessors, which are used for experiments in this paper, and the 'title' and 'text' of the documents are used by all the models.

The effectiveness is assessed by five different measures: Mean Average Precision (MAP), average precision of the top K ($K = 20$) documents, break-even point (b/p), $F_\beta(\beta = 1)$ measure and Interpolated Average Precision (IAP) on 11-points. F_1 is a criterion that assesses the effect involving both precision (p) and recall (r),which is defined as $F_1 = \frac{2pr}{p+r}$. The larger the $top20$, MAP, b/p or F_1 score, the better the system performs. The 11 points measure is the precision at 11 standard recall levels (i.e. recall $= 0, 0.1, \cdots, 1$).

The experiments tested across the 50 collections of independent datasets, which satisfy the generalized cross-validation for the statistical estimation model.

The statistical method, t-test, was also used to verify the significance of the experimental results. If the p-value associated with t is significantly low (< 0.05),

there is evidence to verify that the difference in means across the paired observations is significant.

5.2   Baseline Models and Settings

The experiments were conducted extensively covering all major representations
such as phrases and patterns in order to evaluate the effectiveness of the proposed topic-based IF model. The evaluations were conducted in terms of two
technical categories: topic modelling methods and pattern mining methods. For
each category, some state-of-the-art methods are chosen as the baseline models.
More details about these baseline models are given below.

(1) *Topic modelling based category*

TNG: In the phrase-based topic model, n-gram phrases that are generated by
using the TNG model [14], which can be used to represent user interest needs and
phrase frequency is used to represent topic relevance. Readers who are interested
in the details can refer to [14].

PBTM: We have proposed a topic-based model PBTM_FCP [6] which uses
closed patterns to represent topics and uses patterns' support to represent topic
relevance. PBTM_FCP is chosen as a baseline model for the pattern-based topic
models. The following equation is used to calculate the relevance of a document
d with PBTM_FCP:

$$Rank_C(d) = \sum_{j=1}^{V} \sum_{k=1}^{n_j} |c_{jk}|^{0.5} \times \delta(c_{jk}, d) \times f_{jk} \times \vartheta_{D,j} \qquad (5)$$

where c_{jk} is a closed pattern in PBTM_FCP and n_j is the total number of closed
patterns in topic j.

The parameters for all topic models are set as follows: the number of iterations
of Gibbs sampling is 1000, the hyper-parameters of the LDA model are $\alpha = 50/V$
and $\beta = 0.01$. Our experience shows that filtered results are not very sensitive to
the settings of these parameters. But the number of topics V affects the results
depending on various data collections. In this paper, V is set to 10.

In the process of generating pattern enhanced topic representations, the minimum support σ_{rel} for every topic in each collection is different, because the
number of positive documents in collections of the RCV1 is very different. In order to ensure enough transactions from positive documents to generate accurate
patterns for representing user needs, the minimum support σ_{rel} is set as follows
:

$$\sigma_{rel} = \begin{cases} 1 & n \leq 2 \\ max(2/n, 0.3) & 2 < n \leq 10 \\ max(3/n, 0.3) & 10 < n \leq 13 \\ max(4/n, 0.3) & 13 < n \leq 20 \\ 0.3 & otherwise. \end{cases} \qquad (6)$$

where n is the number of transactions from relevant documents in each transactional database.

(2)*Pattern-based category*

FCP: Frequent closed patterns are generated from the documents in the training dataset and used to represent the user's information needs. The minimum support in the pattern-based models, including phrases, sequential closed patterns, is set to 0.2.

Sequential Closed Patterns(SCP): The Pattern Taxonomy Model [21] is one of the state-of-the-art pattern-based model. It was developed to discover sequential closed patterns from the training dataset and rank the incoming documents in the filtering stage with the relative supports of the discovered patterns that appear in the documents.

n-**Gram:** Most researches on phrases in modelling documents have employed an independent collocation discovery module. In this way, a phrase with independent statistics can be indexed exactly as an word-based representation. In our experiments, we use n-Gram phrases to represent a document collection (i.e. user information needs), where $n = 3$.

5.3   Results

Five different thresholds ($\varepsilon = 0.3, 0.4, 0.5, 0.6, 0.7$) are used in order to find proper SMPatterns in the proposed SPBTM using the 50 human assessed collections and the results are shown in Table 5. Based on the comparison in Table 5, SPBTM achieves the best result when $\varepsilon = 0.5$ for this dataset. Therefore, this result is used to compared with all the baseline models mentioned above. The results are depicted in Table 6 and evaluated using the measures in Section 5.1.

Table 6 consists of two parts. The top and bottom parts in Table 6 provide the results of the topic modelling methods and the pattern mining methods, respectively. The *improvement%* line at the bottom of each part provides the percentage of improvement achieved by the SPBTM which consistently performs the best among all models against the second best model in that part for each measure.

We also conducted the T-Test to compare the SPBTM with all baseline models. The results are listed in Table 7. The statistical results indicate that the

Table 5. Comparison of SPBTM results with different values of threshold ε, using the first 50 collections of RCV1

Threshold ε	MAP	b/p	$top20$	F_1
0.3	0.452	0.436	0.513	0.445
0.4	0.455	0.436	0.521	0.445
0.5	0.456	0.446	0.524	0.446
0.6	0.449	0.433	0.513	0.439
0.7	0.442	0.425	0.515	0.435

Table 6. Comparison of all models using the first 50 collections of RCV1

Methods	*MAP*	*b/p*	*top20*	F_1
SPBTM	**0.456**	**0.446**	**0.524**	**0.446**
TNG	0.374	0.367	0.446	0.388
PBTM_FCP	0.424	0.420	0.494	0.424
improvement%	7.5	6.2	6.1	5.2
SCP	0.364	0.353	0.406	0.390
n-Gram	0.361	0.342	0.401	0.386
FCP	0.361	0.346	0.428	0.385
improvement%	25.3	26.3	22.4	14.4

Table 7. T-Test *p*-values for all modes compared with the SPBTM model

Methods	*MAP*	*b/p*	*top20*	F_1
TNG	0.0003	0.0005	0.0066	0.0002
PBTM_FCP	0.0005	0.0299	0.0267	0.0002
SCP	0.00004	0.0001	0.0002	0.0002
n-Gram	0.0001	0.0001	0.0004	0.0002
FCP	0.00002	0.00004	0.0031	0.0001

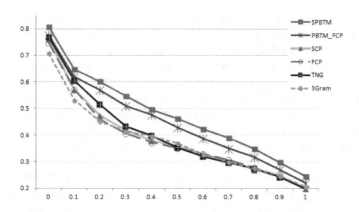

Fig. 2. 11 point results of comparison between the proposed SPBTM and baseline models

proposed SPBTM significantly outperforms all the other models (all values in Table 6 are less than 0.05) and the improvements are consistent on all four measures. Therefore, we conclude that the SPBTM is an exciting achievement in discovering high-quality features in text documents mainly because it represents the text documents not only using the topic distributions at a general level but also using hierarchical pattern representations at a detailed specific level, both of which contribute to the accurate document relevance ranking.

The *11-points* results of all methods are shown in Fig. 2. The results indicate that the SPBTM model has achieved the best performance compared with all the other baseline models.

6 Discussion

As we can see from the experiment results, taking topics into consideration in generating user interest models and also in document relevance ranking can greatly improve the performance of information filtering. The reason behind the SPBTM and the PBTM achieving the excellent performance is mainly because we inventively incorporated pattern mining techniques into topic modelling to generate pattern-based topic models which can represent user interest needs in terms of multiple topics. Most importantly, the topics are represented by patterns which bring concrete and precise semantics to the user interest models. These comparisons can strongly validate the first hypothesis. Moreover, the outstanding performance of the SPBTM over the PBTM_FCP indicates the significant benefit of using the proposed SMPatterns in estimating document relevance over using frequent closed patterns. This result clearly supports the second hypothesis.

6.1 Significant Matched Patterns

In the SPBTM, the patterns which represent user interests are not only grouped in terms of topics, but also partitioned based on equivalence classes in each topic group. The patterns in different groups or different equivalence classes have different meanings and distinct properties. Thus, user information needs are clearly represented according to various semantic meanings as well as distinct properties of the specific patterns in different topic groups and equivalence classes. However, among all matched patterns in each equivalence class, not all of them are useful for estimating the document relevance. The results in Table 5 show that the best performance achieved by the SPBTM is when the threshold ε is 0.5. This result indicates that, selecting more matched patterns as SMPatterns (i.e., $\varepsilon < 0.5$) actually hurts the performance of document relevance ranking. When ε is small, some short matched patterns would be selected. These short patterns are much less specific than longer patterns to represent the documents and also possibly brings bias to the document relevance ranking. Similarly, the performance also deteriorates when selecting less matched patterns (i.e., $\varepsilon > 0.5$). This is because some useful matched patterns will not be selected due to the high threshold, which will negatively affect the quality of the selected SMPatterns.

From Table 6, we can see that the PBTM_FCP achieved better performance than all the other models but SPBTM. SPBTM is the only model which outperforms the PBTM_FCP. This result is an excellent example to show the quality of closed patterns as well as SMPatterns.

6.2 Topic-Based Relevance Estimation

Table 6 shows that all the topic-based models outperform all the other baseline models including the pattern-based and phrase-based models. As we have mentioned above, this is mainly because the topic-based models represent the documents not only using patterns or phrases, but also using topic distributions. Most importantly, the patterns or phrases used by the topic-based models are topics related, which is a key difference from the pattern-based or phrase-based baseline models.

6.3 Complexity

As discussed in Section 4, there are two algorithms in the proposed model, i.e. user profiling and document filtering. The complexity of the two algorithms is discussed below.

For user profiling, the proposed pattern-based topic modelling methods consist of two parts, topic modelling and pattern mining. For the topic modelling part, the initial user interest models are generated using the LDA model, and the complexity of each iteration of Gibbs sampling for the LDA is linear with the number of topics (V) and the number of documents (N), i.e. $O(V * N)$ [15]. For pattern mining, the efficiency of the FP-Tree algorithm for generating frequent patterns has been widely accepted in the field of data mining. It should be mentioned that the user profiling part can be conducted off-line which means that the complexity of the user profiling part will not affect the efficiency of the proposed IF model.

For information filtering, the complexity to determine its relevance to the user needs is linear to the size of the feature space for the pattern-based methods (i.e. SCP, n-Gram, and FCP), $O(S)$ where S is the size of the feature space. For the topic modelling based methods, due to the use of topics, the complexity of determining a document's relevance is $O(V * S)$ where V is the number of topics and S is the number of patterns in each topic representation. Theoretically, the complexity of the topic-based methods is higher than the pattern-based or term-based methods but practically, the number of SMPatterns is much smaller than the number of frequent patterns. Therefore, the complexity of the SPBTM model is very often acceptable.

7   Related Work

Documents can be modelled by various approaches that primarily include term-based models [2,9], pattern-based models [16,21] and probabilistic models [8,10]. Term-based models have an unavoidable limitation on expressing semantics and problems of polysemy and synonymy. Therefore, people tend to extract more semantic features (such as phrases and patterns) to represent a document in many applications. Aiming at representing documents with multiple topics in a more detailed way, topic models are incorporated in the frame of language model and achieve successful retrieval results [15, 18]. Also, topic models [12, 13, 20] can extract user information needs by analysing content and represent them in terms of latent topics discovered from user profiles. But in all of these topic models, a fundamental assumption is a topic can be represented by a word-based multinomial distribution. Thus, it is desirable to interpret topics or documents with coherent and discriminative representations. TNG model generated topical phrases has achieved a slight improvement on IR task [14] and IF task from our experiment results in this paper, which mainly because of too limited occurrences of the discovered phrases to represent the document relevance.

8   Conclusion

This paper presents an innovative pattern enhanced topic model for information filtering including user interest modelling and document relevance ranking. The SPBTM generates pattern-based topic representations to model user's information interests across multiple topics. In the filtering stage, the SPBTM selects SMPatterns, instead of using all discovered patterns, for estimating the relevance of incoming documents. The proposed approach incorporates the semantic structure from topic modelling and the specificity as well as the statistical significance from the SMPatterns. The proposed model has been evaluated by using the RCV1 and TREC collections for the task of information filtering. Compared with the state-of-the-art models, the proposed model demonstrates excellent strength on document modelling and relevance ranking.

The proposed model automatically generates discriminative and semantic rich representations for modelling topics and documents by combining topic modelling techniques and data mining techniques. Moreover, the significant topical patterns for incoming documents can effectively represent user's interests. The technique not only can be used for information filtering, but also can be applied to many content-based user interest modelling tasks.

References

1. Bayardo Jr., R.J.: Efficiently mining long patterns from databases. ACM Sigmod Record 27, 85–93 (1998)
2. Beil, F., Ester, M., Xu, X.: Frequent term-based text clustering. In: KDD 2002, pp. 436–442. ACM (2002)
3. Blei, D.M., Ng, A.Y., Jordan, M.I.: Latent dirichlet allocation. The Journal of Machine Learning Research 3, 993–1022 (2003)
4. Buckley, C., Voorhees, E.M.: Evaluating evaluation measure stability. In: SIGIR 2000, pp. 33–40. ACM (2000)
5. Cheng, H., Yan, X., Han, J., Hsu, C.-W.: Discriminative frequent pattern analysis for effective classification. In: ICDE 2007, pp. 716–725. IEEE (2007)
6. Gao, Y., Xu, Y., Li, Y.: Pattern-based topic models for information filtering. In: Proceedings of International Conference on Data Mining Workshop SENTIRE, ICDM 2013. IEEE (2013)

7. Gao, Y., Xu, Y., Li, Y., Liu, B.: A two-stage approach for generating topic models. In: PADKDD 2013, pp. 221–232 (2013)
8. Lafferty, J., Zhai, C.: Probabilistic relevance models based on document and query generation. In: Language modeling for information retrieval, pp. 1–10. Springer, Heidelberg (2003)
9. Robertson, S., Zaragoza, H., Taylor, M.: Simple bm25 extension to multiple weighted fields. In: CIKM 2004, pp. 42–49. ACM (2004)
10. Sparck Jones, K., Walker, S., Robertson, S.E.: A probabilistic model of infor- mation retrieval: development and comparative experiments: Part 2. Information Processing & Management 36(6), 809–840 (2000)
11. Steyvers, M., Griffiths, T.: Probabilistic topic models. Handbook of Latent Semantic Analysis 427(7), 424–440 (2007)
12. Tang, J., Wu, S., Sun, J., Su, H.: Cross-domain collaboration recommendation. In: KDD 2012, pp. 1285–1293. ACM (2012)
13. Wang, C., Blei, D.M.: Collaborative topic modeling for recommending scientific articles. In: KDD 2011, pp. 448–456. ACM (2011)
14. Wang, X., McCallum, A., Wei, X.: Topical n-grams: Phrase and topic discovery, with an application to information retrieval. In: ICDM 2007, pp. 697–702. IEEE (2007)
15. Wei, X., Croft, W.B.: LDA-based document models for ad-hoc retrieval. In: SIGIR 2006, pp. 178–185. ACM (2006)
16. Wu, S.-T., Li, Y., Xu, Y.: Deploying approaches for pattern refinement in text mining. In: ICDM 2006, pp. 1157–1161. IEEE (2006)
17. Xu, Y., Li, Y., Shaw, G.: Reliable representations for association rules. Data & Knowledge Engineering 70(6), 555–575 (2011)
18. Yi, X., Allan, J.: A comparative study of utilizing topic models for information retrieval. In: Boughanem, M., Berrut, C., Mothe, J., Soule-Dupuy, C. (eds.) ECIR 2009. LNCS, vol. 5478, pp. 29–41. Springer, Heidelberg (2009)
19. Zaki, M.J., Hsiao, C.-J.: CHARM: An efficient algorithm for closed itemset mining. In: SDM, vol. 2, pp. 457–473 (2002)
20. Zhang, Y., Callan, J., Minka, T.: Novelty and redundancy detection in adaptive filtering. In: SIGIR 2002, pp. 81–88. ACM (2002)
21. Zhong, N., Li, Y., Wu, S.-T.: Effective pattern discovery for text mining. IEEE Transactions on Knowledge and Data Engineering 24(1), 30–44 (2012)

A Decremental Search Approach
for Large Scale Dynamic Ridesharing

Ali Shemshadi, Quan Z. Sheng, and Wei Emma Zhang

School of Computer Science
The University of Adelaide, SA 5005, Australia
{ali.shemshadi,michael.sheng,wei.zhang01}@adelaide.edu.au

Abstract. The Web of Things (WoT) paradigm introduces novel applications to improve the quality of human lives. Dynamic ridesharing is
one of these applications, which holds the potential to gain significant
economical, environmental, and social benefits particularly in metropolitan areas. Despite the recent advances in this area, many challenges still
remain. In particular, handling large-scale incomplete data has not been
adequately addressed by previous works. Optimizing the taxi/passengers
schedules to gain the maximum benefits is another challenging issue. In
this paper, we propose a novel system, MARS (Multi-Agent Ridesharing System), which addresses these challenges by formulating travel time
estimation and enhancing the efficiency of taxi searching through a decremental search approach. Our proposed approach has been validated using
a real-world dataset that consists of the trajectories of 10,357 taxis in
Beijing, China.

Keywords: Taxi Ridesharing, Web of Things, Spatio-temporal Data,
Incomplete Data.

1 Introduction

Dynamic ridesharing is known as a promising solution for insufficient transport
supply for people's commutes, particularly at rush hours or in extreme weather in
metropolitan areas [19,14]. The recent emergence of the Web of Things (WoT)
provides unique opportunities for dynamic ridesharing in the sense that it is
possible to acquire, integrate, and analyze real-time data collected from various
sources such as sensors, smart devices, and people [2,13,4]. Indeed, the incentives
of dynamic ridesharing are not limited to economical benefit (e.g., reduced total
mileage and fuel consumption), they include several environmental and social
benefits as well (e.g., less air pollution and passenger waiting time) [1,3].

In recent years, dynamic ridesharing has become a very active research area
and several approaches have been proposed [15,9,12,19]. Unfortunately, many
technical challenges still remain unsolved [8]. One of the major challenges is the
complexity of the ridesharing process. It is shown that this problem is proved
to be an extension of *Traveling Salesman Problem* [9], which is NP-hard. One of
the recent approaches to overcome this problem is incremental search approach

B. Benatallah et al. (Eds.): WISE 2014, Part I, LNCS 8786, pp. 202–217, 2014.
© Springer International Publishing Switzerland 2014

Table 1. Taxi speed estimation based on GPS reading analysis

All Records	Stopped	High Probability	Low Probability	Impossible
	$v \leq 1\ km/h$	$1 < v \leq 90\ km/h$	$90 < v \leq 200\ km/h$	$v > 200\ km/h$
15,784,344	6,419,947	9,280,967	32,283	51,147

[9]. The basic idea of this approach is that the ridesharing application extends the search areas for the origin and destination of a query step by step until the nearest available taxi is found. However, in practice, many other factors such as uncertainty, incomplete data, time, costs, and safety may affect the success of a ridesharing application [11]. In particular, due to technical difficulties such as hardware/software failure, network fluctuations, and erroneous sensor readings, taxi trajectories data can be incomplete or noisy.

Table 1 shows the estimated speeds of taxis in a big city based on their real GPS readings [9]. There are 51,147 records showing taxis traveling with $200km/h$ or more, which are most likely due to GPS malfunction of taxis. Inability to handle uncertainty is actually one of the main reasons that limit the large-scale deployment of dynamic ridesharing systems. In this paper, we particularly focus on technical challenges such as (i) Can we limit the expansion of search areas in the incremental approach? (ii) How can we increase the efficiency and effectiveness of incremental search through incorporating this limit? and (iii) How to manage incoming user queries in the absence of complete data?

To deal with these challenges, we propose a novel framework, MARS (Multi-Agent Ridesharing System). Our main contributions are as follows:

- We introduce the *economic search margin* to limit the expansion of search area. Then we propose a new decremental search approach which is based on the idea of gradually reducing the search area rather than increasing it. Taking advantage of the concept of the economic search margin, we use an early stop mechanism to increase the searching performance. Thus, taxi search algorithm can halt further recursion if no suitable taxi is found.
- We propose a solution for operating dynamic ridesharing process based on incomplete data by using an interval travel time estimate. The travel time estimation itself is not the focus of our research, but we show how to incorporate incomplete travel time estimate index when there is no data available for a specific pair of source/destination.
- We conduct extensive experiments using a large scale real-world dataset collected in Beijing, China. We compare our approach with an enhanced implementation of the incremental search and the experimental results demonstrate the efficiency and benefits of our approach.

The rest of this paper is organized as follows. In Section 2 we analyze the structure of the problem from different aspects including motivation, definition, modeling and discussing our framework. We present the technical details of our

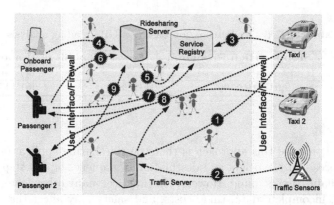

Fig. 1. Illustration of a multi-agent dynamic ridesharing scenario

solution in Section 3 and the results of the experimental evaluation in Section 4. Section 5 reviews related works and Section 6 concludes the paper.

2   Preliminaries

In this section, we first use a scenario to illustrate the dynamic ridesharing problem. Then we briefly introduce some basic concepts related to our work.

2.1   Ridesharing Illustration

Figure 1 illustrates a scenario for dynamic ridesharing. In a normal process, each registered taxi continuously transmits its status to the traffic server when it is active and the traffic server continuously updates taxi position in the database that records GPS readings (1). The traffic server also keeps scanning the traffic status data, and updates its traffic forecast model using a traffic modeling agent (2). Whenever a taxi is ready to receive a ridesharing request, the taxi's agent (3) can register the taxi in the service registry. The ridesharing server receives requests from onboard passengers (4), which in turn registers the corresponding taxis into the service registry (5).

When a new user (passenger) logs into the system, her agent will be launched (6) and registers her into the service registry (7). As the user submits her query with certain specified parameters such as time windows (i.e., pickup and drop off), origin and destination, the ridesharing application offers her the best taxi that minimizes the costs as well as the extra distance (8). The results are determined through her direct query. The generated list of the nearest taxis are passed to her agent. At the end, if the user selects one of the proposed options, it will be registered and monitored until the user gets her booked taxi (9).

2.2   Basic Concepts

In this section, we present some basic concepts that are closely related to our work in this paper. Table 2 summarizes the notations to be used in the following definitions and throughout this paper.

Definition 1 (Segment). *For the sake of simplicity of calculations, the time and space domains are considered discrete in value. Thus, the space range is split into a grid. Each grid cell is called a segment. Each segment is denoted as s_i while function neighbours(s_i) returns the grid cells which have a shared border with s_i. The list of the taxis within the borders of a cell is denoted by $s_i.l_v$.*   □

Definition 2 (Interval). *Denoted by $i = [\underline{i}, \overline{i}]$, $\underline{i} \leq \overline{i}$, is a set of real numbers with the property that any real number n, $\underline{i} \leq n \leq \overline{i}$, is a member of i.*   □

Definition 3 (Distance Matrix). *A distance matrix, denoted as Δ, contains the set of distances of different segments in the grid. Several distance measures such as road network, Manhattan or Euclidean can be used to initialize this matrix. This matrix is as the following:*

$$\Delta = \begin{pmatrix} \varnothing & \delta_{12} & \delta_{13} & \cdots & \delta_{1M} \\ \delta_{21} & \varnothing & \delta_{23} & \cdots & \delta_{2M} \\ \vdots & \vdots & \vdots & \ddots & \vdots \\ \delta_{M1} & \delta_{M2} & \delta_{M3} & \cdots & \varnothing \end{pmatrix}$$

Where δ_{ij} is the distance between segments s_i and s_j.   □

Definition 4 (Travel Time Estimate). *Travel time estimate is a measure given to segments to hold the track of travel time between every two segments in the area of interest (AOI). $\Theta(t)$ denotes the travel time estimation matrix, which is defined as the following:*

$$\Theta(t) = \begin{pmatrix} \varnothing & \theta_{12}(t) & \theta_{13}(t) & \cdots & \theta_{1M}(t) \\ \theta_{21}(t) & \varnothing & \theta_{23}(t) & \cdots & \theta_{2M}(t) \\ \vdots & \vdots & \vdots & \ddots & \vdots \\ \theta_{M1}(t) & \theta_{M2}(t) & \theta_{M3}(t) & \cdots & \varnothing \end{pmatrix}$$

Where

$$\theta_{ij}(t) = \begin{cases} \varnothing & i = j; \\ \lfloor \underline{\theta_{ij}(t)}, \overline{\theta_{ij}(t)} \rceil & otherwise. \end{cases}$$ □

Definition 5 (Query). *A query Q is a passenger's request for a taxi ride which is associated with a timestamp $Q.t$ indicating when the query is submitted, a pickup point (segment) $Q.o$, a delivery point (segment) $Q.d$, a time window $Q.wp$ defining the time interval when the passenger needs to be picked up at the pickup point, and a time window $Q.wd$ defining the time interval when the passenger needs to be dropped off at the delivery point. The early and late bounds of a pickup*

Table 2. List of notations for traffic modeling

ID	Definition
S	The set of segments
s_i	A grid segment i in segments list
Δ	The matrix of distances between segments
δ_{ij}	The distance between segments i and j
$\Theta(t)$	The matrix of historical travel time estimate
$\theta_{ij}(t)$	Travel time estimate interval between segments i and j
Q	A query for a taxi ride
$Q.t$	The birth time of query Q
$Q.p$	Passengers count specified in Q
$Q.o$	The pickup point (segment) of query Q
$Q.d$	The delivery point (segment) of query Q
$Q.wp, Q.wd$	Respectively the pickup and the delivery time window of Q
V	Taxi status
$V.s$	The current schedule of the taxi status V
$V.s.o, V.s.d$	Respectively, the current origin and destination of taxi V
$V.l$	Geographical taxi location
segment$(V.l)$	Corresponding segment of a taxi
Υ	A set of taxi trajectories
R^*	Economic search range
$u.cost$	Ridesharing cost for the user u
$u.exclusiveCost$	Sole (exclusive) ride cost for the user u
$u.wp, u.wd$	Respectively, user u's pickup and delivery time windows

window are denoted by $Q.wp$ and $\overline{Q.wp}$. Likewise, $Q.wd$ and $\overline{Q.wd}$ denote the bounds of the delivery window. To keep the description's simplicity, each query indicates only one passenger's request. □

Definition 6 (Taxi Status). *A taxi status V represents the instantaneous state of a taxi and comprises a taxi identifier $V.id$, a timestamp $V.t$, and a geographical location $V.l$.* □

Definition 7 (Satisfaction). *Given a taxi status V and a query Q, V satisfies Q if and only if (i) new passengers count is smaller than the current seat capacity of the taxi; (ii) Taxi with status V can pick up the passenger of Q at $Q.o$ no longer than $\overline{Q.wp}$ and drop him/her off at $Q.d$ no later than $\overline{Q.wd}$; (iii) Taxi with status V can pick up and drop off existing passengers no later than the late bound of their corresponding pickup and delivery time windows.* □

Figure 2(a) presents a view of a simple scenario where a passenger would like to travel from segment s_3 (i.e., Q_o) at time t_1 to segment s_0 (i.e., Q_d) at time t_3. Although there are four taxis (see the figure), if we limit the search area to $Q.o$ at t_0 and $Q.d$ at t_3, no taxi can satisfy the user with the exact pickup and drop off points, within the pickup and delivery time windows, specified by the user.

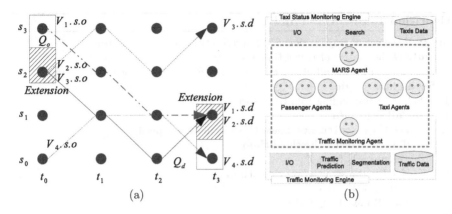

Fig. 2. (a) Search process, (b) MARS framework

The incremental approach expands the search area in each step and repeats the search. In Figure 2(a), after we extend the search area (see shadowed areas), two taxis (i.e., $taxi_1$, $taxi_2$) were found to satisfy the user's query.

2.3 The MARS Framework

First, we briefly present the architecture of our proposed framework (Figure 2(b)). It consists of three main modules, namely the *Ridesharing Multi-Agent System*, the *Traffic Monitoring Engine* and the *Taxi Status Monitoring Engine*.

Multi-Agent System. There are four types of agents in our framework, as shown in Figure 2(b). The *Passenger Agents* take care of both on-board and off-board passengers. The *Taxi Agents* are in charge of transferring data between taxis and engines as well as negotiating with the passenger agents.

Traffic Monitoring Engine. The Traffic Monitoring Engine models and analyzes the traffic status. It is also responsible for indexing roads map and traffic data. This engine provides the ridesharing applications with the latest traffic status predictions and routes data. We will cover the technical details on generating travel time estimate index from raw taxi trajectories data in Section 3.

Taxi Status Monitoring Engine. The Taxi Status Monitoring Engine maintains taxis' data and performs search for incoming user queries. The Search service requires *Segmentation* and the Traffic Prediction service from the *Traffic Monitoring Engine*. A *MARS Agent* will be in charge of handling users' queries as well as performing transactions with the *Traffic Monitoring Agent*. This module is the focus of our work and the technical details are given in Section 3.

Algorithm 1. UPDATE-THETA

Input: $\mathcal{V}$ the set of trajectories, $\theta^*(t)$ and $\theta^-(t)$ limits for travel time
Output: Θ the matrix of travel time estimate
1: Let $\overline{\theta(t)} \leftarrow \theta^*(t)$ for all $\theta(t) \in \Theta(t)$ **and** Let $\underline{\theta(t)} \leftarrow \theta^-(t)$ for all $\theta(t) \in \Theta(t)$
2: **for all** $taxi \in \mathcal{V}$ **do**
3:     Select corresponding taxi trajectories Υ from $\mathcal{V}$
4:     **for all** $V \in \Upsilon$ **do**
5:         Let $V.s \leftarrow$ `calculate-speed`(V) calculate its speed
6:         **if** $V.s <$ moving threshold **then**
7:             remove V from Υ
8:         **if** $V.l$ is the same as previous **then**
9:             remove V from Υ
10:         **for all** next $V' \in \Upsilon$ **do**
11:             **if** `segment`$(V'.l)$ equals `segment`$(V.l)$ **then**
12:                 break for loop
13:             Let $\theta(t) \leftarrow$`diff`$(V.t, V'.t)$
14:             Let $s_i \leftarrow$`segment`$(V.l)$
15:             Let $s_j \leftarrow$`segment`$(V'.l)$
16:             Update $\overline{\theta_{ij}(t)} \leftarrow$`max`$(\theta(t), \overline{\theta_{ij}(t)})$
17:             Update $\underline{\theta_{ij}(t)} \leftarrow$`min`$(\theta(t), \underline{\theta_{ij}(t)})$
18: **return** $\Theta(t)$

---

3   Technical Details

In this section, we present the technical details of our solution. Based on the proposed framework in the previous section, we cover the technical details of *Traffic Prediction* (in the Traffic Monitoring Engine) and *Search* (in the Taxi Status Monitoring Engine).

3.1   Travel Time Estimation

Although travel time estimation itself is not the focus of this paper, we still need to provide an estimation method for it. In the absence of trip based data where the origins and destinations of trips have not been specified, historical travel records can be used to approximate the travel time between two segments. The input for this module is a set of historical raw taxi trajectories and their timestamps denoted as V. Algorithm 1 shows the main procedure for generating travel time estimate index from raw sensor data.

Algorithm 1 requires the minimum and maximum travel times as its input. These numbers can be estimated based on the distance of every segment and the size of the area of interest (AOI). At the first step, the algorithm initializes the travel time matrix with the minimum and maximum travel times. Then for each taxi after filtering irrelevant and unreliable data, the next steps will become more efficient. After calculating the permutations of entries, the matrix can be updated. It happens only if the origin and destination of a trip are not the same because it is rare (or unlikely) that a passenger takes a cyclic route. Finally, the updated travel time estimate matrix is ready. In the next step, whenever a new query is received, the $\Theta(t)$ can be updated based on $Q.t$ timestamp.

3.2   Search

Area extensions are applied to the origin and destination areas in order to find the taxis with the least extra distance for taxis because for most of the queries no taxi is found which exactly matches the specified requirements. As mentioned before, one of the recent taxi search approaches is exploiting an *incremental* search approach. It has been used in designing dynamic ridesharing applications such as T-Share [9]. Algorithm 2 shows a procedure based on this approach.

Algorithm 2. INCREMENTAL-SEARCH

Input: $Q, origin, dest$
Output: T set of taxis which satisfy the query
1: Let $T_1 \leftarrow$ getTaxis($origin, Q.wp$)
2: Let $T_2 \leftarrow$ getTaxis($dest, Q.wd$)
3: **for all** $taxi \in T_1$ **do**
4:    **if** $taxi \in T_2$ **then**
5:       Let $T \leftarrow T \cup taxi$ add taxi to output list.
6: **if** T is empty **then**
7:    Let $origin \leftarrow$ expand($origin, Q.o$) extending the search area.
8:    Let $dest \leftarrow$ expand($dest, Q.d$) extending the search area.
9:    **return** INCREMENTAL-SEARCH($Q, origin, dest$)
10: **else**
11:    **return** T

The set of different steps of the incremental search algorithm takes place in the following order. First, the taxis at extended origin area and extended destination area are queried and stored in T_1 and T_2 respectively. The list of the common taxis in the two sets are stored in the T set. Then, if T is empty, the area is expanded and search is recursively called with the expanded areas and the same query. Otherwise, T is returned. One major problem of this approach is that the search area growth rate is very low (only one segment at a time) and it does not provide any mechanism to stop search and extension if no economically justifiable taxi is available.

In this paper, instead of increasing the search area, we decrease the search area in each step. The details of our decremental search approach is shown in Algorithm 3. The proposed algorithm includes the following steps. First, if each one of the origin and destination areas has only one segment (not an extended area), we cannot split them further and the set of common taxis in the specified time windows is returned. Otherwise, the rest of the process is applied as follows. The set of taxis which appear at the specified time windows are stored in T. Next, if T is not empty, T will be replaced by the search results in the shrunken areas. Otherwise, the taxi search algorithm can stop the further recursion. We view

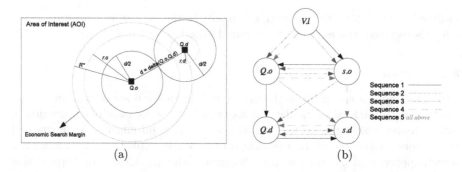

Fig. 3. (a) Symmetric (solid) and asymmetric (dashed) extended search areas (b) Possible schedule sequences for an incoming query

this early-stop feature an important one since it can increase the *efficiency* of the application. Utilizing a limit for expansion, unlike the other approach, there exists a point where the algorithm can stop further recursion.

Algorithm 3. DECREMENTAL-SEARCH

Input: $Q, origin, dest$
Output: T set of taxis which satisfy the query
1: **if** $origin$ equals $Q.o$ **and** $dest$ equals $Q.d$ **then**
2:     **return** $T \leftarrow$ getTaxis($origin, dest, Q.wp, Q.wd$)
3: Let $T_1 \leftarrow$ getTaxis($origin, Q.wp$)
4: Let $T_2 \leftarrow$ getTaxis($dest, Q.wd$)
5: **for all** $taxi \in T_1$ **do**
6:     **if** $taxi \in T_2$ **then**
7:         Let $T \leftarrow T \cup taxi$ add taxi to output list.
8: **if** T is **not** empty **then**
9:     Let $origin \leftarrow$ shrink($origin, Q.o$) shrinking the search area.
10:     Let $dest \leftarrow$ shrink($dest, Q.d$) shrinking the search area.
11:     Let $T' \leftarrow$ DECREMENTAL-SEARCH($Q, origin, dest$)
12: **if** T' is **not** empty **then**
13:     Let $T \leftarrow T'$
14: **return** T

To initialize, our decremental search requires the total extended area that includes all of the economically justifiable taxis. We call it *Economic Search Margin* and denote the radius with R^* (Figure 3(a)). It is worth mentioning that this area is not necessarily a circular area. The shape of this area depends on the distance of each segment from the center ($Q.o$ or $Q.d$). For any segment $s_i \in S$ if the distance of s_i is equal or less than R^*, it will be included in the economic search margin. An important question is that how far we can go to look for economically justifiable taxis (i.e., how to specify the economic search margin)? In order to answer this question, we analyze the possible sequences for new trips. Suppose that a user u wants to register a new trip (with pickup time window $u.wp$

and delivery time window $u.wd$) for a taxi with pre-scheduled trips (with pickup time window $u'.wp$ and delivery time window $u'.wd$) of m users denoted as u'. For the sake of simplicity, we consider two sets of users but the approach can be generalized to support sequences with multiple pickup/drops. The following sequences are possible for the taxi (see Figure 3(b)):

Sequence 1: Respectively, pre-scheduled pickup, new user pickup, new user delivery, pre-scheduled delivery. The riding cost can be formulated as follows:

$$u.cost = \frac{cost(\delta_{od}, \theta_{od}(u.wp))}{(m+1)} \tag{1}$$

$$u'.cost = \frac{cost(\delta_{io}, \theta_{io}(u'.wp))}{m} + \frac{cost(\delta_{od}, \theta_{od}(u.wp))}{m+1} + \frac{cost(\delta_{dj}, \theta_{dj}(u.wd))}{m} \tag{2}$$

Sequence 2: Respectively, pre-scheduled pickup, new user pickup, pre-scheduled delivery, new user delivery. The riding cost can be formulated as follows:

$$u.cost = \frac{cost(\delta_{oj}, \theta_{oj}(u.wp))}{m+1} + cost(\delta_{jd}, \theta_{jd}(u'.wd)) \tag{3}$$

$$u'.cost = \frac{cost(\delta_{io}, \theta_{io}(u'.wp))}{m} + \frac{cost(\delta_{oj}, \theta_{oj}(u.wp))}{m+1} \tag{4}$$

Sequence 3: Respectively, new user pickup, pre-scheduled pickup, new user delivery, pre-scheduled delivery. The riding cost can be formulated as follows:

$$u.cost = cost(\delta_{oi}, \theta_{oi}(u.wp)) + \frac{cost(\delta_{id}, \theta_{id}(u'.wp))}{m+1} \tag{5}$$

$$u'.cost = \frac{cost(\delta_{id}, \theta_{id}(u'.wp))}{m+1} + \frac{cost(\delta_{dj}, \theta_{dj}(u.wp))}{m} \tag{6}$$

Sequence 4: Respectively, new user pickup, pre-scheduled pickup, pre-scheduled delivery, new user delivery. The riding cost can be formulated as follows:

$$u.cost = cost(\delta_{oi}, \theta_{oi}(u.wp)) + \frac{cost(\delta_{ij}, \theta_{ij}(u'.wp))}{m+1} + cost(\delta_{jd}, \theta_{jd}(u'.wd)) \tag{7}$$

$$u'.cost = \frac{cost(\delta_{ij}, \theta_{ij}(u'.wp))}{m+1} \tag{8}$$

Sequence 5: One or more overlapping schedules. Since the level of uncertainty is higher in this case, each ride's cost can be formulated as follows:

$$\underline{u.cost} = \frac{cost(\delta_{od}, \theta_{od}(u.wp))}{(m+1)} \tag{9}$$

$$\overline{u.cost} = \overline{cost}(\delta_{oi}, \theta_{oi}(u.wp)) + \frac{\overline{cost}(\delta_{ij}, \theta_{ij}(u'.wp))}{m+1} + \overline{cost}(\delta_{jd}, \theta_{jd}(u'.wd)) \tag{10}$$

$$\underline{u'.cost} = \frac{cost(\delta_{ij}, \theta_{ij}(u'.wp))}{m+1} \tag{11}$$

$$\overline{u'.cost} = \frac{\overline{cost}(\delta_{io}, \theta_{io}(u'.wp))}{m} + \frac{\overline{cost}(\delta_{od}, \theta_{od}(u.wp))}{m+1} + \frac{\overline{cost}(\delta_{dj}, \theta_{dj}(u.wd))}{m} \tag{12}$$

Based on the above analysis, we have Theorem 1 to find the maximum distance of search area expansion that a ridesharing application can use for searching taxis with economically justifiable additional cost.

Theorem 1: If δ_{od} denotes the distance between $Q.o$ and $Q.d$ (the length of user's travel), then $R^* = \delta_{od}$ is the maximum additional distance that a taxi can undertake to be economically justifiable. Thus, a taxi with scheduled trips further this distance is not within the economically justifiable range.   □

Proof. We prove Theorem 1 by using the costs and benefits analysis for all possible schedule sequences. Based on the mutual benefit principle and by summing the total cost for all stakeholders, we have:

$$u.cost + m * u'.cost < u.soleRideCost + m * u'.soleRideCost \qquad (13)$$

This statement should fit for all of the sequences. For instance, for Sequence 1:

$$u.cost + m * u'.cost = cost(\delta_{oi}, \theta_{oi}(t_1)) + cost(\delta_{ij}, \theta_{ij}(t_2)) + cost(\delta_{jd}, \theta_{jd}(t_3)) \qquad (14)$$

Let us suppose by contradiction that $R^* > \delta_{od}$. Then, if $soleRideCost$ denotes $u.soleRideCost + m * u'.soleRideCost$, we will have:

$$u.cost + m * u'.cost \geq soleRideCost \qquad (15)$$

Which totally denies the fact that everyone who takes part in the ridesharing process, must benefit (economically) from it. As a result, Theorem 1 is proved. □

4   Experimental Results

We have implemented the proposed approach and conducted extensive experiments to study its performance and effectiveness. Due to the space constraint, we will report two experiments. We used Java 1.6 and Java Agent Development Framework (JADE) 3.7 for the experiment. All of the presented results of this paper were executed on a computer with a 2.4 GHz Intel Core 2 Duo processor and 4 GB of memory running Mac OS X 10.9.1.

4.1   Dataset

We used a real-world dataset that contains records of GPS readings for 10,357 taxicabs for the city of Beijing [19,20]. The sampling frequency for each taxi differs from a few seconds to a couple of hours. We used this taxi trajectories data for two main purposes: initiating the system and estimating traveling times. Taxi movements followed successful queries while we hypothetically suppose that each taxi can finish a trip within the maximum bound of travel time estimation. In order to prepare the data to be used in the experiment, a few steps were taken which are described in the following.

We first divided the city into 76 parts in length and 76 parts in height, which means that we have 5,776 segments. Each segment represents approximately 850 meters in length and 1,112 meters in height. All GPS readings with their corresponding segments falling outside the AOI boundaries were removed. Then we got the taxi's velocity through calculating the distance and elapsed time from previous record. Finally, in order to model the moving taxis, we analyzed their previous trips. Since we did not have the previous trips of the taxis, in order to seize every possible trip, we calculated the permutations of successor records and considered each one of them as a possible trip. To get a set of permutations with more reasonable size, possible stops (for example records with velocity less than 1 km/h) as well as cyclic trips with the same itineraries were dismissed.

Figure 4(a) shows the scatter plot of information availability between different segments in the existing database. As it shows, the existing data covers most of the central segments within the map. However, even if no data exists in outer segments, it does not mean that no route can connect them. If no data is specified for two segments s_i and s_j, we suppose that the travel time estimate between them is an interval within the maximum and minimum travel time estimates. The segments count in the figure is the number of segments in S if all of its segments in all rows are enumerated one after the other. As we expected, even the furthest segments can be hardly accessible within the specified maximum time limit (3,600 seconds).

4.2   Performance

In our implementation, the user agents continuously generated and submitted queries to the MARS agent. We generated more than 6,200 random ridesharing requests in our experiments. Travel distance and search expansion ability have a direct relationship with each other. In other words, a bigger travel distance means more workload imposed on the algorithm. Figure 4(b) shows the average distance of users' trips generated from the queries. It should be noted that the average distance of trips is approximately high, which was designed to test the system's capability in handling heavy workload.

The aim of this experiment is to examine the benefits of our proposed decremental approach in terms of performance and query success rate. To analyze the effect on performance factor, we executed the decremental and incremental approaches and analyzed the time taken to process a certain number (6,200) of randomly generated queries. To assure that the system can handle heavier workloads, we leveraged the workload by distributing the queries evenly in the AOI. For the baseline, we used an enhanced version of incremental approach in which in each round, the extension is applied to the search area by adding all the surrounding segments (instead of one by one segment strategy used by e.g., T-Share) in each iteration.

We executed the same set of rideshare queries both of the incremental and decremental approaches. During the experiment, the running times and acceptance results for processing every query were recorded. Figure 4(c) shows the average amount of taxi search time taken to process 100 queries during the experiment for both incremental and decremental search approaches. As shown, after processing 6,000 queries, the optimized incremental approach takes nearly 9 seconds to process 100 queries while the decremental approach only takes up less than 5 seconds. Our proposed decremental search approach shows to be more efficient in terms of process time. Figure 4(e) shows that the decremental

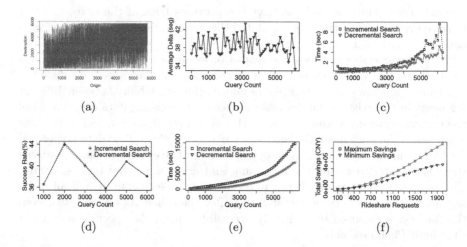

Fig. 4. (a) Scatter plot for available data in the dataset, (b) Statistics of the generated index and queries, (c) Comparison of average search time of incremental and decremental search, (d) Comparison of success rate of incremental and decremental search, (e) Comparison of total search time of incremental and decremental search, (f) The sum of all users' savings for successful queries

search approach overtakes the enhanced incremental approach in terms of the total time spent for search.

We also compare the number of successful search results for both algorithms. Figure 4(d) depicts the ratio of the successful queries to all of all generated queries in each interval. For example, while only less than 10% of the first 100 queries were successful, nearly 40% of the last 100 queries in both of the approaches were successful in finding a rideshare. Although at some points very slight differences can be detected between the two approaches, the total success of both approaches remained very close. In fact, only 3 out of 6,200 queries had different results between the two approaches. This can be reflected by the almost identical plot of the two approaches in the figure.

4.3   Savings

In this experiment, we investigated the effectiveness of the proposed solution via the amount of passengers savings. To assess the amount of cost savings, in the first step we need a pricing scheme. We used the following formula to calculate the taxi fares (in Chinese Yuan)[1]:

$$\overline{cost(\delta, \theta(t))} = 12 + \delta * 2.2 + (\overline{\theta(t)} - \theta(t)) * (0.01)$$
$$\underline{cost(\delta, \theta(t))} = 12 + \delta * 2.2 \tag{16}$$

---

[1] http://www.numbeo.com

Our pricing scheme is composed of three separate fees: a *starting fee*, a *distance fee*, and a *traffic fee*. Using the same results from the previous experiment, we performed a cost analysis for successful queries based on the pricing scheme. For each query and each involved passenger, we estimated and compared the price for a sole ride as well as the price for a rideshare. Figure 4(f) shows the cumulative plot of the minimum and maximum estimations for the total savings of 2,000 successful queries. The upper bound for savings is gained through subtracting the worst case (i.e., sole ride) from the best case of ridesharing. Similarly, the lower bound is gained through subtracting the best cost of sole ride from the worst cost of the ridesharing. The results shows that the application of our approach can be effective in terms of cost savings, saving minimum of 300,000 and maximum of 700,000 Chinese Yuan for its passengers within one hour of operation during rush hours.

5   Related Works

5.1   Dynamic Ridesharing

Ridesharing has been actively studied in past few years. Dynamic ridesharing is generally described as an automated system that facilitates drivers and riders to share one time trips close to their departure times/places and can be characterized with features like dynamic, independent, cost-sharing, non-recurring trips, prearranged and automated matching [1].

Most of the proposed applications have not been tested for either real-life and/or large-scale datasets. These solutions are developed based on a variety of techniques and approaches such as analyzing social connections in SRSS [6], multi source-destination path planning [18] and even cloud computing [5]. Recently, multi-agent systems have been deployed to simulate/implement solutions [12]. Some other very recent solutions such as T-Share [9], Noah [15] and kinetic tree algorithm [7] treat dynamic taxi ridesharing problem as finding k-nearest neighbors (KNN) problem although it does not cover all challenges in minimizing system wide travel time and travel distance. These works are some of the very few works that use large-scale datasets. However, the support for fluctuations in data availability is still limited and can be extended.

Destination/path prediction can be also interesting. DesTeller is a system for destination prediction based on trajectories considering passengers privacy [17,16]. It uses *Sub-Trajectory Synthesis* (SubSyn) to solve data sparsity problem when query trajectories continue to non-terminal links. R2-D2 uses semi-lazy learning to support probabilistic path prediction in dynamic environments. It is composed of an *update* and a *prediction* process and has been applied to a large scale dataset. Another approach uses *Hidden Markov Models* in order to predict the future locations of moving objects [10].

5.2   Travel Time Estimation

Travel time estimation is one of the key steps towards developing more accurate dynamic ridesharing systems [9]. It is a highly challenging problem because it deals with several uncertainty factors such as traffic fluctuations, demand and supply, traffic signals, weather conditions and seasonal changes [19,20].

One of the recent works proposes an approach for travel time estimation using large-scale taxi data with partial information [21]. The proposed model focuses on uncertainty in path choices. It infers the possible paths for each trip and then estimates the link travel times by minimizing the error between the expected path travel times and the observed path travel times. This model uses only the current time data and does not support historical data.

6   Conclusion

Despite recent active research efforts, dynamic ridesharing still remains many challenges. In this paper, we have proposed a framework for dynamic ridesharing using intelligent agent technology. The proposed framework can handle incomplete data and increases the performance of taxi search by developing a decremental approach. We have conducted extensive experiments using a real-world, large-scale dataset of taxi trajectories collected in Beijing, China to evaluate the proposed search approach. The experimental results confirm the effectiveness and the efficiency of the proposed approach. Our proposed decremental search algorithm is faster than the state-of-art approaches, yet being able to maintain query accuracy at the similar level.

The work presented in this paper is the first step to address the challenges of dynamic ridesharing. There are several interesting directions for our future research. We plan to investigate the modeling of different sources of uncertainty and analyze their impacts on taxi ridesharing. We also plan to integrate the user decision making process into our approach for better ridesharing results.

References

1. Agatz, N., Erera, A., Savelsbergh, M., Wang, X.: Optimization for dynamic ridesharing: A review. European J. of Operational Research 223(2), 295–303 (2012)
2. Bhaumik, C., Agrawal, A.K., Sinha, P.: Using social network graphs for search space reduction in internet of things. In: Proceedings of the 2012 Conference on Ubiquitous Computing (Ubicomp), pp. 602–603. ACM (2012)
3. Caulfield, B.: Estimating the environmental benefits of ride-sharing: A case study of dublin. Transportation Research Part D: Transport and Environment 14(7), 527–531 (2009)
4. Crowcroft, J.: Fie: Future internet enervation. In: ACM SIGCOMM Computer Communication Review, vol. 40, pp. 48–52. ACM, New York (2010)
5. Dimitrieski, V.: Real-time carpooling and ride-sharing: Position paper on design concepts, distribution and cloud computing strategies. In: Proceedings of the 2013 Federated Conference on Computer Science and Information Systems (FedCSIS), Kraków, Poland, pp. 781–786 (2013)
6. Gidófalvi, G., Herenyi, G., Bach Pedersen, T.: Instant social ride-sharing. In: Proceedings of the 15th World Congress on Intelligent Transport Systems, p. 8. Intelligent Transportation Society of America, New York (2008)

7. Huang, Y., Jin, R., Bastani, F., Wang, X.S.: Large scale real-time ridesharing with service guarantee on road networks. Computing Research Repository (2013)
8. Lin, Y., Li, W., Qiu, F., Xu, H.: Research on optimization of vehicle routing problem for ride-sharing taxi, Shaoxing, China, pp. 494–502 (2012)
9. Ma, S., Zheng, Y., Wolfson, O.: T-share: A large-scale dynamic taxi ridesharing service. In: Proceedings of 29th International Conference on Data Engineering (ICDE 2013), pp. 410–421. IEEE, Brisbane (2013)
10. Qiu, D., Papotti, P., Blanco, L.: Future locations prediction with uncertain data. In: Blockeel, H., Kersting, K., Nijssen, S., Železný, F. (eds.) ECML PKDD 2013, Part I. LNCS, vol. 8188, pp. 417–432. Springer, Heidelberg (2013)
11. Santos, D.O., Xavier, E.C.: Dynamic taxi and ridesharing: A framework and heuristics for the optimization problem. In: Proceedings of the 23rd International Joint Conference on Artificial Intelligence (IJCAI 2013), pp. 2885–2891. AAAI Press, Beijing (2013)
12. Sghaier, M., Zgaya, H., Hammadi, S., Tahon, C.: A distributed optimized approach based on the multi agent concept for the implementation of a real time carpooling service with an optimization aspect on siblings. International Journal of Engineering 5(2), 217–241 (2011)
13. Sheng, Q.Z., Li, X., Zeadally, S.: Enabling Next-Generation RFID Applications: Solutions and Challenges. IEEE Computer 41(9), 21–28 (2008)
14. Tao, C.C.: Dynamic taxi-sharing service using intelligent transportation system technologies. In: Proceedings of International Conference on Wireless Communications, Networking and Mobile Computing (WiCom 2007), pp. 3209–3212. IEEE, Shanghai (2007)
15. Tian, C., Huang, Y., Liu, Z., Bastani, F., Jin, R.: Noah: a dynamic ridesharing system. In: Proceedings of the 2013 ACM SIGMOD Conference (SIGMOD 2013), pp. 985–988. ACM, New York (2013)
16. Xue, A.Y., Zhang, R., Zheng, Y., Xie, X., Huang, J., Xu, Z.: Destination prediction by sub-trajectory synthesis and privacy protection against such prediction. In: Proceedings of 29th International Conference on Data Engineering (ICDE 2013), Brisbane, Australia, pp. 254–265 (2013)
17. Xue, A.Y., Zhang, R., Zheng, Y., Xie, X., Yu, J., Tang, Y.: Desteller: A system for destination prediction based on trajectories with privacy protection. In: Proceedings of the VLDB Endowment, vol. 6, pp. 1198–1201. VLDB Endowment (2013)
18. Yousaf, J., Li, J., Chen, L., Tang, J., Dai, X., Du, J.: Ride-sharing: A multi source-destination path planning approach. In: Thielscher, M., Zhang, D. (eds.) AI 2012. LNCS, vol. 7691, pp. 815–826. Springer, Heidelberg (2012)
19. Yuan, J., Zheng, Y., Xie, X., Sun, G.: Driving with knowledge from the physical world. In: Proceedings of the 17th International Conference on Knowledge Discovery and Data Mining (SIGKDD), pp. 316–324. ACM (2011)
20. Yuan, J., Zheng, Y., Zhang, C., Xie, W., Xie, X., Sun, G., Huang, Y.: T-drive: driving directions based on taxi trajectories. In: Proceedings of the 18th International Conference on Advances in Geographic Information Systems (SIGSPATIAL), pp. 99–108. ACM (2010)
21. Zhan, X., Hasan, S., Ukkusuri, S.V., Kamga, C.: Urban link travel time estimation using large-scale taxi data with partial information. Transportation Research Part C: Emerging Technologies 33, 37–49 (2013)

Model-Based Search and Ranking of Web APIs across Multiple Repositories

Devis Bianchini, Valeria De Antonellis, and Michele Melchiori

Dept. of Information Engineering University of Brescia
via Branze, 38, 25123 Brescia, Italy
{devis.bianchini,valeria.deantonellis,michele.melchiori}@unibs.it

Abstract. Web API search and reuse for agile Web application development may benefit from selection criteria that combine several perspectives: they can be performed based on features used to describe APIs, or according to the co-occurrence of Web APIs in the same applications, or they can be driven through ratings assigned by designers who used the Web APIs for their own mashups. Nevertheless, different Web API repositories usually focus on a subset of these perspectives, thus providing complementary Web API descriptions. In this paper, we propose a unified model for Web API characterization. The model enables a cross-repository search of Web APIs and mashups, based on different kinds of similarity between them, identified regardless the complementarity of their descriptions. This unified representation improves retrieval results if compared with a Web API search performed over multiple repositories considered separately.

1 Introduction

Web API selection and aggregation, performed for mashup and short-living application development, may benefit from the adoption of criteria that combine different perspectives [1]: a *component* perspective (based on features used to describe Web APIs); an *application* perspective (i.e., information about mashups composed of the Web APIs); an *experience perspective* (including ratings assigned by web designers, who used Web APIs to develop their own mashups). The advantages coming from a multi-perspective Web API search have been confirmed by several approaches, that combined categories, tags and technical features like the adopted protocols and data formats in Web API descriptions with the co-occurrence of APIs in the same applications [2], with a quality-based model for Web APIs [3], with the network traffic around APIs and mashups, as an indicator of their success, and ratings assigned by designers [4]. Existing approaches rely on a single Web API repository. The `ProgrammableWeb` repository[1] is the most common one for sharing Web APIs and mashups. It contains over 11,500 Web APIs, where about 1,200 of them have been registered in the last year. APIs have been used in more than 7,400 mashups, while over 2,800 mashup

---

[1] http://www.programmableweb.com/

B. Benatallah et al. (Eds.): WISE 2014, Part I, LNCS 8786, pp. 218–233, 2014.
© Springer International Publishing Switzerland 2014

owners are registered in the repository. Nevertheless, different repositories emphasize complementary aspects to be considered for Web API search. Although `ProgrammableWeb` constitutes a well-known meeting point for the community of mashup developers, it does not provide a comprehensive Web API model that includes all the perspectives: it is mainly focused on a feature-based description of Web APIs (through categories, tags and technical features) and on the list of mashups that have been developed using the Web APIs. Another repository, `Mashape`[2], a cloud API hub leveraging a twitter-like organization, associated each Web API with the list of developers who adopted or declared their interest for it, denoted as *consumers* and *followers*, respectively. Other public repositories, such as `apigee` or `Anypoint API Portal`[3], focus on different and only partially overlapping aspects as well. This scenario brings to situations where: (i) the same Web APIs or mashups are registered multiple times within different repositories; (ii) Web APIs (resp., mashups) are searched and ranked according to distinct criteria in separate repositories, to meet different Web API (resp., mashup) descriptions (for instance, in `ProgrammableWeb` Web APIs are ranked with respect to the number of mashups they have been used in, Web API ranking performed on `Mashape` repository depends on the number of API followers). As proved in [1], performing Web API search and ranking on a comprehensive API descriptor, that includes different and complementary descriptive aspects, would improve retrieval results. This implies that it is not enough to search for APIs within distinct repositories considered separately and simply merge search results, but a real unified view over the repositories before starting the search is required. In this sense, similarity between Web APIs and mashups across different repositories should be exploited to enrich search results. Current Web API search scenarios lack of a model that provides a unified representation of Web APIs and mashups, to ease the identification of similar resources regardless the complementarity of their descriptions across different repositories [5]. Behind the advantage of avoiding multiple copies of the same API among the search results, although described with different properties depending on the repository from where API has been extracted, such a unified view would improve the retrieval outcomes as expected.

In this paper, we discuss about the definition of this model such that: (i) its unified representation covers the three perspectives mentioned above in Web API description, namely component, application and experience perspectives; (ii) it is part of a framework that enables the identification of different kinds of similarity between Web APIs and mashups, to provide a cross-repository search of these resources; (iii) it is integrated with a Web API and mashup search engine, that exploits similarity measures to properly access complementary information across repositories. A preliminary experimental evaluation confirms the improved search results, obtained by applying our approach, compared with Web API search performed on multiple repositories considered separately.

---

[2] https://www.mashape.com/

[3] https://api-portal.anypoint.mulesoft.com

The paper is organized as follows: Section 2 presents a motivating example for introducing the unified model, that is discussed in Section 3; similarity criteria are presented in Section 4; in Section 5 we describe the Web API and mashup search based on the model; results of the preliminary evaluation are discussed in Section 6; a comparison with related work is provided in Section 7; Section 8 closes the paper.

2   Motivating Example

Let's consider a web designer who aims at including a face recognition functionality to access the private area of his/her own web site. Since developing this kind of applications from scratch would require very specific competencies and could be costly and time-consuming, the designer prefers to look for existing available Web APIs, that implement the desired functionalities, and examples of their use in mashups shared by other designers. Now let's consider the situation depicted in Figure 1. The figure reports some mashups and Web APIs that are relevant for the designer's purpose, obtained from ProgrammableWeb and Mashape repositories, by issuing a query with "face recognition" keywords. For example, the Recognizer mashup, where the LambdaLabs Face and SkyBiometry APIs are used together to provide multiple recognition services based on biometrics features, might fit the designer's goal. The Recognizer mashup can be used by the designer to infer how LambdaLabs Face and SkyBiometry APIs can be fruifully used together. Nevertheless, while the SkyBiometry API has been used in 49 mashups (including Recognizer, Art4Europe, SaveUp applications) and has been positively rated by other designers, the LambdaLabs Face API did not reached the same popularity. However, the latter API is similar to the ReKognition API in a different repository. This API is rated better than the LambdaLabs Face one and has 237 consumers and 372 followers on Mashape. Therefore, a satisfactory search should return ReKognition API ranked better than LambdaLabs Face API to be used, for instance, together with SkyBiometry API for developing a face recognition application.

If we would consider the two repositories separately, the ReKognition API taken from Mashape can not be associated with any mashup that is relevant for the designer's search, since in this repository mashups information are not shared. Similarly, the SkyBiometry API can not be suggested to be used together with the successful ReKognition API on Mashape. Public Web API repositories (e.g., ProgrammableWeb, Mashape, apigee, Anypoint API Portal) provide facilities that enable to search for both Web APIs and mashups (if available) by specifying one or more keywords, that are matched against textual descriptions of APIs and mashups, but they do not enable any advanced search and ranking strategy relying on the component, application and experience perspectives highlighted in the introduction. The only way a designer may combine different viewpoints for Web API ranking is to manually analyze basic sorting facilities provided by existing repositories (such as, the popularity on ProgrammableWeb, meant as the number of mashups where a Web API has been included, and the

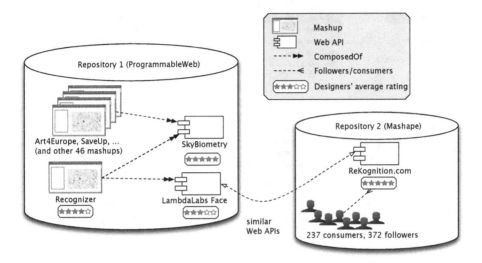

Fig. 1. The scenario considered in the motivating example, where a subset of face recognition APIs and mashups taken from the `ProgrammableWeb` and `Mashape` repositories is shown

number of followers of an API on `Mashape`). Finally, a seamless selection of Web APIs and mashups is necessary to be exploited as suggestions for a designer who aims at developing a new application from scratch (he/she may start from a single API, e.g., `SkyBiometry`) or at completing an existing one (he/she may learn from available mashups, viewed as sets of already aggregated APIs, e.g., `Recognizer`).

To overcome these limitations, in the following, we provide the designer with a unified model that enables a cross-repository search of Web APIs and mashups.

3   Web API and Mashup Unified Model

We introduce a unified *Web Mashup resource Descriptor* (hereafter, WMD), that embraces: (i) Web APIs as extracted from repositories; (ii) Web mashups, composed of one or more APIs. WMDs collect together these two kinds of resources by abstracting the set of their common features, as emerged through the analysis of the most popular public repositories and of state of the art approaches on Web API selection. Web APIs and mashups are basic elements of the *component* and *application perspectives* proposed in [1]. On top of this representation, ratings assigned by designers to Web mashup resources are considered to estimate their popularity (*experience perspective*). The WMD representation, derived from the proper combination of the three perspectives above, enables advanced search and ranking capabilities as described in the next sections. WMDs are collected and stored within a relational database, as shown in Figure 2.

Modeling Web Mashup Resources. A WMD is denoted by a unique identifier, corresponding to the URL of the Web API or the mashup, a human-readable

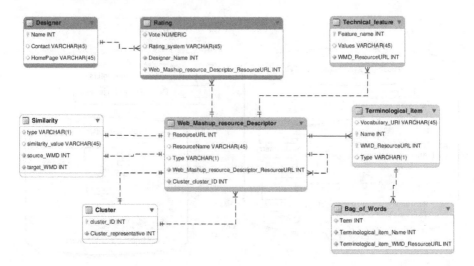

Fig. 2. The relational schema of the database containing the Web Mashup resource Descriptors, represented according to the unified model

name, a resource type, whose values, either M or W, denote the fact that WMD represents a Web API or a mashup. Each WMD is associated with a set of *terminological items*, that correspond to: (a) categories extracted from top-down classifications imposed within a given repository, where the resource is registered; (b) a term with explicit semantics, either a term extracted from WordNet or a concept extracted from an ontology in the Semantic Web context [6]; (c) a simple keyword or tag without an explicit representation of semantics. We distinguish keywords and tags as follows: tags are designer-assigned, bottom-up terms aimed at classifying WMDs in a folksonomy-like style, keywords are recurrent terms extracted from WMD textual descriptions using common IR techniques. A terminological item is in turn described by a representative name, an optional property that denotes the vocabulary, ontology, taxonomy or Word-Net sense where the item is defined (denoted with **Vocabulary_URI** in Figure 2) and a set of other terms (denoted as *bag of words*) used to further characterize the item (optional). In particular, given an item t_i, if t_i is a category, the item name corresponds to the category name, its bag of words is empty and t_i is described by the taxonomy or the classification the category belongs to. If t_i is extracted from WordNet, its bag of words coincides with the list of synonyms of the term, and it is described by a reference to the WordNet sense the term belongs to. If t_i is a concept extracted from an ontology, its bag of words is composed of the names of other concepts related to t_i by semantic relationships in the ontology (in the current version of our approach, we consider OWL/RDF equivalence and direct subsumption relationships); moreover, t_i is described by the URI of the ontology where it is defined. Finally, if t_i is a keyword or a tag without explicit semantics, its bag of words is empty and the item does not refer to any vocabulary or taxonomy where its meaning is properly defined. A WMD is further characterized through the set of technical features (e.g., protocols, data formats, security mechanisms) that have been adopted for the Web API (if the WMD represents a single component) or for the APIs that compose the mashup (if the WMD represents a whole web application). A self-relationship is defined on WMDs, to denote Web API composition into mashups. The abstraction of Web API and mashup descriptions through a single data structure, namely the Web_Mashup_resource_Descriptor table, is meant to support the unified search of Web APIs and mashups as shown in the motivating example (Section 2). The search might start from a Web API (e.g., SkyBiometry in Figure 1), pass through mashups that contain the API (e.g., Recognizer) and find other Web APIs that have been used in the same mashups. Or it might start from a mashup and retrieve all the Web APIs that have been used together in the mashup. The final goal is to retrieve APIs or mashups that are relevant for the keywords, tags or categories specified in the request. In this sense, collecting both Web API and mashup representations as records in a single table ensures the maximum search flexibility. Conceptually, it is equivalent to a pair of entities, representing Web APIs and mashups, and a parent entity that collects common features. Moreover, a unique table, if properly indexed, allows for good performance while inspecting the database during search.

Modeling the Experience Perspective. Each WMD is also associated with *ratings* assigned to it by designers, who may be either Web API providers, Web mashup owners, or they may be WMD consumers, who rate resources according to their personal opinion. In [1] the value of quantitative ratings is selected by the designers according to the NHLBI 9-point Scoring System. In this paper, we assume the same system as well. Since we refer to quantitative ratings uniformly distributed over a continuous range, the mapping from a different scoring system to this one is possible. In our previous work, we discussed how designers can be further characterized by their development skill, that is self-declared, as shown in [1]. Designers' skill can be exploited to properly weight their ratings, considering as more trustworthy the opinions of more expert designers. In [1] designers' skills have been used by the system for Web API search, but have not been published to preserve the anonymity of designers' reputation stored in the database. A designer is allowed to know his/her own skill only. More sophisticated anonymization techniques can be investigated, if necessary, as well as methods to automatically estimate designers' skill based on their experience in mashup development. In our unified model, we did not considered designers' skills yet, since they can not be directly extracted from public repositories we considered. Future efforts will be devoted to the integration of this aspect as well.

Extracting and Organizing Descriptors. WMDs are acquired by means of wrappers, designed to invoke specific methods made available by public repositories to query their contents[4]. Moreover, wrappers may interact with proper modules implemented on the Java platform to extract specific kinds of

---

[4] See, for instance, http://api.programmableweb.com for the ProgrammableWeb repository or http://www.mashape.com/mashaper/mashape#!documentation for the Mashape repository.

terminological items: a module based on `jWordNet` library, used to extract synonyms and senses; a module implemented on `Jena`, to parse concept definitions from OWL/RDF ontologies; an IR-based module, to extract keywords from textual descriptions of resources (when available). Other wrappers can be added to the system according to the modularized architecture, that requires only to create wrappers and connect them to the specific modules listed above. Within the database, *similarity* values between WMDs are stored as well, as detailed in the next section.

4 Cross-Repository Similarity Metrics

Similarity metrics have been defined to compare Web mashup resources, represented according to the unified model. We will distinguish among the following kinds of similarity, that will be detailed in the following:

- *terminological similarity*, based on terminological items;
- *technical similarity*, computed as the number of common values of technical features among the compared resources;
- *compositional similarity*, aimed at measuring the degree of overlapping between two mashups or compositions of Web APIs, evaluated as the number of common or similar Web APIs in the two compositions.

As for the abstraction of Web API and mashup features within the same table, also these different kinds of similarity are abstracted using a single table in Figure 2, that is, `Similarity` table, where the `type` attribute denotes the kind of similarity (among terminological, technical and compositional), similarity value is always normalized in the $[0, 1]$ range and is computed between two WMDs, namely the `source_WMD` and the `target_WMD`. All metrics are symmetric. The aim is at exploiting similarity values to cluster descriptors in order to support their search (see Section 5). Starting from the hypothesis that clustered resources tend to be relevant for the same request [7], similarity-based clustering is exploited to identify a unique resource, that represents a bundle of similar ones (i.e., the *cluster representative*). The request is then compared against the representative resource instead of against each clustered ones, in order to filter out not relevant results, thus improving the resource retrieval effectiveness (see Section 6 on experiments).

Terminological Similarity between Web Mashup Resources. The terminological similarity between two Web mashup resources res_1 and res_2, denoted with $TermSim(res_1, res_2) \in [0, 1]$, is based on the comparison of their terminological items, that is:

$$TermSim(res_1, res_2) = \frac{2 \cdot \sum_{t_1 \in \mathcal{T}_1, t_2 \in \mathcal{T}_2} itemSim(t_1, t_2)}{|\mathcal{T}_1| + |\mathcal{T}_2|} \in [0, 1] \qquad (1)$$

where we denote with $\mathcal{T}_i$ the set of terminological items used to characterize res_i, t_1 and t_2 are terminological items, $|\mathcal{T}_i|$ denotes the number of items in $\mathcal{T}_i$ set and $itemSim(\cdot)$ values are aggregated through the Dice formula. Pairs to be considered for the $TermSim$ computation are selected according to a maximization function that relies on the assignment in bipartite graphs. The point here is how to compute $itemSim(t_1, t_2) \in [0, 1]$ given the different types of involved terminological items. The algorithm for the $itemSim(\cdot)$ calculus is shown in Algorithm 1.

Algorithm 1. The $itemSim(\cdot)$ calculus algorithm

   Input   : Two terminological items t_1 and t_2.
   Output: The calculated $itemSim(t_1, t_2)$ value.

1 **if** *(t_1.type == C) and (t_2.type == C) (categories)* **then**
2     $itemSim(t_1, t_2) = Sim_{cat}(t_1, t_2)$ (using $Sim_{cat} \in [0, 1]$ defined in [1]);
3 **else if** *(t_1.type == WD) and (t_2.type == WD) (WordNet terms)* **then**
4     $itemSim(t_1, t_2) = Sim_{tag}(t_1, t_2)$ (using $Sim_{tag} \in [0, 1]$ defined in [1]);
5 **else if** *(t_1.type == O) and (t_2.type == O) (ontological concepts)* **then**
6     $itemSim(t_1, t_2) = $ H-MATCH(t_1, t_2) (using the H-MATCH$\in [0, 1]$
    function, given in [8]);
7 **else if** *(t_1.type == K) and (t_2.type == K) (keywords)* **then**
8     $itemSim(t_1, t_2) = StringSim(t_1, t_2) \in [0, 1]$ (using the Levenshtein
    measure);
9 **else**
10    $\Upsilon_1 = \{t_1.\text{name}\} \cup t_1.\text{bagOfWords}$;
11    $\Upsilon_2 = \{t_2.\text{name}\} \cup t_2.\text{bagOfWords}$;
12    $itemSim(t_1, t_2) = max_{i,j}\{StringSim(t_1^i, t_2^j)\}$, where $t_1^i \in \Upsilon_1$ and $t_2^j \in \Upsilon_2$;
13 **return** $itemSim(t_1, t_2)$;

When the types of t_1 and t_2 coincide, proper metrics from the literature are used for the comparison. In all the other cases, a comparison between the names of terminological items using the Levenshtein string similarity measure $(StringSim(\cdot))$ is performed, except for the case of WordNet terms and ontological concepts, that are expanded with the bag of words assigned to each item in order to look for a better matching term in the set (in fact, for these kinds of items only, `bagOfWords` is not empty).

Technical Similarity between Web Mashup Resources. The technical similarity between two Web mashup resources res_1 and res_2, denoted with $TechSim()$ $\in [0, 1]$, evaluates how many common feature values the two resources share. This metric is used to quantify the degree of compatibility between the resources in terms of protocols, data formats and other technical features. Feature values are compared only within the context of the same feature. Let $\mathcal{F}_X^{res_1}$ (resp., $\mathcal{F}_X^{res_2}$) the set of values admitted for the technical feature X associated with the Web mashup resources res_1 and res_2, respectively. The technical similarity between the Web mashup resources res_1 and res_2 is computed as follows:

$$TechSim(res_1, res_2) = \frac{1}{N}\Big[\sum_j \frac{2 \cdot |\mathcal{F}_j^{res_1} \cap \mathcal{F}_j^{res_2}|}{|\mathcal{F}_j^{res_1}| + |\mathcal{F}_j^{res_2}|}\Big] \in [0, 1] \qquad (2)$$

where j iterates over the kinds of technical features, $|\mathcal{F}_j^{res_1} \cap \mathcal{F}_j^{res_2}|$ denotes the set of common values for the technical feature j on res_1 and res_2, $|\mathcal{F}_j^{res_k}|$ denotes the number of values admitted for technical feature j on resource res_k, N is the number of kinds of technical features on which the comparison is based. For example, if res_1 presents {XML, JSON, JSONP} as data formats and {REST} as protocol, while res_2 presents {XML, JSON} as data formats and {REST, Javascript, XML} as protocols, the $TechSim()$ value is computed as:

$$\frac{1}{2}\left[\frac{2 \cdot |\{\text{XML, JSON, JSONP}\} \cap \{\text{XML, JSON}\}|}{|\{\text{XML, JSON, JSONP}\}| + |\{\text{XML, JSON}\}|} + \frac{2 \cdot |\{\text{REST}\} \cap \{\text{REST, Javascript, XML}\}|}{|\{\text{REST}\}| + |\{\text{REST, Javascript, XML}\}|}\right] \quad (3)$$

In this example, XML is used both as data format and as XML-RPC protocol and it is considered separately in the two cases. The terminological and the technical similarity measures are equally weighted to compute the overall Web resource similarity, computed as follows:

$$WebResourceSim(res_1, res_2) = 0.5 \cdot TermSim(res_1, res_2) + \\ +0.5 \cdot TechSim(res_1, res_2) \in [0, 1] \quad (4)$$

By contruction, if $res_1 = res_2$, then $WebResourceSim(res_1, res_2) = 1.0$.

Compositional Similarity between Web Mashup Resources. The *compositional similarity* between two Web mashup resources res_1 and res_2, that represent two Web mashups, denoted as $MashupCompSim(\cdot) \in [0, 1]$, measures the degree of overlapping between two mashups as the number of common or similar APIs between them, that is

$$MashupCompSim(res_1, res_2) = \frac{2 \cdot \sum_{i,j} WebResourceSim(res_1^i, res_2^j)}{|res_1| + |res_2|} \quad (5)$$

where res_1^i and res_2^j are two Web APIs, used in res_1 and res_2 mashups, respectively, $|res_1|$ (resp., $|res_2|$) denotes the number of Web APIs in res_1 (resp., res_2). $WebResourceSim(\cdot)$ values are aggregated through the Dice formula. Pairs to be considered for the $MashupCompSim$ computation are selected according to a maximization function that relies on the assignment in bipartite graphs.

5   Web Mashup Resource Search

The similarity metrics introduced in the previous section have been exploited for Web mashup resource search, that relies on the unified representation of resources through the model presented in Section 3. The basic idea of our search approach is to avoid a pairwise comparison of the Web API request $\mathcal{R}$ against each WMD extracted from the multiple repositories. Instead, we provide a WMD clustering based on terminological similarity. The request is compared against a representative WMD for each cluster, in order to identify the most relevant cluster(s) of WMDs. After the identification and the selection of such clusters, we perform a more in depth comparison between $\mathcal{R}$ and each relevant WMD according to all the types of similarities described in the previous section, distinguishing between WMDs that represent Web APIs and WMDs that represent mashups, and we provide the designer with a ranked list of relevant resources (either APIs or mashups) to be selected for his/her purposes. Finally, a further modification of the ranking is based on ratings (if available) assigned by other designers to Web mashup resources. In the following, we present the main phases of the search procedure.

Request Formulation. The request for a resource is formulated by the designer as follows: $\mathcal{R} = \langle \mathcal{K}_\mathcal{R}, \mathcal{F}_\mathcal{R}, \mathcal{M}_\mathcal{R} \rangle$, where $\mathcal{K}_\mathcal{R}$ is a set of keywords, $\mathcal{F}_\mathcal{R}$ is a set of pairs ⟨tech_feature=value⟩ and $\mathcal{M}_\mathcal{R}$ is a mashup (that is, a set of Web APIs) where the Web API to search for will be aggregated. The elements $\mathcal{F}_\mathcal{R}$ and $\mathcal{M}_\mathcal{R}$ in the request are optional. In particular, the latter is used to differentiate the kind of search that is being performed: (i) if $\mathcal{M}_\mathcal{R} = \emptyset$, then the designer is looking for a single Web API, for instance to start a new mashup application from scratch; (ii) otherwise, if $\mathcal{M}_\mathcal{R} \neq \emptyset$, the designer's purpose is to find a Web API to be included in an existing mashup, to complete it or to substitute a Web API within the mashup.

Clustering. The clustering procedure is performed off-line, thus not affecting the performance of the approach. We employ a hierarchical bottom-up clustering algorithm [9]. The term "hierarchical" means that this technique classifies WMDs into clusters at different levels of similarity. Pairwise comparisons between WMDs is performed according to the terminological similarity. In this way, we give more importance first to the terminological items, that are usually adopted to give a functional characterization of Web APIs and mashups in current repositories. Roughly speaking, we agree on the fact that categories, (semantic) tags, ontological concepts and keywords are adopted to categorize or classify the repository contents. The technique operates in a bottom-up way since it places a WMD into its own cluster and then proceeds through a progressive merging of clusters until all WMDs are clustered. Two clusters are merged first if they contain two WMDs, one from each cluster, with the maximum terminological similarity. The result of clustering is a similarity tree, where single WMDs are the leaves and intermediate nodes have an associated value representing the $TermSim()$ value at which a pair of clusters is merged. Only those nodes whose associated $TermSim()$ value is equal or greater than a threshold $\delta \in [0, 1]$ are considered as candidate clusters. Higher values of δ determine higher similarity between cluster members, but also an higher number of clusters with few members. This will impact on performance, as discussed in the experimental results. For each cluster C_k, the centroid is selected as C_k representative, denoted with $\widehat{C_k}$, that is the descriptor closest to all the other descriptors in C_k, considering the terminological similarity.

Search. The search procedure, starting from the set of clusters and the request $\mathcal{R}$ formulated by the designer, is described in Algorithm 2. In the algorithm, the set $\mathcal{W}$ of relevant Web APIs, the set $\mathcal{M}$ of relevant mashups and a buffer set Ω are initialized as empty sets to be further populated (row 1). The request $\mathcal{R}$ is compared against the centroids of clusters according to the $TermSim()$ similarity (rows 2-4). Clustering enables to apply terminological comparison only

---

Algorithm 2. Web mashup resource search algorithm

Input : The set $\{C_k\}$ of clusters; the request $\mathcal{R} = \langle \mathcal{K_R}, \mathcal{F_R}, \mathcal{M_R} \rangle$ formulated by the designer.

Output: The set $\mathcal{W}$ of ranked relevant Web APIs; the set $\mathcal{M}$ of ranked relevant mashups.

1 $\mathcal{W} = \emptyset; \mathcal{M} = \emptyset; \Omega = \emptyset;$

2 **foreach** *Centroid* $\widehat{C_k}$ **do**

3 $\quad$ **if** $TermSim(\mathcal{R},\widehat{C_k}) \geq \gamma_1$ **then**

4 $\quad\quad$ $\Omega = C_k \cup \Omega;$

5 **foreach** $WMD_i \in \Omega$ **do**

6 $\quad$ Compute the WebResourceSim($\mathcal{R}$, WMD$_i$);

7 $\quad$ **if** $WebResourceSim(\mathcal{R},\ WMD_i) \geq \gamma_2$ **then**

8 $\quad\quad$ **if** WMD_i.type == W **then**

9 $\quad\quad\quad$ Add WMD$_i$ to $\mathcal{W}$;

10 $\quad\quad$ **else if** WMD_i.type == M **then**

11 $\quad\quad\quad$ Add WMD$_i$ to $\mathcal{M}$;

12 $\mathcal{M} = \text{Rank}(\mathcal{M}, \rho_1); \mathcal{W} = \text{Rank}(\mathcal{W}, \rho_2);$

13 $\mathcal{W} = \text{ApplyRatings}(\mathcal{W}); \mathcal{M} = \text{ApplyRatings}(\mathcal{M});$

14 **return** $\mathcal{W}$ and $\mathcal{M}$;

---

to cluster centroids, thus avoiding overloading due to the pairwise comparison between the request $\mathcal{R}$ and each WMD extracted from the repositories. Relevant Web mashup resource descriptors are temporarily stored within the Ω buffer set (row 4). At this point, a more in depth comparison between the request and each relevant descriptor is performed according to the $WebResourceSim()$ metric, that takes into account both the terminological and the technical similarity. We note that two thresholds, namely γ_1 and γ_2, are used in rows 3 and 7 to filter out not relevant results. These thresholds are set within the $[0, 1]$ range and must be chosen according to the following considerations: (i) the higher the thresholds, the faster the search, since less resources are marked as relevant, but the search recall is obviously decreased; (ii) according to this viewpoint, the value of γ_1 dominates the one of γ_2, since resources are filtered out according to γ_1 first. We performed preliminary experiments on a training set of resources and we fixed $\gamma_1 \simeq 0.7$ to increase filtering and ensuring best precision; the recall reduction is balanced by the clustering procedure performed off-line, which collect together very close resources. On the other hand, we kept γ_2 low (i.e., $\gamma_2 \in [0.3, 0.5]$) in order to accept as much search results as possible.

Ranking. The ranking procedure applied in the last part of the algorithm ensures that the most relevant results are proposed to the designer first, moving the less relevant ones at the end of the results list. It is worth noting that, if the descriptor extracted from one of the repositories is incomplete (e.g., the technical features are not specified), the overall $WebResourceSim()$ value is lower. This meets our aim of proposing first Web resource descriptors that present a more complete specification, as extracted from available repositories.

Ranking is performed by invoking the $Rank()$ function (row 12), that is differentiated with respect to the type of Web mashup resources. In case of mashups, a ranking function $\rho_1 : \mathcal{M} \mapsto [0, 1]$ is used, that is computed as follows:

$$\rho_1(WMD) = WebResourceSim(\mathcal{R}, WMD)\cdot$$
$$\cdot MashupCompSim(\mathcal{M}_\mathcal{R}, WMD) \in [0, 1] \tag{6}$$

According to this equation, the closer the WMD, that in this case represents a mashup, to $\mathcal{M}_\mathcal{R}$ in the request, according to the compositional similarity between mashups, the better the ranking of WMD in the results list. This means that those mashups, that are more similar to the mashup where the designer will insert the API he/she is looking for, will be suggested to the designer first. In case of APIs, a ranking function $\rho_2 : \mathcal{W} \mapsto [0, 1]$ is computed as a variant of ρ_1, that is:

$$\rho_2(WMD) = WebResourceSim(\mathcal{R}, WMD)\cdot$$
$$\cdot \frac{1}{|\mathcal{M}_{WMD}|} \sum_{k=1}^{|\mathcal{M}_{WMD}|} MashupCompSim(\mathcal{M}_\mathcal{R}, M_k) \in [0, 1] \tag{7}$$

where $\mathcal{M}_{WMD}$ is the set of mashups that contain the resource WMD, that in this case represents a Web API, $M_k \in \mathcal{M}_{WMD}$ is one of these mashups and $|\mathcal{M}_{WMD}|$ denotes the number of mashups. According to this equation, the closer the mashups to MR where WMD is used, according to the compositional similarity, the better the ranking of WMD in the results list.

Finally, a further promotion/penalty mechanism is implemented to take into account the ratings assigned by designers to Web mashup resources (row 13). The mechanism starts from the scoring system we adopted in our approach, that has been widely described in [1]. Here, we further extended this rating system adding ranking promotions/penalties as reported in Table 1. Depending on the rating in which the average score of a Web mashup resource falls, the position of the resource in the results list is increased or decreased as shown in the third column of the table.

Table 1. The 9-point Scoring System for the assignment of ranking promotions and penalties to the Web mashup resources

Rating (additional guidance on strengths/weaknesses)	Score	Ranking promotion or penalty
POOR (completely useless and wrong)	0.2	-4
MARGINAL (several problems during execution)	0.3	-3
FAIR (slow and cumbersome)	0.4	-2
SATISFACTORY (small performance penalty)	0.5	-1
GOOD (minimum application requirements are satisfied)	0.6	0
VERY GOOD (good performance and minimum application requirements are satisfied)	0.7	+1
EXCELLENT (discreet performance and satisfying functionalities)	0.8	+2
OUTSTANDING (very good performance and functionalities)	0.9	+3
EXCEPTIONAL (very good performance and functionalities and easy to use)	1.0	+4

6 Experimental Evaluation

The aim of the preliminary experiments described in this section has been to check the capability of our approach to provide improved search results compared with the separated use of multiple repositories. For the experiments, we considered: (i) the ProgrammableWeb repository, focused on mashups (built with Web APIs), Web API technical features, tags and categories; (ii) the Mashape repository, where APIs are classified through categories and associated with the number of designers interested in the Web APIs; (iii) the Anypoint API Portal repository, where interested designers, categories and technical features are considered for Web API characterization. We considered the application scenario presented in the motivating example. Moreover, we extracted about 1,400 Web APIs and related mashups (if available) from the three repositories, also considering other orthogonal application domains, related to different categories. Experiments have been run on an Intel laptop, with 2.53 GHz Core 2 CPU, 2GB RAM and Linux OS. Experiments have been performed ten times using different requests. In each experiment, we randomly chose a mashup M and we extracted from the mashup a Web API $\mathcal{W}$. We then issued a request using the features of $\mathcal{W}$, given a mashup $M' = M/\{\mathcal{W}\}$. The same request has been issued multiple times using different synonyms. For each request, we manually tagged as relevant the Web API $\mathcal{W}$ itself and all those Web APIs close to $\mathcal{W}$ in terms of protocols, data formats, similarity of mashups where they have been included, functionalities provided by the Web APIs (according to the documentation for the Web APIs provided in the considered repositories). We also asked five expert users to validate our choices. We selected as expert users a set of designers who developed at least ten mashups in the domain of interest, using different kinds of APIs, different data formats and protocols. The idea was to evaluate the search results using the classical IR measures of precision and recall and the average position of $\mathcal{W}$ among the first 10 search results. Precision refers to the number of relevant Web APIs within the set of search results, that is:

$$precision = \frac{|\{relevant\ Web\ APIs\} \cap \{retrieved\ Web\ APIs\}|}{|\{retrieved\ Web\ APIs\}|} \in [0, 1] \qquad (8)$$

Recall refers to the percentage of relevant Web APIs that have been effectively retrieved, that is:

$$recall = \frac{|\{relevant\ Web\ APIs\} \cap \{retrieved\ Web\ APIs\}|}{|\{relevant\ Web\ APIs\}|} \in [0, 1] \qquad (9)$$

Precision and recall measure the effectiveness of the retrieval process and should be maximized. The average position of $\mathcal{W}$ among the first 10 search results is a measure to evaluate Web API ranking. If, among search results, we find mashups, as allowed by our model, we considered as positive those mashups that contain at least a relevant API for the request. The results are shown in Table 2.

We note that, even if we merge the results from the considered repositories, queried separately, our approach presents better precision, recall and overall ranking. Better precision and recall are due to the particular *itemSim*(·) similarity we considered in our approach, that enables to overcome discrepancies due

Table 2. Preliminary evaluation results

	Precision	recall	average $\mathcal{W}$ position
ProgrammableWeb	0.62	0.59	7.9
Mashape	0.58	0.4	-
Anypoint API Portal	0.60	0.51	8.1
Union of results (considering repositories separately)	0.59	0.5	9.3
Our approach	0.91	0.79	2.1

to the adoption of synonyms instead of using the same term. Moreover, different repositories use different categories to classify the same APIs. This limitation cannot be solved simply merging search results coming from distinct repositories, while our approach is able to mitigate it by combining different kinds of similarity measures. Better ranking results are ensured since our approach enables to consider partially overlapping aspects coming from different repositories in a joint way. Given its relative complexity compared with simple keyword-based search and basic ranking facilities provided by available repositories, our approach pays in terms of response times, as shown in the second column of Table 3. However, by applying the clustering procedure, times significantly decrease, as evident in the third column of Table 3, using $\gamma_1 = 0.7$ and $\gamma_2 = 0.5$ (see Section 5).

Table 3. Response times (with and without clustering) in the experimental evaluation

	times without clustering (sec.)	times with clustering (sec.)
ProgrammableWeb	4.464	-
Mashape	3.234	-
Anypoint API Portal	2.360	-
Union of results (considering repositories separately)	~10.058	-
Our approach	19.894	7.091

Of course, the cut-off imposed by thresholds γ_1 and γ_2 has an impact on the precision, recall and ranking of search results. But if we consider these values with and without clustering (see Table 4), they still outperform the values of the same measures if the search is performed on ProgrammableWeb, Mashape or Anypoint API Portal and by simply merging the sets of search results coming from these repositories. The considered public repositories do not present any difference with and without clustering, since a clustering mechanism is not provided on them. By varying γ_1 and γ_2 thresholds, precision, recall, ranking and response times change. For instance, if we decrease γ_1 to 0.5, the recall increases by 3%, but response times increase by 26%. Therefore, the increment of recall values does not justify decreased performance. Better response times are also ensured through the proper setup of δ threshold during cluster identification. If δ increases, more clusters are obtained, thus requiring an higher number of comparisons between the request and cluster centroids. However, also in this case, precision, recall and ranking slightly vary due to the more in-depth comparison between the request and each WMD within a candidate cluster as shown in rows 5-11 of Algorithm 2. Experimental results shown in Tables 2-4 have been obtained with $\delta = 0.6$.

Table 4. Preliminary evaluation results: impact of clustering on precision, recall and ranking of search results ($\gamma_1 = 0.7$, $\gamma_2 = 0.5$)

	Without clustering				With clustering			
	Precision	recall	average $\mathcal{W}$ ranking	times (sec.)	Precision	Recall	average $\mathcal{W}$ ranking	times (sec.)
Our approach	0.91	0.79	2.1	19.894	0.85	0.71	2.1	7.091

7   Related Work

Model-driven Web API selection and mashup development have been addressed by several approaches in the last years. Advanced solutions for mashup development [10] provide technologies, models and CASE tools to ease the designers in aggregating the component Web APIs, setting the interactions between them and generating the glue code required to deploy the mashup application. These models are not targeted at Web API or mashup search and ranking over public or private repositories. In [11] the formal model based on Datalog rules defined in [12] is proposed to search for mashup components (called *mashlets*): when the designer selects a mashlet, the system suggests other mashlets to be connected on the basis of recurrent patterns of components in the existing mashups. In [13] semantic annotations have been proposed to enrich Web API modeling in presence of high heterogeneity and proper metrics based on such annotations have been defined to improve recommendation of Web APIs. These approaches rely on a complex Web API model and a complex request formulation, that are unfeasible for Web designers' expertise, that is mainly focused on Web programming technologies. Different approaches followed, where simpler models have been discussed, not based on formal specifications or semantic annotations, and further extended with other aspects, such as the *collective knowledge* on Web API use, coming from experiences of other designers, and ratings assigned to APIs and mashups [1–4] (see [1] for a detailed survey).

With respect to these most recent approaches for Web API selection, we aim at providing a unified representation of Web APIs and mashups over multiple repositories, aggregating complementary descriptions of resources to search for, and we proposed a more flexible search, that enables seamless selection of Web APIs and mashups. To the best of our knowledge, this is the first attempt to provide a cross-repository search of Web APIs and mashups.

8   Concluding Remarks

In this paper, we discussed a unified model for Web API and mashup characterization and search across multiple repositories, based on selection criteria, that combine several complementary perspectives. As proved in [1], performing Web API search and ranking on a comprehensive API descriptor, that includes different and complementary descriptive aspects, would improve retrieval results. Preliminary experiments have been run to test effectiveness of Web API search in terms of precision and recall. Experiments demonstrated an improved search results, obtained by applying our approach, compared with Web API search performed on multiple repositories considered separately. Evolutions of this approach will be investigated to check how the productivity of Web designers is increased through the use of multiple repositories for Web API selection, where different repositories focus on complementary Web mashup resource descriptions. The possibility of further enriching the model through semantic aspects (e.g., semantic annotation of the unified view over the resources) will be investigated as well.

References

1. Bianchini, D., De Antonellis, V., Melchiori, M.: A Multi-perspective Framework for Web API Search in Enterprise Mashup Design. In: Salinesi, C., Norrie, M.C., Pastor, Ó. (eds.) CAiSE 2013. LNCS, vol. 7908, pp. 353–368. Springer, Heidelberg (2013)
2. Torres, R., Tapia, B., Astudillo, H.: Improving Web API Discovery by leveraging social information. In: Proceedings of the IEEE International Conference on Web Services, pp. 744–745 (2011)
3. Cappiello, C., Matera, M., Picozzi, M., Daniel, F., Fernandez, A.: Quality-Aware Mashup Composition: Issues, Techniques and Tools. In: Proc. of 8th Int. Conference on Quality of Information and Communications Technologies (QUATIC 2012), pp. 10–19 (2012)
4. Gomadam, K., Ranabahu, A., Nagarajan, M., Sheth, A., Verma, K.: A Faceted Classification Based Approach to Search and Rank Web APIs. In: Proc. of International Conference on Web Services (ICWS), pp. 177–184 (2008)
5. Upadhyaya, B., Xiao, H., Zou, Y., Ng, J., Lau, A.: A Framework for Composing Personalized Web Resources. In: Chignell, M., Cordy, J.R., Kealey, R., Ng, J., Yesha, Y. (eds.) The Personal Web. LNCS, vol. 7855, pp. 65–86. Springer, Heidelberg (2013)
6. Bianchini, D., De Antonellis, V., Melchiori, M., Salvi, D.: Semantic-enriched service discovery. In: Proc. of the 22nd International Conference on Data Engineering (ICDE), pp. 38–47 (2006)
7. Trombos, A., Villa, R., van Rijsbergen, C.: The effeeffective of query-specific hierarchic clustering in information retrieval. Information Processing & Management (38), 559–582 (2002)
8. Castano, S., Ferrara, A., Montanelli, S.: Matching Ontologies in Open Networked Systems: Techniques and Applications. Journal on Data Semantics 2, 25–63 (2006)
9. Castano, S., De Antonellis, V., De Capitani di Vimercati, S.: Global Viewing of Heterogeneous Data Sources. IEEE TKDE 13(2), 277–297 (2001)
10. Matera, M., Picozzi, M., Pini, M., Tonazzo, M.: PEUDOM: A mashup platform for the end user development of common information spaces. In: Daniel, F., Dolog, P., Li, Q. (eds.) ICWE 2013. LNCS, vol. 7977, pp. 494–497. Springer, Heidelberg (2013)
11. Greenshpan, O., Milo, T., Polyzotis, N.: Autocompletion for Mashups. In: Proc. of the 35th Int. Conference on Very Large DataBases (VLDB), Lyon, France, pp. 538–549 (2009)
12. Abiteboul, S., Greenshpan, O., Milo, T.: Modeling the Mashup Space. In: Proc. of the Workshop on Web Information and Data Management, pp. 87–94 (2008)
13. Bianchini, D., De Antonellis, V., Melchiori, M.: Semantics-Enabled Web API Organization and Recommendation. In: De Troyer, O., Bauzer Medeiros, C., Billen, R., Hallot, P., Simitsis, A., Van Mingroot, H. (eds.) ER 2011 Workshops. LNCS, vol. 6999, pp. 34–43. Springer, Heidelberg (2011)

Common Neighbor Query-Friendly Triangulation-Based Large-Scale Graph Compression

Liang Zhang, Chen Xu, Weining Qian, and Aoying Zhou

Institute for Data Science and Engineering,
East China Normal University, Shanghai, China
{52101500013,52111500010}@ecnu.cn,{wnqian,ayzhou}@sei.ecnu.edu.cn

Abstract. Large-scale graphs appear in many web applications, and are inevitable in web data management and mining. A lossless compression method for large-scale graphs, named as *bound-triangulation*, is introduced in this paper. It differs itself from other graph compression methods in that: 1) it can achieve both good compression ratio and low compression time. 2) The compression ratio can be controlled by users, so that compression ratio and processing performance can be balanced. 3) It supports efficient common neighbor query processing over compressed graphs. Thus, it can support a wide range of graph processing tasks. Empirical study over two real-life large-scale social networks, which different underlying data distributions, show the superior of the proposed method over other existing graph compression methods.

Keywords: Graph compression, social graph, triangle listing, common neighbor query.

1 Introduction

Graphs are widely used in Web-related applications to represent different kinds of data, such as linkage structure of web pages, social networks, and semantic networks. However, due to their essential characteristics, they are not easy to be queried efficiently. Fitting a large-scale graph, such as a friendship network from a popular social networking website, into smaller space is not an easy task. Graph compression is a natural approach to transform a large-scale graph into some data structures with less storage consumption.

There are several requirements on large-scale graph compression. First, the resulting compressed graph should be compact, so that they can be efficiently transmitted or stored in main memory. Second, the compression and decompression operations should be efficient. Thus, we would not spend too much time before a graph is ready to be stored in main memory or used by a specific application. Furthermore, for some popular query operators, it is welcome that they can be evaluated efficiently over the compressed graph. Last but not the least, a flexible compression method that can achieve different levels of compression ratio is useful for users to balance between storage space and query processing.

B. Benatallah et al. (Eds.): WISE 2014, Part I, LNCS 8786, pp. 234–243, 2014.
© Springer International Publishing Switzerland 2014

This paper aims at the problem of lossless compression of large-scale graphs. An algorithm named as *bound-triangulation* is introduced. It utilizes triangulation structures in graphs to compress them. Intuitively, in a large-scale graph, there are many dense subgraphs, within which there are many triangles. Efficient data structures for representing these triangles lead to compact storage of the whole graph. Meanwhile, our compression method can efficiently support common-neighbor query, which is a basic building block of many graph processing tasks.

The characteristics of bound-triangulation compression method are as follows. It achieves both low compression ratio and compression time, which is demonstrated over real-life social networks. The compression ratio can be controlled by parameters. Thus, users may balance the storage consumption and processing efficiency. For common-neighbor queries, it achieves better query processing performance with equivalent compression ratio.

The remainder of this paper is organized as follows. After the related work, which is introduced in Section 2, the graph compression problem is formalized in Section 3. In Section 4, our triangulation-based graph compression method and the evaluation of common neighbor queries over the compressed graph is introduced in details. The experimental results over two large-scale real-life social graphs are reported in Section 5. The Section 6 is for concluding remarks.

2   Related Work

In this section, we describe several existing compression algorithms. Most of them target to reduce storage cost of graphs, and the results must maintain the certain properties of original graphs.

2.1   Node Merge Schemes

Gilbert proposed a way to combine similar nodes into one to compress the graph. They used node or edge attributes and topological information to measure similarities between nodes[7].

Buehrer proposed the Virtual Node Miner method[4], which was based on the idea of identifying cliques and used virtual nodes as indirection connecting level to link two node sets in order to reduce the number of edges.

2.2   Node Reorder Schemes

U Kang and Christos envisioned a graph as a collection of hubs connecting spokes, with super-hubs connecting the hubs, and so on[10]. They proposed that the real-world graph could be shattered easily by removing hub nodes from them, and proposed the SLASHBURN algorithm which group the non-zero elements and give the hubs that connecting more spokes lower id.

Blandford used *cosine measure* to determine the similarity between a pair nodes in the graph, and proposed the BUILD-GRAPH algorithm to construct a node-node similarity graph which re-index the nodes in order to give the nodes with similar adjacency list closer number[2].

2.3   Adjacency List Schemes

When graphs are represented in adjacency list, each element in adjacency list requires log_n bits to store. As a result, we can reduce the graph storage cost by minimizing the number of bits needed to represent the graph[8].

No encoding stores node IDs that link with the node for each node in the graph. DeltaEncoding records the first element and the difference between the elements in the sorted adjacency list and the previous one[3]. GammaEncoding records twice the absolute difference, when it is a minus, the result minus one in addition. Intervalsencoding is similar to delta encoding, but it encodes intervals of node IDs.

Adler proposed the FIND-REFERENCE algorithm, specifically designed for graph structures with many shared links[1]. They attempted to find nodes that share several common neighbors corresponding to cases where one node might have copied the links of the others.

2.4   Common Neighbor Related Applications

There has been increasingly interest in researching links between objects in networks recently, one of the fundamental researches is link prediction. As a general rule, the greater the amount of common neighbors that two node have, the greater possibility that there will be a link between them in the future.

Cui proposed an common neighbor based approach for link prediction [6]. He used an adjacency matrix to measure the common neighbor of two nodes in the graph. Chaturvedi also studied the common neighbor leading link prediction method in social network[5]. They defined the common neighbors of two nodes in the graph as 'brokers', and proposed a prediction method to concentrates on the strength of links of each common 'broker' of two nodes.

3   Problem Definition

We discuss compression of undirected, unlabeled graphs in this paper. We assume an undirected graph $G(V, E)$ with a set of nodes V and a set of edge E. We denoted the number of nodes with n and the number of edges with m.

We then defined problems exist in graph compression. We considered that the graph consist of a node for each user, and an undirected edge for a relationship between two users. We would like the compressed graph to support same operations as an adjacency list representation of an unlabeled graph, while occupying significantly less space.

The compression ratio is one of the most important indicator to judge the compression effect. The compression ratio means the ratio of the uncompressed graph size and compressed graph size. The compression time is another important indicator, which means the total time the algorithm takes to compress the graph. We also use the response time of common neighbor query to evaluate the compress algorithms, the faster the better.

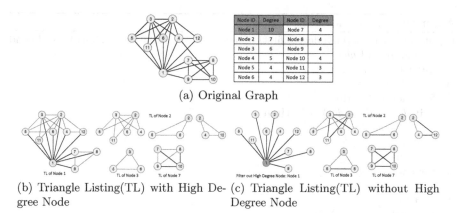

(a) Original Graph

(b) Triangle Listing(TL) with High De-   (c) Triangle Listing(TL) without High
gree Node   Degree Node

Fig. 1. A Graph with High Degree Node (Triangle listing is calculated based on the yellow node in each phase)

4   Compression Algorithm

We presented an algorithm that work on the basis of merging node pairs, which connected to each other and their common neighbors to triangle elements to save the storage cost of the graph with adjacency list representation. Our algorithm tried to list all triangles in the graph and denote all the edges included in these triangles by triangle elements.

4.1   Triangle Listing

A triangle could be defined as a set of three nodes such that each possible edge between them was presented in the graph. In complex network studies, one often deals with huge graphs, typically with several million nodes or edges and up to a few billions. Both time and space required by triangle listing computations are then key issues.

We can found that there might be $\Theta(n^3)$, or $\Theta(m^{\frac{3}{2}})$ triangles in the graph[11]. However, in practice, the graph is often sparse. Two simple algorithms, node-iterator and edge-iterator, which were first introduced by Alon Itai[9], were widely used in this context.

4.2   Triangle Listing Example

Fig. 1 is an example of triangle listing on a graph with high degree nodes. Fig. 1(a) is a graph with 12 nodes. We applied the triangle listing algorithm on the graph, the result is shown in Fig. 1(b). We could get a triangle list of the graph containing 26 triangles. The blue edges means that they have been contained in least two triangles. In addition, the *Node 1*, which possessed the highest degree in the graph, appeared in 15 triangles. It took 27 storage units to denote the

---

Algorithm 1. Bound-triangulation

Input: ordered list (high degree first) of nodes (1...*n*); Adjacencies *Adj*(*v*); low bound
    lb; high bound *hb*
Output: $G'(V, T, E)$
 1: $V' \leftarrow \emptyset, T \leftarrow \emptyset$
 2: **for each** $v \in V$ **do**
 3:    **if** $d(v) < lb \ or \ d(v) > hb$ **then**
 4:       $V' \leftarrow v, V \leftarrow V \setminus v$
 5:    **end if**
 6: **end for**
 7: **for each** $v \in V$ **do**
 8:    $A(v) \leftarrow \emptyset$
 9: **end for**
10: **for each** $u \in (1...n)$ **do**
11:    **for each** $v \in Adj(v)$ **do**
12:       **if** $u < v$ **then**
13:          **for each** $w \in A(u) \cap A(v)$ **do**
14:            $W \leftarrow W \cup w, E \leftarrow E \setminus (u, w), E \leftarrow E \setminus (v, w)$
15:          **end for**
16:         $A(v) \leftarrow A(v) \cup \{u\}$
17:       **end if**
18:    **end for**
19:    $T \leftarrow T \cup \{[u, v], W\}$
20: **end for**
21: $V \leftarrow V \cup V'$
22: **return** $G'(V, T, E)$

---

edges linked to *Node 1* in the triangle list, and it need only 11 storage units
to represent these edges by adjacency list representation. 19 edges in the graph
have appeared at least twice in the triangle list. Apparently, we paid a lot of
time and space cost, but got a worse result.

We tried to filter out part of nodes to improve the performance. We filtered
out the nodes which its degree is less than 2 (a node need at least two neighbors
to form a triangle), and we also filter out the nodes with high degree. In the
example, we filter out *Node 1* from triangle listing calculation. The result is
shown in Fig. 1(c). There are only 11 triangles in the result triangle list. The
algorithm took much less time and space on computation, and got a compressed
graph with lower compression ratio, there are only 7 redundancy edges.

4.3 Bound-Triangulation Algorithm

Schank proposed the *forward* algorithm which could list all triangles in a graph
in $\Theta(m^{\frac{3}{2}})$ time and need $O(m)$ space, which was better than other triangle
listing algorithms[12]. However, when we applied the *forward* algorithm on a
large graph with millions of nodes and edges, it didn't work well. It took long
time to list all triangles in the graph.

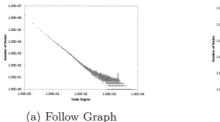
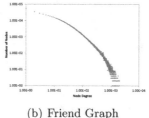

(a) Follow Graph   (b) Friend Graph

Fig. 2. Node Degree Distribution

It has been observed that most real-world graphs have an important characteristic in common, their degrees are very heterogeneous. The majority of nodes have very low degrees while the minority have high degrees. We have to take an excessive quantity of resources to list the triangles that formed by a high degree node and its neighboring low degree nodes. As a result, we tried to ignore these high degree nodes when we listing the triangles in the graph and proposed *bound-triangulation* algorithm with better time complexity.

Algorithm 1, *bound-triangulation* algorithm, takes an adjacency list, *Adj*, as input. Lines 2 to 6 set the value of low bound and high bound, and filtered out all nodes that didn't satisfy the value from the triangle listing computation. Line 7 to 9 create an empty array, A, for the rest nodes. Line 10 to 20 iteratively processed following steps. For each node satisfying the bound, v, and for each neighboring node of v which the degree was less than v, we set it as u. If we found a nonempty common node set W in $A|u| \cap A|v|$, we output the triangle set $\{[u, v], W\}$ and remove the edge set $\{(u, w) \cup (v, w)|w \in W\}$ from E. Then we add node v to array $A[u]$. Finally, Line 21 put all node which filtered out before back to V.

4.4 Common Neighbor Query over Compressed Graph

We use common neighbor query to satisfy the need of discovering common neighbors between two nodes in a graph in many situations. In the graph compressed by our *bound-triangulation* algorithm, we do not need to decompress the graph to get the common neighbor set in many cases. Therefore, we can signally reduce the query time when we can get the set we query from the triangle elements in the compressed graph, while the graph compressed by other adjacency list based algorithms need to be decompressed to get the result.

5 Experiments

5.1 Experimental Setup

In this section, we present experimental result to our *bound-triangulation* algorithm, other adjacency list based algorithms and ID reorder algorithms. We

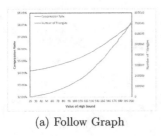
(a) Follow Graph

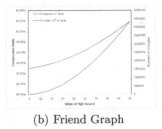
(b) Friend Graph

Fig. 3. Tradeoff between Compression Ratio and the Number of Triangles

experimented the algorithms on two social graphs, which fetched from Sina Weibo[1]. We experimented the algorithms on a single PC with a 3.06 GHz Intel "Core 2 Duo" processor and 4 GB of RAM (1066 MHz DDR3 SDRAM).

We ran the experiments on two undirected social graphs with different node degree distribution, as shown in Fig. 2. The Follow Graph was a user followed social graph, when one node followed another node, there would be an undirected link between these two nodes. It contains *2675412* nodes and *11857851* edges. The nodes follow the power-law distribution in Follow Graph. The Friend Graph is a friend social graph. There will be an undirected link between two nodes, only if they both follow each other. It contains *784882* nodes and *28316226* edges. Therefore, we experimented our algorithm on these two graphs with different degree distribution to prove the efficiency of our algorithm.

Our algorithm had two parameters: low bound and high bound of node degree. They had direct control on the compression time, compression ratio and the number of triangles in the compressed graph. We conducted an extensive experimental study on different values of these two parameters for two graphs in our experiment.

We measured the compression performance using the compression ratio and compression time. The performance of common neighbor query on the compressed graph is another indicator we care about. We processed common neighbor query on the graph compressed by our algorithm and other adjacency list based algorithms to compare the average query time.

5.2 Tradeoff between Compression Ratio and the Number of Triangles

First, we observed the tradeoff between the compression ratio and the number of triangles with different value of bound. We observed that the low bound generated little impact on the compression ratio. In the experiment, we ran our algorithm with the value of high bound from 20 to 200 on Follow Graph, and set the low bound as 2. We also ran the experiment with the high bound value from 5 to 50 on Friend Graph, because there were more nodes in Friend Graph that contained a high degree value. Through observing the variation in quantities

---

[1] http://weibo.com

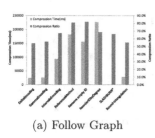

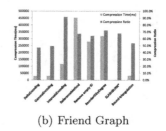

(a) Follow Graph   (b) Friend Graph

Fig. 4. Comparison of Compression Ratio and Compression Time (Since SLASHBURN performed badly on compression time, we ran SLASHBURN on a subset of experiment graphs and only record the compression ratio.)

of triangles in the compressed graph, shown in Fig. 3, we discovered that the number of triangles would increase, while the value of high bound increased.

As we discussed above, a user in social network may be interested in discovering if there are any common neighbors that shared by him and his friends. It's necessary to maintain a certain amount of triangle elements in the compressed graph. Therefore, we need to find a balance of compression ratio and the number of triangles to provide quick response to common neighbor query.

5.3   Comparison on Compression Ration and Compression Time

In this experiment, we set the low bound as 2 and high bound as 90 on Follow Graph which reach the similar compression ratio to DeltaEncoding and GammaEncoding. We also set the low bound as 2 and high bound as 15 in our algorithm on Friend Graph. The performance in the compression ratio and compression time is shown in Fig. 4.

The results showed that our method took less time than all node reorder methods. Most ID reorder methods need to scan the whole graph before it assign a new ID each time, so it takes much more time than adjacency list methods which only scan the graph once. In Fig. 4, our *bound-triangulation* algorithm got a compressed graph with similar compression ratio and compression time as DeltaEncoding and GammaEncoding easily. In addition, the IntervalsEncoding and ReferecnceMethod didn't work well on both graphs.

5.4   Comparison of the Performance of Common Neighbor Query

We produced 1000 random common neighbor queries, and recorded the average query time on the compressed graphs, the result is show in Fig. 5. The DeltaEncoding, GammaEncoding and IntervalsEncoding need to decompress the graph before running query on it, so they took much time on decompression. The ReferencMethod also need to scan other rows which each element referenced. The ID reorder methods need to regain the original IDs before running query on the graph, it took additional processing time and additional storage space to store the mapping table.

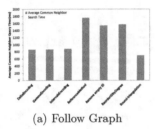
(a) Follow Graph

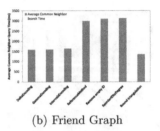
(b) Friend Graph

Fig. 5. Comparison on Common Neighbor Query Time with other Algorithms

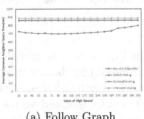
(a) Follow Graph

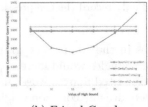
(b) Friend Graph

Fig. 6. Comparison on Common Neighbor Query Time with Different Value of High Bound

Our *bound-triangulation* algorithm maintain some triangles which composed by node pairs and their neighbors. It can answer common neighbor query without scan the whole compressed graph or decompress it in many cases. It responses more quickly than other algorithms.

We ran common neighbor query on both graphs with different value of high bound and compare the average query time with other adjacency list based algorithms, the result in show in Fig. 6. The query time of the compressed graph by our algorithm would decrease, while the value of high bound increased. However, when the value of high bound is too high, the query time does not decrease anymore and increases with the high bound, but it's still better than other adjacency list based algorithms in most situations.

6 Conclusion

The problem of lossless compression of large-scale graphs is studied in this paper. The algorithm of *bound-triangulation* is introduced. The basic idea of it is to group common neighbors of node pairs, and form triangle elements, so that the storage cost is reduced. We show that by listing majority part of triangles in the graph, our algorithm is effective in graph compression that both low compression ratio and compression time can be achieved. By controlling the number of triangles to be listed, users can balance between the storage cost and processing performance. Furthermore, this compression method is friendly to common neighbor queries, which is a basic building block for many graph processing tasks. Experimental results over real-life social graphs show the efficiency of our proposed method.

Our future work include the revision of the compression method to support more queries over graphs, and online compression of large-scale graphs.

Acknowledgments. This work is partially supported by National Basic Research (973) Program under grant number 2010CB731402 and National Science Foundation of China under grant number 61170086.

References

1. Adler, M., Mitzenmacher, M.: Towards compressing web graphs. In: Proceedings of the Data Compression Conference, DCC 2001, pp. 203–212. IEEE (2001)
2. Blandford, D., Blelloch, G.E.: Index compression through document reordering. In: Proceedings of the Data Compression Conference, DCC 2002, pp. 342–351. IEEE (2002)
3. Boldi, P., Vigna, S.: The webgraph framework i: compression techniques. In: Proceedings of the 13th International Conference on World Wide Web, pp. 595–602. ACM (2004)
4. Buehrer, G., Chellapilla, K.: A scalable pattern mining approach to web graph compression with communities. In: Proceedings of the 2008 International Conference on Web Search and Data Mining, pp. 95–106. ACM (2008)
5. Chaturvedi, A., Acharjee, T.: An efficient modified common neighbor approach for link prediction in social networks. IOSR Journal of Computer Engineering (IOSR-JCE) 12, 25–34 (2013)
6. Cui, H.: Link prediction on evolving data using tensor-based common neighbor. In: 2012 Fifth International Symposium on Computational Intelligence and Design (ISCID), vol. 2, pp. 343–346. IEEE (2012)
7. Gilbert, A.C., Levchenko, K.: Compressing network graphs. In: Proceedings of the LinkKDD Workshop at the 10th ACM Conference on KDD (2004)
8. Hannah, D., Macdonald, C., Ounis, I.: Analysis of link graph compression techniques. In: Macdonald, C., Ounis, I., Plachouras, V., Ruthven, I., White, R.W. (eds.) ECIR 2008. LNCS, vol. 4956, pp. 596–601. Springer, Heidelberg (2008)
9. Itai, A., Rodeh, M.: Finding a minimum circuit in a graph. SIAM Journal on Computing 7(4), 413–423 (1978)
10. Kang, U., Faloutsos, C.: Beyond'caveman communities': Hubs and spokes for graph compression and mining. In: 2011 IEEE 11th International Conference on Data Mining (ICDM), pp. 300–309. IEEE (2011)
11. Latapy, M.: Main-memory triangle computations for very large (sparse (power-law)) graphs. Theoretical Computer Science 407(1), 458–473 (2008)
12. Schank, T., Wagner, D.: Finding, counting and listing all triangles in large graphs, an experimental study. In: Nikoletseas, S.E. (ed.) WEA 2005. LNCS, vol. 3503, pp. 606–609. Springer, Heidelberg (2005)

Continuous Monitoring of Top-*k* Dominating Queries over Uncertain Data Streams

Guohui Li[1], Changyin Luo[1], and Jianjun Li[1]

School of Computer Science and Technology, Huazhong University of Science and Technology, Wuhan, P.R. China
{Guohuili,luochangyin}@hust.edu.cn, jianjunwh@gmail.com

Abstract. In many scenarios, e.g., environmental monitoring using multiple sensors, the uncertain data objects arrive continuously (online) and need to be processed in a streaming manner. We first formally define the problem of continuous probabilistic top-*k* dominating (*PTOPK*) query processing over uncertain data streams based on a count-based sliding window model. Based on the observation that *PTOPK* does not change dramatically in consequent sliding window and most uncertain data objects not in *PTOPK* cannot be inserted in *PTOPK* in a certain period of time, an efficient postponed examination algorithm (PEA) is proposed. With PEA, the scores calculation for some uncertain data objects not in *PTOPK* can be postponed and the computation cost can be saved. Extensive experiments have been conducted to demonstrate the efficiency of our approaches.

Keywords: uncertain data, top-*k* dominating query, uncertain stream.

1 Introduction

Top-*k* dominating query [12,15,9] assimilates the advantages of top-*k* and skyline queries and eliminates their limitations. It uses a ranking function to rank points just as in top-k query, and uses the dominating relationship as in skyline query. This query returns the *k* records with the higest domination scores from the dataset. The existing works [12,15] explored the top-*k* dominating query in the *certain databases*. Uncertain data are inevitable in many applications due to various factors such as limitations of measuring equipments, delays in data updates and data randomness and incompleteness. Recently, query processing over uncertain data has attracted much attention. In many real application scenarios, the collected uncertain data are produced in a streaming fashion, such as financial data trackers, sensor networks, environmental surveillance, and location based services, etc. Uncertain streaming data processing has drawn considerable attention from the database research community recently [3,6,13]. A few studies have addressed the evaluation of continuous query over uncertain data streams, such as top-k [7,2] and skyline [17,4].

Probabilistic top-*k* dominating query [10,11] is firstly designed to retrieve *k* uncertain objects, which are expected to dynamically dominate the largest number of uncertain objects under the semantics of dynamic dominance. Zhang et al. [16] studied the threshold-based probabilistic top-*k* dominating query, which is based on a user-defined threshold to retrieve *k* uncertain data with the highest score. Santoso et al. [14] explores

B. Benatallah et al. (Eds.): WISE 2014, Part I, LNCS 8786, pp. 244–255, 2014.
© Springer International Publishing Switzerland 2014

the continuous top-k dominating query over certain data streams and proposes a unique index structure, a Close Dominance Graph (CDG), to support the processing of query. owever, their methods based on CDG cannot be used to tackle our problem. The reason is that the dominance relationship between uncertain data objects cannot be expressed by CDG.

The existing works [10,11,16] retrieve *PTOPK* from the whole databases, while ours concentrate on monitoring *PTOPK* under the uncertain data streaming environments, the methods developed in previous cannot be used directly. Comparing with the certain streaming data, it is much more complicated and expensive to compute scores (dominating ability) of uncertain data in streaming environments. Because it first needs to compute score of each instance, then, deduces score of uncertain data based on their instance scores. So we'd better avoid computing dominating scores whenever possible. How to conduct probabilistic top-k dominating query over uncertain data streams remains an open problem. To the best of our knowledge, there is no similar work studying this query problem.

Our contributions can be summarized as follows.

– We define the problem of probabilistic top-k dominating query over uncertain data streams by employing a count-based sliding window model.
– An effective PEA algorithm is proposed to continuously compute *PTOPK*.
– A thorough experimental evaluation is carried out based on synthetic and real-life data sets, providing evidence regarding the scalability and efficiency of the proposed algorithm.

2 Related Work

Top-k dominating query is a relatively new topic in the field of data management, which was first introduced in [12]. Yiu and Mamoulis [15] further explored the top-k dominating query problem on certain data, and proposed several algorithms based on aggregate R-tree index structure. However, they are not applicable to retrieveing *PTOPK* on uncertain data directly. In [10,11,16], the problem of *PTOPK* query was introduced to address uncertain datasets. In [8], Kontaki et al. proposed a grid-based indexing algorithm to answer continuous top-k dominating query in subspace. Later, they developed an event-based approach to monitor top-k dominating queries over certain data streams [9]. However, their approaches cannot be used to tackle our problem directly. The key reason is that the dominance relationship between uncertain data is different from that of certain data. The most related work [5] proposes a dynamic programming approach based on the Poisson-Binomial Recurrence method to compute *PTOPK* over sliding windows, which is different from our work. Because it focuses on *tuple-level uncertainty*. However, our paper focuses on attribute-level uncertainty [10].

3 Problem Definition

An uncertain data object U is represented by a set of instances u_i ($1 \leq i \leq m$) such that each instance $u_i \in U$ is a point in a d–dimensional space $D = \{D_1, ..., D_d\}$, and is associated with a probability $P(u_i)$, where $0 < P(u_i) \leq 1$ and $\sum_{u_i \in U} P(u_i) = 1$. For an instance u, we use $u.D_i$ to denote its i-th dimension. Note that, any two instances u_i and u_j from the same uncertain data U are mutually exclusive.

Definition 1. *(Instance Dominance). Given two uncertain data instances u and v, instance u dominates instance v (denoted by $u \prec v$), iff (1) $u.D_k \leq v.D_k$ for $1 \leq k \leq d$ and, (2) there is at least one dimension j $(1 \leq j \leq d)$ such that $u.D_j < v.D_j$.*

Example 1. In Figure 1, there are three uncertain data A, B and C with their instances located in a 2-dimensional space. A has three instances a_1, a_2 and a_3. According to Definition 1, we can see that a_1 dominates b_1, b_2, b_3, c_1 and c_2, while a_3 can be dominated by all the other instances. In this paper, we use **score** to measure dominating ability. Below, we first define the dominating ability of instance.

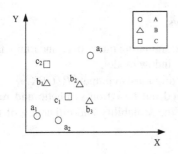
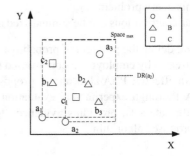

Fig. 1. A set of uncertain data   **Fig. 2.** Spatial aggregate query for a instance

Definition 2. *Given an instance (u_i) of an uncertain data object U, its dominating ability is defined as the sum of the probabilities of the instances which are dominated by u_i in a sliding window, i.e.,*

$$score(u_i) = \sum_{\forall v_j \in V \wedge u_i \prec v_j} P(v_j) \tag{1}$$

where V is the set of uncertain data instances in the sliding window.

Definition 3. *Given two uncertain data objects U and V, the probability of U dominating V is defined as follows:*

$$P(U \prec V) = \sum_{i=1}^{|U|} \left(P(u_i) \cdot \sum_{\forall v_j \in V \wedge u_i \prec v_j} P(v_j) \right) \tag{2}$$

where $|U|$ is the number of instances in U.

Now we can give the definition of the dominating ability (score) of an uncertain data object.

Definition 4. *The score of an uncertain data object weighs how outstanding an uncertain data object is compared to other uncertain data in a sliding window $\mathcal{W}$. Given an uncertain data object U, $score(U)$ is defined as follows:*

$$score(U) = \sum_{V \in \mathcal{W} \wedge V \neq U} P(U \prec V) \tag{3}$$

We now further introduce the notation of *dominance region* for an instance in the following.

Definition 5. *Given an instance u_i, its dominance region, denoted as $DR(u_i)$, is the rectangular region with diagonal corners coordinate of u_i and* space$_{max}$, *where* space$_{max}$ *is the top-right corners of the data space.*

Figure 2 illustrates an example of $DR(a_1)$ and $DR(a_2)$. In $DR(a_1)$, since a_1 dominates two uncertain data B and C, it is easy to obtain $score(a_1) = 1/3 \times (1+1) = 2/3$. a_2 dominates part of the instances of B and C (b_2, b_3, c_1), we can get that $score(a_2) = 1/3 \times (2/3 + 1/2) = 7/18$. Because instance a_3 does not dominate any instance, its score is zero. Therefore, $score(A) = score(a_1) + score(a_2) = 2/3 + 7/6 = 19/18$. This process is named *window query method*.

Definition 6. *(Sliding Window Probabilistic Top-k Dominating Query, SWPTD). Given a dynamic uncertain data stream UDS containing N uncertain data $U_1, U_2, \ldots, U_N$, and a sliding window $\mathcal{W}$, **SWPTD** query monitors the **k** uncertain data with the highest dominating score over sliding window **continuously**.*

4   Baseline Method

A naive approach to retrieving *PTOPK* in stream environments is to perform all domination checking among uncertain data in the sliding window $\mathcal{W}$ every time $\mathcal{W}$ slides. Specifically, when a new uncertain data U_{new} arrives, it leads the oldest one to expire. $score(U_{new})$ is computed based on Definition 4, and the scores of other uncertain data in $\mathcal{W}$ should be updated. Finally, it chooses k uncertain data objects with maximal score values. As $\mathcal{W}$ slides, it has to re-compute the scores of uncertain data objects in $\mathcal{W}$ from scratch. We observe that this process can be accelerated.

In the following section, we design an efficient and effective approach, which can avoid re-computing score of every uncertain data when $\mathcal{W}$ moves in each step and make use of the computation results in the previous window. The reason is that, the set of uncertain data in the consequent $\mathcal{W}$ almost remain the same and the *PTOPK* cannot change dramatically. Thus, we can postpone the examination of uncertain data for inclusion in *PTOPK* for a certain period of time. It is unnecessary to calculate the exact score of each uncertain data in each step. This method is named as *postpone examination algorithm* (denote as PEA for short).

5   PEA Algorithm

In this section, we first introduce the motivation of PEA, and then give an important observation and theorem, finally present the details of PEA algorithm.

5.1 Motivation

This paper focuses on the count-based window, meaning that a new uncertain data object arrival leads to an oldest one expiration, and the number of data objects in a sliding window remains unchanged. Figure 3 shows a sliding window with 2-dimensional uncertain data at time point now=8, $W = \{U_1, U_2, \ldots, U_8\}$. We can see that U_2 partially dominates U_4 and fully dominates all other uncertain data. U_4 fully dominates U_1, U_3, U_6, U_7, U_8 and partially dominates U_5, we assume that $score(U_2) = 6.5$ and $score(U_4) = 5.4$ respectively. U_5 fully dominates U_1, U_6, U_7, U_8 and partially dominates U_3, we assume $score(U_5) = 4.5$. Furthermore, U_6 fully dominates U_1 and partially dominates U_7, so we assume $score(U_6) = 1.8$. However, U_7 just partially dominates U_1, so $score(U_7) = 0.6$. If we set k=2, U_2 and U_4 (shaded MBB in Figure 3) belong to *PTOPK*, and all the other uncertain data in W are not in *PTOPK*. Some uncertain data which are not in *PTOPK* cannot become a member of *PTOPK* in a short period of time. Let us take U_7 as an example, at now=8, $score(U_7) = 0.6$. We assume that the sliding window moves two steps, U_{new1} and U_{new2} are two new incoming uncertain data. We suppose both U_{new1} and U_{new2} are dominated by U_7, $score(U_7)$ will be increased by two, so $score(U_7) = 0.6 + 2 = 2.6$. Furthermore, we suppose that U_{new1} and U_{new2} are not dominated by U_4, which has the k-th largest score in *PTOPK* (here,K=2). On the other hand, two new uncertain data arrival must cause two oldest ones expiration, and we assume that the two oldest ones are all dominated by U_4, so $score(U_4) = 5.4 - 2 = 3.4$. That is, after two updates in W, $score(U_7)$ are still smaller than $score(U_4)$. In other words, U_7 still has no opportunity to join *PTOPK* even after the sliding window moves two steps.

When now=10, the real update in W is shown in Figure 4, we find that U_7 is not inserted in *PTOPK*. From this example, we find that each uncertain data $U_i \notin PTOPK$ has a corresponding period of time (named it as **minimum time interval**:MTI), in which U_i can not be inserted in *PTOPK*, so we can postpone judgement for uncertain data whether it can be inserted in *PTOPK*. If we can find MTI for each uncertain data not in *PTOPK*, a large number of computation can be saved. The following subsection discusses how to calculate MTI, and how to make use of it to improve algorithm performance.

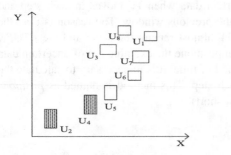 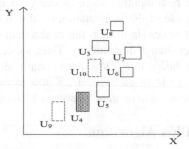

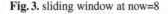

Fig. 3. sliding window at now=8   **Fig. 4.** sliding window at now=10

5.2 Postponing Calculating the Scores of Uncertain Data

Uncertain data in a sliding window can be classified into two types: uncertain data in *PTOPK* and not in *PTOPK*.

Table 1. Dominance Relationships

ID	dominance	score change	Δ	
	U_{new} arrival		$\Delta score_k.arrival$	$\Delta score(U_i).arrival$
a	$U_i \not\prec U_{new} \wedge U_k \not\prec U_{new}$	$score(U_i) \rightarrow, score_k \rightarrow$	0	0
b	$U_i \not\prec U_{new} \wedge U_k \prec U_{new}$	$score(U_i) \rightarrow, score_k \uparrow$	1	0
c	$U_i \prec U_{new} \wedge U_k \prec U_{new}$	$score(U_i) \uparrow, score_k \uparrow$	1	1
	U_{old} expiration		$\Delta score_k.expiration$	$\Delta score(U_i).expiration$
d	$U_i \not\prec U_{old} \wedge U_k \not\prec U_{old}$	$score(U_i) \rightarrow, score_k \rightarrow$	0	0
e	$U_i \not\prec U_{old} \wedge U_k \prec U_{old}$	$score(U_i) \rightarrow, score_k \downarrow$	-1	0
f	$U_i \prec U_{old} \wedge U_k \prec U_{old}$	$score(U_i) \downarrow, score_k \downarrow$	-1	-1

Table 2. Relative Score Difference

combination	ad	ae	af	bd	be	bf	cd	ce	cf
RSD	0	-1	0	1	0	1	0	-1	0

Let U_i be an uncertain data object not in *PTOPK*, thus, $score(U_i) < score_k$ ($score_k$ is used to denote the k-th best score, corresponding to uncertain data U_k in *PTOPK*) . When a new uncertain data (U_{new}) arrives, there are three kinds of dominance relationships: (a) $U_i \not\prec U_{new} \wedge U_k \not\prec U_{new}$, so $score(U_i)$ and $score_k$ remain unchanged. (b) $U_i \not\prec U_{new} \wedge U_k \prec U_{new}$, so $score(U_i)$ remains unchanged and $score_k$ is increased by one. (c) $U_i \prec U_{new} \wedge U_k \prec U_{new}$, both scores are increased by one. When an oldest uncertain data (U_{old}) expires, there are also three kinds of dominance relationships: (d) $U_i \not\prec U_{old} \wedge U_k \not\prec U_{old}$, so $score(U_i)$ and $score_k$ remain unchanged. (e) $U_i \not\prec U_{old} \wedge U_k \prec U_{old}$, so $score(U_i)$ remains unchanged and $score_k$ is decreased by one. (f) $U_i \prec U_{old} \wedge U_k \prec U_{old}$, so both $score(U_i)$ and $score_k$ are decreased by one. These dominance relationships are listed in Table 1. The symbol $\rightarrow$ is used to denote that the value of score remains unchanged. The symbol $\downarrow$ ($\uparrow$) is used to denote that the value of score is decreased (increased) by one.

We use the *relative difference* (Δ) of score to denote *score change* in comparison with its original score. $\Delta score_k.arrival$ ($\Delta score_k.expiration$) is used to denote the relative difference of $score_k$ when U_{new} arrives (U_{old} expires). In the same way, we can define $\Delta score(U_i).arrival$ and $\Delta score(U_i).expiration$ for $U_i \notin PTOPK$. For example, when U_{new} arrives, in case of dominance relationship **b** listed in Table 1, $score(U_i)$ remains unchanged and $score_k$ is increased by one. Correspondingly, we have $\Delta score(U_i).arrival = 0$ and $\Delta score(U_k).arrival = 1$. When U_{old} expires, in case of dominance relationship **e**, $\Delta score(U_i).expiration = 0$ and $\Delta score_k.expiration = -1$. Based on the definition of relative difference (Δ), it is easy to obtain Δ value in the last two columns of Table 1.

We use the *score difference (SD)* to denote score change after both U_{new} arrival and U_{old} expiration are taken into consideration. Based on the concept of relative difference (Δ), score difference of $score_k$ ($SD(score_k)$) is defined in Equation 4, which is sum of both $\Delta score_k.arrival$ and $\Delta score_k.\exp iration$. Similarly, $SD(score(U_i))$ is defined in Equation 5. We use *relative score difference (RSD)* defined in Equation 6 to denote the final relative score change between $score_k$ and $score(U_i)$.

$$SD(score_k) = \Delta score_k.arrival + \Delta score_k.\exp iration \tag{4}$$

$$SD(score(U_i)) = \Delta score(U_i).arrival + \Delta score(U_i).\exp iration \tag{5}$$

$$RSD = SD(score_k) - SD(score(U_i)) \tag{6}$$

As listed in Table 1, when U_{new} arrives and U_{old} expires, there are three domination cases respectively. So when a sliding window moves in a step, there are $3 \times 3 = 9$ kinds of dominance relationships, the corresponding RSD values are listed in Table 2. *RSD* has three possible values: 0, -1, 1.

Observation 1. *When a sliding windows moves in each step, the **minimum** relative score difference between $score_k$ and the score of $U_i \notin PTOPK$ ($score(U_i)$) is -1. So obviously we have the following observation.*

If $RSD = -1$, which indicates that $score_k - score(U_i)$ is become smaller in this moving step. As shown in Table 2, there are only two combinations (**ae**, **ce**) satisfying $RSD = -1$. We find that the dominance relationship **e** listed in Table 1 is a necessary condition that can make $RSD = -1$. That is, if and only if $U_i \not\prec U_{old} \wedge U_k \prec U_{old}$, which can make the value of $score_k - score(U_i)$ become smaller in the end. Therefore, we can come to the conclusions: i) dominance relationship **e** is a necessary condition for $RSD = -1$. ii) after a time interval with the length of $score_k - score(U_i)$ (it is a *MTI*), $score(U_i)$ can have an opportunity to equal to $score_k$. Furthermore, each uncertain data not in *PTOPK* has a MTI. From Definition 4, we know that the score of uncertain data is not always an integer, in order to obtain a safe MTI, we address it in a conservative way. That is, we set $MTI = \lfloor score_k - score(U_i) \rfloor$.

From *score difference* aspect, U_i has an opportunity to be inserted in *PTOPK* only if $score(U_i) \geq score_k$. On the other hand, if an uncertain data object U_j in *PTOPK* expires during time interval of $\lfloor score_k - score(U_i) \rfloor$ time units, U_i also has an opportunity to join *PTOPK*. Note that all the data in *PTOPK* will expire with the passage of time, and $\exp_1$ is used to denote the *minimum* expiration time for all uncertain data in *PTOPK*. The following theorem gives us a hint on when to calculate score for an uncertain data object $U_i \notin PTOPK$.

Theorem 1. *(minimum time interval: MTI). Given the minimum expiration time $\exp_1$ of the top-k dominating uncertain data and the current time point now, an uncertain data object U_i cannot be inserted in PTOPK in less than MTI time units where*

$$MTI(U_i) = \min\{\exp_1 - now, \ \lfloor score_k - score(U_i) \rfloor\} \tag{7}$$

Proof. There are two possible cases in which $U_i \notin PTOPK$ can become a member of *PTOPK* after a sliding window moves a step: (i) if an uncertain data object in *PTOPK* expires, or (ii) if $score(U_i) \geq score_k$. In the first case, since a data in *PTOPK* expires, it is possible for U_i to be added into *PTOPK*. For the second case, when an U_{new} arrives, since $U_k \prec U_i$, if $U_i \prec U_{new}$, according to dominance transitivity, we have $U_k \prec U_{new}$. In other words, we have $score(U_i) < score_k$ when U_{new} arrives. On the other hand, when an U_{old} expires, if and only if $U_i \nprec U_{old} \wedge U_k \prec U_{old}$, $score_k = score_k - 1$. However $score(U_i)$ is unchanged. We assume *the best case scenario*, in which the expired U_{old} is always dominated by U_k, but not dominated by U_i. That's to say, $U_i \nprec U_{old} \wedge U_k \prec U_{old}$ always holds when a sliding window moves in each step and U_i can have the opportunity to become a member of the *PTOPK* after at least $(\lfloor score_k - score(U_i) \rfloor)$ time instances. From above discussion, we can draw an important conclusion that U_i cannot be inserted into *PTOPK* in less than MTI time instances, the theorem has been proved.

MTI in Theorem 1 is just a conservative lower bound, indicating that during this interval, U_i is impossible to be in *PTOPK*. Therefore, it is unnecessary to calculate $score(U_i)$ during the time period of MTI. The larger the MTI is, the more computation can be saved. However, after a sliding window moves MTI time instances, we have to check whether U_i can be inserted in *PTOPK*. That's to say, if *now* is the current time, U_i will be checked again as a possible candidate for *PTOPK* until $\min\{\exp_1, \lfloor score_k - score(U_i) \rfloor + now\}$, and this timestamp is named as *checking time* (*CT* for short). Hereafter we use $U_i.CT$ to denote the checking time for U_i.

The checking time for uncertain data objects not in *PTOPK* are organized into a queue and they are sorted in the increasing order of their values. The head of the queue contains the checking time that will be processed next. When $U_i.CT = now$, meaning that it is time to check whether U_i can be inserted in *PTOPK*, $score(U_i)$ should be computed at this moment.

5.3   The Postponed Examination Algorithm (PEA)

The framework of PEA algorithm based on aR-tree index [16] is outlined in algorithm 1, which is composed of two parts: the first one is that treating the first $\mathcal{W}$ as a static database D and we use the PTD method in [10] to retrieve an initial *PTOPK* in $\mathcal{W}$. For each $U_j \notin PTOPK$ in the first $\mathcal{W}$, we compute its MTI and then $U_j.CT$ is obtained. Finally, these $U_j.CT$ are organized in a queue (line 2).

As $\mathcal{W}$ moves in each steap, the algorithm modifies the results by removing the outdated U_{old} and adding an U_{new}. The previous calculation information is used to delay checking uncertain data left in $\mathcal{W}$ which have no opportunity to join the *PTOPK*. When U_{new} arrives (line3), there is an oldest U_{old} expired, if U_{old} is in *PTOPK*, *PTOPK* should be updated first, meaning that it has to find an uncertain data object to be inserted in *PTOPK*. We process it as follows: we find an uncertain data object $U_T \notin PTOPK$, whose score $(score(U_T))$ has a largest value in the queue, and then inserts U_T into *PTOPK* (Line 6). This step ensures that *PTOPK* is fresh enough whenever $\mathcal{W}$ slides. On the other hand, the information of *PTOPK* are needed to compute the MTI for each $U_i \notin PTOPK$. Therefore, updating *PTOPK* must be done firstly. If U_{old} is not in

---

Algorithm 1. PEA(uncertain data streams)

Input: uncertain data streams
Output: probabilistic top-k dominating query on uncertain data steams
 1: call PTD algorithm to retrieve an initial $PTOPK$ in the first W
 2: Compute MTI and checking time for $U_j \notin PTOPK$ in the first W
 3: **while** each new incoming U_{new} arrive **do**
 4:     delete oldest U_{old}
 5:     **if** $U_{old} \in PTOPK$ **then**
 6:         find an uncertain data object U_T satisfying $score(U_T) = \max\{score(U_i)|U_i \notin PTOPK\}$, insert U_T in $PTOPK$
 7:     **else**
 8:         update scores of uncertain data in $PTOPK$
 9:     **end if**
10:     insert U_{new} into aR-tree, update index and update scores of uncertain data in $PTOPK$
11:     compute $score(U_{new})$ using window query method
12:     **if** $score(U_{new}) \geq score_k$ **then**
13:         insert U_{new} in $PTOPK$, update $PTOPK$
14:     **else**
15:         Compute MTI for U_{new}
16:     **end if**
17:     **for** $U_i.CT \leftarrow Queue.pop(); U_i.CT = now \wedge queue \neq \varphi; Queue.pop()$ **do**
18:         Checking($U_i.CT$, now)
19:         **if** $|PTOPK| \geq K$ **then**
20:             compute $score(U_i)$ using window query method
21:             **if** $score(U_i) \geq score_k$ **then**
22:                 insert U_i in $PTOPK$ and update $PTOPK$
23:             **else**
24:                 Compute the MTI for U_i
25:             **end if**
26:         **end if**
27:     **end for**
28: **end while**

$PTOPK$, it has to update scores for uncertain data objects in $PTOPK$ (line 8). After processing U_{old} expiration, we insert U_{new} into the aR-tree and update scores of uncertain data objects in $PTOPK$. Note that since $k \ll |W|$, the cost in this score update is affordable. $score(U_{new})$ are calculated using the window query method (line 11). If $score(U_{new}) \geq score_k$, meaning that U_{new} can be inserted in $PTOPK$ (line13). Otherwise, U_{new} has no opportunity to join $PTOPK$ at now, we should compute MTI for U_{new} (line 15). Finally, for all the $U_i.CT$ in the queue, if $U_i.CT = now$, we should check whether U_i can be inserted in $PTOPK$ at present (line 17-27).

6   Performance Evaluation

Experiments are run on PCs with 2.0GHz Intel Core 2 Duo CPU and 2G memory under Windows XP operating system. All algorithms are implemented in C++.

The computational cost is measured by the running time per 1000 steps. Two type of datasets are used in our evaluation process.

Synthetic datasets are generated using methodologies in [1] with respect to the following parameters listed in Table 3. The default values are in bold font. Particularly, number of instances per uncertain data follows a uniform distribution in$[1,\mathcal{M}]$. Each MBB to bound an uncertain data object is a hype-cube with the edge length (denoted *len*) following a normal distribution in the range $[0, len]$.

The edge length (*len*) follows either *uniform* or *Gaussian* (with mean $\frac{len_{min}+len_{max}}{2}$ and variance $\frac{len_{max}-len_{min}}{5}$) distribution. Centers of uncertain data (data's MBBs) follow either *anti-correlated* or *independent* distribution. Therefore, in all we have four types of synthetic datasets combining uncertain data centers and edge length: anti-uniform(A-U), anti-Gaussian(A-G), inde-uniform(I-U) and inde-Gaussian(I-G).

Real datasets. we use 2D geographical data sets *California Streams* (*CAS*) and *Tiger Streams* (*TS*), available at http://www.chorochronos.org/. CAS is an anti-correlated dataset. TS is an independent dataset.

Table 3. Parameter values

parameter	values
dimensionality d	**2**,3,4,5,6,7
$\|\mathcal{W}\|$ (million)	**0.01**,0.02, 0.03,0.04,0.05
$\mathcal{M}$	**100**,150,200,250,300
len	5,10,15,20,25
k	8,**32**,64,128,256
data types	**I-U**, I-G, A-U, A-G, CAS, TS

Since we are the first to study top-k dominating query over uncertain data streams, no algorithm can be used to compare with our approaches directly. However, Zhang et al. [16] proposed *EXACT* algorithm to retrieve *PTOPK* in database. We can consider each sliding window $\mathcal{W}$ as a small database, thus, *EXACT* algorithm can be invoked continuously to monitor *PTOPK*. We refer to the modified *EXACT* algorithm as **EIC** (EXACT invoked continuously). In order to provide a fair comparison between EIC and our approach, we have to set the threshold $q = 0$ in EIC.

Only a few experiment results are presented here due to space limitation.

Figure 5(a) reports the result of performance evaluation over datasets I-U by varying the dimensionality, while other parameters are setting in default values. Specially, BFA stands for baseline method. As the dimensionality grows, the CPU times of three algorithms increase. It is evident that BFA is not appropriate for streaming environments and it is omitted from subsequent experiments. PEA has a better performance than EIC. The reason are twofold. On one hand, computing dominating score is based on the *possible worlds* in EIC, Which is more costly than our method. On the other hand, PEA algorithm can avoid computing *PTOPK* from scratch.

Figure 5(b),5(c), 5(d) report the possible impacts against window size in I-G and two real datasets. Although the processing time of two algorithms both increase as $|\mathcal{W}|$

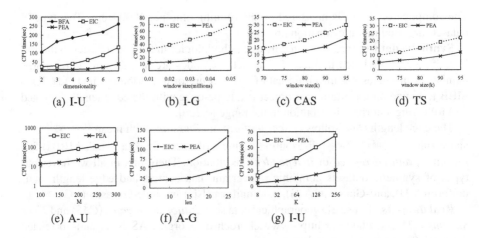

Fig. 5. Running time vs.different parameters

becomes larger. PEA has overall better performance than EIC. Figure 5(e) reports the experiment results against A-U with different average instance numbers. The y-axis in figure is illustrated in logarithmic scale. The CPU times of both algorithms are increased when $\mathcal{M}$ becomes larger. This is reasonable, as the number of instances become larger, computation of score for each uncertain data is more costly.

Figure 5(f) reports experiment results against A-G with different edges range of MBBs. The CPU times of both algorithms increase when the edges length become larger. This is as expected. Since the larger of edges length, the larger uncertainty regions of uncertain data would cause more uncertain data to intersect with each other, which makes the pruning approaches designed in EIC hard to distinguish from each other.

Figure 5(g) reports running time vs number of answers (K) in I-U dataset. The values of other parameters are set to default values. The CPU times of both algorithms increase as K becomes larger. However, PEA has a better performance than EIC. This is because that more candidates of uncertain data need to be checked as K increases in EIC, while this has a limited influence in PEA.

7   Conclusion

In order to retrieve *PTOPK* in uncertain data streaming environments, it is unwise to re-execute from scratch as the sliding window slides. Therefore, PEA algorithm is proposed, which computes the *minimum time interval* for each uncertain data not in *PTOPK* so as to avoid computing its score in that period, and makes use of *PTOPK* information computed in previous sliding window to accelerate the processing. The efficiency and effectiveness of PEA are extensively investigated in experimental studies.

Acknowledgments. This work was substantially supported by National Natural Science Foundation of China under Grant 61300045, and China Postdoctoral Science Foundation under GrantNo.2013M531696.

References

1. Borzsony, S., Kossmann, D., Stocker, K.: The skyline operator. In: Proceedings of the 17th International Conference on Data Engineering, pp. 421–430. IEEE (2001)
2. Chen, T., Chen, L., Ozsu, M., Xiao, N.: Optimizing multi-top-k queries over uncertain data streams. IEEE Transactions on Knowledge and Data Engineering (2012)
3. Cormode, G., Garofalakis, M.: Sketching probabilistic data streams. In: Proceedings of the 2007 ACM SIGMOD International Conference on Management of Data, pp. 281–292. ACM (2007)
4. Ding, X., Lian, X., Chen, L., Jin, H.: Continuous monitoring of skylines over uncertain data streams. Information Sciences 184(1), 196–214 (2012)
5. Feng, X., Zhao, X., Gao, Y., Zhang, Y.: Probabilistic top-k dominating query over sliding windows. In: Ishikawa, Y., Li, J., Wang, W., Zhang, R., Zhang, W. (eds.) APWeb 2013. LNCS, vol. 7808, pp. 782–793. Springer, Heidelberg (2013)
6. Jayram, T., McGregor, A., Muthukrishnan, S., Vee, E.: Estimating statistical aggregates on probabilistic data streams. In: Proceedings of the Twenty-Sixth ACM SIGMOD-SIGACT-SIGART Symposium on Principles of Database Systems, pp. 243–252. ACM (2007)
7. Jin, C., Yi, K., Chen, L., Yu, J.X., Lin, X.: Sliding-window top-k queries on uncertain streams. Proceedings of the VLDB Endowment 1(1), 301–312 (2008)
8. Kontaki, M., Papadopoulos, A.N., Manolopoulos, Y.: Continuous top-k dominating queries in subspaces. In: Panhellenic Conference on Informatics, PCI 2008, pp. 31–35. IEEE (2008)
9. Kontaki, M., Papadopoulos, A.N., Manolopoulos, Y.: Continuous top-k dominating queries. IEEE Transactions on Knowledge and Data Engineering 24(5), 840–853 (2012)
10. Lian, X., Chen, L.: Top-k dominating queries in uncertain databases. In: Proceedings of the 12th International Conference on Extending Database Technology: Advances in Database Technology, pp. 660–671. ACM (2009)
11. Lian, X., Chen, L.: Probabilistic top-k dominating queries in uncertain databases. Information Sciences (2012)
12. Papadias, D., Tao, Y., Fu, G., Seeger, B.: Progressive skyline computation in database systems. ACM Transactions on Database Systems (TODS) 30(1), 41–82 (2005)
13. Papapetrou, O., Garofalakis, M., Deligiannakis, A.: Sketch-based querying of distributed sliding-window data streams. Proceedings of the VLDB Endowment 5(10), 992–1003 (2012)
14. Santoso, B.J., Chiu, G.M.: Close dominance graph: An efficient framework for answering continuous top-k dominating queries. IEEE Transactions on Knowledge and Data Engineering, 1 (2013)
15. Yiu, M.L., Mamoulis, N.: Multi-dimensional top-k dominating queries. The VLDB Journal 18(3), 695–718 (2009)
16. Zhang, W., Lin, X., Zhang, Y., Pei, J., Wang, W.: Threshold-based probabilistic top-k dominating queries. The VLDB Journal 19(2), 283–305 (2010)
17. Zhang, W., Lin, X., Zhang, Y., Wang, W., Zhu, G., Xu Yu, J.: Probabilistic skyline operator over sliding windows. Information Systems (2012)

Keyword Search over Web Documents Based on Earth Mover's Distance

Jiangang Ma[1], Quan Z. Sheng[2], Lina Yao[2], Yong Xu[3], and Ali Shemshadi[2]

[1] Center for Applied Informatics, Victoria University, Australia
jiangang.ma@vu.edu.au
[2] School of Computer Science, The University of Adelaide, Australia
{michael.sheng,lina.yao,ali.shemshadi}@adelaide.edu.au
[3] School of Economics and Commerce, South China University of Technology, China
xuyong@scut.edu.cn

Abstract. Keyword search is widely used in many practical applications. Unfortunately, most keyword-based search engines compute the similarity distance between two Web documents by only matching the keywords at the same positions in both the query and the document vectors, without considering the impact of the keywords at neighbouring positions. Such approach usually results in incompleteness of search results. In this paper, we exploit the Earth Mover's Distance (EMD) as a distance function, which is more flexible against other distance functions such as Euclidean distance. To overcome the limitation of EMD-based computation complexity, we use the filtering techniques to minimize the total number of actual EMD computations. We further develop a novel lower bound as a new EMD filter for partial matching technique that is suitable for searching Web documents. The experimental results demonstrate the efficiency of EMD-based search with filtering techniques.

Keywords: Web-based similarity search, EMD, partial matching.

1 Introduction

Keyword search is a technique which aims to find objects that are close to a query object. It is an active area for both research and development [5,11,3,4,14]. An example of Web documents is Web services [3], which are loosely coupled software component published on the Web. These Web documents include textual information such as titles, headings, names and service operation descriptions. Represented in HTML and/or XML documents, it becomes feasible to search such Web documents through keyword search [1,5]. Other examples of Web documents include blogs, product review sites, cloud services, and so on.

A challenge is that the growing number of Web documents makes it difficult for search engines to find desired Web documents. Unfortunately, traditional keyword-based document search engines still have several limitations. Firstly, due to the heterogeneous naming conventions used by different document providers, the keyword-based model may result in a large term dictionary especially when

B. Benatallah et al. (Eds.): WISE 2014, Part I, LNCS 8786, pp. 256–265, 2014.
© Springer International Publishing Switzerland 2014

scaling to massive number of Web documents. Secondly, as Web document description is typically comprised of limited terms, the resultant term vectors will be extremely sparse, and matching documents based on traditional similarity measures (e.g., cosine similarity) will lead to poor results as the large sparse term vectors are less likely to coincide on common terms. Finally, traditional keyword-based search engines compute a similarity score by only matching the keywords at the same positions in both the query and the document vectors, without considering the impact of the keywords at neighboring positions.

To overcome the limitations mentioned above, we propose an Earth Mover's Distance (EMD)-based similarity search algorithm for finding similar Web documents based on the partial match of any of the Web document or a combination of them. EMD has been widely employed in multimedia search [12] to better approximate human visual perception, and speech recognition, because EMD can effectively capture the differences between the distributions of two objects' main features, and allow for partial matching. This means if keywords in a query only match partially the keywords in a document record, EMD-based similarity search is still able to find the most similar matching pairs of words. However, one limitation of EMD lies on its computation complexity. To overcome this issue, we exploit filtering techniques to minimize the total number of actual EMD computations. Furthermore, we develop a *generalized independent minimization lower bound* as a new EMD filter for partial matching that is suitable for searching Web documents. The filter is then incorporated into a k-NN *algorithm* for producing top-k results. Our key contributions are as the following. Firstly, we propose a novel model to represent Web documents. Each Web document is formalized as a set of weighted points that are stored in a database. This model can be used to capture partial similarity relationship between the documents and facilitate keyword search. Secondly, we develop a *generalized independent minimization lower bound* (LB_{GIM}) as a new EMD filter for partial matching. It is flexible for keyword-based content searches where words are always compared with different lengths. For this reason, LB_{GIM} also has great potential to be used in other application domains. We compute LB_{GIM} by proposing a *unified greedy algorithm*. We also conduct an extensive experimental evaluation on a set of real-world datasets of Web documents. The experimental results show that our approach produces query answers with high recall/precision and low response time.

The remainder of this paper is organized as follows. In Section 2, we present the model for Web documents and the EMD-based search algorithm. In Section 3, we report our experiments to show the effectiveness and efficiency of the proposed approach. Finally, in Section 4, we provide some concluding remarks.

2   The Proposed Methodology

2.1   Web Document Model

A Web document usually includes two kinds of key information: *content* and the *structure data*. The content of a Web document consists of a set of keywords.

The structure data refer to hyperlink information among Web documents. In this paper, we focus on the content information of Web documents.

Specifically, each Web document contains its content that is represented by a *keyword sequence*, which consists of ordered keywords separated by spaces where a keyword is a string of contiguous characters. Specifically, we model a Web document as *a record* stored in a database, where each record usually has some keywords to describe the basic content. Each record $r \in \mathbf{DB}$ is described by *keyword sequences* $\mathbf{wd}^r = \{wd_1^r, wd_2^r, ...\}$. Such records could not only be searched but also mined by proper algorithms.

Definition 1. *(Web Document Model) Each record $r \in \mathbf{DB}$ is a set of weighted points $r = \{(wd_1^r, t_1^r), (wd_2^r, t_2^r), ..., (wd_n^r, t_n^r)\}$, where wd_i^r is a single keyword in r and t_i^r is the length of the keyword wd_i^r. The length of the keyword is the number of characters consisting of the keyword.*

Based on this definition, a query q is also a specific Web document that is represented as a set of weighted points $q = \{(wd_1^q, t_1^q), (wd_2^q, t_2^q), ..., (wd_m^q, t_m^q)\}$.

2.2   EMD-Based Search Process

EMD Calculation. EMD describes the normalized minimum amount of work required to transform one distribution to the other. Computing the exact EMD requires solving the famous *transportation problem* [6] in operations research. In the context of our work, the *subtask* is to find EMD between a keyword and an attribute word sequences $\mathbf{kw}^q$ and $\mathbf{aw}^{r,M^q}$ from a query q and a record r, which describes their similarity. Then we perform this subtask for all filtered records, and the top-k ones with smaller EMDs are returned as final results to q.

For the sake of simplicity, let any subtask to involve two word sequences $\mathbf{kw}$ and $\mathbf{aw}$ instead of $\mathbf{kw}^q$ and $\mathbf{aw}^{r,M^q}$, where $|\mathbf{kw}| = n_1$ and $|\mathbf{aw}| = n_2$. Finding the minimum work of the subtask to transform $\mathbf{kw}$ to $\mathbf{aw}$ through the flow $\mathbf{f} = \{\forall 1 \leq i \leq n_1, 1 \leq j \leq n_2 : f_{ij}\}$ is equivalent to computing the optimal solution to the following linear program (LP) with variable f_{ij}:

$$
\begin{aligned}
&\text{minimize} : \sum_{i=1}^{n_1} \sum_{j=1}^{n_2} f_{ij} d_{ij} \\
&\text{subject to} : \\
&\forall 1 \leq i \leq n_1 : \sum_{j=1}^{n_2} f_{ij} \leq w_{kw_i} & (1.1) \\
&\forall 1 \leq j \leq n_2 : \sum_{i=1}^{n_1} f_{ij} \leq w_{aw_j} & (1.2) \\
&\forall 1 \leq i \leq n_1, 1 \leq j \leq n_2 : f_{ij} \geq 0 & (1.3) \\
&\sum_{i=1}^{n_1} \sum_{j=1}^{n_2} f_{ij} = \min\left(\sum_{i=1}^{n_1} w_{kw_i}, \sum_{j=1}^{n_2} w_{aw_j}\right) & (1.4)
\end{aligned}
\tag{1}
$$

where $\mathbf{d} = \{\forall 1 \leq i \leq n_1, 1 \leq j \leq n_2 : d_{ij}\}$ is the ground distance matrix between $\mathbf{kw}$ and $\mathbf{aw}$; $\mathbf{w_{kw}}$ and $\mathbf{w_{aw}}$ are weight vectors of keywords and attribute words respectively. These distances and weights are necessary *input values* to LP (1) that will be defined later. Furthermore, constraint (1.1) restricts the total outgoing flow from any keyword not to exceed the corresponding weight; (1.2) limits the total incoming flow to any attribute word to be no larger than the weight; (1.3) ensures the positiveness of all flows; (1.4) defines the amount of total flow which is equal to the minimum of keyword and attribute word sequences' total weights. Note that unlike in other works [2], the flow system we consider here is more general, which means the total weights of word sequences may not be equal. Also the system can be modeled as a directed *complete bipartite graph* with n_1 keyword nodes and n_2 attribute word nodes as two parts. For simplicity, let the total number of nodes $n = n_1 + n_2$ and total number of edges (representing flows) $m = n_1 \times n_2$.

Now if we assume the optimal flow $\mathbf{f}^*$ is found for the above LP, the EMD is then the corresponding work *normalized* by the amount of total flow:

$$EMD\,(\mathbf{kw},\,\mathbf{aw}) = \frac{\sum_{i=1}^{n_1} \sum_{j=1}^{n_2} f_{ij}^* d_{ij}}{\sum_{i=1}^{n_1} \sum_{j=1}^{n_2} f_{ij}^*} \tag{2}$$

which has the capability to avoid favoring shorter queries in our partial words matching context.

Conditional Lower Bound. We now discuss the generation of the EMD filter LB_{GIM} for word sequences partial matching. There are many ways to compute exact EMD with *arbitrary* distance metric. The streamlined approach we adopted is the transportation-simplex method due to its supercubic, between $\Omega\,(n^3)$ and $O\,(n^4)$, empirical performance [8,12]. This algorithm exploits the special structure of the transportation problem and hence has better performance. It should be noted that there exist other theoretically polynomial time algorithms like the interior-point method [7]. The flow problem can then be solved by the Orlin's algorithm in $O\,(n^3 \log n)$ time which is the *similar* to the transportation-simplex method's empirical runtime complexity. For EMD with special distance metrics like L_p-norms, more efficient solutions [8,13,10] exist.

Even though the algorithms for calculating the exact EMD are deemed to be efficient, as the amount of Web document contents continue to increase, the framework's response time for top-k queries may be driven down quickly. Hence we use the filtering idea similar to [12,2,9] to minimize the total number of actual EMD computations. However, their lower bound filters are only for EMD with equal weights ($\sum_{i=1}^{n_1} w_{kw_i} = \sum_{j=1}^{n_2} w_{aw_j}$ in our case), which is clearly not suitable for the comparison of word sequences that requires solving the general EMD (including the unequal total weights case). For this reason, we develop the LB_{GIM} filter generalized from LB_{IM} in [2] for word sequences partial matching. We will also see in the next subsection how this filter is incorporated into the k-NN algorithm for producing the final results.

Our LB_{GIM} is a *conditional lower bound* depending on three cases of LP (Equation (1)): 1)$\sum_{i=1}^{n_1} w_{kw_i} < \sum_{j=1}^{n_2} w_{aw_j}$; 2)$\sum_{i=1}^{n_1} w_{kw_i} > \sum_{j=1}^{n_2} w_{aw_j}$; and 3) $\sum_{i=1}^{n_1} w_{kw_i} = \sum_{j=1}^{n_2} w_{aw_j}$. In particular, case 3) is the same as LB_{IM}. For cases 1) and 2), we develop new lower bound LPs (4) and (3) respectively in the following and LB_{GIM} can be either of their normalized optimal solutions.

$$\text{minimize } \sum_{i=1}^{n_1} \sum_{j=1}^{n_2} f_{ij} d_{ij}$$

$$\text{subject to } \forall 1 \le i \le n_1 : \sum_{j=1}^{n_2} f_{ij} = w_{kw_i} \quad (2.1)$$

$$\forall 1 \le i \le n_1, 1 \le j \le n_2 : f_{ij} \le w_{aw_j} \quad (2.2)$$

$$\forall 1 \le i \le n_1, 1 \le j \le n_2 : f_{ij} \ge 0 \quad (2.3)$$

(3)

$$\text{minimize } \sum_{i=1}^{n_1} \sum_{j=1}^{n_2} f_{ij} d_{ij}$$

$$\text{subject to } \forall 1 \le i \le n_1, 1 \le j \le n_2 : f_{ij} \le w_{kw_i} \quad (3.1)$$

$$\forall 1 \le j \le n_2 : \sum_{i=1}^{n_1} f_{ij} = w_{aw_j} \quad (3.2)$$

$$\forall 1 \le i \le n_1, 1 \le j \le n_2 : f_{ij} \ge 0 \quad (3.3)$$

(4)

Although LB_{GIM} involves 3 possible cases, we can still provide a fast *unified greedy algorithm* to resolve them. Moreover, the algorithm enables us to avoid explicitly solving these LPs which takes much longer time. The main idea for developing the greedy algorithm is to look at LB_{GIM} from an algorithmic view rather than restricting ourselves to its mathematical view (the LP formulations). The algorithmic problem are then depicted in Figure 1. Specifically, we are asked to find the minimum cost flow subject to the constraints in LPs. If case 1) happens (the upper figure), we can consider each keyword in turn due to the relaxed constraint (2.2). For each keyword, we first ascendingly sort all its edges by distances. Then the *greedy strategy* is always trying to assign a *larger flow* value along the edge with the *shorter distance* until constraint (2.1) is fulfilled. Afterwards the unassigned edges associated with the keyword will have zero flow value. LB_{GIM} can then be easily calculated from assigned flows and edge distances. If case 2) happens (the lower figure), without loss of generality, we can reverse the direction of flow, consider each attribute word in turn and do the same thing as case 1). Finally, if case 3) happens, we consider both case 1) and case 2) and take the larger LB_{GIM} value which is closer to the exact EMD value. Figure 2 shows an example where $\mathbf{kw} = \{\,'HOLDEN', 'CAR', 'SERVICES'\}$ and $\mathbf{aw} = \{\,'CITY', 'HOLDEN'\}$. If we use SED as the distance metric, $\mathbf{d} = \begin{pmatrix} 6 & 0 \\ 3 & 6 \\ 8 & 8 \end{pmatrix}$, $\mathbf{w_{kw}} = \{6, 3, 8\}$ and $\mathbf{w_{aw}} = \{4, 6\}$ and the problem falls into the case 2) of LB_{GIM}. So we reverse the flow direction with $\mathbf{d}^T = \begin{pmatrix} 6 & 3 & 8 \\ 0 & 6 & 8 \end{pmatrix}$ and follow the greedy algorithm to get the total cost of assigned flows $TC = 3 \times 3 + (4 - 3) \times 6 + 6 \times 0 = 15$, where the total flow is $TF = 4 + 6 = 10$. Hence $LB_{GIM} = \frac{TC}{TF} = 1.5$.

EMD Filter Process. With the definitions of search query and database record, we can now describe the query processing. For every basic query q and its k required answers, we are able to operate the search process in database with a *boolean function* $\phi^{q,k} : r \to \{0, 1\}$, where $\phi^{q,k}(r) = 1$ if $EMD(\mathbf{kw}^q, \mathbf{aw}^{r,M^q}) \le \mathbf{EMDs}^{q,\mathbf{DB}}(k)$[1]. EMD function computes EMD between keyword and attribute word sequences, and the *ascendingly sorted* set of all EMD values is defined as: $\mathbf{EMDs}^{q,\mathbf{DB}} = \{\forall r \in \mathbf{DB} : EMD(\mathbf{kw}^q, \mathbf{aw}^{r,M^q})\}^+$. This process can then be translated into the following SQL query:

---

[1] The k-th element in the set $\mathbf{EMDs}^{q,\mathbf{DB}}$.

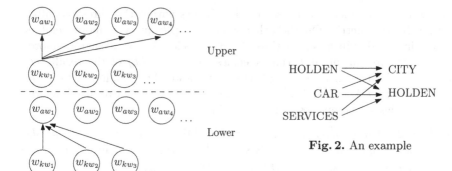

Fig. 1. Algorithmic view of LB_{GIM}

Fig. 2. An example

$$SELECT \ * \ FROM \ \textbf{DB}$$
$$WHERE \ \phi^{q,k}(r)$$
$$ORDER \ BY \ E\bar{M}D$$

where $\phi^{q,k}(r)$ is a boolean predicate in SQL and $E\bar{M}D$ denotes the extra created attribute name storing contents of $\textbf{EMDs}^{q,\textbf{DB}}$.

Generating the set $\textbf{EMDs}^{q,\textbf{DB}}$ may take long time for a large number of exact EMD computations. Fortunately, our LB_{GIM} filter uses $E\hat{M}D$ function to approximate EMD, and $\textbf{E}\hat{\textbf{M}}\textbf{Ds}^{q,\textbf{DB}} = \left\{\forall r \in \textbf{DB} : E\hat{M}D(\textbf{kw}^q, \textbf{aw}^{r,M^q})\right\}^+$ can be computed in a much shorter time. This set together with range and top-k queries is able to generate a filtered set $\textbf{DB}' \subseteq \textbf{DB}$ for processing the final top-k query. Therefore, the boolean function which describes the search process can be revised as: $\phi^{q,k} : r \in \textbf{DB}' \rightarrow \{0, 1\}$, where $\phi^{q,k}(r) = 1$ if $EMD\left(\textbf{kw}^q, \textbf{aw}^{r,M^q}\right) \leq$ $\textbf{EMDs}^{q,\textbf{DB}'}(k)$. For every query q and k required answers for q, to maximize the efficiency, $\textbf{QoS}^q$ is compared with $\textbf{QoS}^r$ first to generate a filtered set $\textbf{DB}'' \subseteq$ $\textbf{DB}$. Afterwards, the basic query process continues to generate $\textbf{DB}''' \subseteq \textbf{DB}''$ as described previously for processing the final top-k query. Note that both basic query and advanced query processes can be translated into a series of more complicated SQL queries.

2.3 Top-k Records Retrieval

In the previous sections, we have investigated the LB_{GIM} filter approach for word sequences partial matching. This approach can get an optimal solution by using a greedy strategy. In this section, we would like to study how this filter can be incorporated into the k-NN algorithm for producing the final results.

Two common query types exist in any retrieval systems, namely the *range query,* and the *top-k query* that is also known as the k nearest neighbor (k-NN) query. A range query is associated with a specific *metric* and a *threshold.*

The answer to this query is the set of the objects within the threshold after measuring by the metric. On the other hand, a top-k query is more flexible, since the threshold, which is sometimes hard to decide, is not needed anymore. Instead, it requires an input k that specifies the *cardinality* of the result set. In our work, the two metrics are EMD and its lower bound filter, which are calculated by the EMD and $\widehat{EMD}$ functions respectively as introduced before. Formally, we define the queries as the following:

Definition 2. *A range query q to a domain $\mathbb{D} \subseteq \mathbf{DB}$ with metric c and a threshold ϵ asks for a range set of records* $\mathbf{RS}_{c,\epsilon}^{q,\mathbb{D}} = \left\{ \forall r \in \mathbb{D} \mid c\left(\mathbf{kw}^q, \mathbf{aw}^{r,M^q}\right) \leq \epsilon \right\}$, *where c is either obtained from EMD or $\widehat{EMD}$.*

Definition 3. *A top-k query q to a domain $\mathbb{D} \subseteq \mathbf{DB}$ with metric c asks for a top-k set of records* $\mathbf{TKS}_c^{q,\mathbb{D}} = \left\{ \forall r \in \mathbb{D} \mid c\left(\mathbf{kw}^q, \mathbf{aw}^{r,M^q}\right) \leq \epsilon' \right\}$, *where c is either from EMD or $\widehat{EMD}$, and $\epsilon' = \mathbf{EMDs}^{q,\mathbb{D}}(k)$.*

The *domain* $\mathbb{D}$ for a query in our approach is either $\mathbf{DB}$ or its subset and $\mathbf{EMDs}^{q,\mathbb{D}}(k)$ is defined similarly in Section 3. A set of records are returned for a range query and a top-k set of records for a top-k query. In order to correctly utilize an $\widehat{EMD}$ filter to reduce the number of exact EMD calculations, it is necessary to combine the range and top-k queries together as shown in Algorithm 1. In the algorithm, step 1 and 2 issue a top-k query to $\mathbf{DB}$ with the result from $\widehat{EMD}$ as the metric. Assume we use LB_{GIM} with LD for $\widehat{EMD}$ and there are totally R records in domain $\mathbf{DB}$, then for a query these two steps take $O\left(RN^2 \log N + R \log R\right)$ time where N is the maximum number of words ($\geq$ any n) contained in a pair of matching word sequences. Step 3 and 4 then compute the exact EMDs for the set of k records obtained in step 2, and set the maximum one as the threshold to issue a range query to $\mathbf{DB}$ with $\widehat{EMD}$ as the metric. These steps take $O\left(kN^3 \log N + R\right)$ time. Finally, step 5 and 6 issue another top-k query to the remaining records ($\mathbf{DB}'$ obtained in step 4) with EMD as metric for getting the result set. Assuming there are totally R' records left after filtering from step 3 and 4, the final steps take $O\left(R'N^3 \log N + R' \log R'\right)$ time.

3   Experiments

3.1   Experimental Setting

The experiments were conducted using over 4,500 real-world Web documents that we collected from the Web. These documents represent a wide range of sources include normal web pages, as well as Web services collected from some public accessible sources[2,3]. To evaluate the quality of the search, all these Web documents are classified into 30 different categories (e.g., business). Each document includes three parts: i) the keywords that are included in that document,

---

[2] http://www.uoguelph.ca/~qmahmoud/qws/dataset
[3] http://wiki.cse.cuhk.edu.hk/user/zbzheng

Algorithm 1. k-nn Algorithm with LB_{GIM}

Input: query q with $\mathbf{kw}^q$ and M^q, input number k
Output: set $\mathbf{TKS}_{EMD}^{q,\mathbf{DB}}$.

1. $\forall r \in \mathbf{DB}$: compute $E\hat{M}D\left(\mathbf{kw}^q, \mathbf{aw}^{r,M^q}\right)$.

2. Construct the set $\mathbf{I} = \mathbf{TKS}_{E\hat{M}D}^{q,\mathbf{DB}}$.

3. $\forall r \in \mathbf{I}$: compute $EMD\left(\mathbf{kw}^q, \mathbf{aw}^{r,M^q}\right)$. Set $\epsilon = \max_r EMD\left(\mathbf{kw}^q, \mathbf{aw}^{r,M^q}\right)$.

4. Construct the set $\mathbf{DB}' = \mathbf{RS}_{E\hat{M}D,\epsilon}^{q,\mathbf{DB}}$.

5. $\forall r \in \mathbf{DB}'$: compute $EMD\left(\mathbf{kw}^q, \mathbf{aw}^{r,M^q}\right)$.

6. Construct the result set $\mathbf{TKS}_{EMD}^{q,\mathbf{DB}'}$ which is the required output $\mathbf{TKS}_{EMD}^{q,\mathbf{DB}}$.

---

ii) the URL of the Web document indicating the source of the Web document, and iii) the description explaining the contents of that document. A Java program was developed to perform preprocessing these Web documents. We used C++ programming language to implement the EMD-based search algorithms described in this paper. The query workload we selected includes a set of documents randomly selected from our collection. Those queries will allow us to test the top-k query values. In addition, final performance of search results was obtained through averaging the results used from all the queries. PHP (v5.3.2) was employed to implement a GUI that allows users to input queries, upload the data and display research results. As users typically appreciate a search system that is able to return the relevant documents (good precision) and rarely miss relevant documents (good recall), in our experiments, we used the widely-adopted *precision, recall,* and F measures to measure the search performance. All the experiments were conducted on a PC running Windows 7 with an Intel (R) Core(TM) i5-2410M 2.30GHz CPU and 4Gb memory.

3.2 Experimental Results

We first compared the precision of similarity search with EMD-based search algorithms combining with top-k query. For top-k query, the precision measures the value p/k, where p is the relevant documents out of the k Web documents retrieved. Figure 3 compares the effect of the database size on the precision of EMD-based search. To check the effect of the database size on the precision, we varied different values of the database and calculated the corresponding precision. Here k was set as 40. As shown in the figure, the precisions are all under 0.7 for different categories (here we picked up three categories, other categories had the similar results) for ground distance LD (Figure 3.a) while the precision are around 0.65 for ground distance SED (Figure 3.b). This experiment shows that the precision with LD is higher than that of SED due to the advantage of edit distance.

We also measured the recall of EMD-based similarity search. We selected a subset of test queries used in the precision experiments for this recall experiment.

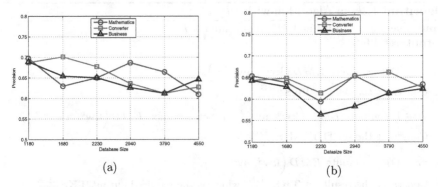

Fig. 3. Precision vs database size: (a) LD, (b) SED

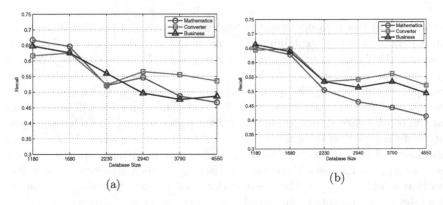

Fig. 4. Recall vs database size: (a) LD, (b) SED

For each test query, we manually checked the number of relevant services (P) in the database for that query. The recall was calculated as A/P where A is the number of relevant services retrieved. Figure 4 shows the impact of database sizes and ground distances. Overall, the impact of the sizes on the recall of EMD-based search for Web documents is not significant, which indicates a good property of EMD approach. Also, recall values in both diagrams are well above 0.4 for all DB sizes.

4   Conclusions

Effectively and efficiently searching desired Web documents remains a challenging task. In this paper, we have studied the main problem related to the current keyword-based search approaches: leaving partial similarities between queries and Web documents uncovered, leading to incompleteness in the search results. To overcome the problem, we have proposed a novel approach which can capture partial similarity relationship between a query and the Web documents. In particular, we have developed an earth mover's distance (EMD)-based filtering approach which is the first effective and easy-to-compute lower bound for the general EMD. Our approach is flexible and suitable for keyword-based content searches where words are compared with different lengths. We have also conducted experiments over a set of real datasets of Web documents. Our experimental result show the efficiency and effectiveness of the approach. Our future work includes integrating Semantic Web technique for more accurate search. Performance optimization of our proposed search method is another focus of the future research work.

References

1. Al-Masri, E., Mahmoud, Q.: Investigating web services on the world wide web. In: Proc. of the 17th Intl. World Wide Web Conf., WWW 2008 (2008)
2. Assent, I., Wenning, A., Seidl, T.: Approximation techniques for indexing the earth mover's distance in multimedia databases. In: Proc. of the 22nd Intl. Conference on Data Engineering, ICDE 2006, pp. 11–22 (2006)
3. Dong, X., et al.: Similarity search for web services. In: Proc. of the 30th Intl. Conf. on Very Large Data Bases, VLDB 2004 (2004)
4. Fu, A., Liu, W., Deng, X.: Detecting phishing web pages with visual similarity assessment based on earth mover's distance (EMD). IEEE Trans. on Dependable and Secure Computing 3(4), 301–311 (2006)
5. Fujii, A.: Modeling anchor text and classifying queries to enhance web document retrieval. In: Proc. of the 17th Intl. World Wide Web Conf., WWW 2008 (2008)
6. Hitchcock, F.: The distribution of a product from several sources to numerous localities. J. Math. Phys. 20(2), 224–230 (1941)
7. Karmarkar, N.: A new polynomial-time algorithm for linear programming. In: Proc. of the 16th Annual ACM Symposium on Theory of Computing, pp. 302–311 (1984)
8. Ling, H., Okada, K.: An efficient earth mover's distance algorithm for robust histogram comparison. IEEE Trans. on Pattern Analysis and Machine Intelligence 29(5), 840–853 (2007)
9. Ljosa, V., Bhattacharya, A., Singh, A.K.: Indexing spatially sensitive distance measures using multi-resolution lower bounds. In: Ioannidis, Y., et al. (eds.) EDBT 2006. LNCS, vol. 3896, pp. 865–883. Springer, Heidelberg (2006)
10. Pele, O., Werman, M.: Fast and robust earth mover's distances. In: 2009 IEEE 12th International Conference on Computer Vision, pp. 460–467. IEEE (2009)
11. Poblete, B., Baeza-Yates, R.: Query-sets: using implicit feedback and query patterns to organize web documents. In: Proc. of the 17th Intl. World Wide Web Conf., WWW 2008 (2008)
12. Rubner, Y., Tomasi, C., Guibas, L.: The earth mover's distance as a metric for image retrieval. International Journal of Computer Vision 40(2), 99–121 (2000)
13. Shirdhonkar, S., Jacobs, D.: Approximate earth mover's distance in linear time. In: Proc. of Intl. Conf. on Computer Vision and Pattern Recognition, CVPR 2008 (2008)
14. Wan, X.: A novel document similarity measure based on earth mover's distance. Information Sciences 177(18), 3718–3730 (2007)

iPoll: Automatic Polling Using Online Search

Thin Nguyen, Dinh Phung, Wei Luo, Truyen Tran, and Svetha Venkatesh

Deakin University, Australia
{thin.nguyen,dinh.phung,wei.luo,truyen.tran,
svetha.venkatesh}@deakin.edu.au

Abstract. For years, opinion polls rely on data collected through telephone or person-to-person surveys. The process is costly, inconvenient, and slow. Recently online search data has emerged as potential proxies for the survey data. However considerable human involvement is still needed for the selection of search indices, a task that requires knowledge of both the target issue and how search terms are used by the online community. The robustness of such manually selected search indices can be questionable. In this paper, we propose an automatic polling system through a novel application of machine learning. In this system, the needs for examining, comparing, and selecting search indices have been eliminated through automatic generation of candidate search indices and intelligent combination of the indices. The results include a publicly accessible web application that provides real-time, robust, and accurate measurements of public opinions on several subjects of general interest.

Keywords: web search, information extraction, opinion polls.

1 Introduction

Public sentiment is everything, as claimed by Abraham Lincoln. The same holds true one and half centuries later, although Internet and social media have made population opinions easier and quicker to change. Accurate real-time measure of public opinions has become increasingly challenging.

For decades, population opinions and behaviors are studied through polls. Gallup, one of the world's largest pollsters, conducts regular polls on diverse issues. For example, one poll aims to estimate the rate of diabetes in every state within a specific time and monitor its change over time. Polls are costly and time-consuming. For example, to track the approval rate of the US president, Gallup phones about 1,500 adults daily. Compiling the results causes delay. And not everyone can afford to purchase the results.

Alternatively, trends and drift of population opinions and behavior might be discovered through examining the community's online activities, such as web search. Everyday, people look for information online and their activities are recorded by the search engines. Online search trends have been found to be predictive of some important social-demographic indicators. For example, the validity of using online search queries to estimate public opinion trends was examined in [18]. It was found that conventional surveys can be replaced by the

B. Benatallah et al. (Eds.): WISE 2014, Part I, LNCS 8786, pp. 266–275, 2014.
© Springer International Publishing Switzerland 2014

query trends. A recent study assessed the validity of weekly Google Trends data against Gallup's issue salience surveys [13], asking "What do you think is the most important problem facing this country today?"[1] The author found that the search data can be a good surrogate of the salience of public opinion. In these studies, a single or a small number of search terms were selected to approximate a polling result or an issue of interest. Such a keyword needs to have a stable meaning that accurately reflects the target issue. Finding such a keyword is often difficult and often involves nontrivial human involvement.

In this study, we present an automatic polling system, based on online search activity, that can return the outcome similar to conventional surveys. We propose a way to eliminate the need for manual search-term selection. This study differs from existing studies in that we use a large number of keywords for each target issue. Such keywords can be populated using existing online tools and each keyword is only required to be loosely related to the target issue. Although the keywords may have varying and undetermined degree of relevance to the target issue, through novel use of modern machine learning techniques, we are able to automatically combine piece-meal information scattered around all these search terms. In the same fashion of ensemble learning, the aggregated information from all keywords provides a more accurate and more robust measurement of the target issue. This represents a novel data-driven approach to reduce human involvement polling approximation. Our system is evaluated against the historical records Gallup polls. We also deployed an interactive online website that provides real-time polling results freely accessible to the general public.

The main contribution of this study is to propose a novel approach to build automatic polling systems using online behavior data, validated in a large number of surveys. To address the noisy nature of usage of certain terms in the online community, we used a large number of terms, combined through a machine learning model, to reduce the uncertainty of the polling results. The significance of this work lies in using machine learning methods to relate conventional polls with online search activities, with automatic search term selection. The system has been successfully deployed online. With such implementation, for example, public opinions for each region or community can be assessed in real time and decision makers can make or adjust policies promptly for each community. While Gallup will not release its polls for 2014 until 2015, our website already contains similar information that is being updated in real-time. Also, a huge cost for expensive traditional polls would be saved. The cost of running a website like ours is negligible compared to the Gallup. Therefore, our polling results are available to the public free of charge.

2   Background

People's interests in different topics change from time to time, and the change is reflected in their web search activities. Google search related to fitness, diet,

---

[1] www.gallup.com/poll/1675/most-important-problem.aspx, retrieved April 2014.

weight loss and smoking is highest in January and declined throughout the remainder of the year [2]. Search related to pornography, prostitution and mate-
seeking has peaks in winter and summer months and troughs in spring and fall
months [12]. Seasonality was also found in online searching for mental health
terms, such as anxiety, depression and suicide [1].

Such ups and downs of web search volume have been exploited in many application domains. For example, in economics, Google Trends was used to predict
short-term values of economic indicators, namely unemployment claims and consumer confidence [3]. Similarly, the trend of web searches was used to predict
consumer behavior of cultural products, including movies and songs [7].

Web search activity has been explored in several areas of health studies. In
public health, an example is Google Flu Trends [6], which aggregates Google
flu-related searches[2]. The online search volume for flu-related terms was found
to be predictive of flu spread in the real world [4]. In drug safety, web search
logs can help detecting adverse drug events [17]. In suicide studies, the search
volume of suicide-related terms in a state was found to strongly correlate with
the suicide rate for that state [8], providing the base for an alarm system.

Despite the wide use of web search activities, most existing studies considered
the correlation between a target variable with a small number of pre-selected
search terms. In mental health, a positive relationship was found between the
Google Trends data on "depression" and "anxiety" with unemployment [14]. A
strong correlation was found between the suicide rates in each state of the United
States and the Google search volume in that state for "commit suicide" and "how
to suicide" [8]; or "hydrogen sulfide suicide" [9]. Similarly, the rate of Google
searches for "HIV" was found to be strongly correlated with the incidences of
HIV reported by the Centers for Disease Control and Prevention [10].

To build an automatic polling system, the problem for automatic search term
generation and selection remains to be solved.

3   Method

In this section we describe a procedure to build a polling system that captures
online search behavior and estimates a number of poll indices. Figure 1 is a
high-level flowchart showing the steps for this procedure. We focus on three
main poll categories from Gallup: Politics, Religion and Well-being. We define
a *poll index* as a sub-category within a *poll category*. Our aim is to estimate
the prevalence of a poll index for every state in the United States in a target
year using online search activity data. Thus a poll index is the outcome (or the
dependent variable) and the search activity data of its relevant terms are the
features (or the independent variables) in a prediction model.

In the training phase, for a target year and a poll index, data for dependent
and independent variables in the previous years are used to fit a prediction
model. In the testing phase, data for independent variables (web search activity)

---

[2] http://www.google.org/flutrends/

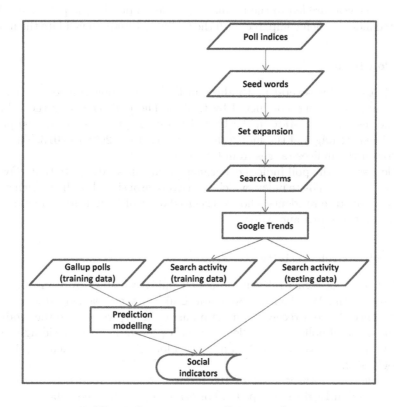

Fig. 1. An automatic polling procedure

Table 1. Annual polls conducted by Gallup from 2008 to 2013

Poll category	Poll index	Description
Politics	Democrat lean	The percentage of state residents who identify as Democrats or who identify as independents but say they lean Democratic
	Republican lean	The percentage of state residents who identify as Republicans or who identify as independents but say they lean Republican
Religion	Catholic	The percentage of state residents who report they are Catholic
	Protestant	The percentage of state residents who report they are Protestant
	Very religious	The percentage of state residents who say religion is important in their lives and say they attend church weekly or nearly weekly
Wellbeing	Diabetes	The percentage of state residents who say a physician or nurse told them they have diabetes
	Obese	The percentage of state residents whose Body Mass Index score is 30 or higher
	Overall wellbeing	An average of six sub-indexes, which individually examine life evaluation, emotional health, physical health, healthy behavior, work environment and basic access

in the target year are fed to the prediction model to predict the poll index. This predicted index is then compared with the Gallup estimate to validate the model.

3.1 Poll Indices

We collect data for several poll indices in three main poll categories: religion, politics and well being, conducted by Gallup. The pollster tracks the value of these indices for every state in the United States. It provides the annual prevalence (the percentage) of the indices for each state from 2008 to 2013. Thus, the value for all the indices ranges from 0 to 100.

Table 1 shows the poll indices experimented in this study. Note that "Overall wellbeing" is a composite index, which is an average of six sub-indices, such as the percentage of state residents who experienced a lot of happiness and enjoyment without a lot of stress and worry.

3.2 Web Search Activity

To select a set of terms relevant to a poll index, we start by using its name as the seed term. We then use Semantic Link[3], a set expansion tool, to grow the set. Since the tool receives a unigram as its input, we remove the modifiers from the name of polls, such as 'lean', 'very' or 'overall', when deciding seeding words. Other set expansion tools can be used for this purpose, such as [11], [16], or Google Sets.

Semantic Link returns a set of 100 terms that most frequently occurs with the given seed in English Wikipedia. For example, for the seed "obese", the tool returns "bmi", "calories", "carbohydrate" and "cholesterol". Then each of these terms, including the seed, is input to Google Trends, limited to Web search queries in the United States for each year from 2008 to 2013. For a year, Google Trends returns a vector, where each element corresponds to the relative search volume for a given term in a state.

For a poll index, not all of the relevant terms have sufficient online search volume in the country for the period. Only "Overall wellbeing" has all of its 101 relevant terms whose enough search volume. Relevant terms whose sufficient online search volume for each poll index are provided in the supplemental resource[4], consisting of 534 unique terms in total.

3.3 Prediction Modeling

Given a poll index, we describe a regression model to predict the value of the index for a state in the United States in a year. Let $\mathcal{D}$ be the training dataset consisting of n observations

$$\mathcal{D} = \{(\mathbf{x}_i, y_i) \mid \mathbf{x}_i \in \mathbb{R}^m, y_i \in \mathbb{R}\}_{i=1}^{n}$$

---

[3] http://semantic-link.com

[4] https://s3-us-west-2.amazonaws.com/wise2014/supplement.pdf

where y_i denotes the value of a Gallup index for a state i in a year, m is the number of terms chosen to predict the value and $\mathbf{x}_i := \{x_{i1}, x_{i2}, ..., x_{im}\}$ is the search data, where x_{ij} is the relative search volume for term j in state i in the year.

The regression model for the prediction is defined as

$$\hat{y}_i = \beta_0 + \sum_{j=1}^{m} \beta_j x_{ij} \tag{1}$$

where $\hat{y}_i$ is an estimate of a poll index in state i in a year, β_0 is the intercept and β_j is the regression coefficient associated with term j. When $m=1$, Equation 1 is a *univariate* linear regression. This model will be used as baseline in this work.

When $m>1$, Equation 1 is a *multivariate* linear regression. We use mean squared error (*mse*) as the loss function for this model

$$mse = \frac{1}{n} \sum_{i=1}^{n} (y_i - \hat{y}_i)^2 \tag{2}$$

A low *mse* model learned from training data will be used to predict polls for testing data.

When the number of features m is large, the model given in Equation 1 can fit the training data well, resulting in a small *mse*, but performs poorly when applied to new data, for example predicting polls for future years. This is also known as overfitting. To avoid this issue, we use a penalized regression approach, Lasso [15], to select independent variables for the multiple linear regression. The loss function is then defined as

$$J(\beta) = \frac{1}{n} \left[\sum_{i=1}^{n} (y_i - \hat{y}_i)^2 + \lambda \sum_{j=1}^{m} |\beta_j| \right] \tag{3}$$

where λ is the penalty parameter. To minimize the loss, this procedure can shrink some coefficients to zero, which is also a feature selection process. An implementation of this penalized regression model, *glmnet* [5], is used in the study. Of the models returned by the package, the one-standard-error (*1se*) model is chosen. This model has largest value of the penalty parameter λ whilst its performance is still within 1 standard error of the best accuracy (or the lowest *mse*). This practice may prevent overfitting since not too many features are included in the model while the prediction performance is also assured.

The performance of a prediction model is evaluated using Pearson correlation between the Gallup outcome and the value predicted by the model.

4   Result and Discussion

4.1   Single Search Term as the Predictor

The online search activity data for the relevant terms were used to estimate the value of the poll indices for each state in 2012 and 2013. The estimation model

Table 2. Pearson correlation between the Gallup outcome and machine prediction using the volume of online search for relevant terms

Poll index	2012					2013				
	Single term		Multiple terms			Single term		Multiple terms		
	Seed	Best	M1	M2	M3	Seed	Best	M1	M2	M3
Democrat lean	-0.011	0.683	0.792	0.812	0.924	0.006	0.714	0.804	0.805	0.947
Republican lean	-0.161	0.718	0.758	0.847	0.959	0.16	0.743	0.737	0.827	0.949
Catholic	0.453	0.717	0.907	0.918	0.966	0.404	0.647	0.879	0.899	0.95
Protestant	0.181	0.847	0.929	0.953	0.978	0.213	0.784	0.929	0.943	0.975
Very religious	0.576	0.781	0.752	0.944	0.973	0.455	0.713	0.737	0.938	0.969
Diabetes	0.45	0.769	0.898	0.861	0.862	0.394	0.727	0.863	0.831	0.867
Obese	0.236	0.769	0.797	0.91	0.88	0.211	0.706	0.753	0.868	0.872
Overall wellbeing	0.14	0.73	0.869	0.822	0.895	0.1	0.759	0.812	0.782	0.836
Average	0.233	0.752	0.838	0.883	0.930	0.243	0.724	0.814	0.862	0.921

for an index in a year is built using data from previous years back to 2008 as training. For example, to predict "*Very religious*" in 2013, data from 2008 to 2012 is used as training.

A *univariate* regression model was used as a baseline to predict the prevalence for a target poll index. Firstly the search volume for a seed was used as the predictor. As shown in Table 2, the result of this prediction is very poor: on average, 23.3% for 2012 and 24.3% for 2013. The highest correlation these seed predictors gain is for "*Very religious*" in both 2012 (57.6%) and 2013 (45.5%), using the Web search volume for "*religious*" for the prediction.

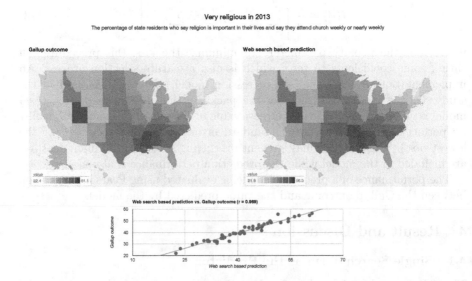

Fig. 2. Gallup's "*Very Religious*" 2013 and machine prediction using web search data

Secondly, to find out the best performance these simple regressions can gain, each of all 534 search terms was used as a predictor in the task. For a poll index, the highest correlation that these single models gain was reported. The highest correlation is with "*Protestant*" in both 2012 (84.7%) and 2013 (78.4%), using Web search volume for "*beliefs*" and "*churches*" as the predictor respectively.

4.2   Multiple Search Terms as the Predictors

When using *multivariate* regression models to predict the prevalence for a poll index, three types of the independent variables in the initial model (matrix **x** in Eq. 1) were experimented. They are (**M1**) the search terms relevant to the poll index; (**M2**) the search terms relevant to the poll category the poll index belongs to; and (**M3**) all the search terms. The features in the final model were then selected by Lasso [5].

As shown in Table 2, the prevalence estimated by the multivariate models was very strongly correlated with Gallup's, ranging from *Overall well being* 2013 ($r=0.836$) to *Protestant* 2012 ($r=0.978$). An example for the comparison of Gallup outcome and machine estimation is shown in Figure 2.

On average, the best performance is with **M3**, the model of all terms included in the initial, followed by **M2**,with terms in the poll category and then by **M1**, with terms in the poll index. As shown in Table 2, almost all multiple variable models are better than the single variable models in the prediction of the poll indices in 2012 and 2013.

Of the three poll categories, on average, *religion* indices are best predicted, at 97.2% in 2012 and 96.5% in 2013, followed by *politics* (94.2% in 2012 and 94.8% in 2013) and then *wellbeing* (87.9% in 2012 and 85.8% in 2013).

Fig. 3. Regression model of search terms to predict "*Very Religious*" 2013. The size of variables indicates the absolute magnitude of corresponding coefficients; the color indicates sign of the coefficients: blue for positive and red for negative.

Feature selection. As seen above, model **M3** whose all 534 search terms were initially included works best in the prediction of all the poll indices. However, not all terms were included in the final prediction model. For example, Figure 3 shows the *1se* model learned by Lasso to predict *Very religious* for 2013. About 15% of search terms were chosen into the prediction model of the poll index. A majority of the chosen features are religion related, such as *priesthood, churches,* and *religious*, whose the largest positive weights in predicting of *Very religious*. The resulting model gains an accuracy of 96.9%, demonstrating an automatic and efficient way to select search terms as features into a prediction model of polls.

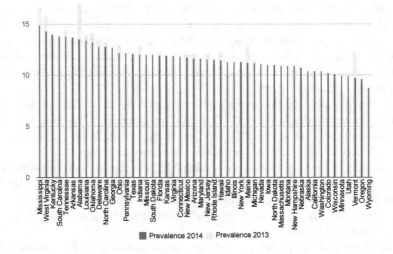

Fig. 4. Prediction of "*Diabetes*" 2014 based on web search for terms relevant to the poll. The shadow chart shows Gallup's "*Diabetes*" in 2013.

4.3   Polling System

A polling system using Google search volume is implemented and placed at http://ipoll.heroku.com. Figure 4 shows an example of machine prediction on the prevalence of *Diabetes* in 2014. The website also contains the poll indices for 2012 and 2013. Snapshots of the system are provided at https://s3-us-west-2.amazonaws.com/wise2014/supplement.pdf.

5   Conclusion

We have investigated a novel problem of leveraging the activity of online search as a sensor of public opinion on several politics, religion and well-being indices. It was found that the online activities was predictive of the indices. The result should be of interest to those wishing to gather public opinion on their targeting subjects. One possible application would be building automatic polling systems, supporting officials to make or adjust policies timely and at the right place.

References

1. Ayers, J.W., Althouse, B., Allem, J.-P., Niels Rosenquist, J., Ford, D.E.: Seasonality in seeking mental health information on Google. American Journal of Preventive Medicine 44(5) (2013)
2. Carr, L.J., Dunsiger, S.I.: Search query data to monitor interest in behavior change: Application for public health. PLoS ONE 7(10), e48158 (2012)
3. Choi, H., Varian, H.: Predicting the present with Google Trends. Economic Record 88(s1), 2–9 (2012)
4. Dugas, A.F., Hsieh, Y.-H., Levin, S.R., Pines, J.M., Mareiniss, D.P., Mohareb, A., Gaydos, C.A., Perl, T.M., Rothman, R.E.: Google Flu Trends: Correlation with emergency department influenza rates and crowding metrics. Clinical Infectious Diseases 54(4), 463–469 (2012)
5. Friedman, J., Hastie, T., Tibshirani, R.: Regularization paths for generalized linear models via coordinate descent. Journal of Statistical Software 33(1), 1 (2010)
6. Ginsberg, J., Mohebbi, M.H., Patel, R.S., Brammer, L., Smolinski, M.S., Brilliant, L.: Detecting influenza epidemics using search engine query data. Nature 457(7232), 1012–1014 (2008)
7. Goel, S., Hofman, J.M., Lahaie, S., Pennock, D.M., Watts, D.J.: Predicting consumer behavior with web search. Proceedings of the National Academy of Sciences 107(41), 17486–17490 (2010)
8. Gunn III, J.F., Lester, D.: Using Google searches on the Internet to monitor suicidal behavior. Journal of Affective Disorders (2012)
9. Hagihara, A., Miyazaki, S., Abe, T.: Internet suicide searches and the incidence of suicide in young people in Japan. European Archives of Psychiatry and Clinical Neuroscience 262(1), 39–46 (2012)
10. Jena, A.B., Karaca-Mandic, P., Weaver, L., Seabury, S.A.: Predicting new diagnoses of HIV infection using Internet search engine data. Clinical Infectious Diseases (2013)
11. Letham, B., Rudin, C., Heller, K.A.: Growing a list. Data Mining and Knowledge Discovery 27(3), 372–395 (2013)
12. Markey, P.M., Markey, C.N.: Seasonal variation in Internet keyword searches: A proxy assessment of sex mating behaviors. Archives of Sexual Behavior, 1–7 (2012)
13. Mellon, J.: Internet search data and issue salience: The properties of Google Trends as a measure of issue salience. Journal of Elections, Public Opinion and Parties 24(1), 45–72 (2014)
14. Tefft, N.: Insights on unemployment, unemployment insurance, and mental health. Journal of Health Economics 30(2), 258–264 (2011)
15. Tibshirani, R.: Regression shrinkage and selection via the Lasso. Journal of the Royal Statistical Society. Series B (Methodological), 267–288 (1996)
16. Wang, R.C., Cohen, W.W.: Iterative set expansion of named entities using the web. In: Proceedings of the IEEE International Conference on Data Mining (ICDM), pp. 1091–1096 (2008)
17. White, R.W., Tatonetti, N.P., Shah, N.H., Altman, R.B., Horvitz, E.: Web-scale pharmacovigilance: Listening to signals from the crowd. Journal of the American Medical Informatics Association (2013)
18. Zhu, J., Wang, X., Qin, J., Wu, L.: Assessing public opinion trends based on user search queries: Validity, reliability, and practicality. In: The Annual Conference of the World Association for Public Opinion Research, Hong Kong, (2012)

Comparing the Predictive Capability of Social and Interest Affinity for Recommendations

Alexandra Olteanu[1], Anne-Marie Kermarrec[2], and Karl Aberer[1]

[1] Ecole Polytechnique Federale de Lausanne (EPFL), Lausanne, Switzerland
[2] INRIA Rennes-Bretagne Atlantique, Rennes, France

Abstract. The advent of online social networks created new prediction opportunities for recommender systems: instead of relying on past rating history through the use of collaborative filtering (CF), they can leverage the social relations among users as a predictor of user tastes similarity. Alas, little effort has been put into understanding when and why (e.g., for which users and what items) the *social affinity* (i.e., how well connected users are in the social network) is a better predictor of user preferences than the *interest affinity* among them as algorithmically determined by CF, and how to better evaluate recommendations depending on, for instance, what type of users a recommendation application targets. This overlook is explained in part by the lack of a systematic collection of datasets including both the explicit social network among users and the collaborative annotated items. In this paper, we conduct an extensive empirical analysis on six real-world publicly available datasets, which dissects the impact of user and item attributes, such as the density of social ties or item rating patterns, on the performance of recommendation strategies relying on either the social ties or past rating similarity. Our findings represent practical guidelines that can assist in future deployments and mixing schemes.

Keywords: Social affinity, Interest affinity, Recommender systems, Collaborative Filtering, Evaluation.

1 Introduction

Recommender systems are inescapable in a wide range of web applications, e.g. Amazon or Netflix, to provide users with books or movies that match their interest. Accurate recommendations generate returns of investments up to 30% due to increased sales [24]. Many such systems rely on collaborative filtering (CF) approaches that recommend items based on user rating history. Concomitantly, the rising popularity of social networks has provided new opportunities to filter out relevant content for users. For instance, recommendation services like Epinions, Last.fm or BeerAdvocate are enhanced with virtual social networks.

As a result, existing works have proposed both pure social recommenders (SR)[1] that only leverage the social ties among users [33], and hybrid approaches

---

[1] For readability, social refers to both trust and social.

B. Benatallah et al. (Eds.): WISE 2014, Part I, LNCS 8786, pp. 276–292, 2014.
© Springer International Publishing Switzerland 2014

that either augment the CF recommendation engine with social guidelines [39,31] or incorporate CF mechanisms into a social recommendation engine [20].

A common practice in evaluating such approaches is to resort to *(i)* one [42,45,35,22,20,25,30], sometimes two [39,31] datasets and *(ii)* global averages for the metrics of choice. Alas, this has made it difficult to draw generalizable conclusions on the effectiveness of leveraging the social ties for recommendations compared with CF across datasets of different nature.

Furthermore, the use of global metrics[2] to evaluate and compare the recommendation approaches may be inconclusive as they provide little insight into *when* and *why* the approaches succeed or fail [13]. Although the impact of the parameters of a recommendation strategy has been often inspected [9,39,22,41,29,20,7], little systematic effort has been devoted into understanding how various user or item attributes are affecting the performance [2], and none of such analyses, to our knowledge, have included SR approaches.

Orthogonal to designing better hybrid approaches that combine SR and CF features, our goal is to gain insight into the relative benefits of each of these approaches that, in turn, can guide future deployments and mixing schemes. To do so, we perform an extensive empirical analysis that dissects the recommendation performance, measured by precision and coverage, and does a fine grained comparison across various user and item classes on six *publicly available* datasets including both the ratings information and the social network among users (§3). All datasets are medium to large-scale and exhibit various properties regarding user social ties and items ratings. We focus on the two ends of the problem spectrum, which places on the one side the *interest affinity* among users (resp. items), as algorithmically determined by CF from user rating history, and at the other side the *social affinity* as inferred from users social network by *pure* SR (§2). Our analysis addresses two main questions:

(1) Are global metrics able to reflect the performance of a given recommendation strategy across various settings? Our analysis shows that one cannot rely on global metrics to assess a given recommender performance not only across all datasets but also within each dataset, across different classes of users or items. Even a slight change in the global average might hide important changes in the performance distribution across a dataset demographics. One may thus need to understand and optimize the performance on a specific demographic subset depending on the application specifics (e.g., for a beer recommendation service, it might be more important to be accurate in the recommendations to experienced and, thus, harder to please users [34]).

(2) Are there user or item attributes that hint at the CF (interest affinity) performance with respect to SR (social affinity)? In our results, we find that when the basis of formulating connections among users stems from *plain* friendship, rather than from sharing interests, SR leads to less more precise recommendations. Further, items likeability (the rating they received on average) and user selectiveness are good predictors of the recommendation performance: relying on *interest affinity* for items similarity leads to more precise predictions for highly

---

[2] Metrics that are computed or averaged for all predictions.

liked items, while for indulgent users (that typically give high ratings) leveraging the *social affinity* is best. More results are discussed in (§3).

2   Problem Definition

Typically, a recommender task is to predict ratings for unseen items to users. To do so, a set of items I, a set of users U, and a set of items $I_u \subseteq I$ rated by each user u with a rating $r_{u,i}$ on a Likert scale from 1 to 5 is considered. If the recommender system exploits the social ties among users, for each user u a set of friends F_u is assumed. This paper looks at the predictive capability of social ties (SR) compared to the one of items or users rating similarity (CF) for items recommendation.

2.1   Comparison Framework

We conduct our study using a comparison framework that implements a rec-ommendation template under which, to make a recommendation for user u on target item i, two main steps are performed[3]: (1) identify the set of similar users (resp. items) with u (resp. i) and (2) compute weighted aggregates of their rat-ings on i (resp. from u) according to the similarity with u (resp.i). On top of it, we implement the main building blocks of SR and CF as used for comparison in literature [3,20,21,35,25,23]. Specifically, we implement (a) item- and user-based CF variants as often used as reference point by previous work [20,35,38,25], and (b) a SR approach that aggregates the ratings similarly with CF, yet, instead of deriving users affinity based on how similar they rated items in the past, it does so based on their social ties. Next we describe each approach and motivate our choices.

Collaborative Filtering (CF) approaches are usually grouped in two main classes: *neighborhood-* and *model-based* [12]. Model-based variants have received lot of attention as their accuracy was considered superior, yet neighborhood-based CF, though simpler, remains competitive [11]. Further, they exploit dif-ferent patterns in data, none of them consistently out-perform the other: model-based CF is typically effective at estimating the overall model related to all items simultaneously, while neighborhood-based CF better captures local associations in data [6]. This trait makes neighborhood-based CF suitable for our purpose to compare the predictive capability of *interest affinity* (inferred based on implicit similarity links as determined by CF) and *social affinity* (computed based on ex-plicit social links among users). Further, neighborhood-based CF offers a simple and intuitive template for recommendation to easily implement a pure SR-based approach on top of it and fairly compare the two under the same setting.

We use common variants of the two main types of neighborhood-based CF: user- and item-based CF. Briefly, for each user u (resp. item i) a neighborhood

---

[3] As in neighborhood-based CF [18].

UN_u (resp. IN_i) of users (items) similar with u (resp. i) is built and their ratings on the target item i (resp. from active user u) are aggregated as:

$$p_{u,i} = \frac{\sum_{v \in UN_u} sim(u,v) r_{v,i}}{\sum_{v \in UN_u} sim(u,v)} \quad (1)$$

for user-based CF, where $sim(u,v)$ is the similarity between users u and v, as estimated by the Pearson correlation of the ratings given by u and v on the same items[4]; respective, $p_{u,i} = \frac{\sum_{j \in IN_i} sim(i,j) r_{u,j}}{\sum_{j \in IN_i} sim(i,j)}$ for item-based CF, where $s(i,j)$ is the Pearson correlation of the ratings received by i and j from the same users.

Social Recommendation (SR). In contrast to CF[5], when the ratings received by target item i are aggregated according to Eq. (1), SR weights them based on the social affinity between the active user (i.e., the user for which we want to make a prediction) and the users that have rated item i in the past.

Social Affinity (relatedness) of two nodes in a social graph can be estimated using random walks (RWs) [28], which have been used for both friend [5,26] and item recommendations [43,20,14]. In short, for each prediction, we run RWs on the social graph that start at user u needing a recommendation on item i, and stops when they either reach a user v that have rated the target item i, or have performed a maximum number of steps k_{max}[6]. We denote a RW stopping condition with $s_{v,i,k}$, which is *true* if $i \in I_v$ or $k >= k_{max}$, meaning that the RW stops at v. Then, the social affinity between u and user v that rated the target item i is the probability to reach v using different paths and number of steps: $P(X_{u,i} = v) = \frac{\sum_k P(X_{u,i,k} = v)}{\sum_{w \in U} \sum_k P(X_{u,i,k} = w)}$, where the random variable $X_{u,i}$ represents the nodes that rated item i and can be reached at any step of the RW starting at node u, while $X_{u,i,k}$ represents only the subset of nodes reachable at step k:

$$P(X_{u,i,k} = v) = \sum_{w \in U} P(X_{u,i,k-1} = w) P(X_w = v) \quad (2)$$

where $P(X_{u,i,0}) = 1$ and X_w the random variable to pick a friend of node w. For unweighted graphs (as those used in our evaluation), we have: $P(X_w = v) = \frac{1}{|F_w|}$.

Thus, the probability to step on node $v \in F_w$ at step $k+1$ after being at node w at step k is $P(X_{u,i,k+1} = v | X_{u,i,k} = w, \bar{s}_{w,i,k}) = P(X_w = v)$, where $X_{u,i,k}$ is the random variable for nodes that can be reached at step k when looking for i, $\bar{s}_{w,i,k}$ is the negation of $s_{w,i,k}$, and $P(X_{u_i,k+1} = v | X_{u,i,k} = u, s_{w,i,k}) = 0$ to complete the probability distribution. To also complete the specification of the probability distribution in Eq. (2), we define a final state $\perp$, to which the RW goes when it terminates: $P(X_{u,i,k} = \perp) = 1 - \sum_{v \in U} P(X_{u,i,k} = v)$.

To determine if we performed enough RWs to make an admissible prediction, after each RW we compute the variance $\sigma^2 = \frac{\sum_{j=1..T} (r_j - \bar{r})}{T}$ in the results of all the

---

[4] Note that we also consider only positive correlations [20].

[5] For brevity, when referring to both user-based and item-based CF, we use only CF.

[6] Set to 6 based on the "six-degree of separation" assumption [36] that most of the nodes are reachable within 6 hopes [20].

walks [20], where T is the number of successful walks[7], r_j is the result returned by the j-th RW, and $\bar{r}$ is the mean of the results return by the RWs. If the variance σ^2 converges to a constant (i.e., the variance after $j+1$ walks varies with less than $\epsilon = 0.0001$ from the variance after j walks), or the total number of (successful and unsuccessful) walks reaches the maximum number of walks $T_{max} = 1000$, we stop from running more RWs. Then, to make a prediction, in Eq. (1), we replace the similarity between active user u and user v which have rated item i with their relatedness in the social network: $p_{u,i} = \sum_{\{v \in U | i \in R_v\}} P(X_{u,i} = v) r_{v,i}$.

3   Empirical Analysis

In this section we perform an extensive analysis that juxtaposes the SR (social affinity) and CF (interest affinity) as predictors for item recommendation, structured in three parts. First, we present a comprehensive characterization of the datasets. Second, we apply global metrics to evaluate the recommendation strategies, and examine if they capture the performance variation across various settings. Finally, we do a fine grained analysis of the impact of user and item properties on the performance, organized as a set of questions about CF and SR properties. These questions are largely inspired by admitted properties of CF or SR, such as, CF performs better on users for which it has more information [15,4,8], the recommendation accuracy decreases towards the long-tail items (i.e., less popular items) [40], or SRs are superior on *cold start* users [20,33].

3.1   Metrics and Experimental Setup

To evaluate the recommendation performance, we use the well-known *leave one out* strategy. Specifically, we remove from the dataset only the rating we want to predict and leave the other ratings and social network unchanged. Then, we compare CF and SR along two popular metrics: (1) The *coverage* measures a recommendation strategy ability to make predictions, and it is the number of ratings the system succeeded to make divided by the total number of ratings that it tried to predict. (2) The *Root Mean Square Error (RMSE)* captures the average error between the predictions and the real ratings, measuring the recommendation precision: $RMSE = \sqrt{\frac{1}{N} \sum (r_{u,i} - p_{u,i})^2}$, where N is the number of predictions, $r_{u,i}$ the real rating given by u to item i, while $p_{u,i}$ is the prediction. Note that the smaller the RMSE is, the more precise the recommendations are.

Albeit RMSE ability to gauge the performance for pervasive top-k recommendations is debated [10], it best fits our purpose to measure performance shifts across classes of items/users. The accuracy metrics deemed suitable to evaluate top-k performance, are biased towards the performance on preferred items (i.e., high ratings) [19]. Moreover, many recommender systems that leverage the social ties optimize for RMSE [44], making our analysis convenient to compare with.

---

[7] A random walk is successful if it encounters a user that have rated the target item.

Table 1. Datasets Figures

Dataset	Users	Items	Ratings	Social Links	Links Type
Ciao	12,375	99,762	284,086	237,350	direct
Epinions1	49,290	139,738	664,824	487,181	direct
Epinions2	22,166	296,277	922,267	355,813	direct
Epinions	132,000	755,760	13,668,319	841,372	direct
Flixster	786,936	48,794	8,196,077	7,058,819	symmetric
Douban	129,490	58,541	16,830,839	1,692,952	symmetric

Two approaches are used to report RMSE and coverage values for a set of users/items: (1) compute the RMSE (resp. coverage) over all the predictions to users (or for items) in the set; or (2) compute the RMSE (resp. coverage) for each item/user separately and average the results over all users (resp. items) in the set. While the first measures the overall performance on estimating the ratings, the second weights each user (resp. item) equally measuring how good the predictions are on average for each user (resp. item) in the set. We measured both, yet, due to space limitations, when the two variants lead to similar conclusions we show only results with the second one; both are included otherwise. Finally, when measuring how a certain user (resp. item) property impacts the results, we group the users (resp. items) by logarithmically binning them regarding the property value, and then compute the performance for each bin[8].

3.2   Datasets Characterization

We conduct our analysis on 6 real world publicly available datasets including both ratings and a social network (figures are summarized in Table 1):

Epinions is a popular product review site where people rate products and build lists of trusted users whose reviews they find useful. We use two rating datasets from Epinions: one is collected by the authors of [32] around 2006 (noted *epinions1*), and one is collected in May 2011 by the authors of [42] (noted *epinions2*). In addition to product ratings, in Epinions, users can also rate product reviews. We also use a dataset, made available by Epinions.com to the authors of [32] containing ratings on product reviews, instead of ratings on products (noted *epinions*). In all datasets the ratings are on a scale from 1 to 5.

Douban is a Chinese product review site that represents one of the largest online communities in China. As in Epinions, users rate and review products in order to receive recommendations. In addition, at the date of crawling, it provided a Facebook-like social networking service [31].

Ciao defines itself as a *multi-million-strong online community* in which users critically review and rate millions of products. It provides the same functionality as Epinions (i.e., users can both rate products and indicate the trusted users)[41].

Flixster is a large social movie rating service that allows users to create Facebook-like friendship relations and share ratings [22], which are from

---

[8] We use logarithmic binning (in base 4) to account for the fact that some values in the degree, popularity, or activity distributions are frequent while others are not. A linear binning leads to bins with few or no points.

Table 2. Dataset Statistics. Bold marks the highest value per column, while italic the lowest.

Dataset	Ratings Per User	Ratings Per Item	Avg. Degree	Mean Rating	Median Rating
Ciao	22.9	*2.8*	**19.1**	4.16	4
Epinions1	13.4	4.7	9.8	3.99	4
Epinions2	41.6	3.1	16	3.97	4
Epinions	103.5	18.0	*6.3*	**4.67**	5
Flixster	*10.4*	167.9	8.9	*3.8*	4
Douban	**129.9**	**287.5**	13.0	3.84	4

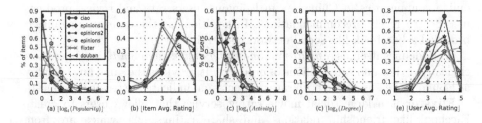

Fig. 1. Distribution of ratings as function of: (a) user activity; (b) item popularity; (c) user degree; (d) rating value

0.5 to 5 (with a step of 0.5). To ensure uniformity across the analyzed datasets, we round the ratings to the next integer so as to obtain ratings on a 1 to 5 scale.

Data Statistics. We want to understand the properties of the datasets we analyze, the resemblance among them, as they might explain the performance variations across them. Table 2 highlights basic statistics for each dataset.

Rating Distributions. Fig. 1 shows the rating distributions across user and item properties, and the rating value. In Fig. 1(a) we notice similar patterns across datasets with only little variation (for larger datasets, the level of user activity at which the peak number of ratings is produced is shifted towards higher ranges). In contrast, the rating distribution according to item popularity, Fig. 1(b), varies greatly: while in some datasets (*ciao, epinions1, epinions2*) the highest fraction of ratings is given to unpopular items, in others (the largest ones) this is accounted for popular items. Fig. 1(c) also shows that while in *flixster* and

Fig. 2. The distribution of items as a function of (a) item popularity and (b) average rating per item, and the distribution of users as a function of (c) user activity, (d) user (out-)degree and (e) average rating per user

epinions most ratings are given by moderately social connected users, in other datasets a higher number of ratings is credited to lower degree users. Looking at rating distributions according to the rating value, Fig. 1(d), we see that in all datasets the values are skewed towards higher ranges (peaking around 5).

Item Distributions. We observe similar patterns across all datasets: Fig. 2(a) illustrates that with only one exception (*epinions*) the cold start items (with only few ratings) represent a significant fraction of all items. Fig. 2(b) shows that in all datasets most of the items received on average a rating of 3 or 4.

User Distributions. SR is believed to address *cold start* users, as it does not require them to rate items for making predictions, but only to be connected in the social network. Given that in some datasets the number of *cold start* users is significant (roughly 50% [20]), improving on this set of users might significantly impact the overall performance. Thus, on average such approaches were found to outperform CF [20,33]. Yet, when the percentage of cold start users is not significant, this might not be the case. Fig. 2(c) shows that while in some datasets (*epinions, flixster*) cold start users are a significant percentage, this is clearly not the case in others (*douban, ciao*). Additionally, regardless of their fraction, cold start users always produce a minor fraction of ratings (see Fig. 1). In Fig. 2(d), we notice that, except *flixster*, the number of low degree users is larger than the number of cold start users, which in turn might affect SR overall performance. Finally, Fig. 2(e) shows that, on average, users tend to give higher rating values.

Correlations. We also check the correlation among item and user properties (item popularity, user activity and degree, and the average rating received by an item or given by a user). As in general we found low or no correlation, we report only on statistically significant ($p < 0.01$) moderate Pearson correlations ($|r| \geq 0.2$). We found moderate and positive correlations among users degree and their level of activity in *ciao* ($r = 0.59$), *epinions1* ($r = 0.45$) and *epinions*($r = 0.36$). In *flixster* ($r = 0.43$), *douban* ($r = 0.35$) and *epinions* ($r = 0.30$) there is a positive correlation between items popularity and the ratings they got, i.e, popular items tend to obtain higher ratings. Item popularity also correlates negatively with users level of activity in *flixster* and *douban* ($r = -0.20$ in both datasets), i.e., active users are more inclined to rate unpopular items. While in *douban* there is a negative correlation ($r = -0.29$) between users level of activity and the ratings they give on average, indicating that active users are more likely to give lower ratings; in *epinions* popular items tend to get higher ratings ($r = 0.31$).

We will see in the next sections how these varying data properties explain the different performance numbers obtained when aggregating the results differently (e.g., user-oriented vs. item-oriented evaluation) within and across datasets.

3.3   Overall Performance Characterization

A common practice in recommender systems evaluation is to show how their performance varies with approach-dependent parameters. Yet, even when there are correlations between the parameter values and performance level, it is difficult to know, for instance, if the improvements hold for the entire population, or only

Table 3. Overall performance. In each cell we report RMSE (Coverage) computed over all the ratings in the dataset. Bold highlights the best value on each row.

Dataset	User CF	Item CF	Social
Ciao	**1.144** (0.410)	1.285 (0.318)	1.252 (**0.626**)
Epinions1	**1.186** (0.512)	1.428 (0.463)	1.362 (**0.663**)
Epinions2	**1.164** (**0.483**)	1.361 (0.395)	1.406 (0.365)
Epinions	**0.466** (0.930)	0.602 (0.579)	0.559 (**0.951**)
Flixster	1.013 (0.969)	**0.889** (**0.991**)	1.349 (0.985)
Douban	**0.784** (0.996)	0.809 (**0.997**)	1.037 (0.894)

for some subgroups. Thus, we want to observe if there is a trivial relationship between the experimental results obtained through globally computed metrics that summarize the performance, typically used to evaluate recommendation systems [20,22,35,39], and the averaged performance at user (resp. item) level. Table 3 reports the globally computed metrics (*rating-oriented evaluation*) per dataset and approach. For error rates, with only one exception (i.e., *flixster*), user-based CF performs best across all the datasets. In terms of coverage, there is no clear winner: SR performs best for *ciao*, *epinions1* and *epinions*, while user-based CF for *douban* and *epinions2*, and item-based CF for *flixster*. Next, we check if these results are also confirmed by the *user (resp. item)-oriented evaluations* (§3.1) which measures how well an approach does on average per user (resp. item). In Fig. 3 the boxplots show the shape of the average performance distribution for users (resp. items), its central value, and variability.

User-Oriented Evaluation. Fig. 3(a) shows the per-user performance variation across datasets. Though it mostly confirms the overall results (in terms of winners) for most datasets, there are exceptions in which SR, resp. item-CF, fares better than the globally computed metrics indicate in Table 3: e.g., the coverage on *flixster*, where the fraction of unsocial users is lower than that of cold start users, and RMSE on *epinions*, where there is a higher fraction of items with similar ratings, than of users giving similar ratings.

Item-Oriented Evaluation. Similarly, barring the coverage on *flixster* and *douban*, Fig. 3(b) also confirms (in terms of winners) the figures in Table 3. Yet, we notice that except *epinions* and *flixster*, in all the other datasets both the distributions and the average coverage values are significantly shifted towards lower ranges regarding the user-oriented evaluation, which is explained in part by the much higher fraction of unpopular items than of cold start users that these datasets exhibit.

This demonstrates that it is difficult to rely on global metrics to assess or explain a given recommender performance, a finer granularity has to be applied; and that indeed no general conclusion can be drawn regarding the relative superiority of a given recommendation method over another, not only across datasets but also within each dataset.

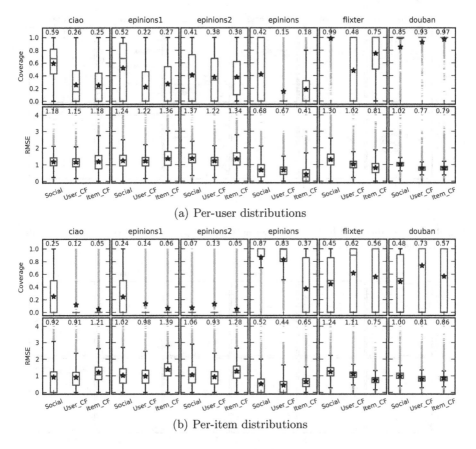

(a) Per-user distributions

(b) Per-item distributions

Fig. 3. Results Distribution: The boxplots divide the data, except outliers (the blue lines), in four equal buckets. A data point displays the performance on a particular user (resp. item). The redline splitting the boxplot is the median, while the star is the average performance (also plotted above each boxplot).

3.4 In-Depth Performance Characterization

We aim to understand the benefit of each approach under a variety of settings. In this regard, we address a set of questions about the properties of CF and SR, some of which are well embedded in the conventional wisdom:

Does CF Fare Better for Users (resp. items) with More Ratings? The belief is that CF does better when a user has rated more items [15,8]. To test it, we analyze how CF performs as users are more active (have rated more items). Fig. 4 shows that users' level of activity impacts the ability to make predictions (the coverage) similarly across all approaches: being more active helps only until some threshold after which rating more items either does not help (*epinions, douban*) or can even be harmful (*epinions2*). Further, while rating more items

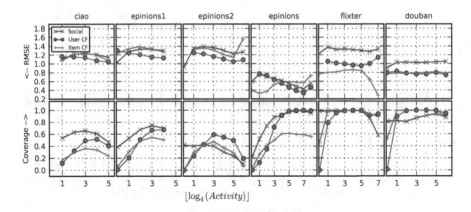

Fig. 4. Performance as a function of user activity: (top) average RMSE per user; (bottom) average coverage per user

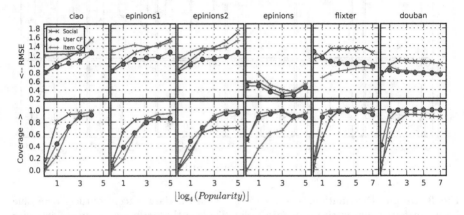

Fig. 5. Performance as a function of item popularity: (top) average RMSE per item; (bottom) average coverage per item

tends to help user-based CF to make precise prediction (in *epinions* and *flixster* after slightly improving for a while, the error increases again), item-based CF has a more inconsistent pattern. Looking at the relative performance of CF regarding SR (barring cold start users, i.e., the first bin on the $\log_4$ scale), we notice that users level of activity impacts user-based CF and SR similarly in terms of both coverage and RMSE. Exceptions are the coverage results on the datasets that exhibit no correlation among users social degree and their level of activity (*douban, flixter*).

As with more ratings per user, the belief is that more ratings per item help CF [40]. To challenge it, we look how CF performs with the number of ratings per item. Fig. 5 shows that the average coverage per item is improving as items are more popular only until some threshold when they plateau. In contrast, for *ciao, epinions1, epinions2* (datasets with a small number of ratings per item, Table 2)

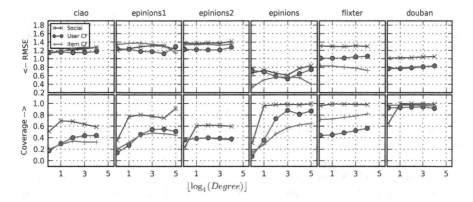

Fig. 6. Performance as a function of node degree: (top) average RMSE per user; (bottom) average coverage per user

the predictions are less precise as the items are more popular, *invalidating* the belief. Checking the relative performance of CF regarding SR, we notice that more ratings per item helps CF to increase its precision regarding SR. The only exception is *epinions* (to easily spot the patterns, follow on y-axis the distance between points corresponding to the same bin but with distinct approaches).

Does SR fare better for cold star users? The belief is that SR deals better with cold start users [20] (with less than 5 items rated [16]) as it only requires them to be connected to other users to make predictions. Indeed, Fig. 4 shows that SR achieves better coverage for these users (leftmost bins) across datasets. Yet, this is not always the case when it comes to precision (RMSE). For instance, we observe that for *flixster* and *douban* (when the social ties stem from friendship), CF attains a better precision for all users, including cold start ones.

Does SR fare better for users with more social connections? Intuitively, more social information available should help SR. To check this, we study how SR performs across users with various social degrees. Fig. 6 shows that higher degrees help improve the coverage only until users are moderately connected (have at least 5 connections), after which linking to more users seems to bring little or no benefit for SR, even declining on *ciao*. Neither SR's precision improves as users are more socially active: it either slightly decreases, or plateau. This means that having too many friends might also introduce noise. This hints that many social ties might not reflect as much friendship, similarity or trust. However, on most datasets higher degrees tend to have a weak to no impact on SR's precision. Further, as with the level of activity, baring the low degree users, the social degree impacts user-based CF and SR in a similar way, in particular for those datasets in which the degree correlates with the level of activity.

Is CF doing better on low degree nodes? Since CF does not leverage the social links to make predictions, it should not be affected by their absence, and, thus, should perform better on *unsocial* (low-degree) users. Yet, Fig. 6 shows that CF succeeds to obtain a better coverage on unsocial users only for *douban* and *epinions2*. For RMSE, while on some datasets CF does better on *unsocial* users, when there is a correlation between user degrees and how many items they rate (*ciao, epinions1,* and *epinions*), it performs comparable with SR.

Is SR's Precision re. CF Smaller on Facebook-like Networks? The process of creating connections primarily based on "plain" friendship (Facebook-like) does not necessarily correlate with one's opinions as it is orthogonal to a product recommendation task. Yet, when the basis of forming connections is to connect with people whose opinions one shares, there might be more agreement in how users rate the same items. Indeed, this distinction is clearly visible in our results (Fig. 6 to 8): while SR fares comparable with CF in terms of RMSE in Epinions datasets and *ciao*, in Facebook-like *flixster* and *douban* CF significantly outperforms SR. In addition, being more socially active has little to no impact on the results obtained for *flixster* and *douban* (Fig. 6). Thus, this indicates that the underlying nature of the network and whether or not the connections are related or orthogonal to the recommendation task is an important factor as well.

Is the performance independent of users selectiveness or items likeability? Only few studies hint at the relation between user selectiveness [34] or items likeability and recommendation performance. Yet, in Fig. 7 we notice consistent patterns across datasets, in particular, for RMSE. In all datasets item-based CF is more precise when items are either liked (received high ratings), or disliked (received low ratings), while SR and user-based CF are more precise for users that are either very selective (giving mostly low ratings) or indulgent (offering mostly high ratings). Also note how similarly both the user and item average rating impacts the precision across all datasets (i.e., leading to similar curves for all datasets). This is surprising as it indicates that the users (resp. items) average rating is predictive for the recommendation approach precision. It is also worth noting that user-based CF and SR precision (although with slightly different values) follow almost identical curves. Yet, as Fig. 7 illustrates, for coverage the patterns are not consistent across all datasets.

4   Related Work

Collaborative Filtering (CF) has been widely used by major commercial applications such as Amazon, Movielens, or Netflix [24,27,1]. These methods leverage users rating history and predict the rating of a target item and a source user by looking at the ratings on the target item given by similar users, *user-based approaches* [17], or at what ratings items similar to the target one have received from the source user, *item-based approaches* [38]. Yet, relying solely on CF is ineffective when dealing with large numbers of items, given the sparsity of the user-item ratings matrix. *Cold start* users and items are particularly affected, CF often failing to make predictions in such cases (i.e., leading to a low *coverage*).

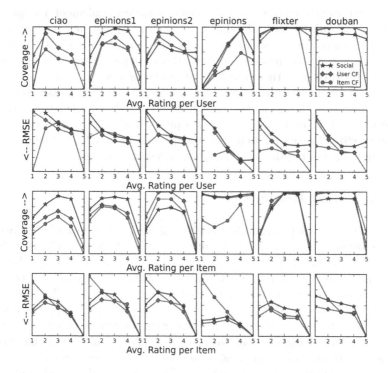

Fig. 7. Performance as a function of average rating value per item and per user

Social recommender systems(SR). In contrast, SR systems leverage users social ties to make predictions [46,33,37,16,35], assuming that these reflect common tastes or interests. SR systems deal better with *cold start* users, as they require users only to be connected to other users in the social network, and do not have to wait for users to grow a rating history to make predictions. Alas, while these systems tend to achieve better coverage, they can also suffer due to sparse ratings and sparse trust relations. Thus, in order to consider the ratings of users that are not directly connected, various approaches propagate the trust among their users [46,16,33,35]. Yet, in these cases the recommender might end up considering ratings of weakly trusted users, thus affecting the precision [20].

Social-enhanced collaborative approaches incorporate social factors to the collaborative framework by tailoring the rating similarity based on the social ties [25]; making predictions based on friend ratings weighted by the level of trust, and integrating them in the CF framework [29]; adding social regularization factors to matrix factorization recommendation techniques by constraining a user inferred taste (her feature vector) with the average taste of her friends, and the similarity with each of them [31], thus making her feature vector depend on those of her friends [22], or by accounting for the social ties heterogeneity [39].

In contrast, *collaborative-enhanced social approaches* implement a social-based framework that falls back on CF when trusted users did not rate the target item. TrustWalker enhances a social-based approach with item-based CF [20], and employs a random walk model that first tries to exploit the social network by looking for the ratings on the target item at trusted nodes (*trust-based approach*). Yet, as the random walk advances, if a rating on this item is not found, the likelihood to return the rating of a similar item (*item-based approach*) increases. TrustWalker acts in extreme settings as a pure SR approach when the random walk never stops for similar items, and as pure item-based CF when the walk never starts (navigating the same problem spectrum with us).

5 Concluding Remarks

We conducted an in-depth empirical analysis on six *publicly available* datasets to study the respective merits of the *interest affinity*, as derived by CF, and the *social affinity*, reflecting how well connected users are in the social graph, for items recommendations. We focused on the building blocks of the analyzed strategies, without aiming to exhaustively inspect all possible implementations, as we argue that their understanding can better guide more complex deployments. Our study conveys that the level of user activity, item popularity or the density and nature of the underlying social network are as many characteristics that can impact the performance of recommendation systems. One needs to understand the dataset demographics and optimize the performance based on each application specificities. We make a case for hybrid approaches, that dynamically adapt as the system evolves and the properties of user and item change over time.

Acknowledgements. We thank Stefan Bucur, Yelena Mejova and Saket Sathe for their valuable feedback on earlier versions of this work. This work was partially supported by the grant Reconcile: Robust Online Credibility Evaluation of Web Content from Switzerland through the Swiss Contribution to the enlarged European Union.

References

1. Adomavicius, G., Tuzhilin, A.: Toward the next generation of recommender systems: A survey of the state-of-the-art and possible extensions. IEEE Trans. on Knowl. and Data Eng. (2005)
2. Adomavicius, G., Zhang, J.: Impact of data characteristics on recommender systems performance. ACM Trans. Manage. Inf. Syst. (2012)
3. Adomavicius, G., Zhang, J.: Stability of recommendation algorithms. ACM Trans. Inf. Syst. (2012)
4. Amatriain, X.: Mining large streams of user data for personalized recommendations. SIGKDD Explor. Newsl. (2013)
5. Backstrom, L., Leskovec, J.: Supervised random walks: predicting and recommending links in social networks. In: WSDM (2011)

6. Bell, R.M., Koren, Y.: Lessons from the netflix prize challenge. SIGKDD Explor. Newsl. (2007)
7. Bellogín, A., Cantador, I., Díez, F., Castells, P., Chavarriaga, E.: An empirical comparison of social, collaborative filtering, and hybrid recommenders. ACM Trans. on Intel. Sys. and Tech. (2013)
8. Burke, R.: Integrating knowledge-based and collaborative-filtering recommender systems. In: Workshop on AI and Electronic Commerce (1999)
9. Chen, W., Hsu, W., Lee, M.L.: Making recommendations from multiple domains. In: KDD (2013)
10. Cremonesi, P., Koren, Y., Turrin, R.: Performance of recommender algorithms on top-n recommendation tasks. In: RecSys (2010)
11. de Campos, L.M., Fernandez-Luna, J.M., Huete, J.F., Rueda-Morales, M.A.: Measuring predictive capability in collaborative filtering. In: RecSys (2009)
12. Desrosiers, C., Karypis, G.: A comprehensive survey of neighborhood-based recommendation methods. In: Recommender Systems Handbook. Springer (2011)
13. Ekstrand, M., Riedl, J.: When recommenders fail: predicting recommender failure for algorithm selection and combination. In: RecSys (2012)
14. Fouss, F., Pirotte, A., Renders, J.-M., Saerens, M.: Random-walk computation of similarities between nodes of a graph with application to collaborative recommendation. IEEE Trans. on Knowl. and Data Eng. (2007)
15. Golbandi, N., Koren, Y., Lempel, R.: Adaptive bootstrapping of recommender systems using decision trees. In: WSDM (2011)
16. Golbeck, J.: Computing and Applying Trust in Web-based Social Networks. PhD thesis, University of Maryland (2005)
17. Goldberg, D., Nichols, D., Oki, B.M., Terry, D.: Using collaborative filtering to weave an information tapestry. Communications of the ACM (1992)
18. Herlocker, J.L., Konstan, J.A., Borchers, A., Riedl, J.: An algorithmic framework for performing collaborative filtering. In: SIGIR (1999)
19. Herlocker, J.L., Konstan, J.A., Terveen, L.G., Riedl, J.T.: Evaluating collaborative filtering recommender systems. ACM Trans. Inf. Syst. (2004)
20. Jamali, M., Ester, M.: Trustwalker: a random walk model for combining trust-based and item-based recommendation. In: KDD (2009)
21. Jamali, M., Ester, M.: Using a trust network to improve top-n recommendation. In: RecSys (2009)
22. Jamali, M., Ester, M.: A matrix factorization technique with trust propagation for recommendation in social networks. In: RecSys (2010)
23. Kermarrec, A.-M., Leroy, V., Moin, A., Thraves, C.: Application of random walks to decentralized recommender systems. In: Lu, C., Masuzawa, T., Mosbah, M. (eds.) OPODIS 2010. LNCS, vol. 6490, pp. 48–63. Springer, Heidelberg (2010)
24. Konstan, J., Riedl, J.: Recommended for you. IEEE Spectrum (2012)
25. Konstas, I., Stathopoulos, V., Jose, J.M.: On social networks and collaborative recommendation. In: SIGIR (2009)
26. Liben-Nowell, D., Kleinberg, J.: The link-prediction problem for social networks. Journal of the American Society for Information Science and Technology (2007)
27. Linden, G., Smith, B., York, J.: Amazon.com recommendations: Item-to-item collaborative filtering. IEEE Internet Computing (2003)
28. Lovász, L.: Random walks on graphs: A survey. Combinatorics, Paul Erdos is Eighty 2(1), 1–46 (1993)
29. Ma, H., King, I., Lyu, M.R.: Learning to recommend with social trust ensemble. In: SIGIR (2009)

30. Ma, H., Yang, H., Lyu, M.R., King, I.: Sorec: social recommendation using probabilistic matrix factorization. In: CIKM (2008)
31. Ma, H., Zhou, D., Liu, C., Lyu, M.R., King, I.: Recommender systems with social regularization. In: WSDM (2011)
32. Massa, P., Avesani, P.: Trust-aware bootstrapping of recommender systems. In: ECAI Workshop on Recommender Systems (2006)
33. Massa, P., Avesani, P.: Trust-aware recommender systems. In: RecSys (2007)
34. McAuley, J.J., Leskovec, J.: From amateurs to connoisseurs: modeling the evolution of user expertise through online reviews. In: WWW (2013)
35. Meyffret, S., Médini, L., Laforest, F.: Trust-based local and social recommendation. In: RSWeb (2012)
36. Milgram, S.: The small world problem. Psychology Today (1967)
37. Pitsilis, G., Knapskog, S.J.: Social trust as a solution to address sparsity-inherent problems of recommender systems. In: Recommender Systems and the Social Web (2009)
38. Sarwar, B., Karypis, G., Konstan, J., Riedl, J.: Item-based collaborative filtering recommendation algorithms. In: WWW (2001)
39. Shen, Y., Jin, R.: Learning personal + social latent factor model for social recommendation. In: KDD (2012)
40. Steck, H.: Item popularity and recommendation accuracy. In: RecSys (2011)
41. Tang, J., Gao, H., Liu, H.: mTrust: Discerning multi-faceted trust in a connected world. In: WSDM (2012)
42. Tang, J., Liu, H., Gao, H., Das Sarmas, A.: eTrust: understanding trust evolution in an online world. In: KDD (2012)
43. Yang, S.-H., Long, B., Smola, A., Sadagopan, N., Zheng, Z., Zha, H.: Like like alike: joint friendship and interest propagation in social networks. In: WWW (2011)
44. Yang, X., Steck, H., Guo, Y., Liu, Y.: On top-k recommendation using social networks. In: RecSys (2012)
45. Yang, X., Steck, H., Liu, Y.: Circle-based recommendation in online social networks. In: KDD (2012)
46. Ziegler, C.-N.: Towards Decentralized Recommender Systems. PhD thesis, Albert-Ludwigs-Universitat Freiburg (2005)

End-User Browser-Side Modification
of Web Pages

Oscar Díaz[1], Cristóbal Arellano[1], Iñigo Aldalur[1],
Haritz Medina[1], and Sergio Firmenich[2]

[1] University of the Basque Country (UPV/EHU), San Sebastián, Spain
{oscar.diaz,cristobal.arellano,inigo.aldalur}@ehu.es
[2] LIFIA, Universidad Nacional de La Plata and CONICET, Argentina
sergio.firmenich@lifia.info.unlp.edu.ar

Abstract. The increasing volume of content and actions available on
the Web, combined with the growing number of mature digital natives,
anticipate a growing desire of controlling the Web experience. Akin to the
Web2.0 movement, webies' desires do not stop at content authoring but
look for controlling how content is arranged in websites. By content, we
mainly refer to HTML pages, better said, their runtime representation:
DOM trees. The vision is for users to "prune" (removing nodes) or "graft"
(adding nodes) existing DOM trees to improve their idiosyncratic and
situational Web experience. Hence, Web content is no longer consumed
as canned by Web masters. Rather, users can remove content of no
interest, or place new content from somewhere else. This vision accounts
for a post-production user-driven Web customization (referred to as
"Web Modding"). Being user driven, appropriate abstractions and tools
are needed. The paper introduces a set of abstractions (formalized in
terms of a domain-specific language) and an IDE (realized as an add-on
from *Google Chrome*) to empower non-programmers to achieve HTML
rearrangement. The paper discusses the technical issues and the results
of a first validation.

Keywords: Web Modding, Web Widget, End User Programming,
Visual Programming, Domain Specific Languages, WebMakeUp.

1   Introduction

Modding is a slang expression that is derived from the verb "modify". Modding
refers to the act of modifying hardware, software, or virtually anything else,
to perform a function not originally conceived or intended by the designer
[19]. The rationales for modding should be sought in the aspiration of users
to contextualize to their own situation the artefact at hand. This ambition is
not limited to video games, cars or computer hardware. The need also arises
for the Web. As an example, consider a TV-guide website (e.g. *www.tvguia.es*).
For a given user, favourite channels might be scattered throughout the channel
grid, hence, forcing frequent scrolling. In addition, users might move to
other websites (e.g. *www.filmaffinity.com*) to get more information about the

B. Benatallah et al. (Eds.): WISE 2014, Part I, LNCS 8786, pp. 293–307, 2014.
© Springer International Publishing Switzerland 2014

scheduled movies. If *tvguia* is recurrently visited, this results in a poor user experience. Traditionally, this is addressed through Web Personalization, i.e. a set of techniques for making websites more responsive to the unique and individual needs of each user [3]. Similar to other software efforts, traditional personalization scenarios prioritize the most demanded requirements while minority requests are put aside. However, as a significant portion of our social and working interactions are migrated to the Web, we can expect an increase in "long-tail" personalization petitions. These idiosyncratic petitions might be difficult to foresee or too residual to be worth the effort. **"Web modding"** moves the power to the users. Web modding (hereafter, just modding) aims at Web content being consumed in ways other than those foregone by Web masters. Rather, users are empowered to rearrange Web content "after manufacture", e.g. removing content of no interest (leading to less cluttered pages while reducing scrolling) or placing new content obtained from somewhere else (reducing moving back and forth between sites so that a single viewing context is provided). The research question is how to achieve this empowerment.

This question admits different answers depending on the target audience. We frame our work along three main requirements: available time (30'), available expertise (no programming experience), and sparking motivation (improving the Web experience). This rules out fine-grained, absorbing programmatic approaches, and demands more declarative and abstract means. This is what Domain-Specific Languages (DSLs) are good for. DSLs are full-fledged languages tailored to specific application domains by using domain-specific terms. Domain abstractions are closer to how users conceive the problem, facilitating engagement, production and promptness. This work's contribution rests on the three pillars of DSLs applied to Web modding, i.e. ascertaining the right concerns (Section 3), finding appropriate DSL constructs to capture those concerns (Section 4), and finally, developing suitable editors that ease the production of DSL expressions (Section 5). The later is realized through *WebMakeup*, a *Google Chrome* extension that turns this browser into an editor for defining Web mods. *WebMakeup* is available at the *Chrome Web Store*: https:// chrome.google.com/webstore/detail/ alnhegodephpjnaghlcemlnpdknhbhjj. Mods are exported as *Google Chrome* extensions that once installed, will transparently customize the page next time is visited. A first evaluation is provided in Section 6. We start by characterizing Web Modding.

2   Characterizing Web Modding through Related Work

Web Modding sits in between Web Personalization [17] and Web Mashup [20]. As a personalization technique, modding aims at improving the user experience by customizing Web content. There are also important differences. In Web Personalization, the website master (the "who") decides the personalization rules (the "how"), normally at the inception of the website (the "when"), preferentially using a server-centric approach (the "where"). By contrast, modding aims at

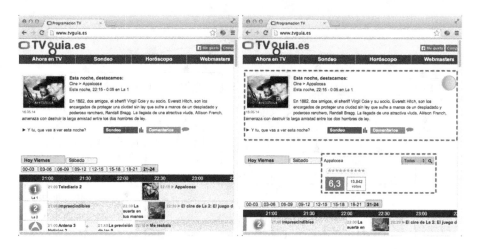

Fig. 1. *www.tvguia.es* before (left) and after (right) being modded: channel *"La 1"* is removed & *filmaffinity* ratings are introduced

empowering end-users (the "who") to rearrange Web content once in operation (the "when") by acting on the DOM tree (runtime realization of HTML pages) (the "how") at the client side (the "where"). Nevertheless, modding also shares similitudes with mashups: both tap into external resources. However, and unlike mashups, modding does not create a bright new website. Rather, it sticks with the modded website. Just like modding a car does not build a new car, modding a website does not create a new website but just operates on the browser side to change its DOM tree.

Web modding pays off for websites frequently visited but unsatisfactory Web experience. As an example, consider *www.tvguia.es*. This website provides the channel grid plus the-movie-of-the-day recommendation (see Figure 1 (left)). A user might just focus on some few channels, hence a thorough channel grip becomes a nuisance. In addition, content from other websites about the recommended movie might be of interest. Figure 1 (right) depicts a modded version: channel *"La 1"* is removed whereas additional content about the recommended movie is obtained from *www.filmaffinity.com*. The fragment extracted from *filmaffinity* is referred to as a *widget*, in this case, the *filmAffinity* widget.

The bottom line is that mod scenarios are characterized as being idiosyncratic, situational, and, potentially, short-lived, aiming not so much at synergistically combining third-party data (as mashups do) but improving the user experience of existing websites. Since these scenarios are very dependent on Web consumption habits and user interests, modding necessarily has to be do-it-yourself (DIY). This implies keeping the modding effort on a scale within the time and the skills of end users. This scale is a main driver in finding a balance between expressiveness (what can be modded) and effort (the cost of developing the mod).

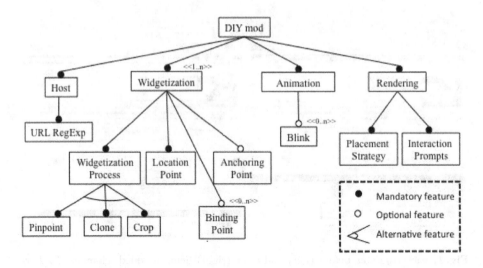

Fig. 2. Feature diagram for DIY Web Modding

Our target is for *Web Modding to be conducted by users with no programming skills in around 30 minutes.*

Implementation wise, modding implies browser-based programming. Modding is already possible for skilful JavaScript programmers but certainly outside the scope of end users [16]. This rules out fine-grained, absorbing programmatic approaches (e.g. *Chickenfoot* [1], *Co-Scripter* [12]), and calls for coarser grained, light-weight component-based standpoints. Unfortunately, most works on Web components (e.g. widgets) favours a programmer perspective, addressing the definition [18], implementation [6,9] and cloning of Web components [13,7]. A higher-level of abstraction is needed. Domain-Specific Languages (DSLs) come to the rescue. DSLs are full-fledged languages tailored to specific application domains by using domain-specific terms [8]. To increase the chances for DSLs to be adopted, three main landmarks stand out: ascertaining the right concerns, finding appropriate constructs to capture those concerns, and finally, developing appropriate editors that intuitively permit users to come up with DSL expressions. Next sections address each of these landmarks for modding.

3 Ascertaining the Right Concerns

Web Modding sits within the field of Web Augmentation [2], i.e. conducting changes upon the runtime representation of HTML pages (i.e. DOM trees) at the time the page is loaded into the browser. Those changes can affect the content, rendering, layout or dynamics of the page. Among the numerous uses of Web Augmentation, modding focuses on performing a function not originally conceived or intended by the host designer [5]. Finally, DIY modding addresses the empowerment of end-users to mod by themselves. As in other areas of end-user design, more (expressiveness) can be less (usage). Therefore, DIY modding is necessarily going to be less expressive (i.e. more domain-specific) than general modding. We focus on improving the user experience through content rearrangement, i.e. content removal (leading to less cluttered pages) and content cloning, i.e. taking content from somewhere else (providing a single viewing context while cutting down moving back and forth between browser tabs). This sets the domain.

Along DSL good practices [14], concerns raised during DIY modding are captured as a feature diagram [10]. A feature diagram represents a hierarchical decomposition of the main concepts (i.e. features) found in the domain. The diagram also captures whether features are mandatory, alternative or optional. Figure 2 depicts the feature diagram for the domain "DIY modding". Issues include, **hosting** (i.e. setting the ambit of the modding), **widgetization** (i.e. the definition of widgets whose addition and removal shape the modding), **animation** (i.e. defining possible dynamics among the widgets), and finally, the **rendering directives** for the mod. Next paragraphs delve into the details (bold font is used for the features).

3.1 Hosting

A *mod* is a set of changes conducted upon the runtime representation of an HTML page at the time the page is loaded. Therefore modding does not happen in a vacuum but within the setting of an existing website, i.e. the **host**. The host can be characterised by a URL expression or a regular expression (e.g. www.amazon.com/*) so that all pages meeting the expression are subject to the mod. The expressiveness much depends on the target audience. For our purpose, we limit **url regexp** to those ending by "*". More complex expressions are not supported.

3.2 Widgetization

Modding is about customizing HTML content. HTML pages are conceived as DOM documents. The granularity at which HTML customization happens influences complexity. A finer-grained approach will certainly improve expressiveness but at the cost of complexity and learnability. Therefore, we opt for a coarser grained approach: *widgets*. For the purpose of this work, a widget is a coarse-grained DOM node (a.k.a. fragment), which accounts for a meaningful mod unit.

A widget can be defined *from scratch* through HTML and JavaScript. This is not possible for non-programmers. Alternatively, 3rd parties can help. But this also contradicts our setting that is characterized as being idiosyncratic, situational, and, potentially, short-lived, hence, the introduction of 3rd parties does not payoff. We are then forced to explore a different approach: *widget mining*. That is, users do not create widgets on their own but extract them from existing pages at the time the need arises. We then do not talk about widget creation but *widgetization* of existing code. To this end, we support tree variants: pinpoint, crop and clone.

Pinpoint supports inside-the-host widgetization, i.e. the widget is obtained from the host. In this case, extraction points hold the host's URL and a structure-based coordinate, i.e. an XPath expression that pinpoints the DOM node to be turned into a widget (see later). Widget *movie-of-the-day* is a case in point. It singularizes the DOM node that holds the content for the recommended movie. However, outside-the-host widgetization is more complex. A naive approach to extract existing functionality from a web page is just copy&paste. However, since HTML, CSS, and JavaScript are all "context-dependent", moving fragments from their original scope is rarely feasible. This moves us to the other two variants.

Clone is used for outside-the-host widgetization when the fragment to be extracted is "static", i.e. it holds content and style but not functionality (no JS scripts associated). The aim is for the widget to look like the raw content in the original page. Here, widgetization is achieved through cloning. Since style needs to be replicated, cloning is not limited to the selected DOM node but also its ancestors' CSS styles are inherited[1]. Since code is replicated, what if the original is upgraded? How are changes propagated to the replica? To this end, we introduce *refreshTimer,* a parameter that sets the refresh polling time to four possible values: onload (i.e. the widget is calculated every time the host page is loaded), daily, weekly or never.

So far, we assume widgets to be obtained from a single HTML fragment (**singleCloned**). However, the content of interest might be spread across different nodes. An interesting case is that of the Deep Web. Deep Web sources store their content in searchable databases that only produce results dynamically in response to a direct request. Here, the "meaningful functional unit" (i.e. the node to be widgetized) includes two fragments (**complexCloned**): the request fragment and the response fragment. The *filmAffinity* widget illustrates this situation. The "functional unit" includes not only the ranking table (i.e. the output) but also the search entry form to type the movie title. Hence, creating *filmAffinity* implies two extractions: one to collect the ranking table; another to obtain the entry form[2]. Last but not least, so-created widgets are parameterized by the form entries. This permits to fix some form entries (e.g. set "Gone with the wind" as the movie title) or even better, bind the entry to some data which is dynamically extracted from the hosting page at runtime (so called "binding points", see later).

Crop is used for outside-the-host widgetization when the fragment is "dynamic", i.e. it holds scripts. In this scenario, cloning does not work. Functionality is difficult to extract in an automatic way (refer to [13] for the difficulties on extracting JS code). Here, we resort to pixel-based cropping. Using iframes, it is possible to load the source webpage on the background. Next, the desired fragment can be addressed by referencing the height and width w.r.t the cropping start coordinates.

---

[1] *HTMLClipper* (http://www.betterprogramming.com/htmlclipper.html) is used to propagate replication from content to the associated CSS-like directives.

[2] Labelling a newly created *widget* with an existing name, makes the extraction engine glue them together and be offered as a unit (provided they come from the same page).

Once DOM nodes are turned into widgets, they start exhibiting some additional characteristics. Widgets can have parameters and a state (i.e. visible or collapsed). But most importantly, widgets might hold reference points, i.e. directives that refer to some location in terms of Web coordinates. We distinguish tree kind of reference points:

- **Location points**, which indicate from where the widget was obtained. They contain a Web coordinate plus the framing page.
- **Anchoring points**, which refer to the new setting where the widget is to be rendered, i.e. the position (i.e. before or after) w.r.t a given Web coordinate.
- **Binding points**, which denote how widget parameters can be bound to content from the host. It holds the name of the parameter and the host's Web coordinate. As an example, consider *filmAffinity*. This widget needs to be recalculated every time *guiaTV*'s movie-of-the-day changes. To this end, *filmAffinity* holds the *title* parameter. This parameter holds a binding point to the DOM node in *guiaTV* that keeps the title of the recommended movie. At runtime, the movie-of-the-day is recovered, and *filmAffinity* is dynamically computed after the current title.

Previous paragraphs refer to Web coordinates. A Web coordinate is a means to address content within a DOM tree (a.k.a. locators). For considerations about locators refer to [11].

3.3   Animation

Modding is about rearranging content. But this rearrangement does not need to happen in a single shot. Specifically, *widgets* can be in two states: visible or collapsed. When visible, widgets have the capacity to respond to events, such as keystrokes or mouse actions. When collapsed, widgets leave no trace in the screen. A *widget* has an initial state, i.e. the state at the time the hosting page is loaded (e.g. if visible, the widget is rendered as soon as the page is loaded). This state might be amenable to be changed by interacting with other *widgets*. A common approach for describing GUI dynamics is through statecharts [4]. However, statecharts are far too complex for our target audience. A simpler mechanism is needed.

Broadly, state changes can be described as event-condition-action rules. First studies, however, demonstrate that rules were a too fine-grained specification. Needed are higher abstractions that permit to capture recurrent patterns as a single construct. Based on previous evaluations, we noticed a recurrent animation pattern. Let's illustrate it with two widgets: *movieOfTheDay* and *filmAffinity*. Consider the later is to be made visible or collapsed upon mouse in/mouse out *movieOfTheDay*. This can be captured through a pair of rules:

ON mouse-in *movieOfTheDay* **WHEN** *filmAffinity*.state = "collapsed" **DO** *filmAffinity*.state = "visible"
ON mouse-out *movieOfTheDay* **WHEN** *filmAffinity*.state = "visible" **DO** *filmAffinity*.state = "collapsed"

We found this pattern so common that decided to introduce a DSL primitive for it: the blink. A **blink** accounts for a directed relationship between two widgets W1 and W2. We say *"W1 blinks W2"*, if acting upon W1 (e.g. clicking) causes W2 to change its state (from visible to collapsed or vice versa, depending on the W2 current state). Previous example can now be expressed as *"movieOfTheDay blinks filmAffinity on clicking"*. So far, we limit animation to *blinks. Blink* events are limited to *mouse-in* (being *mouse-out* its *blink* counterpart) and *click* (being *click* also its *blink* counterpart).

3.4   Rendering

Inlaying new widgets into an existing DOM structure can make the host's layout be disrupted. Specifically, HTML introduces some attributes to describe the rendering strategies for DOM nodes, namely: the layout strategy (HTML's "display" attribute) which can be arranging the content horizontally (inline) or vertically (block); minimum and maximum size intervals (HTML's attributes *minHeight, minWidth, maxHeight, maxWidth*); and the overflow strategy (HTML's "overflow" attribute) that indicates what to do in case the content exceeds the size intervals (i.e. make container scrollable, show the overflowed content or hide the overflowed content). Widget inlaying might disturb the page layout, causing one-dimension distortion or even worse, two-dimension distortion. We decide this concern to be hardwired within the DSL engine. Better said, the engine supports contingency actions to alleviate this situation (e.g. if container is 80% full then, WA overflow strategy is set to "warn"; if container is 90% full and the widget fits inside then, WA overflow strategy = "resize", etc.).

4   Finding Appropriate Constructs

Previous feature diagram captures main concerns to be solved during DIY modding. Next, these abstractions are realized in a language by looking into variabilities and commonalities in the feature diagram [14]. Variable parts must be specified directly in or be derivable from DSL expressions. In the first case, the variants become DSL constructs. However, some alternatives can be hardwired into the DSL engine as heuristics. Being heuristics, they might fail and hence, they are not as reliable as if provided by the user. The upside is that they simplify the user's life, hence, improving learnability and development. We decided *rendering* to be hardwired into the engine. That is, widget placement is to be assisted by the DSL engine. The rest of features are set by the user through the DSL. This section introduces the DSL metamodel.

Figure 3 provides the metamodel for mod description. A *mod* is a set of changes conducted upon the runtime representation of an HTML page (i.e. the host). These changes are described in terms of widgets. Widgets are characterized by a **locationPoint** (i.e. how to obtain it), an **anchoringPoint** (i.e. where to locate it), and, optionally, distinct **bindingPoints** (i.e. how widget parameters

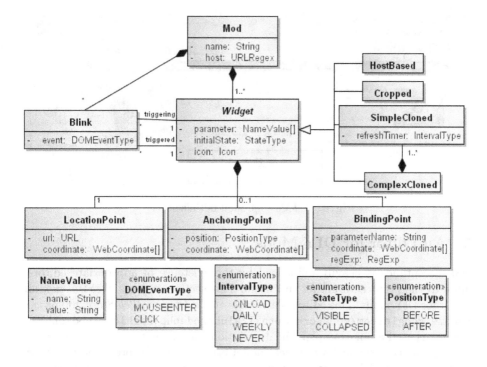

Fig. 3. A DSL for Web Modding: abstract syntax

can be obtained from the host's content). Each widget stands for a rearrangement operation as follows:

- If *locationPoint* exists without *anchoringPoint*, this accounts for content removal (only for host-based widgets).
- If *locationPoint* differs from *anchoringPoint*, this accounts for content displacement (only for host-based widgets).
- Otherwise, the widget captures content addition.

But not all contents need to be added/removed at loading time. *Blinks* permit to hand this decision over to the current user. This makes content rearrangement dependent upon user interactions. For instance, *"movieOfTheDay blinks filmAffinity on clicking"* permits to postpone till runtime the decision of rendering *filmAffinity*. If you click, you get *filmAffinity*. If complementary outside-the-host widgets exists (e.g. *filmIMDB* extracts the ratings from the IMDB website), then this content can be shown either simultaneously (e.g. *"movieOfTheDay blinks filmIMDB on clicking"*) or in a cascade way (*"filmAffinity blinks filmIMDB on clicking"*). But not only additions, also removals can be left pending until interaction time: *"movieOfTheDay blinks movieOfTheDay on clicking"* permits current users decide whether they want to delete (i.e. collapse) *movieOfTheDay* by clicking on it. Next, we address how to make mods affordable to end users.

Fig. 4. *WebMakeup*: mod initialization

5   An Editor for DIY Mods

DSL acceptance is heavily influenced by the existence of appropriate editors, more to the point if targeting end users. This section outlines *WebMakeup*, an editor for DIY mods. This editor is available at the *Chrome Web Store*: https:// chrome.google.com/webstore/detail/alnhegodephpjnaghlcemlnpdknhbhjj. *TVguia* is used as an example. The description goes along the creation of a *mod*, i.e. a model conforming to the metamodel presented in the previous section. A demo video is available at http://onekin.org/downloads/public/WebMakeup/ video.mov.

Mod creation (Figure 4). *WebMakeup* is a plugin for *Google Chrome* browser. Its installation is reflected by the *WebMakeup* button at the right of the address bar. On clicking this button, a scrollable menu pops up. By clicking *"New makeup"*, the user initializes the mod model (Figure 4 (bottom)). *WebMakeup* turns the current page into the editor canvas: the pointer is turned into a camera, a grid-like structure is interspersed on top of the current DOM tree, and the *piggyBank tab* pops up.

Mod populating (Figure 5). A widget is a DOM node but not all DOM nodes are widgets. We need to singularize the selected DOM node that accounts for a meaningful HTML fragment. Meaningfulness is not inferred by the tool but indicated by the user. To this end, and, as the user moves the cursor around the screen, the DOM node under the current cursor location is highlighted. By clicking, the user singularizes this node as a meaningful HTML fragment, i.e. a widget. A nuisance is the handling of "hidden nodes". These nodes are those that do not have a graphical counterpart and hence, they cannot be pinpointed through the cursor. For instance, a table row ($<tr>$) is graphically hidden if its graphical space is totally taken by its content. If the row does not explicitly have some graphical counterpart (e.g. a border), then all the space is occupied by the row's content so that the cursor will always select the row's content rather than the row element itself. To overcome this problem, we resort to the keyboard. Keys *"w"*, *"s"*, *"a"* and *"d"* help to move up, down, left and right along the DOM tree, respectively, w.r.t to the node being pinpointed by the cursor.

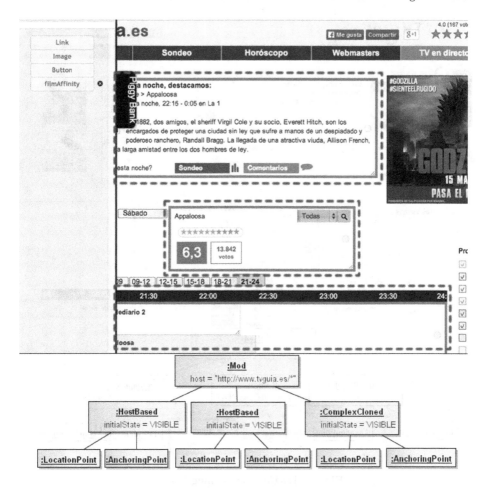

Fig. 5. *WebMakeup*: mod filling up. The *piggyBank* tab is displayed.

No matter the selection mechanism (i.e. cursor vs. keyword), the selected node is surrounded by a decorator. This decorator permits to set the initial widget state by clicking on the "eye" icon (decorators' upper left-hand side corner): visible (open eyes) & collapsed (closed eyes). The example contains two inside-the-host widgets (i.e. *movieOfTheDay* and *TVE1channel*) and outside-the-host widget (i.e. *filmAffinity*). The latter is dragged&dropped from *piggyBank*[3]. Click on this tab to expose the widgets collected from other pages (see it in display in

---

[3] Outside-the-host widgets can be obtained at any time. To this end, the right-click contextual menu is extended with the *widgetizeIT* item. At any time, select it for a grid-like structure to be interspersed on top of the page you are looking at. As the user moves the cursor around the screen, the DOM node under the current cursor location is highlighted. By clicking, the selected node is turned into a widget and kept in the extension's variable: *piggyBank*.

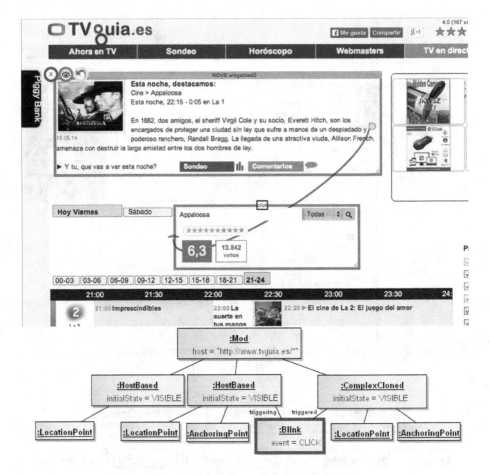

Fig. 6. *WebMakeup*: defining *blinks*

Figure 5). Placement heuristics will warn or prevent from dropping widgets in certain places. In all cases, *WebMakeup* works out the Web coordinates.

Mod enhancement (Figure 6). At any time during editing, widgets can be:

- Deleted. Widget removal is achieved by clicking upon the X icon on the widget decorator. In the example, we remove *TVE1channel*. Model wise, this is reflected by deleting its anchoring point. An important remark: banners cannot be removed. Though this is a common desire among users, up to 84% of the top 100 websites rely on advertising to generate revenue [15]. Though adverts can be a nuisance, they are the ones that pay the bill. So for the time being, we take the decision of making *WebMakeup* ad-friendly.
- Rearranged. This is conducted through drag&drop once the widget is selected. Model wise, this is reflected as an update on the anchoring point.

– "Blinked". *Blinks* are graphically represented through pipes. Widget decorators have in their right-hand side a yellow circle. This circle denotes a
pipe start. Click and drag from this point to expand till reaching another
widget. This sets a blink from the triggering widget (the pipe's start) to the
triggered widget (the pipe's end). An entry field on top of the pipe serves
to indicate the blink's event. Figure 6 illustrates the case *"movieOfTheDay
blinks filmAffinity on clicking"*.

Once the edition finishes, the mod can be exported as a Chrome extension. Once
the extension is installed, the mod will be automatically enacted next time the host
page is loaded. For our running example, the generated extension is available at
`http://onekin.org/downloads/public/WebMakeup/extension.zip`.

6   Usability Evaluation

ISO definition of usability (ISO 9241-11, Guidance on Usability (1998)) refers
to the extent to which a system (e.g. *WebMakeup*) can be used by specified
users (e.g. end users) to achieve specified goals (e.g. content re-arrangement)
with effectiveness (e.g. mod completion), efficiency (e.g. 30') and satisfaction in
a specified context of use (e.g. browsing sessions). This section provides first
insights not only about *WebMakeup* but also about the satisfaction of users on
the result of the mod.

Research Method. The study was conducted in a laboratory of the
Computer Science Faculty of San Sebastián. Before the participants started, they
were informed about the purpose of the study and were given a brief description
of it (5 minutes). Then, a *WebMakeup* sample was presented to illustrate the
main functionality of the tool. The sample mod adapts a conference website by
removing the logo of the conference and adding information about the weather
forecast and information about the authors obtained from the DBLP. Next,
participants were handed out a sheet with the instructions to create a new mod
similar to the one used here as a running example. Participants were asked to
write down the time when *"New WebMakeup"* button is clicked and again when
they saw the augmentation. Last, participants were directed to a *Google Forms*
online questionnaire.

Ten students participated in the study. The majority of participants were
male (80.0%). Regarding age, 80.0% were in the 20-29 age range and all
participants were below thirty five years old. Concerning the participants'
browsing behaviour, 80% accesses to more than 10 websites every day and in the
last year participants had installed between 2 and 20 applications/plugins/add-
ons, with a mean of 6.5.

An online questionnaire served to gather users' experience. It consisted of
four parts, the first one to gather the participants' background, another one to
measure the satisfaction, other one to effectiveness and the last one to measure
the productivity. In order to evaluate effectiveness, the questionnaire contained
the proposed tasks so that participants could indicate if they had performed
them, while productivity was measured using the minutes taken in such tasks.

Table 1. Satisfaction results from 1 (completely disagree) to 5 (completely agree)

Item	Mean	St. Dev.
1. I found the tool easy to use	3.4	0.966
2. I have made all the things that I wanted	3.1	0.994
3. I have always known how to do the things	2.1	0.738
4. There was no errors	4.0	0.817
5. It is fast	3.9	0.850
6. I am satisfied with the things I made	4.1	1.197
7. Removing content improves my Web experience	3.2	1.174
8. Adding content in a single view improves my experience	4.5	0.699
9. Demo is interesting to be told to friends	3.9	1.229

Satisfaction was measured using 9 questions, respectively, with a 5-point Likert scale (1=completely disagree, 5=completely agree). Descriptive statistics were used to characterize the sample and to valuate the participants' experience using *WebMakeup*.

Results. All participants but one successfully created the proposed augmentation. Those who successfully ended the sample took between 19 and 35 minutes with a mean of 24.2 minutes to fulfil the task. Table 1 shows scores for the satisfaction survey. As for the tool itself (items 1 to 5), subjects were reasonably happy. A shortcoming detected during the experiment was the lack of facilities to store work-in-progress mods. So far, *WebMakeup* forces to obtain the mod in a single session. Also, two subjects found the *blink* relationship misleading. On the upside, most of users finished under 30'. As for the notion of modding itself, subjects found content deletion and content rearrangement effective means to improve their web experience (items 7 and 8). Interesting enough, providing a single viewing context was found more interesting than content removal. In general, users found the experience rewarding (items 6 and 9).

7   Conclusions

Webies 2.0 no longer take the Web as it is but imagine fancy ways of customizing it for their own purposes. This work presents our vision for DIY modding along three main requirements: available time (30'), available expertise (no programming experience), and spark motivation (improving the Web experience). These requirements ground a coarse-grained, light-weight approach to DIY modding that is so far limited to content rearrangement. A fully-working editor, *WebMakeup*, demonstrates the feasibility of this vision. First evaluation is encouraging about the potentiality of Web Modding to improve the Web experience, and hence, the need for tools that make this vision possible.

Acknowledgments. This work is co-supported by the Spanish Ministry of Education, and the European Social Fund under contract TIN2011-23839. Aldalur has a doctoral grant from the Spanish Ministry of Science & Education.

References

1. Bolin, M., Webber, M., Rha, P., Wilson, T., Miller, R.C.: Automation and Customization of Rendered Web Pages. In: UIST 2005, pp. 163–172 (2005)
2. Bouvin, N.O.: Unifying Strategies for Web augmentation. In: HyperText 1999, pp. 91–100 (1999)
3. Cingil, I., Dogac, A., Azgin, A.: A Broader Approach to Personalization. Communications of the ACM 43(8), 136–141 (2000)
4. Daniel, F., Furlan, A.: The interactive API (iAPI). In: Sheng, Q.Z., Kjeldskov, J. (eds.) ICWE 2013 Workshops. LNCS, vol. 8295, pp. 3–15. Springer, Heidelberg (2013)
5. Díaz, O., Arellano, C.: The Augmented Web: Rationales, Opportunities & Challenges on Browser-side Transcoding. ACM Transactions on the Web (2014)
6. Ennals, R., Brewer, E.A., Garofalakis, M.N., Shadle, M., Gandhi, P.: Intel Mash Maker: Join the Web. SIGMOD Record 36, 27–33 (2007)
7. Firmenich, S., Winckler, M., Rossi, G., Gordillo, S.E.: A Crowdsourced Approach for Concern-Sensitive Integration of Information across the Web. Journal of Web Engineering 10(4), 289–315 (2011)
8. Fowler, M.: Domain-Specific Languages. Addison-Wesley Professional (2010)
9. Han, H., Tokuda, T.: A Method for Integration of Web Applications Based on Information Extraction. In: ICWE 2008, pp. 189–195 (2008)
10. Kang, K.C., Cohen, S.G., Hess, J.A., Novak, W.E., Peterson, A.S.: Feature-Oriented Domain Analysis (FODA) Feasibility Study. Technical report, Carnegie-Mellon University (1990)
11. Leotta, M., Clerissi, D., Ricca, F., Tonella, P.: Visual vs. DOM-Based Web Locators: An Empirical Study. In: Casteleyn, S., Rossi, G., Winckler, M. (eds.) ICWE 2014. LNCS, vol. 8541, pp. 322–340. Springer, Heidelberg (2014)
12. Leshed, G., Haber, E.M., Matthews, T., Lau, T.: CoScripter: Automating & Sharing How-To Knowledge in the Enterprise. In: CHI 2008, pp. 1719–1728 (2008)
13. Maras, J., Stula, M., Carlson, J., Crnkovic, I.: Identifying Code of Individual Features in Client-Side Web Applications. IEEE Transactions on Software Engineering 39(12), 1680–1697 (2013)
14. Mernik, M., Heering, J., Sloane, A.M.: When and How to Develop Domain-Specific Languages. ACM Computing Surveys 37, 316–344 (2005)
15. PageFair. The Rise of Adblocking (2013), http://blog.pagefair.com/2013/the-rise-of-adblocking/
16. Pilgrim, M.: Greasemonkey Hacks: Tips & Tools for Remixing the Web with Firefox. In: Getting Started, 12. Avoid Common Pitfalls, ch. 1, pp. 33–45. O'Reilly (2005)
17. Rossi, G., Schwabe, D., Guimarães, R.: Designing Personalized Web Applications. In: WWW 2010, pp. 275–284 (2001)
18. W3C. Requirement For Standardizing Widgets (2006), http://dev.w3.org/2006/waf/widgets-reqs/
19. Wikipedia. Modding (2014), https://en.wikipedia.org/wiki/Modding
20. Yu, J., Benatallah, B., Casati, F., Daniel, F.: Understanding Mashup Development. IEEE Internet Computing 12, 44–52 (2008)

Mobile Phone Recommendation
Based on Phone Interest

Bozhi Yuan[1,2], Bin Xu[1,2], Tonglee Chung[1,2], Kaiyan Shuai[3], and Yongbin Liu[1,2]

[1] Department of Computer Science and Technology, Tsinghua University, China
[2] Tsinghua National Laboratory for Information Science and Technology, China
[3] Computer School, Beijing Information Science and Technology University, China
{lawby1229,13636157238}@163.com, xubin@tsinghua.edu.cn,
{tongleechung86,yongbinliu03}@gmail.com

Abstract. As cellular users change mobile phone frequently, mobile phone recommendation system is of great importance for mobile operator to achieve business benefit. There are essential challenges for researchers to design such system. Among them, a critical one is how to obtain and model user's interest of mobile phone. So far, recommendation approaches based on phone's hardware features or personalized web behavior could not achieve satisfactory results. In this paper, we propose phone interest for mobile phone recommendation. Phone interest is a latent level concept which is extracted from a group of users' web log data, who have the same mobile phone. We propose a novel probabilistic model named "Phone Interest Model" only based on mobile web log data. All the log data are from cellular operators server, not from mobile phone's application. The model proves its effectiveness on large scale of station cellular data from real cellular operator. In experiments, we validated the model against 1.3 billion of mobile Web logs for 4 million distinct users in Beijing metropolitan areas, and show that the model achieves a good performance in the phone recommendation, also outperforms the baseline methods and offers significantly high fidelity.

Keywords: phone recommendation, phone interest, mobile data, cellular data, user behavior.

1 Introduction

Mobile phone recommendation system is of great importance for mobile operator to achieve business benefit. For example, mobile service providers always competed in the existing market space, strived for market share of products or services and customers. It's best to leverage recommendation of mobile phone as an opportunity to take the lead and push things forward. In a user-derived market, the number of contract user and contract phone is especially significant for mobile service operators.

So far, recommendation approaches based on phone's hardware features or personalized Web behavior could not achieve satisfactory results. Because each mobile phone is different from hardware, appearance, price etc., hence we have to model the mobile recommendation from the mobile phone's perspective. It's inevitable to lead to a specific mobile phone being suitable for some particular Websites or applications. The

B. Benatallah et al. (Eds.): WISE 2014, Part I, LNCS 8786, pp. 308–323, 2014.
© Springer International Publishing Switzerland 2014

character of mobile phone should be reflected from the crowd of users who have the same phone. And how to establish a model to extract the potential phone interest from the mobile phone has become a very intractable problem. The cellular data is the log data from the station of the operator, which describes the Web-access of all users who have used the cellular network. By using cellular network data, it is tempting to think that by analyzing the crowd of users' Web behavior and latent phone interest for mobile phone recommendation should be easy.

This paper proposes a modeling approach which takes as input cellular data. we explore mobile Web log analysis with user cellular data to build a comprehensive model describing phone interest. The model is named "Phone Interest Model" (PIM), which includes three parts: phone, Website behavior and phone interest. A particularly good data source comes from Event Detail Records (EDRs) maintained by a cellular network operator. EDRs contain information such as the IP address and time of each http link, network volume of uplink and downlink, as well as the identity of the cellular tower with which the phone was associated at that time. The learning process of "Phone Interest Model" (PIM) is as follows:

First, we manually labeled the access intent of the each application as "App-Behavior" via the "Useragent" field of the raw cellular data of the operator. Second, Websites are extracted from raw Web log for each user. Associated with access intent (App-Behavior) of each user access, the features of each mobile phone are generated. Third, a probabilistic topic model is proposed for extracting latent phone interest layer between mobile phone and Websites-behavior. We validate the PIM model against the EDRs from Beijing. The data lasts for 2 weeks, and covers 2 main urban districts, with an area of approximately 40 square kilometers. And more than 4 million user IDs appeared in the estimated 1.3 billion EDRs.

The contributions of the paper are: We propose a novel probabilistic model Phone Interest Model (PIM) to analyze phone interest, based only on the usage history of cellular data. And we proves its effectiveness for mobile phone recommendation on large scale of EDRs from real cellular data.

The paper is organized as follows.In Section 2, we describe recent related work on Web user behavior analysis and mobile recommendation.In Section 3, we give an overview of the data we use, and observe some important characteristics and definitions of the data in several aspects, and raise the remaining problem. Section 4 describes the discovery of "App-Behavior" from raw HTTP log, also introduces two baseline method "K-means Interest Learning" (KIL) and "K-SVM Interest Learning" and formally describes the "Phone Interest Model" (PIM) in detail. Section 5 describes the experimental results applying the framework on our dataset, and compares performance between the different method and gives detailed analysis on the result in several different angels. Finally, we conclude the paper in Section 6.

2   Related Work

User interest and behavior mining based on Web usage has long been a hot topic[1][2]. White, et, al.[3] consider 5 different contextual information to model user interest, and then do recommendation based on it. Nasraoui, et, al. [4] study user behavior of a particular Website based on tracking user profiles and their evolving. Some researchers use clustering methods to extract types of users.[5][6] But they either do clustering on the users' perspective and cluster user into different types, or on the Websites' perspective and make URL groups. Extracting user types as well as Website topics in a unified model with hierarchical clustering methods is still rarely seen.

The mobile recommendation and human behavior play an important role in many fields.Chen, Deng-Neng, et al.[7] built a recommendation system via the AHP. Soe-Tsyr Yuan, Y.W[8] presented a personalized contextualized mobile advertising infrastructure for the recommendation of advertisement. Fan.Y, Zhimei.W[9] built a scalable personalized mobile information pushing platform, which can recommend the location-based services to users. Tsao.Kowatsch T, Maass W.[10] investigated the use of mobile recommendation agents and they developed a model to better understand the impact of MRAs on usage intentions, product purchases and store preferences of consumers. Do, Gatica-Perez [11] mine user pattern using mobile phone app usage, including mobile Web usage on mobile phone.Zheng V W, Cao B, Zheng Y, et[12] mined useful knowledge from many users' GPS trajectories based on their partial location and activity annotations to provide targeted collaborative location and activity recommendations for each user.Huang K, Zhang C, Ma X, et[13] use a variety of contextual information, such as last used App, time, location, and the user profile, to predict the user's App whether will be open. Pinyapong S, Kato T.[14] proposed the relationship between 3 factors which are time, place and purpose. In consequence, they have summarized the basic rules to analyze essential data and algorithms to query processing. Ricci F[15].has done a nice survey of the mobile recommender systems, who has illustrated the overview of major techniques supported functions, and specific computational models.

Our method mines phone Interest patterns in mobile Web usage from the station cellular data of mobile operator's perspective, which is both comprehensive and large scale.

3   Problem Definition

We present required definitions and formulate the problem of the mobile phone recommendation based on the station cellular data of mobile operator. Without loss of generality, we assume there are two sets of the mobile users, source user set and the target user set. Our goal is to recommend one or more specific mobile phone to the target user from the model which is trained by the source user set. First of all, we give some formal definitions of the concepts used.

Table 1. Field details of the dataset

Filed name	Data type	Description
User Id	String	IMSI, the unique identifier of a user
Phone Id	String	IMEI, the unique identifier of a phone
Host	String	The domain name of the host requested
Content Type	String	The ContentType attribute in HTTP header
User agent	String	The app infomation in HTTP header

Table 2. Application Behavior

1	SEARCHING	11	PLAYING GAME
2	DO RECORD	12	MANAGING PHONE
3	BROWSING WEIBO	13	INTERNET SURFING
4	DO SHOPPING	14	READING
5	CHATTING	15	PHOTOGRAPHING
6	WATCHING STREAM	16	SENDING/RECIEVING MAIL
7	BROWSING RENREN	17	SEARCHING MAP
8	READ NEWS	18	LISTENING MUSIC
9	FLASHSLIGHTING	19	MESSAGING
10	BROWSING BLOG/ZONE	20	COMMUNICATION

Definition 1. *Cellular data. The cellular data is the log data from the station of the operator towers. Each tower records the cellular data, which can describes the Web-access of the users who are covered by a tower. One line in the dataset corresponds to a HTTP request/response pair occurred when using cellular network. All of follow framework and the method in this paper are based on cellular data. The main structure of the cellular data is shown in Table 1.*

Definition 2. *User. A user if uniquely identified by "IMSI" (the unique id of a SIM card) in the dataset. It corresponds to a real person using the mobile Web, regardless of what devices are used.*

Definition 3. *Mobile Phone/Device. A mobile device (phone or pad) which is uniquely identified by "IMEI" (the unique id of a phone) in the dataset. It corresponds to a real device (a cellphone serial number) using the mobile Web, regardless of its phone number.*

Definition 4. *Useragent. The useragent is the domain name of the HTTP request which acts as a client in a network protocol during communications within a client server distributed computing system. Some specific useragents can reflect the details of users' application.*

Definition 5. *Website. A Website is the domain name of the HTTP request. It may or may not be the address that is directly requested by the user / app. The Website reflects the host address of the HTTP request server.*

Definition 6. *App-Behavior. When the user uses web-applications, the applications have to send one or more request to the service host. Each request contains a intention from the application or user. For example, a request requires the game service, so the intention of the request is "gaming". We define the intention of each request as app-behavior. We can recognize the application's intention via the "Useragent" domain from the cellular data, and give the each app-behavior a label of natural language. The 20 types App-Behavior is shown in Table 2.*

Definition 7. *Phone Interest. Phone Interest is the distribution of the Website and App-Behavior, which reflects the interest of users who have device (mobile phone or pad)* with the same type are interested in. The phone interest may contain several different Websites and App-Behaviors. Each tuple of Website and App-Behavior is associated with a weight, representing for the degree of users' fondness. On the other hand Website and App-Behavior may serve different phone interest.

Problem 1. **Mobile Phone Recommendation**. (1) Given a complete cellular data in a period of time of an area. (2) Find out the latent phone interest of each type of phone from the cellular data, and build an interest model. (3) According to the "Phone Interest Model" (PIM) and the records of user, recommend the suitable phones to the user.

4   Phone-Interest Model (PIM)

In this section, we proposed a method to build a model, which can represent the real property and phone interest (defined in Definition 7) of different mobile phone. We attempt to extract latent phones' interests from their Web usage cellular data (defined in Definition 1); we propose two baseline methods K-means[16][17] and SVM (Support Vector Machine)[18][19] and also propose a probabilistic topic modeling method based on LDA (Latent Dirichlet Allocation)[20]. In our model, extracted latent layer represents the phone interest.

Let us briefly introduce notations below. T is the set of the phone interest; z is the index of phone interest; U is the set of user; u is the index of user; D is the set of the mobile devices; d is the index of mobile device; w is the index of Website of each request server address; b is the index of App-Behavior of each request want to express;

4.1   App-Behavior Discovery

Since raw HTTP log may not truly reflect user behavior. We do some additional work for the each HTTP request to express the intention of requester. We define the intention of requester as App-Behavior. First, we extract the Useragent domain from HTTP cellular data, and then we use natural language recognition to analyze the purpose of the request, and tag a formal label as the App-Behavior for this request. The label set contains 20 distinct labels, and each of the word can express a kind of attempt of the user. Based on the cellular data analysis above, the 20 different App-Behavior labels are given in Table 2.

4.2   Website and App-Behavior Represent of the Mobile Phone Feature

Since same Website address can appear in different request cases, and each host of those request may not have the same Web intention, we can't only use the Website address as the feature of the mobile phone. And we should add the user purpose part to the mobile phone feature. For each mobile device d, visited Website w, and each App-Behavior b, we define the transformation $f(w, b)$ as the feature of mobile phone. The set of mobile device D is represent as $D = \{\langle f(w, b)_{di}, n_{f(w,b)_{di}} \rangle\}$, where $f(w, b)_{di}$ is the ith feature dimension $f(w, b)$ of the mobile device d, and $n_{f(w,b)_{di}}$ is the number of times that $f(w, b)_{di}$ is visited by mobile device d. According to feature expression above, it can describe the purpose of the user, and also can reflect the different Website. For simplicity, we use symbol F to represent the feature space of $f(w, b)$, and use f to represent dimension index of $f(w, b)$ in feature space F.

4.3 K-Means Interest Learning (KIL)

We can use an easy method to extract "phone interest". We assume that the user always try to use or access the applications or websites overtime, even if the hardware or device is not perfectly suitable for those services. And most of the users can be divided into the limited number of groups, so we can use the tuple of host, App-Behavior as each phone's feature field, and run a cluster method to get each centroid of all clusters, which can represent the favorite hosts and App-Behaviors of each specific mobile device. Then the features of user's web behavior from the user log data is extracted, to match which kinds of mobile devices are similar to them.

For benchmark testing, we use K-means[16][17] method to cluster the mobile devices D into several clusters, with the tuple of host, App-Behavior F as the device feature. The vectors of the centroid which we calculate in the last step, are considered as "user interest". We can use those vectors to judge the test user which cluster they belong to, and recommend mobile devices which exist in the cluster to user.

The specific steps of "K-means Interest Learning" (KIL) are shown in Algorithm 1. The input is the set of mobile device D and the interest number K. The algorithm returns the set of interest centroid C. The goal is to calculate the distance between the formal cellular data of user u and interest center c to decide which c_j a user u belongs to. Finally, with the user u who is clustered by centroid c_j, the mobile devices d which are assigned to c_j will be recommended to this user.

Algorithm 1. K-means Interest Learning

Require:
  1: The set of mobile device D
  2: The interest number K
Ensure:
  3: Initialize $\{c_1, c_2...c_K\}$ be the cluster centers of interest set C.
  4: **while** each $\{c_1, c_2...c_K\}$ are not convergence to a stable value **do**
  5:     **for** each mobile device $d \in D$ **do**
  6:         assign the device d to the closest interest cluster c_j
  7:     **end for**
  8:     **for** each interest center $c_i \in C$ **do**
  9:         update c_i by averaging all of the device d that have been assigned to it
10:         **if** c_i don't contain any device d **then**
11:             $c_i \leftarrow randomd$
12:         **end if**
13:     **end for**
14: **end while**
15: **return** $\{c_1, c_2...c_K\}$;

4.4 K-SVM Interest Learning (KSIL)

"K-SVM Interest Learning" (KSIL) is modified based on the "K-means Interest Learning" (KIL). KSIL retained all steps of the KIL, and introduced "Support Vector Machine" (SVM)[18][19] for training and predicting. In the training step, the mobile devices d which are assigned by interest centroid c from KIL should be trained as the instance of c. We use F as the device feature, and the identify of interest centroid as the label. In the prediction stage, we use the user u as the instance,and use formal cellular data f as the feature, then use the SVM model to predict which class could be the user u was belonged to. Finally, recommend the mobile devices d which exist in this class to the user u.

4.5 Phone-Interest Model

Topic Model is commonly used in text mining for discovering abstract "topics" in a set of documents. LDA (Latent Dirichlet Allocation) is a commonly used topic model currently, and it has also been applied in discovering user behavior patterns[21]. LDA is a unsupervised, generative model, which models the generation of a document into a two-step process: choosing a topic based on topics distribution over a document; and choosing a word based on words distribution over a topic. We use the specific mobile device type representation of all users who use this type as a document, and use each user's formal log data as the feature to express the words' vector space, and propose a probabilistic topic model for phone interest modeling. Table 3 summarizes the notations used in the PIM.

We also propose a generative model of record set of the mobile devices. Assume that the generating scheme of cellular records are as follows:

There are many hosts/app-behaviors $F = \{f_1, f_2...\}$ that belong to each mobile device d,and each f may belong to different latent phone interest $T = \{z_1, z_2...\}$ with different probability, so there is a probability vector $\theta_d = \{p_{f1}, p_{f2}...\}$ to represent the probability of each host/app-behavior of each device which belong to different phone interest, meanwhile the probability is not unique. Accordingly, we can sample the phone interest z for each specific position of a specific device d by θ_d. And we can unite a probability matrix ϕ of z given f to generate the f for each position by Equation 1.

$$p(f) = \sum_z p(z/\theta_d)p(f/z, \phi) \tag{1}$$

z denotes the phone interest, θ_d denotes the probability distribution of a specific mobile device d over phone interest z and ϕ denotes the probability matrix z over all host/app-behavior f. In each device, there are n independent feature f, and the different θ_d can produce a same f, also f can be produced by different z, so the probability of a specific device d record can be generated by:

$$p(d) = \int_{\theta_d} p(\theta/\alpha) \prod_n \sum_z p(z/\theta_d)p(f/z, \phi) \tag{2}$$

α denotes the hyper parameter of the multinomial distributions and n denotes the host/app-behavior set of each device . If we want to generated all records of the device set we must accumulate the product of every device d, so the generation function of all record set is:

$$p = \prod_{d \in D} p(d) \tag{3}$$

Our goal is to calculate the 2 probability distributions θ and ϕ via phone interest model for clustering the mobile devices and recommending the mobile phones to user, also we have to find out the latent phone interest T from the set of mobile devices D. θ is a multinomial distribution over T specific to mobile device, and the dimensions of θ is $|D| \times |T|$. ϕ is a multinomial distribution over F specific to T, and the dimensions of ϕ is $|T| \times |F|$. We analyse phone interest z each mobile device d belongs to via the θ matrix. In the matrix θ, each row represents a probability distribution of a mobile device d to the latent phone interest T. We use the ϕ matrix and a specific user u formal log data F to predict which phone interest z the user u belong to, and recommend the suitable mobile devices to the user u, meanwhile we extract some phone features f with high probability value on the each topic z to represent the every $PhoneInterest$.

Table 3. Symbol description

T	the set of phone interests.
K	the number of phone interests.
D	the set of mobile devices.
B	the set of App-Behavior.
U	the set of users.
F	the set of formal cellular data feature.
z	the index of phone interest.
d	the index of mobile device.
u	the index of user.
w	the Website.
b	the App-Behavior.
$f(w,b)$	the formal cellular data feature, which is represented by tuple of Website and App-Behavior .
f	the index of formal cellular data feature.
x_{di}	the ith $f(w,b)$ attribute (word) in mobile device d.
z_{di}	the phone interest(topic) assigned to attribute x_{di}.
s_{di}	if x_{di} is a word from a single domain or a cross domain.
θ_d	multinomial distribution over phone interest(topic) specific to mobile device d.
ϕ_z	multinomial distribution over tuple of host and app-behavior specific to phone interest(topic) z.
α, β	Dirichlet priors to multinomial distributions θ and ϕ.

We combine mobile device D, tuple of Website as well as App-Behavior F and phone interest T in a unified generative model PIM. The generative process of "Phone Interest Model" (PIM) is as follows Figure 1.

1. For each mobile device d, draw θ_d from Dirichlet prior α;
2. For each phone interest z, draw ϕ_z, from Dirichlet priors β_z;
3. For each feature i of d appearance $f(w,b)_{di}$ in mobile device d:
   - draw a phone interest (topic) z_{di} from a multinomial distribution θ_d.
   - draw a term for $f(w,b)_{di}$ from multinomial distribution ϕ_{zdi}.

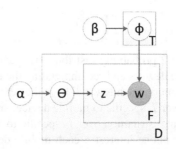

Fig. 1. Plate representation of Phone-Interest Model (PIM)

We use Gibbs sampling to estimate the model parameters, following [6]. For simplicity, we take fixed values for hyper parameters α and β (i.e. $\alpha = 50/|T|$, $\beta = 0.01$). We use Gibbs sampling to estimate the posterior distribution on $f(w, b)$ and z, then use the result to estimate θ and ϕ. The posterior probability can be calculated by Equation 4.

$$P(z_{di}|T_{-di}, F, \alpha, \beta) \propto \frac{m_{d,z_{di}}^{-di} + \alpha_{z_{di}}}{\sum_{z=1}^{K}(m_{d_{di},z}^{-di} + \alpha_z)} \frac{n_{z_{di},f_{di}}^{-di} + \beta_{f_{di}}}{\sum_{f=1}^{F}(n_{z_{di},f}^{-di} + \beta_f)} \tag{4}$$

$m_{d,z}$ denotes the number of times that phone interest z occurs in mobile device d, and $n_{z,f}$ denotes the number of times that feature f of the mobile device generated by phone interest z. Superscript $-di$ means that the quantity is counted excluding the current instance by the ith dimension of the dth mobile device. After the sampling convergences, the multinomial distribution parameters θ and β can be estimated as follows:

$$\theta_{d,z} = \frac{m_{d,z} + \alpha_z}{\sum_{z'}(m_{d,z'} + \alpha_{z'})} \tag{5}$$

$$\phi_{z,f} = \frac{n_{z,f(w,b)} + \beta_{f(w,b)}}{\sum_{f(w,b)'}(n_{z,f(w,b)'} + \beta_{f(w,b)'})} \tag{6}$$

Also, if a user u with his formal cellular data feature f is given,and then we inference the probability distribution over the phone interest T of the user u by:

$$P(T_i = j|T_{-i}, f(w,b)_i, d_i, \cdot) \propto \frac{n_k^{(t)} + n_k^{(t)'} + \beta_t}{\sum_{f=1}^{F}(n_k^{(f)} + n_k^{(t)'} + \beta_f)} \cdot \frac{n_m^{(k)} + \alpha_k}{\sum_{z=1}^{K}(n_m^z + \alpha_z)} \tag{7}$$

Finally, we get the $\theta_{d,z}$ and $\phi_{z,f}$ cross the Equation 5 and 6. The characteristic of the phone interest z can be reflected by the f items which own the higher probability in the ϕ_z. So the phone interest z of each mobile device is expressed by several top f with maximum probability in the ϕ_z. For mobile phone recommendation, we have to infer which phone interest a new user's Web behavior most possibly belongs to using Equation 7 and the cellular data. Then according to θ_z, the user will be recommended the mobile device d, which already belongs to this phone interest z.

5 Experimental Results and Evaluations

In this section, we will demonstrate the experiment results of applying the framework on our dataset, and give analysis based on the experiment results.

5.1 Data Description

The data we use is the mobile Web usage log (HTTP request log) of the cellular network (2G, GPRS) of a mobile operator. The dataset covers the geographical range of Beijing, the capital of China. The data we use covers 2 urban districts, with an area of approximately 40 square kilometers. The area contains a major set of attractions, as well as one of the busiest business districts of the city.

Overview. The time range of the dataset is from June.8 to June.20, 2013. There are totally 1,344,654,257 complete records in the dataset. There are 4,084,230 distinct users in the raw dataset. We clean the raw data by 1) As we focus on tuple of Website, App-Behavior and phone interest, so we use Useragent field to filter out all supporting requests with inner joined by the App-Behavior table. 2) Records with meaningless (i.e. full of question marks) Host field are removed or null. After cleaning, there are 34.262% (460,713,102) records and 1,244,272 distinct users left.

We extracted the top 2,000 users, whose records are over 500 from the data, with 145 distinct phone types (IMEI). We retained the mobile phone records which keep over 10 distinct users in the data. Finally, we have got the data which hold a total of 18,908,715 records, 44 distinct mobile device types of the phone, 19,994 distinct hosts and 1,602 users as our source data. The distribution of mobile device in count of users is shown in Figure 2.

5.2 Phone-Interest Discovery

We run the PIM with the above source data, and use the Equation 6 to calculate a $T \times F$ ϕ parameter matrix, and follow Equation 5 to calculate a $|D| \times |T|$ θ matrix, where D is the number of the mobile devices in the source data. We use PIM-k to donate the PIM model which runs with k phone interests, and use ϕ_k, θ_k to donate the ϕ and θ which calculated by the PIM-k.

Evaluation metrics. We evaluate the PIM models based on the source data with mobile device $|D| = 44$, Website host $|W| = 19994$, App-Behavior $|B| = 20$, phone interest $|T| = \{5, 6, 7, 8\}$. To quantitatively evaluate the proposed methods, we use other cellular data as the testing data. All of the mobile devices in the testing data are known, to verify the accuracy of PIM model. In evaluation, since we have filtered the users whose records are more than others as the testing data, we consider those users

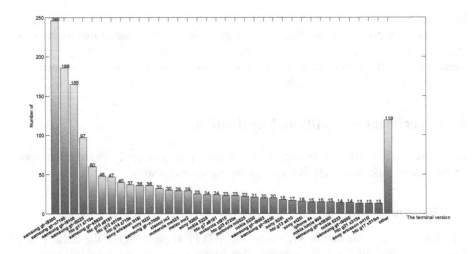

Fig. 2. Distribution of each mobile phone types

already have a suitable phone, which can perfect match their phone interest. According to the phone interest, the PIM model will predict some specific mobile device types for each user, if the result of prediction included the mobile device type which can perfect match the user's, and then we say the prediction is correct, otherwise we say the prediction is wrong. Based on this, we evaluate the prediction performance in terms of P@1 (Precision of the mobile phone recommendation based on the first prediction phone interest), P@2 (Precision of the mobile phone recommendation based on the both first and second prediction phone interest), Recall (Recall for the prediction phone interest) and MAP (Mean Average Precision).

All codes are implemented in Java, and all the experiments are conducted on an x64 server with E5-2609 2.4GHz Intel Xeon CPU and 32G RAM. The operation system is Microsoft Windows Sever 2008 R2 Enterprise. For training the PIM-6, PIM-7 and PIM-8 models, it takes about 2 hours respectively on the entire data set (18908715 records, 44 distinct mobile phone types). Recognizing the computation complexity of LDA style models, we are currently looking into developing more efficient computation mechanism to speed up the process.

5.3 Performance Analysis

Table 4 lists the performance of PIM on different test cases. In the table, MI0 means the first phone interest in the PIM, and each row means the sub-performance of the different phone interest. The last row shows us the average of the performance in the PIM. We also run the "K-means Interest Learning" KIL and "K-SVM Interest Learning" KSIL method on our dataset. The performance of those methods are shown in the Table 5. The proposed PIM method clearly gets an outstanding performance, and outperforms the baseline methods (KIL and KSIL) with different phone interests.

Table 4. The performance with diffrent phone interest

(a) PIM-5

	P@1	P@2	Recall	MAP
MI0	0.4939	0.5783	0.6507	0.8450
MI1	0.3212	0.8339	0.3886	0.9010
MI2	0.6855	0.7860	0.4523	0.8477
MI3	0.1396	0.1955	0.3472	0.7522
MI4	0.3750	0.3750	0.6774	0.8000
Average	**0.4496**	**0.6510**	**0.5032**	**0.8292**

(b) PIM-6

	P@1	P@2	Recall	MAP
MI0	0.5967	0.6290	0.925	0.8852
MI1	0.3571	0.3571	0.7407	0.8203
MI2	0.3181	0.4090	0.4375	0.7428
MI3	0.3939	0.4545	0.4482	0.8468
MI4	0.1923	0.8076	0.3333	0.9618
MI5	0.8070	0.8070	0.3458	0.8382
Average	**0.4878**	**0.5709**	**0.5384**	**0.8492**

(c) PIM-7

	P@1	P@2	Recall	MAP
MI0	0.3275	0.3448	0.6333	0.6966
MI1	0.1923	0.2307	0.2777	0.9389
MI2	0.3666	0.3666	0.3859	0.8414
MI3	0.125	0.125	0.3125	0.8527
MI4	0.5967	0.6290	0.9024	0.9016
MI5	0.8125	0.875	0.6842	0.6829
MI6	0.6923	0.6923	0.3082	0.8002
Average	**0.4464**	**0.4617**	**0.5006**	**0.8163**

(d) PIM-8

	P@1	P@2	Recall	MAP
MI0	0.5806	0.5806	0.7659	0.8965
MI1	0.2068	0.2241	0.1666	0.9520
MI2	0.3253	0.4216	0.1516	0.7875
MI3	0.1791	0.4701	0.4800	0.6772
MI4	0.6058	0.6569	0.2150	0.7832
MI5	0.1724	0.1896	0.2173	0.6933
MI6	0.8125	0.9375	0.5652	0.4864
MI7	0.2226	0.2226	0.7794	0.9237
Average	**0.3065**	**0.4119**	**0.4176**	**0.7750**

Phone Interest Topic Analysis. How many topics are enough for the phone recommendation? We perform an analysis by varying the number of phone interest topics in the proposed PIM method. Figure 3 shows its P@1, P@2, Recall and MAP performance with the number of phone interest topics varied. We see when the phone interest number is up to 6, decreasing the number often obtains a performance improvement. The precision trend becomes best when the number is at 6. This demonstrates the stability of the PIM method with respect to the number of topics. On the other hand, the phone interest number is defined by the how many mobile devices you want to recommend to the user. If is used k to express the number of mobile phone types will recommend to user, and then we roughly calculate the phone interest number $|T|$ via the expression $|T| = |D|/k$.

According to PIM-5, we displayed all 5 phone interests, also we listed top 5 frequent mobile devices with their weights from each phone interest. Their topic Websites/App-Behavior and mobile devices are listed in Table 6. The main mobile devices of the interests are listed, and top 5 tuple of Websites and App-Behavior of each interest are shown with their domains and weights within the interest.

It can be seen that the topic Websites of the interests are quite centralized. Analyst can draw the conclusions from the observation, even if give some meaningful tags for the users of different phone interest cluster, which can present the difference between the different crowd.

Table 5. Recommendation performance by different methods: KIL(K-means), KSIL (K-SVM), PIM

PI	Method	P@1	P@2	Recall	MAP
	KIL	0.3995	**0.6777**	0.3752	0.4125
5 Interests	KSIL	0.387	-	0.382	-
	PIM	**0.4496**	0.6510	**0.5032**	**0.8292**
	KIL	0.3323	0.5195	0.3229	0.3427
6 Interests	KSIL	0.344	-	0.334	-
	PIM	**0.4878**	**0.5709**	**0.5384**	**0.8492**
	KIL	0.3430	**0.5254**	0.2905	0.3741
7 Interests	KSIL	0.370	-	0.354	-
	PIM	**0.4464**	0.4617	**0.5006**	**0.8163**

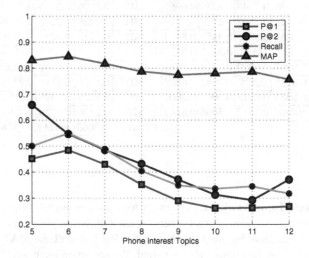

Fig. 3. Performance of PIM with different topic

Interest 0 is mostly about chatting, the top 2 Websites are related to live chat, but the chat clients are supported by different providers, and it is shown that Nokia phone is dominant in this interest, which is proven by the fact that all top 5 the most frequent mobile device are produced by Nokia phone.

Interest 1 is centralized by news reading, where about 35% of usage is used for browsing or reading news from the news application of Netease 163. In addition, mail service and browsing blog service also take up a significant part .

Interest 2 is great centralized by send/recieving mail service, since more than half of its usages are related to mail service. And we can observe that a lot of Samsung devices had very good support for this function.

Interest 3 is mostly about sending/recieving the mail service, also searching service and news service have played main roles in this interest. The listed devices indicate that sony ericsson lt18i is perfect suitable for those Web usages.

Table 6. Latent "phone interests" and clusters of mobile phone from the PIM5

	Mobile Phone Type	Weight of	Websites App-Behavior	Weight of
Phone Interest 0	nokia 5228	0.9993	wx.qlogo.cn CHATTING	0.3483
	nokia c7-00	0.9988	mmsns.qpic.cn CHATTING	0.3084
	nokia c5-03	0.9986	short.weixin.qq.com SENDING/RECEIVING MAIL	0.1638
	nokia 5230	0.9775	mobilemaps.clients.google.com SEARCHING	0.0692
	nokia 5233	0.9755	api.baiyue.baidu.com READ NEWS	0.0231
Phone Interest 1	htc g14 710e	0.9788	p.3g.163.com READ NEWS	0.3528
	motorola me525	0.9778	short.weixin.qq.com SENDING/RECIEVING MAIL	0.1803
	sony lt26ii	0.9619	m.qpic.cn BROWSING BLOG/ZONE	0.1010
	htc g13 a510	0.9464	ugc.qpic.cn BROWSING BLOG/ZONE	0.0464
	samsung gt-s5660	0.8645	qzonestyle.gtimg.cn BROWSING BLOG/ZONE	0.0387
Phone Interest 2	motorola xt910	0.9987	short.weixin.qq.com SENDING/RECIEVING MAIL	0.5143
	samsung gt-s7500	0.9914	m.qpic.cn BROWSING BLOG/ZONE	0.0795
	sony lt26i	0.9908	www.google-analytics.com MANAGING PHONE	0.0319
	samsung gt-n7000	0.9774	in1.feed.uu.cc SEARCHING	0.0251
	samsung gt-s5670	0.9224	api.mobile.360.cn MANAGING PHONE	0.0189
Phone Interest 3	sony ericsson lt18i	0.9930	short.weixin.qq.com SENDING/RECIEVING MAIL	0.4139
	htc g10 a9191	0.8894	p.3g.163.com READ NEWS	0.2918
	samsung gt-i9000	0.7955	m.api.dianping.com SEARCHING MAP	0.0913
	samsung gt-s5830i	0.7909	gomarket.goapk.com SEARCHING	0.0885
	nokia lumia 900	0.7778	218.205.179.22:8002 SEARCHING	0.0340
Phone Interest 4	iphone 4s	0.9106	api.weibo.cn BROWSING WEIBO	0.3194
	htc g12 s510e	0.8197	ww3.sinaimg.cn BROWSING WEIBO	0.0867
			ww1.sinaimg.cn BROWSING WEIBO	0.0775
			ww2.sinaimg.cn BROWSING WEIBO	0.0684
			wbapp.mobile.sina.cn BROWSING WEIBO	0.0580

Interest 4 is centralized by browsing Weibo (English name micro blog), a popular micro blog app in China, many users use Iphone 4s or HTC g12 s510e for Weibo.

6   Conclusion

In this paper, we have presented a framework for recommending the mobile device to user. We have proposed a modeling method that models the Web cellular data of the operator. We have also presented a probabilistic topic modeling method to extract latent phone interest from mobile Web usage cellular data. We have applied the proposed framework to a real world large scale dataset from Beijing, capital of China, with more than 18 million data records, and the output shows outstanding mobile phone recommendation accuracy. We have analyzed city-level collective behavior patterns in mobile Web usage based on the model output, and discussed mobile phone clustering and performance using phone interest distribution. For future work, one possibility to extend the work is to take social connections between users in consideration. Mobile Web usage behaviors can spread over social networks, thus establishing relationships between users according to social connections would help enrich context information of a user and improve accuracy of phone interest modeling.

Acknowledgement. This work is supported by China National Science Foundation under grant No.61170212, China National High-Tech Project (863) under grant No. SS2013AA010307, and Ministry of Education-China Mobile Research Fund under grant No.MCM20130381. Beijing Key Lab of Networked Multimedia also supports our research work.

References

1. Srivastava, J., Cooley, R., Deshpande, M., Tan, P.N.: Web usage mining: discovery and applications of usage patterns from web data. SIGKDD Explor. Newsl. 1(2), 12–23 (2000)
2. Kosala, R., Blockeel, H.: Web mining research: a survey. SIGKDD Explor. Newsl. 2(1), 1–15 (2000)
3. White, R.W., Bailey, P., Chen, L.: Predicting user interests from contextual information. In: Proceedings of the 32nd International ACM SIGIR Conference on Research and Development in Information Retrieval, SIGIR 2009, pp. 363–370. ACM, New York (2009)
4. Nasraoui, O., Soliman, M., Saka, E., Badia, A., Germain, R.: A web usage mining framework for mining evolving user profiles in dynamic web sites. IEEE Transactions on Knowledge and Data Engineering 20(2), 202–215 (2008)
5. Xu, J., Liu, H.: Web user clustering analysis based on kmeans algorithm. In: 2010 International Conference on Information Networking and Automation (ICINA), vol. 2, pp. V2-6 –V2-9 (October 2010)
6. Mobasher, B., Cooley, R., Srivastava, J.: Creating adaptive web sites through usage-based clustering of urls. In: Proceedings of the 1999 Workshop on Knowledge and Data Engineering Exchange (KDEX 1999), pp. 19–25 (1999)
7. Chen, D.N., Hu, P.J.H., Kuo, Y.R., Liang, T.P.: A web-based personalized recommendation system for mobile phone selection: Design, implementation, and evaluation. Expert Systems with Applications 37(12), 8201–8210 (2010)
8. Yuan, S.T., Tsao, Y.W.: A recommendation mechanism for contextualized mobile advertising. Expert Systems with Applications 24(4), 399–414 (2003)
9. Yang, F., Wang, Z.: A mobile location-based information recommendation system based on gps and web 2.0 services. Database 7, 8 (2009)
10. Kowatsch, T., Maass, W.: In-store consumer behavior: How mobile recommendation agents influence usage intentions, product purchases, and store preferences. Computers in Human Behavior 26(4), 697–704 (2010)
11. Do, T.M.T., Gatica-Perez, D.: By their apps you shall understand them: mining large-scale patterns of mobile phone usage. In: Proceedings of the 9th International Conference on Mobile and Ubiquitous Multimedia, MUM 2010, pp. 27:1–27:10. ACM, New York (2010)
12. Zheng, V.W., Cao, B., Zheng, Y., Xie, X., Yang, Q.: Collaborative filtering meets mobile recommendation: A user-centered approach. In: AAAI, vol. 10, pp. 236–241 (2010)
13. Huang, K., Zhang, C., Ma, X., Chen, G.: Predicting mobile application usage using contextual information. In: Proceedings of the 2012 ACM Conference on Ubiquitous Computing, pp. 1059–1065. ACM (2012)
14. Pinyapong, S., Kato, T.: Query processing algorithms for time, place, purpose and personal profile sensitive mobile recommendation. In: 2004 International Conference on Cyberworlds, pp. 423–430. IEEE (2004)
15. Horozov, T., Narasimhan, N., Vasudevan, V.: Using location for personalized poi recommendations in mobile environments. In: International Symposium on Applications and the Internet, SAINT 2006, p. 6. IEEE (2006)

16. Hartigan, J.A., Wong, M.A.: Algorithm as 136: A k-means clustering algorithm. Applied Statistics, 100–108 (1979)
17. Wagstaff, K., Cardie, C., Rogers, S., Schrödl, S., et al.: Constrained k-means clustering with background knowledge. In: ICML, vol. 1, pp. 577–584 (2001)
18. Suykens, J.A., Vandewalle, J.: Least squares support vector machine classifiers. Neural Processing Letters 9(3), 293–300 (1999)
19. Tong, S., Koller, D.: Support vector machine active learning with applications to text classification. The Journal of Machine Learning Research 2, 45–66 (2002)
20. Blei, D.M., Ng, A.Y., Jordan, M.I.: Latent dirichlet allocation. J. Mach. Learn. Res. 3, 993–1022 (2003)
21. Farrahi, K., Gatica-Perez, D.: Discovering routines from large-scale human locations using probabilistic topic models. ACM Trans. Intell. Syst. Technol. 2(1), 3:1–3:27 (2011)

Two Approaches to the Dataset Interlinking Recommendation Problem

Giseli Rabello Lopes[1], Luiz André P. Paes Leme[2],
Bernardo Pereira Nunes[1], Marco Antonio Casanova[1], and Stefan Dietze[3]

[1] Department of Informatics, Pontifical Catholic University of Rio de Janeiro,
Rio de Janeiro, RJ – Brazil, CEP 22451-900
{grlopes,bnunes,casanova}@inf.puc-rio.br
[2] Computer Science Institute, Fluminense Federal University,
Niterói, RJ – Brazil, CEP 24210-240
lapaesleme@ic.uff.br
[3] L3S Research Center, Leibniz University Hannover, Appelstr. 9a,
30167 Hannover, Germany
{dietze}@l3s.de

Abstract. Whenever a dataset t is published on the Web of Data, an exploratory search over existing datasets must be performed to identify those datasets that are potential candidates to be interlinked with t. This paper introduces and compares two approaches to address the dataset interlinking recommendation problem, respectively based on Bayesian classifiers and on Social Network Analysis techniques. Both approaches define rank score functions that explore the vocabularies, classes and properties that the datasets use, in addition to the known dataset links. After extensive experiments using real-world datasets, the results show that the rank score functions achieve a mean average precision of around 60%. Intuitively, this means that the exploratory search for datasets to be interlinked with t might be limited to just the top-ranked datasets, reducing the cost of the dataset interlinking process.

Keywords: Linked Data, data interlinking, recommender systems, Bayesian classifier, social networks.

1 Introduction

Over the past years there has been a considerable movement towards publishing data on the Web following the Linked Data principles [1]. According to those principles, to be considered 5-star, a dataset must comply with the following requirements: (i) be available on the Web; (ii) be available as machine-readable structured data; (iii) be in a non-proprietary format; (iv) use open standards from W3C (i.e. RDF and SPARQL) to identify resources on the Web; and (v) be linked to other people's data to provide additional data. This paper addresses the last requirement.

Briefly, in the context of Linked Data, a *dataset* is a set of RDF triples. A resource identified by an RDF URI reference s *is defined in* a dataset t iff s occurs as the subject of a triple in t.

B. Benatallah et al. (Eds.): WISE 2014, Part I, LNCS 8786, pp. 324–339, 2014.
© Springer International Publishing Switzerland 2014

A *feature* of a dataset is a vocabulary URI, a class URI or a property URI used in triples of the dataset. One may then represent the dataset by one or more of its features.

Let t and u be two datasets. A *link* from t to u is a triple of the form (s, p, o) such that s is defined in t and o is defined in u. We say that t is *linked to u*, or that u is *linked from t*, iff there is at least one link from t to u. We also say that u is *relevant* for t iff there is at least one resource defined in u that can be linked from a resource defined in t.

The *dataset interlinking recommendation problem* can then be posed as follows:

> *Given a finite set of datasets D and a dataset t, compute a rank score for each dataset $u \in D$ such that the rank score of u increases with the chances of u being relevant for t.*

To address the dataset interlinking recommendation problem, this paper proposes and compares two approaches respectively based on Bayesian classifiers and on Social Network link prediction measures. Both approaches define rank score functions that explore the dataset features and the known links between the datasets. The experiments used real-world datasets and the results show that the rank score functions achieve a mean average precision of around 60%. Intuitively, this means that a dataset interlinking tool might limit the search for links from a dataset t to just the top ranked datasets with respect to t and yet find most of the links from t.

The rest of the paper is organized as follows. Section 2 discusses related work. Section 3 introduces our proposed approaches based on Bayesian classifiers and on Social Network Analysis techniques. Section 4 presents the experiments conducted to test and compare the approaches. Finally, Section 5 contains the conclusions and directions for future work.

2   Related Work

In this paper, we extend previous work [2,3] that introduced preliminary versions of the rank score functions respectively based on the Bayesian and the Social Network approaches. This paper contains significantly new results over our previous work in so far as it (i) explores different sets of features to compute rank score functions; (ii) uses modified rank score functions to interlink new datasets without known links; and (iii) provides a comprehensive comparison of the approaches using different feature sets.

In more detail, the paper improves previous results as follows. As for the Bayesian ranking definition, the paper formally defines how to manage the lack of observations of co-occurrences between features and links. Without this new definition, null probabilities could lead the score function to a discontinuity region ($log(0)$). In the SN-based ranking definition, we propose a new score function (not defined in [3]) which combines preferential attachment and resource allocation measures. The definition of the similarity set of the target dataset is also novel.

Furthermore, we do not assume that one knows the existing links of the dataset to which one wants to generate recommendations for. This assumption is realistic for new datasets and tackles one of the core challenges of the Linked Data principles. Indeed, for new datasets, the approach proposed in [3] will not work and that presented in [2] will generate recommendations based only on the popularity of the datasets (i.e., the recommendations will be the same for all datasets).

We explored different sets of features - properties, classes and vocabularies - to compute the rank score functions. Moreover, we thoroughly compared the performance of the improved approaches using different feature sets.

Nikolov et al. [4,5] propose an approach to identify relevant datasets for interlinking, with two main steps: (i) searching for potentially relevant entities in other datasets using as keywords a subset of labels in the new published dataset; and (ii) filtering out irrelevant datasets by measuring concept similarities obtained by applying ontology matching techniques.

Kuznetsov [6] describes a linking system which is responsible for discovering relevant datasets for a given dataset and for creating instance level linkage. Relevant datasets are discovered by using the *referer* attribute available in the HTTP message header, as described in [7], and ontology matching techniques are used to reduce the number of pairwise comparisons for instance matching. However, this work does not present any practical experiments to test the techniques.

When compared with these approaches, the rank score functions proposed in this paper use only metadata and are, therefore, much simpler to compute and yet achieve a good performance.

The next set of papers aim at recommending datasets with respect to user queries, which is a problem close, but not identical to the problem discussed in this paper. Lóscio et al. [8] address the recommendation of datasets that contribute to answering queries posed to an application. Their recommendation function estimates a degree of relevance of a given dataset based on an information quality criteria of correctness, schema completeness and data completeness. Wagner et al. [9] also propose a technique to find relevant datasets for user queries. The technique is based on a contextualization score between datasets, which is in turn based on the overlapping of sets of instances of datasets. It uses just the relationships between entities and disregards the schemas of the datasets. Oliveira et al. [10] use application queries and user feedback to discover relevant datasets. Application queries help filter datasets that are potentially strong candidates to be relevant and user feedback helps analyze the relevance of such candidates.

Toupikov et al. [11] adapt the original PageRank algorithm to rank existing datasets with respect to a given dataset. The technique uses the Linksets descriptions available in VoID files as the representation of relationships between datasets and the number of triples in each Linkset as the weight of the relationships. Results show that the proposed technique performs better than traditional ranking algorithms, such as PageRank, HITS and DRank. As the rank score functions defined in this paper, the version of the PageRank algorithm the authors propose depends on harvesting VoID files.

3 Ranking Techniques

Sections 3.1 and 3.2 introduce two approaches to compute rank score functions, leaving a concrete example to Section 3.3.

3.1 Bayesian Ranking

This section defines a rank score function inspired on conditional probabilities. However, we note that the rank score is not a probability function. We proceed in a stepwise fashion until reaching the final definition of the rank score function, in Equation 9.

Let D be a finite set of datasets, d_i be a dataset in D and t be a dataset one wishes to link to datasets in D. Let T denote the event of selecting the dataset t, D_i denote the event of selecting a dataset in D that has a link to d_i, and F_j denote the event of selecting a dataset that has feature f_j (recall that a *feature* of a dataset is a vocabulary URI, a class URI or a property URI used in triples of the dataset).

We tentatively define the rank score function as a conditional probability:

$$score_0(d_i, t) = P(D_i|T) \tag{1}$$

that is, $score_0(d_i, t)$ is the conditional probability that D_i occurs, given that T occurred. As required, this score function intrinsically favors those datasets with the highest chance of defining links from t.

We then rewrite $score_0$, using Bayes's rule, as follows:

$$score_1(d_i, t) = \frac{P(T|D_i)}{P(T)} P(D_i) \tag{2}$$

As in Bayesian classifiers [12,13], by representing t as a bag of features $F = \{f_1, ..., f_n\}$, one may rewrite $score_1$ as:

$$score_2(d_i, t) = \frac{P(\{f_1, ..., f_n\}|D_i)}{P(\{f_1, ..., f_n\})} P(D_i) \tag{3}$$

By the naive Bayes assumption [12,13], $P(\{f_1, f_2, ..., f_n\}|D_i)$ can be computed by multiplying conditional probabilities for each independent event F_j (the event of selecting datasets with just the feature f_j). Moreover, $P(\{f_1, ..., f_n\})$ does not change the rank order because it is the same for all d_i. Hence, we remove this term. The new score function becomes:

$$score_3(d_i, t) = \left(\prod_{j=1..n} P(F_j|D_i) \right) P(D_i) \tag{4}$$

The final score function is obtained from $score_3$ by replacing the product of the probabilities by a summation of logarithms, with the help of auxiliary functions p and q that avoid computing $log(0)$.

Intuitively, the definitions of functions p and q penalize a dataset d_i when no dataset with feature f_j is linked to d_i or when no dataset is linked to d_i. The definitions depend on choosing a constant C that satisfies the following restriction (where m is the number of datasets in D and n is the number of features considered):

$$C < min(C', C'') \tag{5}$$
$$C' = min\{P(F_j|D_i) \in [0,1] \; / \; P(F_j|D_i) \neq 0 \land j \in [1,n] \land i \in [1,m]\}$$
$$C'' = min\{P(D_i) \in [0,1] \; / \; P(D_i) \neq 0 \land i \in [1,m]\}$$

Then, p is defined as follows:

$$p(F_j, D_i) = \begin{cases} C, & if\, P(F_j|D_i) = 0 \\ P(F_j|D_i), & \text{otherwise} \end{cases} \tag{6}$$

Intuitively, p avoids computing $log(P(F_j|D_i))$ when $P(F_j|D_i) = 0$, that is, when no dataset with feature f_j is linked to d_i. In this case, d_i is penalized and $p(F_j, D_i)$ is set to C.

Likewise, q is defined as follows:

$$q(D_i) = \begin{cases} C, & if\, P(D_i) = 0 \\ P(D_i), & \text{otherwise} \end{cases} \tag{7}$$

Intuitively, q avoids computing $log(P(D_i))$ when $P(D_i) = 0$, that is, when no dataset is linked to d_i. In this case, d_i is also penalized and $q(D_i)$ is set to C.

We define the final rank score function in two steps. We first define:

$$score(d_i, t) = \left(\sum_{j=1..n} log(p(F_j, D_i)) \right) + log(q(D_i)) \tag{8}$$

and then eliminate $p(F_j, D_i)$ from Equation 8 :

$$score(d_i, t) = c\,|N_i| + \left(\sum_{f_j \in P_i} log(P(F_j|D_i)) \right) + log(q(D_i)) \tag{9}$$

where

- $c = log(C)$
- $N_i = \{f_j \in F / P(F_j|D_i) = 0\}$
- $P_i = F - N_i$

In particular, we note that, when t does not have any feature (i.e., when $n = 0$), the score function takes into account only the unconditional probability $P(D_i)$. In this case, the most popular datasets, such as DBpedia[1] and Geonames[2], will be favored by the score function at the expenses of perhaps more

---

[1] http://dbpedia.org/
[2] http://www.geonames.org/

appropriate datasets. The ranking may not be accurate in such borderline cases, but a popularity-based ranking is preferable to no ranking at all, when nothing is known about t.

Equation 9, therefore, defines the final score function that induces the ranking of the datasets in D (from the largest to the smallest score). Section 3.3 illustrates how the score is computed.

Based on the maximum likelihood estimate of the probabilities [13] in a training set of datasets, the above probabilities can be estimated as follows:

$$P(F_j|D_i) = \frac{count(f_j, d_i)}{\sum_{j=1}^{n} count(f_j, d_i)} \qquad (10)$$

$$P(D_i) = \frac{count(d_i)}{\sum_{i=1}^{m} count(d_i)} \qquad (11)$$

where $count(f_j, d_i)$ is the number of datasets in the training set that have feature f_j and that are linked to d_i, $count(d_i)$ is the number of datasets in the training set that are linked to d_i, disregarding the feature set. Thus, for any dataset t represented by a set of features, the rank position of each of the datasets in D can be computed using Equations 7, 9, 10 and 11.

Note that Equation 10 depends on the correlation between f_j and d_i in the training set. This means that the higher the number of datasets correlating feature f_j with links to d_i, the higher the probability in Equation 10. Moreover, as Equation 4 depends on the joint probability of the features f_j of t, the higher the number of features shared by t and the datasets linked to d_i with high probability, the higher $score(d_i, t)$ will be. That is, if the set of features of t is very often correlated with datasets that are linked to d_i and t is not already linked to d_i, then it is recommended to try to link t to d_i.

Finally, we stress that, if a dataset t exhibits a set of features F, one can choose any subset of F as the representation of t. Thus, each possible representation may generate different rankings with different performances and one cannot predict in advance which representation will generate the best ranking. Section 4 then compares the results obtained for several different feature sets.

3.2   Social Network-Based Ranking

In Social Networks Analysis (SNA), the network is typically represented as a graph, where the nodes are the entities (e.g., users, companies) and the edges are the relationships between them (e.g., follows, shares, befriends, co-authorships). In SNA, the *link prediction problem* refers to the problem of estimating the likelihood of the existence of an edge between two nodes, based on the already existing edges and on the attributes of the nodes [14]. We propose to analyze the dataset interlinking recommendation problem in much the same way as the link prediction problem.

As in Section 3.1, let D be a finite set of datasets, d_i be a dataset in D and t be a dataset one wishes to link to datasets in D. Recall again that a *feature* of a dataset is a vocabulary URI, a class URI or a property URI used in triples of the dataset.

The *Linked Data network* for D is a directed graph such that the nodes are the datasets in D and there is an edge between datasets u and v in D iff there is a link from u to v.

The *similarity set* of a dataset t, denoted S_t, is the set of all datasets in D that have features in common with t. The *popularity set* of a dataset $d_i \in D$, denoted P_{d_i}, is the set of all datasets in D that have links to d_i.

Among the traditional measures adopted for link prediction [15,14], we will use Preferential Attachment and Resource Allocation. Indeed, the results reported in [16], which analyzed the dataset interlinking recommendation problem using just the existing links, indicate that these two measures achieved the best performance.

Preferential Attachment. The Preferential Attachment score estimates the possibility of defining a link from t to d_i as the product of the cardinality of the similarity set of t, denoted $|S_t|$, and the cardinality of the popularity set of d_i, denoted $|P_{d_i}|$, and is defined as follows:

$$pa_0(t, d_i) = |S_t| \times |P_{d_i}| \tag{12}$$

However, since $|S_t|$ is independent of d_i, this term does not influence the rank score of the datasets. Thus, we may ignore it and define pa as follows:

$$pa(t, d_i) = |P_{d_i}| \tag{13}$$

Resource Allocation. Let d_j be a dataset in D, distinct from d_i. Intuitively, if there are links from t to d_j and from d_j to d_i and there are many other datasets that have links to d_j, then d_j must be a generic dataset (eg. DBpedia, Geonames, etc.). Therefore, d_j does not necessarily suggest any possible link from t to d_i. On the other hand, if there are not many datasets that have links to d_j, then this might be a strong indication that d_j is a very particular dataset for both t and d_i and, therefore, a link from t to d_i might as well be defined. Thus, the strength of the belief in the existence of a link from t to d_i increases inversely proportional to the number of datasets which have links to d_j, i.e., depends on the cardinality of the popularity set of d_j, again denoted $|P_{d_j}|$.

The Resource Allocation score estimates the possibility of defining a link from t to d_i as a summation of the inverse of the cardinality of the popularity set of the datasets in the intersection of the datasets linked from t, which is the similarity set S_t of t, and the datasets linked to d_i, which is the popularity set P_{d_i} of d_i. It is defined as follows:

$$ra(t, d_i) = \sum_{d_j \in S_t \cap P_{d_i}} \frac{1}{|P_{d_j}|} \tag{14}$$

Combined Score. To obtain more accurate results, we combine the two previous scores into a new score, defined as follows:

$$score(t, d_i) = ra(t, d_i) + \frac{pa(t, d_i)}{|\boldsymbol{D}|} \tag{15}$$

This final score gives priority to the ra score; the pa score, normalized by the total number of datasets to be ranked ($|\boldsymbol{D}|$), will play a role when there is a tie or when the ra value is zero. Section 4.3 comments on the adequacy of defining a combined score function.

3.3   Example of Rank Score Computations

We illustrate how to compute rank score functions, using both approaches, with the help of a schematic example. We selected a subset of the datasets indexed by the DataHub[3], using the *Learning Analytics and Knowledge*[4] dataset [17], referred to as *lak* in what follows, as the target of the recommendation.

As features of *lak*, we used three classes, *swc:ConferenceEvent*, *swrc:Proceedings* and *swrc:InProceedings*, obtained from the LinkedUp project Web site[5].

As the candidates to be ranked, we selected the datasets *webscience*, *webconf*, *wordnet*, *dblp* and *courseware*. They were chosen because we considered all datasets that share at least one feature with *lak* (*webscience* and *webconf*) and all datasets linked from them (*wordnet* and *dblp*). In addition, to better illustrate the computation of the rank scores, we also considered *courseware*, one of the datasets linked to *wordnet*.

The similarity set of *lak* consists of the datasets *webscience* and *webconf*, since they share at least one feature with *lak*. The datasets *webscience* and *webconf* shares respectively the *swc:ConferenceEvent* class and the *swc:ConferenceEvent*, *swrc:Proceedings* and *swrc:InProceedings* classes with *lak*.

Table 1 and Table 2 respectively list the URIs of all such datasets and classes. Figure 1 depicts these objects, where the directed thin arrows represent the existing links among the datasets, the thick arrows denote links from *lak* to datasets in its similarity set (used only by Social Network-based approach) and the dashed lines indicate which datasets have what features. The dashed cylinders refer to groups of datasets (the number of datasets grouped is indicated inside the cylinder).

The rank score functions have to rank the datasets *webscience*, *webconf*, *wordnet*, *dblp* and *courseware* according to the chances of defining links from resources in *lak* to resources in each of these datasets. The datasets in the similarity set of *lak* (*webscience* and *webconf*) are included in the list of candidates to be ranked because they are not yet linked from *lak*.

---

[3] http://datahub.io/
[4] http://lak.linkededucation.org
[5] http://linkedup-project.eu/

Table 1. The dataset acronym and the corresponding URI

Dataset	URI
lak	http://lak.linkededucation.org
webscience	http://webscience.rkbexplorer.com
webconf	http://webconf.rkbexplorer.com
dblp	http://knoesis.wright.edu/library/ontologies/swetodblp/
wordnet	http://www.w3.org/TR/wordnet-rdf
courseware	http://courseware.rkbexplorer.com

Table 2. The class feature acronym and the corresponding URI

Class	URI
swc:ConferenceEvent	http://data.semanticweb.org/ns/swc/ontology#ConferenceEvent
swrc:Proceedings	http://swrc.ontoware.org/ontology#Proceedings
swrc:InProceedings	http://swrc.ontoware.org/ontology#InProceedings

The Social Network-based rank score function (shown in Equation 15) ranks *wordnet* in the first position (the largest score value), *dblp* in the second position, *courseware* in the third position and *webscience* and *webconf* (with tied scores) in the last two positions. Recall that the Social Network-based score function is the sum of two terms, *ra* and *pa*. The first two best ranked datasets have scores determined by *ra* greater than zero because they are linked from *webconf*, which is in the similarity set of *lak*. The remaining datasets are ranked only by the *pa* term, including *webconf* and *webscience*, because they are in the similarity set of *lak*.

Using the Bayesian approach, the rank score function ranks *dblp* in the first position, *wordnet* in the second position, *courseware* in the third position and *webscience* and *webconf* (with tied scores) in the last two positions. It is not possible to adequately estimate probability values for *webscience* and *webconf* because they are both not linked from any other dataset. Thus, in this example, their score values will be the minimum, determined in this case by $c*4 = -60$ (omitted from the table in Figure 1 for convenience). Intuitively, the top ranking positions assigned to *wordnet* and *dblp* are justified because both datasets are linked from datasets that share some feature with *lak* and the popularity of both can be estimated.

A manual inspection performed in the two best ranked datasets by both approaches indicated that the recommendation of *dblp* is justified because the DBLP digital library[6] indexes the papers published in the LAK and EDM conferences, as does the *lak* dataset. Then, resources of *lak* can be linked to resources in *dblp* (e.g., using *owl:sameAs* property). The recommendation of *wordnet* is also justified because resources of *lak* can be linked to the corresponding concepts defined in *wordnet*.

Both approaches presented in the paper (SN-based and Bayesian) are related to the correlation between features and links. Therefore, our approaches could

---

[6] http://www.informatik.uni-trier.de/~ley/db/

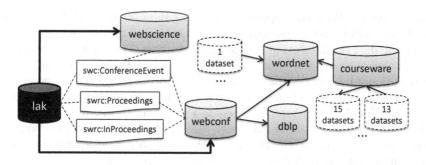

Bayesian Scores (using c=-15)

Partial Result	d_1	d_2	d_3
count(f_1,d_i)	1	1	0
count(f_2,d_i)	1	1	0
count(f_3,d_i)	1	1	0
$\Sigma_{j=1...n}$count(f_j,d_i)	3	3	0
count(d_i)	3	1	13
$P(F_2|D_i)$	0.33	0.33	-
$P(F_2|D_i)$	0.33	0.33	-
$P(F_3|D_i)$	0.33	0.33	-
$P(D_i)$	0.004	0.001	0.019
$|N_i|$	0	0	3
$\log_2(P(F_1|D_i))$	-2	-2	-
$\log_2(P(F_2|D_i))$	-2	-2	-
$\log_2(P(F_3|D_i))$	-2	-2	-
$\log_2(P(D_i))$	-7.86	-9.45	-5.74
score(d_i,t)	**-12.61**	**-14.20**	**-50.74**

f_1=swc:ConferenceEvent, f_2=swrc:Proceedings,
f_3=swrc:InProceedings
d_1=wordnet, d_2=dblp, d_3=courseware
sum(count(d_i))=697

Social Network-based

score(lak, *webscience*) $= 0 + \dfrac{1}{295} = 0.0034$

score(lak, *webconf*) $= 0 + \dfrac{1}{295} = 0.0034$

score(lak, *wordnet*) $= \dfrac{1}{1} + \dfrac{3}{295} = 1.0102$

score(lak, *dblp*) $= \dfrac{1}{1} + \dfrac{1}{295} = 1.0034$

score(lak, *courseware*) $= 0 + \dfrac{13}{295} = 0.0441$

Fig. 1. Example including the datasets links, associated features and the score computation

recommend two datasets that do not share any feature (vocabulary, class and property) as candidates to be interlinked. Considering the example, *lak* and *dblp* have completely different feature sets and yet could be interlinked. As there is *webconf* (that has common features with *lak*) linked to *dblp*, then our approaches can recommend to try to interlink *lak* to *dblp*.

4 Experiments

4.1 Notation and Performance Measures

To motivate how we define the performance measure, recall that the goal of the rank score functions is to reduce the effort required to discover new links from a dataset t. With the appropriate ranking, datasets more likely to contain links from t will be better positioned in the ranking so that the search may be concentrated on the datasets at the top of the ranking. Thus, in the experiments, we evaluated the rank score functions using the Mean Average Precision, which is a traditional Information Retrieval measure [18,19]. Furthermore, we remark that, since the rank score functions induce a ranking of all datasets, the recall is always 100% and is, therefore, not used as a performance measure.

To define the Mean Average Precision (MAP), we adopt the following notation (recall that a dataset u is *relevant* for a dataset t iff there is at least one resource defined in u that can be linked from a resource defined in t):

- D is a set of datasets
- T is a set of datasets, disjoint from D, one wishes to link to datasets in D
- $t \in T$
- G_t is the set of datasets in D with known links from t (the *gold standard* for t)
- $Prec@k_t$ is the number of relevant datasets obtained until position k in a ranking for t, divided by k (the *precision at position* k of a ranking for t)
- $AveP_t = (\sum_k Prec@k_t)/|G_t|$, for each position k in a ranking for t in which a relevant dataset occurs (the *average precision at position* k of a ranking for t)

The *Mean Average Precision* (MAP) of a rank score function over the datasets in T is then defined as follows:

$$MAP = average\{AveP_{t_j} \ / \ t_j \in T \ \wedge \ |G_{t_j}| > 0\} \tag{16}$$

Moreover, in order to evaluate whether the improvements are statistically significant, a paired statistical *Student's T-test* [18,19] was performed. According to Hull [20], the T-test performs well even for distributions which are not perfectly normal. We adopted the usual threshold of $\alpha = 0.05$ for statistical significance. When a paired T-test obtained a p-value (probability of no significant difference between the compared approaches) less than α, there is a significant difference between the compared approaches.

4.2   Dataset

We tested the rank score functions with metadata extracted from the DataHub catalog, a repository of metadata about datasets, in the style of Wikipedia. DataHub is openly editable and can be accessed through an API provided by the data cataloguing software CKAN[7]. The set of data used in our experiments is available at http://www.inf.puc-rio.br/~casanova/ Publications/Papers/2014-Papers/interlinking-test-data.zip and was extracted in April 2013.

---

[7] http://ckan.org

We adopted as features the properties, classes and vocabularies used in the datasets, in different combinations. From the DataHub catalog, we managed to obtain 295 datasets with at least one feature and 697 links between these datasets. The number of distinct features was 12,102, where 10,303 were references to properties, 6,447 references to classes and 645 references to vocabularies; the number of relations between datasets and features was 17,395.

We conclude with brief comments on how we extracted metadata from the DataHub catalog.

Let t be a dataset and V be a set of VoID descriptions [21] for t, available through the catalog. We extracted classes and properties used in t from dataset partitions defined in V, using the *void:class* and the *void:property* properties. We obtained vocabularies used in t from the *void:vocabulary* property. We uncovered links of t from Linkset descriptions associated with t that occur in V. A *void:Linkset* describes a set of triples (s, p, o) that link resources from two datasets through a property p. The *void:subjectsTarget* property designates the dataset of the subject s and the *void:objectsTarget* property indicates the dataset of the object o.

We also extracted links via the catalog API, which exposes a multivalued property, *relationships*, whose domain and range is the complete set of catalogued datasets. In this case, assertions of the form "$t[relationships] = \_node$" and "$\_node[object] = u$" indicate that t is linked to a dataset u.

4.3 Testing Strategy

To evaluate the performance of the rank score functions, we adopted the traditional 10-fold cross validation approach, where a *testing set* is randomly partitioned into 10 equally-sized subsets and the testing process is repeated ten times, each time using a different subset as a *testing partition* and the rest of the objects in the testing set as a *training partition*.

In our experiments, the 295 datasets obtained from the DataHub catalog played the role of the testing set. The 10-fold cross validation then generated 10 different pairs (T_i, D_i), for $i = 1, ..., 10$, of testing and training partitions. The known links between datasets in D_i were preserved, those between datasets in T_i were ignored, and those from datasets in T_i to D_i were used as the gold standard for the datasets in T_i. Each test consisted of computing the MAP for the pair (T_i, D_i). Then, we computed the overall average of the MAPs for the 10 tests, referred to as the *overall MAP* in Section 4.4.

We used the training partition to estimate probabilities, using Equations 10 and 11, when testing the Bayesian approach, and to construct the Linked Data network, when testing the Social Network-based approach.

4.4 Results

This section describes the experiments we conducted to evaluate the rank score functions generated by the two approaches presented in Section 3, referred to as the Bayesian approach and the Social Network-based (SN-based) approach

Table 3. Overall Mean Average Precision

Approach	Feature set			
	properties	classes	vocabularies	all
SN-based	48.46%	**57.18%**	48.27%	51.57%
Bayesian	59.18%	55.31%	51.20%	**60.29%**

(using the rank score function defined in Equation 15). We combined each of the approaches with the following feature sets: (i) only properties; (ii) only classes; (iii) only vocabularies; and (iv) all these three features.

Table 3 depicts the overall MAP results obtained by each combination of approach and feature set. The Bayesian approach using all three features achieved the best performance; the Bayesian approach using properties obtained the second best result; and the SN-based approach using classes was the third best result. In fact, the Bayesian approach obtained better results than the SN-based approach using properties or vocabularies as single features. The worst results obtained by both approaches used vocabularies as a single feature. This probably happened because, in our experiments, we have a restrict number of references to vocabularies in the datasets.

We also calculated the overall MAP of the rank score functions based only on preferential attachment (pa) and resource allocation (ra), using classes as single features. We respectively obtained 43.64% and 44.75%, which are lower than the overall MAP for the rank score function defined in Equation 15.

Finally, we applied a paired T-test to investigate whether there are statistically significant differences between the overall MAP results of the different approaches and selected feature sets. Table 4 shows the p-values obtained by all T-tests performed, where the results is boldface represent differences which are not statistically significant.

The T-test of the SN-based approaches indicate that the SN-based approach using the rank score function defined in Equation 15 and using classes as features outperforms the SN-based approaches using preferential attachment (pa) or resource allocation (ra) and classes as features.

A T-test was also performed for overall MAP results of the SN-based approaches using classes and using the other feature selections. The T-tests indicate that the SN-based approach using the rank score function defined in Equation 15 and classes achieved a statistically significant improvement when compared to all others (using properties, vocabularies and all features). Thus, there are evidences that classes are the best feature selection to be used with the SN-based approach.

For the Bayesian approach, we compared the results obtained by using all features (the configuration with the best overall MAP) with the results obtained using all other feature selections. The T-tests indicate that the overall MAP results of the Bayesian approach using all features and using only properties do not present a statistically significant difference. This suggests that using only properties is an adequate strategy to be adopted with the Bayesian approach.

Table 4. The p values applying T-test

SN-based with *classes*	SN-based with			pa	ra
	properties	*vocabularies*	*all* features		
	5.26E-05	0.00195	0.03683	5.46E-08	1.35E-05
Bayesian with *all* features	Bayesian with			SN-based with *classes*	
	properties	*classes*	*vocabularies*		
	0.10641	0.00408	0.00022	**0.07275**	

We also used a paired T-test to investigate whether there is a statistically significant difference between the overall MAP values obtained by the best configuration for the SN-based approach (using classes) and the best configuration for the Bayesian approach (using all features). The T-tests indicate that there is no statistical difference between the overall MAP results of both approaches.

In conclusion, these observations indicate that the SN-based approach using classes or the Bayesian approach using properties induce the best rank score functions, since they achieve the best results and are simple to compute. This is the main result of the paper.

5 Conclusions

This paper compared two approaches respectively based on Bayesian classifiers and on Social Network Analysis techniques to address the dataset interlinking recommendation problem. Both approaches define rank score functions that explore only metadata features - vocabularies, classes and properties - and the known dataset links. The results show that the rank score functions achieve a mean average precision of around 60%. This means that a dataset interlinking tool might use the rank score functions to limit the search for links from a dataset t to just the top ranked datasets with respect to t and yet find most of the links from t. Thus, the rank score functions are potentially useful to reduce the cost of dataset interlinking.

The computation of the rank score functions depends on harvesting metadata from Linked Data catalogs and from the datasets themselves, a problem shared by other Linked Data techniques, but they are not restricted using only VoID descriptions. This limitation in fact calls attention to the importance of harvesting metadata, that can be carried out in different ways, including the inspection of the datasets by crawlers, a problem we address elsewhere [22], to fulfill the Linked Data promises.

Finally, we plan to further improve the definition of the rank score functions. One generic strategy is to improve the network analysis-based score by considering the frequency of the schema elements. Often two datasets share similar classes and properties, but they strongly differ on the number of instances. Another aspect to explore would be feature similarity (e.g., string similarity between two features), rather than just considering the intersection of the feature sets.

Acknowledgments. This work was partly funded by the LinkedUp project (GA No:317620), under the FP7 programme of the European Commission, by CNPq, under grants 160326/2012-5, 303332/2013-1 and 557128/2009-9, by FAPERJ, under grants E-26/170028/2008, E-26/103.070/2011 and E-26/101.382/2014, and by CAPES, under grant 1410827.

References

1. Berners-Lee, T.: Linked Data. In: Design Issues. W3C (July 2006)
2. Leme, L.A.P.P., Lopes, G.R., Nunes, B.P., Casanova, M.A., Dietze, S.: Identifying candidate datasets for data interlinking. In: Daniel, F., Dolog, P., Li, Q. (eds.) ICWE 2013. LNCS, vol. 7977, pp. 354–366. Springer, Heidelberg (2013)
3. Lopes, G.R., Leme, L.A.P.P., Nunes, B.P., Casanova, M.A., Dietze, S.: Recommending tripleset interlinking through a social network approach. In: Lin, X., Manolopoulos, Y., Srivastava, D., Huang, G. (eds.) WISE 2013, Part I. LNCS, vol. 8180, pp. 149–161. Springer, Heidelberg (2013)
4. Nikolov, A., d'Aquin, M.: Identifying Relevant Sources for Data Linking using a Semantic Web Index. In: WWW2011 Workshop on Linked Data on the Web, Hyderabad, India. CEUR Workshop Proceedings, vol. 813. CEUR-WS.org (March 29, 2011)
5. Nikolov, A., d'Aquin, M., Motta, E.: What Should I Link to? Identifying Relevant Sources and Classes for Data Linking. In: Pan, J.Z., Chen, H., Kim, H.-G., Li, J., Wu, Z., Horrocks, I., Mizoguchi, R., Wu, Z. (eds.) JIST 2011. LNCS, vol. 7185, pp. 284–299. Springer, Heidelberg (2012)
6. Kuznetsov, K.A.: Scientific data integration system in the linked open data space. Programming and Computer Software 39(1), 43–48 (2013)
7. Mühleisen, H., Jentzsch, A.: Augmenting the Web of Data using Referers. In: WWW2011 Workshop on Linked Data on the Web, Hyderabad, India. CEUR Workshop Proceedings, vol. 813. CEUR-WS.org (March 29, 2011)
8. Lóscio, B.F., Batista, M., Souza, D.: Using information quality for the identification of relevant web data sources. In: The 14th International Conference on Information Integration and Web-Based Applications & Services, IIWAS 2012, Bali, Indonesia, December 3-5, pp. 36–44. ACM, New York (2012)
9. Wagner, A., Haase, P., Rettinger, A., Lamm, H.: Discovering related data sources in data-portals. In: Proceedings of the First International Workshop on Semantic Statistics, Co-located with the the International Semantic Web Conference (2013)
10. de Oliveira, H.R., Tavares, A.T., Lóscio, B.F.: Feedback-based data set recommendation for building linked data applications. In: I-SEMANTICS 2012 - 8th International Conference on Semantic Systems, I-SEMANTICS 2012, Graz, Austria, September 5-7, pp. 49–55. ACM (2012)
11. Toupikov, N., Umbrich, J., Delbru, R., Hausenblas, M., Tummarello, G.: Ding! dataset ranking using formal descriptions. In: Proceedings of the WWW2009 Workshop on Linked Data on the Web, LDOW 2009, Madrid, Spain. CEUR Workshop Proceedings, vol. 538. CEUR-WS.org (April 20, 2009)
12. Witten, I.H., Frank, E., Hall, M.A.: Data Mining: Practical Machine Learning Tools and Techniques. Morgan Kaufmann (January 2011)
13. Manning, C.D., Schütze, H.: Foundations of Statistical Natural Language Processing. MIT Press (2002)

14. Lü, L., Jin, C.H., Zhou, T.: Similarity index based on local paths for link prediction of complex networks. Physical Review E 80(4), 046122 (2009)
15. Liben-Nowell, D., Kleinberg, J.: The link-prediction problem for social networks. J. Am. Soc. Inf. Sci. Technol. 58(7), 1019–1031 (2007)
16. Caraballo, A.A.M., Nunes, B.P., Lopes, G.R., Leme, L.A.P.P., Casanova, M.A., Dietze, S.: Trt - a tripleset recommendation tool. In: Proceedings of the ISWC 2013 Posters & Demonstrations Track, Sydney, Australia. CEUR Workshop Proceedings, vol. 1035, pp. 105–108. CEUR-WS.org (October 23, 2013)
17. Taibi, D., Dietze, S.: Proceedings of the LAK Data Challenge, Leuven, Belgium, April 9. CEUR Workshop Proceedings, vol. 974. CEUR-WS.org (2013)
18. Baeza-Yates, R.A., Ribeiro-Neto, B.A.: Modern Information Retrieval - the concepts and technology behind search, 2nd edn. Pearson Education Ltd., Harlow (2011)
19. Manning, C.D., Raghavan, P., Schütze, H.: Introduction to Information Retrieval. Cambridge University Press (July 2008)
20. Hull, D.: Using statistical testing in the evaluation of retrieval experiments. In: Proceedings of the 16th Annual International ACM SIGIR Conference on Research and Development in Information Retrieval, SIGIR 1993, pp. 329–338. ACM, New York (1993)
21. Alexander, K., Cyganiak, R., Hausenblas, M., Zhao, J.: Describing Linked Datasets with the VoID Vocabulary. W3C (March 2011)
22. do Vale Gomes, R., Casanova, M.A., Lopes, G.R., Leme, L.A.P.P.: A metadata focused crawler for linked data. In: Proceedings of the 16th International Conference on Enterprise Information Systems, ICEIS 2014, Lisbon, Portugal, April 27-30, vol. 2, pp. 489–500. SciTePress (2014)

Exploiting Perceptual Similarity:
Privacy-Preserving Cooperative Query Personalization

Christoph Lofi and Christian Nieke

Technische Universität Braunschweig
Mühlenpfordtstr. 23, 38114 Braunschweig, Germany
{lofi,nieke}@ifis.cs.tu-bs.de

Abstract. In this paper, we introduce privacy-preserving query personalization for experience items like movies, music, games or books. While these items are rather common, describing them with semantically meaningful attribute values is challenging, thus hindering traditional database query personalization. This often leads to the use of recommender systems, which, however, have several drawbacks as for example high barriers for new users joining the system, the inability to process dynamic queries, and severe privacy concerns due to requiring extensive long-term user profiles. We propose an alternative approach, representing experience items in a perceptual space using high-dimensional and semantically rich features. In order to query this space, we provide query-by-example personalization relying on the perceived similarity between items, and learn a user's current preferences with respect to the query on the fly. Furthermore, for query execution, our approach addresses privacy issues of recommender systems as we do not require user profiles for queries, do not leak any personal information during interaction, and allow users to stay anonymous while querying. In this paper, we provide the foundations of such a system and then extensively discuss and evaluate the performance of our approach under different assumptions. Also, suitable optimizations and modifications to ensure scalability on current hardware are presented.

Keywords: Personalized Query Processing, Privacy in Information Systems.

1   Introduction

Effective personalization techniques have grown to be an integral and indispensable part of current information systems, and are essential to support users when faced with a flood of different choices. Here, two major approaches are common: a) Using SQL-style personalized queries on meta-data, which unfortunately require users to have extensive domain knowledge in order to formulate precise and efficient queries. Additionally, SQL-style queries are difficult for the large domain of *experience items* like movies, books, or music, as the commonly available meta-data is often not describing the items in a suitable fashion (e.g. if they are suspenseful, funny, or romantic.) b) Adapting recommender systems which proactively suggest items to users based on their user profile, and which became particularly popular in systems like

B. Benatallah et al. (Eds.): WISE 2014, Part I, LNCS 8786, pp. 340–356, 2014.
© Springer International Publishing Switzerland 2014

Amazon or Netflix [1]: While many recommender systems provide recommendations of high quality [2], they have several shortcomings. Especially, for *each user* an elaborate user model needs to be built and stored, requiring up to hundreds of ratings until a user can get meaningful recommendations. This creates a high barrier for new users to join the system. But more severely, this user model contains exhaustive personal information on a user's preferences, her reaction to different items, or her general likes and dislikes. In order to query or use the system, this information must be *clearly associated* with the respective user and needs to be *stored long-term*. Such profiles are highly valuable, and can easily be commercialized, abused, or even stolen. This situation raises many privacy concerns, and repels privacy-conscious users.

In this paper, we therefore present an alternative approach combining advantages of both recommender systems and SQL-based database personalization techniques, while at the same time avoiding many of the associated privacy risks. We realize this with privacy-preserving *query-by-example personalization*, which allows users to query for items fitting their current preferences easily without providing explicit feedback on attributes or their values. In order to obtain meaningful attribute values for each database object, we rely on *perceptual spaces* [3] which encode the implicitly perceived properties of each item, allowing to measure the consensually *perceived similarity* between given items. Similar to recommender systems, this perceptual information is mined from user-item ratings. However, we avoid the drawbacks of recommender systems: no user profiles are necessary to query the system, allowing situative, personalized, and anonymous ad-hoc queries.

In summary, our contributions in this paper are as follows:

- We present privacy-preserving queries for experience products using example-based queries on *perceptual spaces*. Our query-centered personalization is based on an adaption of *Bayesian navigation*.
- We discuss the advantages of our approach over SQL-style-based personalization and recommender systems focusing specifically on *privacy concerns* and *ease-of-use*
- We address the *performance and scalability* issues resulting from adapting Bayesian Navigation with *significant improvements* to the query execution process
- We evaluate the effectiveness of our approach in an *extensive user study*
- We evaluate the effects of our proposed algorithmic optimizations on the system's *performance, scalability* and *result quality*, and show that our approach can indeed be scaled to satisfy the demands of modern information systems.

2   Foundations and System Design

Experience products like movies, games, but also restaurants or hotels are items which are mostly characterized by non-hard attributes, i.e. they are subjectively experienced by their consumers and it is hard to find explicit attributes describing this resulting experience properly (e.g., a movie is considered humorous in a dry sense, as opposed to slapstick). Our approach is intended to allow users to easily explore a database with experience products by using personalized and privacy preserving

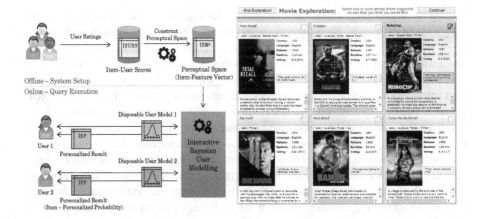

Fig. 1. Basic System Design

Fig. 2. Screenshot of Prototype 2nd display after user provided "The Terminator (1984)" as start example

query-by-example (QBE) navigation in order to avoid directly interacting with attribute values.

The intended semantics of our approach are complementary to both SQL-style personalization as well as to recommender systems: SQL-style personalization offers powerful queries using the usually available meta-data (like actors or genres for movies) and is suitable for users *who know exactly what they are looking for* (i.e. users need good domain knowledge and must be able to formalize a precise query). SQL queries will provide exact results and do not require any form of user profiling. In contrast to that, recommender systems or cross-selling systems focus on the case that *a user does not know what she is looking for*, and the system proactively presents suitable database items. In order to realize this, these systems need to "get to know" each user, i.e. they need to learn each user's likes, and dislikes. A scenario where a group of friends spontaneously decides to watch a movie would require to create an account for the group and vote on up to hundreds of movies, while our system allows to create a temporary profile on-the-fly while browsing. Additionally, the information aggregated in permanent user profiles can become very extensive and poses a serious threat to privacy. Furthermore, systems run into the risk of over-personalization by recommending only those items the system believes a user will like, which in turn makes it more likely that users consume and provide feedback only on similar items, locking them tightly into a "filter bubble" [4].

Our approach is in the middle-ground between both: the user poses a simple query by giving a vague example of what she is looking for, and can navigate through items selected and displayed due to their perceptual similarity to the example, by simply pointing out good suggestions in the display. It adopts a browsing behavior similar to physical book or video stores with "I know what I am looking for when I see it"-semantics. No direct interaction with attribute values is necessary. This type of query occurs quite often naturally, i.e. "There is a movie which I found interesting (because

I liked it, or it was interesting in a way, or me and my friend both liked it, etc.), and now I am interested in more like that".

Accordingly, two major challenges are discussed in this paper: a) How can experience items be represented in a high-dimensional feature space such that meaningful similarity measurements and QBE navigation are possible? b) How can we personalize an example-based query in such a way that it respects the user's feedback actions and privacy?

Most popular QBE approaches in multimedia databases tried to operate on features extracted from the actual multimedia file itself, which could be low-level features (e.g. color histograms or pattern-based features), or so-called high-level features as for example in scene composition [5] or content-based semantic features [6] (e.g., presence of explosions or a flag, etc.). Here, our approach takes a completely different route, as our features result from external user ratings instead of being extracted from the media. Such information has been shown to be very informative, and semantically more meaningful to users than traditional meta data as, e.g. information about the director or actors (as shown in e.g. [7] for movies). In this paper, we demonstrate how such semantically rich rating data can represent each item of an experience product database within a high-dimensional feature space. The idea is that the resulting space implicitly encodes how users perceived a movie, e.g., if it was funny, or if certain plot elements or tropes were present. For this task, we adapt *perceptual spaces*. Perceptual spaces have been introduced in [3], and are built on the basic assumption that each user who provides ratings on items has certain personal interests, likes, and dislikes, which steer and influence her rating behavior [8]. The resulting general system design of our approach is shown in Figure 1: in an offline system initialization phase, a large number of user ratings is processed into a perceptual space, and then our adapted Bayesian Navigation approach is used to personalize user queries by eliciting short term user profiles which are discarded after the query.

2.1   Personalization and Privacy

Privacy concerns severely impact a user's overall satisfaction with a Web-based system (as argued in [9]), and might even prevent them from using it altogether, if the balance between privacy concerns and perceived system utility becomes unfavorable. The focus of our system in terms of privacy is to allow all users to use the personalized query capabilities without requiring a user profile or pre-query preference elicitation. Especially, this means that browsing or querying our system requires no *long term user profiles* (unlike recommender systems), but only *temporary query profiles*, thus removing the need to store and protect this sensitive information. A single query profile will usually not be enough to extract a sufficiently distinctive pattern to identify a user, as it is not connected to other profiles (or a user id) and is only of the form: "an (anonymous) user wanted to start his query with 'The Terminator', and then selected 'Conan' out of a small set of movies proposed by the system" – it provides little insight into the wider preferences of a user.

But still, our system will require a small group of enthusiast users to provide identifiable rating data in order to construct the perceptual space. However, this construction process is completely decoupled from executing queries, and the perceptual space itself does not contain any user related information, not even in an anonymized or masked form. Even just the number of users that participated in its creation are not included. It is basically just a matrix of movie ids and their major perceptual dimensions (n=100 in our case). Therefore, approaches de-anonymizing ratings similar to the ones detailed in [10] cannot be applied. This fact could also allow a "trusted platform" like MovieLens to use its users' ratings to construct a perceptual space, which then could be used by another system like ours. In contrast to publishing anonymized rating data, publishing a perceptual space carries only minimal risks to the user's privacy. But in any case, even if users did decide to contribute ratings to build the space, all users can use the query capabilities of our system without leaving trails of personal information in an ad-hoc fashion.

2.2   Related Work

Content-based retrieval [11] and Query-by-example-based approaches [12, 13] have been very popular during the late 90s in the context of multi-media databases. The paradigm's main selling point is that querying high-dimensional data becomes very simple as it only requires the user to give a starting example and some easy-to-elicit feedback. However, QBE became less popular in recent years as, according to [14], the features used were often unsufficient, as discussed before. Also the Bayesian Retrieval approach [15] adapted to our system   was originally designed to work on image color histograms. Here, our approach takes a completely different route, as our feature space results from the reactions of users after consuming the media instead of being extracted from the media itself. Feature spaces constructed from ratings, like our perceptual space [3], have also been explored, as for example in [16]. However, in contrast to these works, we complement such feature spaces with personalized query capabilities, and discuss and address the design issues resulting from integrating such an approach into an information system. Approaches based on different flavors of Bayesian modeling have also been employed in recommender systems. In [17], users rate a small set of movie trailers in order to generate a user mood profile using Bayesian modeling, which later is integrated into a long-term user profile. However, they do not offer QBE functionality.

Approaches for privacy preservation in recommender systems mostly focus on the protection of the potentially vulnerable rating data. Here, many different approaches have been developed to either anonymize or encrypt the rating data to protect it from misuse by either the recommender platform itself, or from malicious attacks by 3rd parties. In most of these solutions, there is an inherent trade-off between privacy, accuracy, and efficiency [18]. For example in [19], user data is perturbed by adding fixed-distribution random values to each user rating, therefore hindering subsequent user identification, but also decreasing the quality of recommendations. A similar notion is followed by differentially-private recommender systems like [20], which add noise to the item similarity matrix in order to obfuscate the ratings originally provided.

2.3   Perceptual Spaces

Perceptual spaces have been introduced and formalized in [3], we therefore only briefly summarize the most relevant aspects in this section. Perceptual Spaces exploit the bias of users when rating items, which is influenced by personal likes and dislikes towards properties of the rated item. They heavily rely on factor models, a technique popular in recommender systems research [2]. Factor models have originally been developed to estimate the value of non-observed ratings for the purpose of recommending new (yet unrated) items to existing users, but can also be beneficial beyond recommendation tasks [21]. We assume that a perceptional space is a d-dimensional space as follows: each user and each item is represented as a d-dimensional numeric vector. The vector of a user represents her personality, i.e., the degree by which she likes or dislikes certain characteristics. Likewise, item vectors represent the degree to which an item shows the same characteristics. Items which are perceived similarly in some aspect have similar vectors. As we are only interested in items, user vectors are not stored and are discarded in later stages. Furthermore, each user rating can be seen as a function of the user vector and item vector. This assumption reflects established models of human preferences and is well-accepted in recommender systems research [2]. All model parameters are estimated by minimizing a cost function measuring the deviation between the actual observed ratings and those predicted by the model. By formulating this as an optimization problem, user and item vectors can be found that fit the given rating data best.

Formally, a large and sparse user-item-rating matrix is given, containing only rating for around 1-2% of all user-item pairs. The goal is to find a matrix $A = (a_{m,k}) \in \mathbb{R}^{n_M \times d}$ representing movies as d-dimensional coordinates. To achieve this, we also need a helper matrix $B = (b_{u,k}) \in \mathbb{R}^{n_U \times d}$, representing user in the same space. Then, we use a factor model representing a rating function $f : \mathbb{R}^d \times \mathbb{R}^d \to \mathbb{R}$. Basically, this function can predict missing ratings given user and item vectors. We approximate this function and the involved vectors/matrices, we use Euclidian Embedding (as in [22]), and we want the distance between a movie vector a_m and user vector b_u to be small if user u likes movie m; otherwise, it should be large. To account for general effects independent of personal preferences, for each movie m and user u, we introduce the model parameters δ_m and δ_u, which represent a generic movie rating bias relative to the average rating μ. Then, a rating of a movie m by a user u can be predicted by $\hat{r}_{m,u} = \mu + \delta_m + \delta_u - dis_E^2(a_m, b_u)$, i.e. the average rating of all movies (e.g., μ=6.2 out of 1..10) plus the user bias (e.g., δ_u=-1.6 representing a critical user always rating worse than others) and the movie bias (e.g., an overall good movie with an average rating of 8.4, so δ_m=2.2). The last term, $dis_E(\cdot, \cdot)$, represents the distance of the movie vector and the user vector in a d-dimensional space. Finally, all movie vectors (and therefore the matrix A) are approximated by solving a least squares optimization problem with all instances of the above equation for which a rating is known (including a correction for noise). The sult represents our perceptual space.

Unfortunately, the resulting features in this space are implicit and have no direct real-world interpretation, and therefore not suitable for SQL-style queries. However, they allow for measuring perceived similarity effectively (i.e. the distance between the feature vectors). This now allows using the query-by-example paradigm, which provides simple query formulization without the need to explicitly refer to any features.

2.4   Basic Bayesian Retrieval

For approaching query-by-example personalization, we will adapt Bayesian Retrieval as shown in e.g. [15]. Such approaches have been successfully used in multimedia databases research, but have also been adapted to general database retrieval [23]. In short, our approach aims at computing for each database object the probability of the user being interested in it considering the feedback she has provided on a selection of items during a cooperative preference elicitation. We will only briefly summarize the basic theory of Bayesian retrieval and user modeling in this subsection, before highlighting the modifications necessary to adapt the model to our context in the next section, and discussing techniques for improving the computational and memory performance of the approach in section 4. From a user's perspective the interaction style of this approach is similar to non-personalizing similarity navigation (i.e. repeatedly navigating from one item to similar ones), but we will show that incorporating Bayesian personalization will lead to significantly better and quicker interaction performance, outshining similarity navigation.

Elicitation of user preferences is performed interactively over several *steps*. In each step $t = 1,2,...$, the user will be shown a set of objects D_t (the so-called *display*) from the database containing our previously computed perceptual space with n_M objects. The choice of objects in D_t depends on a *selection strategy* (discussed later), which relies on analyzing the probability estimates for each database object representing the belief that it is the one the user is looking for given the current interaction history. This is formalized as follows: the database objects are denoted as $O_1, O_2, ..., O_n$. Each object O_i is annotated with a probability of being the user's target O. Here, the target is the best suited object in the database to fulfill the user's current needs, which is of course yet undefined.

After reviewing the objects, the user will provide feedback on the display's items in form of a user action A_t. In our case, this is simply selecting any items that "look right" (similar to judging covers in a video store). The initial *a priori estimate* before starting the user interaction for each object O_i will be denoted as $P(O = O_i)$. Again, several strategies are viable for providing this *startup distribution* of the a priori estimates (discussed later). After the interaction has started, the probabilities are updated to *a posteriori estimates* respecting the current *interaction history* denoted as: $H_t = A_0, D_0, A_1, D_1, ..., A_t, D_t$. Since D_t is deterministically given by the selection strategy and H_{t-1} is known, we arrive at the following formula (see [15] for details):

Simplified Bayesian Update: For each O_i in the perceptual space:

$$P(O = O_i \mid H_t) = P(O = O_i \mid D_t, A_t, H_{t-1})$$
$$= \frac{P(A_t \mid O = O_i, D_t, H_{t-1}) \, P(O = O_i \mid H_{t-1})}{\sum_{j=1}^{n_M} P(A_t \mid O = O_j, D_t, H_{t-1}) \, P(O = O_j \mid H_{t-1})}.$$

After each user feedback, this update operation has to be performed for all items in the perceptual space, thus forming the *user model* representing in which items the current user is likely interested. This is represented in the model by all a-posteriori probabilities of all items $P(O = O_i \mid H_t)$ with $1 \leq i \leq n_M$.

The term $P(O = O_j|H_{t-1})$ can be computed recursively until the a priori approximation given by the start distribution is reached. But the central term within the previous equitation $P(A_t \mid O = O_i, D_t, H_{t-1})$ remains difficult, i.e. the probability that the user will actually perform the current action A_t given that the current database object O_i is indeed the target O given the history H_{t-1}. This term predicts a user's action considering all currently known information, and is provided via a *user prediction model* (next section).

3   Adapting Bayesian Retrieval

For adapting Bayesian retrieval to our usage scenario, we need to create a suitable user prediction model, startup distribution, and selection strategy.

The **user prediction model** provides an implementation for evaluating the term $P(A_t \mid O = O_i, D_t, H_{t-1})$ from the previous section. This makes up the "semantic core" of the calculation and is essential in determining whether the calculated probabilities will actually correctly represent the users' preferences. To allow an easy and intuitive interaction with the system, we opt for item-based feedback during a user feedback cycle, meaning that a user simply selects any number of items from the display $D_t = \{X_1, \ldots, X_{n_D}\}$ she wants to use as positive examples for further personalization. n_D is a system parameter denoting how many items are shown in each display, and can be adjusted to the type of items and display device which is used (we used $n_D = 9$ in our Web-based prototype). Then, following [15] we also take a soft-min approach for item-based feedback, assuming that the user behaves time invariant (i.e. her decisions are based on her implicitly known target object and we can drop H_{t-1}). Accordingly, the probability for each decision on each single object X_a from the display can be modeled as:

$$P_{soft}(A = a|X_1, \ldots, X_{n_D}, O_j) = \frac{\exp(-d(X_a, O_j))}{\sum_{i=1}^{n_D} \exp(-d(X_i, O_j))}$$

In this formula, $d(X_a, O_j)$ denotes the *distance* between item X_a and the target object O_j and the approach yields highest values for those items X_a closest to the target O_j. The quality of this approach is of course highly dependent on choosing a meaningful metric for the distance, which corresponds to the similarity of the objects. We use the *Euclidian distance measured in the perceptual space*, which is a significant contribution as it provides a semantically meaningful high-dimensional feature space for augmenting regular database objects, and distances measured in this space represent the *consensual subjective semantic similarity* between the objects elicited from a large number of users. These measures are significantly more meaningful than similarity measured on typical meta-data usually available in information systems [3]. Using the assumption of independent decisions, a combined decision can be calculated by multiplying the probabilities of each single decision, e.g. $P_{soft}(A = 1, 2) = P_{soft}(A = 1) * P_{soft}(A = 2)$.

The **selection strategy** decides which objects are included in a display D_i for each feedback cycle t_i and is important to allow a satisfying interaction with the system. Basically, two major approaches are possible here:

a) *Most-probable strategies* select objects with the highest a posteriori ty $P(O = O_i | H_t)$. These selection strategies tend to favor similar objects at the beginning of the interaction, and at a first glance might resemble similarity search for the first one or two displays. However, after few feedback cycles, this strategy will be able to cross larger distances in the space depending on user input. But still, in its naïve form, this strategy is prone to getting stuck in clusters repeating the same objects over and over.

b) *Most-informative strategies* aim at maximizing the information gain in each feedback step [15] by diversifying the display (i.e. displaying those items which would impact the a posteriori probabilities most). This usually leads to higher navigation speeds, and users can traverse the space quickly with only few feedback steps.

While the most-informative strategy shows superior performance (i.e. it needs less user interactions to find a certain target in the space), we found in our pre-study that users were easily confused by the resulting displays, as the "long distance links" to yet untouched areas of the space were perceived as mistakes of the algorithm (e.g. presenting the family movie "Finding Nemo" after a user selected the action movie "The Terminator").

We therefore opted for adapting the most-probable strategy to our needs, sacrificing some interaction speed for increased user satisfaction. In order to avoid being caught in local maxima or clusters, we only consider objects which have not yet been displayed to the user, resulting in the *most-probable unseen strategy*.

The **startup distribution** defines the a-priori probability $P(O = O_i)$ for each database object O_i of being the target object O before the user starts interacting, which can alleviate the cold start problems of the not yet personalized system. Without any additional assumptions, the naïve approach is a uniform startup distribution with $P(O = O_i) = 1/n$ for each $O_i \in \{O_1, ..., O_n\}$. Alternatively, additional information on the database objects could be incorporated into the startup distribution as for example average user ratings or popularity measures. In our prototype implementation, we initialized the system with a simple uniform startup distribution, modified by an explicit, user-provided example. For this, a single feedback-step is transparently executed during startup (i.e. the user does not see this first feedback step). Here, we evaluated two approaches:

Free Example: The user freely provides an example to start the query and we simulate the first user action A_0 as selecting the example from a display D_0 which contains only two items: the example and the movie in the dataset with the maximal distance to that selected movie. Therefore, the first display the user will see is D_1 which is already strongly influenced by the start example.

Supported Free Example: In our pre-studies, we found that some users had a hard time thinking of a good example for starting the query. We therefore presented a set of 12 popular example movies hand-picked from different genres, allowing the user to either provide her own example or pick one of ours. D_0 is then made up of these 12 examples plus the optional freeform example.

4   Mastering Performance Issues

For performance evaluation, we focus only on costs for executing individual user queries, as the costs for maintaining the perceptual space are negligible. The perceptual space is not impacted enough by adding single ratings to justify the effort for continuous updates.

In contrast, query performance raises several demanding issues: Bayesian retrieval as introduced in the last sections requires storing the a posteriori probabilities for each database object individually in the user model of each user. Furthermore, for each user interaction, all probabilities in that user's model need to be updated, and each update of a single probability requires considering all other probabilities. Clearly, this situation presents severe challenges to scalability with respect to the number of objects in the perceptual space n_M, but also to the number of concurrent system users n_{UC}. Both memory space for storing the user models as well as computation time are negatively impacted. Here, the required memory is in $O(n_{UC} * n_m)$, while the computation effort per query is in $O(n_{UC} * n_m^2)$. By simply caching the denominator term in the Bayesian update formula in 2.4, the resulting algorithm can be brought to linear complexity in $O(n_{UC} * n_m)$.

Therefore, we introduce *locality-restricted Bayesian updates* in this section. While these efforts do not improve the theoretical worst-case complexity, they drastically improve the actual time and memory needed for executing a query, making our approach feasible on currently available hardware, even for larger perceptual spaces and many concurrent users. The basic idea is to restrict the probability updates to the *relevant* parts of the perceptual space, i.e. those objects that are close to those selected before. If a user started her query with a family comedy, we can ignore horror movies unless her feedback indicates otherwise by steering toward that direction. This allows us to extend the area under observation slowly and in a directed fashion, without affecting semantics too much.

We implement this by introducing, for each user, the set M_j for memorizing the relevant part of that user's user model, i.e. the objects that are close to formerly chosen objects, which therefore have a high a posteriori probability after interaction step t_j. As the remaining objects are far from the objects that are currently considered to be likely, they will have a low probability and can be ignored until the exploration leads to their section of the space. The set M_{-1} is initialized with the example object selected by the user and its δ_N nearest neighbors in the perceptual space. The parameter δ_N can be freely chosen during system setup, and we will show the effects for different δ_N's in the evaluation section. After each user interaction t_j, the previous set M_{j-1} is expanded by adding the δ_N nearest neighbors in the perceptual space of each display object X that was selected by the user in t_j. As the probabilities of the newly added objects are still unknown, we will heuristically assume that their probability is similar to the closest neighbor already in the set M_{j-1} and use the known probability of this neighbor to initialize the new object.

After all new objects were added to the set M_{j-1}, the locality-restricted Bayesian update computes the new set M_j by calculating the a posteriori probabilities of each object and their respective probabilities in M_{j-1} similar to its original version, but ignoring all objects not being in M_{j-1}. After that, M_{j-1} can be discarded and we continue with the extended set M_j. *Locality-Restricted Bayesian Update at* t_j:

For each $O_i \in M_{j-1}$: $P(O = O_i \mid H_t) =$

$$\frac{P(A_t \mid O = O_i, D_t, H_{t-1}) \, P(O = O_i \mid H_{t-1})}{\sum_{O_k \in M_{j-1}} P(A_t \mid O = O_k, D_t, H_{t-1}) \, P(O = O_k \mid H_{t-1})}$$

with $M_{-1} = \{\delta_N NN(selected(A_0))\}$ and $M_i = M_{i-1} \cup \delta_N NN(selected(A_i))$

In our locality-restricted variant, the actual effort needed to store or update the user model after each interaction is clearly reduced as the set M_i is significantly smaller than the whole perceptual space. But this also incurs a penalty on the semantics of the approach, as relevant objects, which would have been assigned a high probability using a non-restricted approach, might not yet be part of the latest set M_i, and are therefore ignored.

Of course finding nearest neighbors can be very expensive as well, but in our current prototype, we simply materialize a full index of the δ_N-nearest neighbors of each object offline when importing a new perceptual space, which even for 100k objects is easily feasible even on low-end machines. This provides us with near instant access to the nearest neighbors independent of the size of the perceptual space.

5   Evaluations

In this section, we present extensive evaluations of our privacy-preserving query-by-example personalization on perceptual spaces. In the first set of evaluations focusing on usability and semantics, we use a real-life dataset with movie ratings for initializing the perceptual space, and have real users interact with a prototype implementation. In addition to this study, simulations focusing on performance aspects of the approach under different assumptions and parameters were performed. Also, these simulations are run on increasingly larger datasets to analyze scalability issues. We close the section with a discussion of the results and their adaptability to other domains.

The perceptual space used in our real-word data experiment is based on the dataset released during the Netflix Prize challenge [1], and consists of 103M ratings given by 480k users on 17k video and movie titles. This dataset is still one of the largest user-item-rating datasets which are available to the community. All titles are from 2005 and older. We filtered out all TV series and retained only full feature movies for our evaluation, leaving 11,976 movies. The initial construction of the 100-dimensional space took slightly below 2 hours on a standard notebook computer.

For simulations on synthetic data, we randomly generated spaces with 100-dimensions and a varying amount of database items assuming an independent uniform distribution of the attribute values. Although real perceptual spaces are usually not uniformly distributed, our experiments comparing the real perceptual space with the uniformly generated ones showed that the results are close enough to allow drawing some conclusions about performance from simulations to real-world behavior (see section 5.2). Our prototype was implemented in Java 7, and the perceptual space as well as all probabilities computed for the user models were held using the H2 Main Memory DBMS. Our prototype ran on a notebook computer with a Core-i7 at 2.4GHz and 10GB of main memory usable for Java.

5.1   User Study

Our user study was performed using the CrowdFlower.com crowdsourcing platform to recruit participants. The users were redirected to our Web-based prototype implementation (screenshot in Figure 2 in section 2), answered a survey afterwards, and were finally paid for their efforts. 179 users participated, and as we allowed users to perform additional (unpaid) interactions, this resulted in 288 completed queries. In the survey concluding the evaluation, we asked users for feedback on a number of statements, on a scale from 1: strongly disagree, 3: neutral, to 5: strongly agree, and we counted 4 and 5 as explicit agreement and 1 and 2 as explicit disagreement. Our test users were fairly evenly distributed over different age classes, gender, movie knowledge and experience with existing movie subscription system like Netflix or Amazon Instant.

We asked the users to test the system in an explorative fashion with the task of deciding on a suitable movie for an imaginative movie night in mind. We allowed them to start the interaction with an example of their choice, but additionally supported this with displaying a selection of 12 manually selected examples for inspiring the choice of a start example. Users were then able to explore the movie space using Bayesian Navigation as they saw fit, and could end the interaction whenever they felt they had seen enough.

After querying, 74% of all users explicitly claimed to have found a suitable movie, but only 6% explicitly disagreed (the other users were neutral). However, this statement has to be taken carefully as our users often had certain expectations of what movie they should discover when certain examples and feedback were provided. As our dataset was restricted to movies released in or before 2005, many users complained that they expected newer movies in the display which were simply not part of our dataset (e.g. they expected to find "The Hobbit (2012)" when providing "The Lord of the Rings 1 (2001)" as a start example).

To assess the semantic quality of each feedback cycle's display, we asked our test users if the screens presented by the system were a good fit for the feedback they had given, and got 60% explicit agreement and only 12% explicit disagreement to this statement. While this does not allow to quantify how close our displays and query results are to the "best possible ones", it shows that users felt that they were "good enough".

When asked if they prefer our approach over the regular query approaches as used by e.g. Amazon Instant or Netflix (recommendations with SQL-style querying and most-popular-in-category lists), 54% of all participants stated that they explicitly prefer our approach, while only 9% explicitly disagreed. However, when being asked whether they think if this approach is a valuable addition to current state-of-the-art

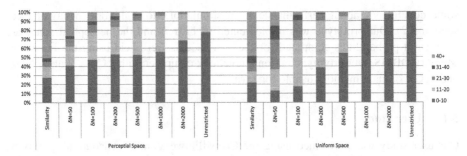

Fig. 3. Simulated Query Runtimes for $n_m = 11,976$ in number of displays that needed feedback to reach a randomly selected target

interaction paradigms, the explicit agreement increased to 73% of all users, while only 9% think our approach is not a useful extension. This result underlines our notion that personalized example-based queries are a very valuable addition to SQL-style filtering, categorization, and recommender systems, but of course are not a full-fledged replacement. SQL-style queries cover the case where a user knows exactly what she is looking for, while recommender systems proactively recommend an item, often even without a query. Our approach is in the middle ground, where users have some intuition about their preferences, but still cannot precisely formulate them as required for an SQL query.

5.2 Performance Simulation

In this section, we evaluate our approach using simulated user interactions on the Netflix perceptual space and on different artificially created ones. As it is rather hard to evaluate the semantic usefulness of an explorative query paradigm in simulations, the main purpose of this section is to showcase the impact of different factors on the systems performance. Especially, we are interested in the scalability of our approach and the impact of our proposed heuristic on memory and CPU consumption to show the expected runtime performance when applied to big datasets.

Therefore, as natural browsing behavior is hard to model, we chose a similar approach as in [23]. In this type of simulation, a particular movie has to be found as quickly as possible, starting from the initial example. While this setting does not represent a typical user's behavior, in addition to assessing performance and scalability, this setup is also able to provide some insight into the speed in which the space could theoretically be traversed.

In order to perform the experiments, we randomly select a movie from the database to be a user's target movie, i.e. the intended perfect match for the user's current preferences which has to be found. The simulation algorithm then picks the best choice of a given feedback display, which is the movie with the smallest distance to the target in the perceptual space, and continues until the target appears in the feedback display. The simulation only had a limited set of start example movies to choose from, depending on the experiment either the 12 handpicked movies we used for our

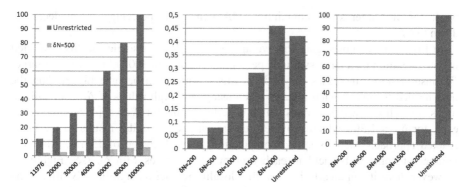

Fig. 4. Memory for User Profiles y: #1000 items stored; x: size n_M

Fig. 5. Time per Update with $n_M = 100ky$: time in sec

Fig. 6. Memory for User Profiles $n_M = 100ky$: #1000 items stored

supported free example startup when using the Netflix perceptual space, or the 12 centroid objects of the clusters resulting from a k-means-clustering of the artificial spaces. For all experimental results presented in the following, we performed at least 1,000 independent simulation runs.

To assess the effectiveness of our approach in this simulation setting, we measured how many displays (with 9 choices each) a simulated user needed to asses before finding the target. In Figure 3 we show this for the Netflix perceptual space, and a uniformly generated one of the same size, each with locality-restricted updates with varying δ_N and an unrestricted version. Furthermore, we included an approach based only on similarity search without Bayesian modeling, i.e. a user provides a start example and is shown a display of the 9 yet-unseen nearest neighbors in the perceptual space, selects one, and obtains a new display with that movie's yet-unseen closets neighbors. Here, we can see that Bayesian retrieval approaches outperform similarity search, as they need significantly less displays, while the locality restricted Bayesian update deteriorates towards similarity search for very small δ_N and converges towards unrestricted Bayesian retrieval for sufficiently large δ_N.

For $\delta_N = 1,000$ the target is found after 10.7 displays on average in the perceptual space, while additional experiments using a uniform space with 100k items resulted in a slightly higher average of 17.7 displays. The parameter δ_N affects the uniform space more strongly than a real perceptual space, but it can be seen that the general behavior of the algorithm in a perceptual space can roughly be approximated by the uniform space. Especially for values of $\delta_N = 500$ the observed query behavior is similar; therefore we use this value when evaluating the scalability of our algorithm in the next experiments.

Figure 4 shows the memory usage of our locally restricted Bayesian update approach for storing the final user model after a query found its intended target for increasing dataset sizes. It can clearly be seen that the locality restriction scales favorably: the user model requires only about 16% the size of the unrestricted approach for the uniform dataset with 11,976 items (the size of the original perceptual space) while for the larger space containing 100k items, this percentage even decreases down to 6%.

In Figure 5, we investigate the average time needed per single locality-restricted update operation on a large synthetic perceptual space with 100k objects. Using larger δ_N unsurprisingly increases the updates times, which can become higher than the time needed for unrestricted updates when using very high values of δ_N (but still, all times are well below 0.5 seconds which is perfectly fine for Web applications). However, for $\delta_N = 1,500$ which provides semantically very similar results to the unrestricted approach, there is still a computation time advantage of 138ms. Also, the number of items to be held in each user model increases with δ_N (see Figure 6). However, here it can clearly be seen that locality restriction significantly affects the amount of memory needed, and even for $\delta_N = 2,000$, only 11.7% of the memory required by the unrestricted variant is consumed.

5.3   Adapting to Different Domains

While we used movies as a running example to demonstrate our system design, our approach can be adapted to any experience product which is frequently consumed and rated. Unfortunately, the free availability of user-item-rating information is limited, especially after the scandals following the release of the Netflix dataset (which is still the most extensive dataset up to now). Therefore, we resorted to rating data crawled from the Web in 2012 [3] to show the applicability in other domains. The first data set contains restaurant ratings in the San Francisco area obtained from yelp.com (3,811 restaurants; 128,486 users; 626,038 ratings). It is only a small item set with a large but rather inactive user base (i.e. only few ratings per user). In contrast, the second dataset consists of ratings of board games from boardgamesgeek.com (32,337 games; 73,705 users; 3,536,455 ratings). Here, a rather small user base is a highly active, rating a large number of items each.

For both datasets, we performed simulations similar to those described in the last subsection and found that the yelp.com data set could be traversed in 8.5 screens on average, while the board games data set needed 11.1 screens on average. No further user studies have been performed so far, as users recruited from crowd sourcing platform most likely lack the domain knowledge to evaluate a system using such data, but real user experiments with these datasets will be part of future works. While we currently do not provide experimentally supported insight into the resulting user experience, we can claim that the adaptation to different domains is at least technically possible and the results are comparable to movies for the simulations.

6   Summary and Discussion

In this paper, we demonstrated privacy-preserving query personalization using Bayesian query-by-example techniques on perceptual spaces. This allows users to query an information system with experience products in an anonymous and ad-hoc fashion, not requiring profiling or explicit preference elicitation. Our system is in the middle ground between SQL-style queries, which need the user to be able to formulize a precise query, and recommender systems which proactively suggest items to users. It allows explorative queries with "I know what I am looking for when I see it" semantics, which also helps users to break out from the "filter bubble" created by recommender systems.

We have shown how to build such a system, and how perceptual spaces can be combined with, and adapted to our proposed Bayesian query-by-example framework. Furthermore, we demonstrated how the resulting performance and scalability issues can be overcome by introducing locality-restricted Bayesian updates which provide tremendous advantages in terms of memory consumption, and saves a significant amount of computation time. Finally, in our extensive evaluations we have presented a user study with over 175 participants, and obtained their opinion on the effectiveness of the system. Furthermore, we have performed several experiments using simulations on synthetic data, and showed that the approach is indeed scalable and adaptable even on lower-end hardware like a laptop.

While we achieved semantically meaningful and simple to formulate queries without requiring user profiling during the query process, our current approach still relies on aggregating user-item-ratings from a group of enthusiast users to construct the perceptual space, which still carries certain privacy concerns. One of the future directions of research is hence the development of representations of experience items which are similarly expressive as perceptual spaces, but can be constructed using data not exhibiting any privacy concerns, like for example using Linked Open Data sources or anonymized product reviews. Furthermore, while system usability is already quite good as is, it could be further improved by explaining the system behavior to users. This especially covers describing why the presented items are a good match to the query, e.g., "The items shown have been selected because they have settings similar to movie X, but share the humor of movie Y – and you used both movies as positive examples".

References

1. Bell, R.M., Koren, Y., Volinsky, C.: All together now: A perspective on the Netflix Price. Chance 23, 24–24 (2010)
2. Koren, Y., Bell, R.: Advances in Collaborative Filtering. In: Recommender Systems Handbook, pp. 145–186 (2011)
3. Selke, J., Lofi, C., Balke, W.-T.: Pushing the Boundaries of Crowd-Enabled Databases with Query-Driven Schema Expansion. In: Proc. VLDB, vol. 5, pp. 538–549 (2012)
4. Pariser, E.: The Filter Bubble: What the Internet Is Hiding from You. Penguin Books (2011)
5. Sundaram, H., Chang, S.-F.: Computable scenes and structures in films. IEEE Trans. Multimed. 4, 482–491 (2002)
6. Neo, S.-Y., Zhao, J., Kan, M.-Y., Chua, T.-S.: Video Retrieval Using High Level Features: Exploiting Query Matching and Confidence-Based Weighting. In: Sundaram, H., Naphade, M., Smith, J.R., Rui, Y. (eds.) CIVR 2006. LNCS, vol. 4071, pp. 143–152. Springer, Heidelberg (2006)

7. Pilászy, I., Tikk, D.: Recommending new movies: even a few ratings are more valuable than metadata. In: ACM Conf. on Recom. Systems (RecSys), New York, USA (2009)
8. Kahneman, D., Tversky, A.: Psychology of Preferences. Sci. Am. 246, 160–173 (1982)
9. Knijnenburg, B.P., Willemsen, M.C., Gantner, Z., Soncu, H., Newell, C.: Explaining the user experience of recommender systems. UMUAI 22, 441–504 (2012)
10. Narayanan, A., Shmatikov, V.: Robust De-anonymization of Large Sparse Dataset. In: IEEE Symposium on Security and Privacy, Oakland, USA (2008)
11. Yoshitaka, A., Lchikawa, T.: A survey on content-based retrieval for multimedia databases. IEEE Trans. Knowl. Data Eng. 11 (1999)
12. Kato, T., Kurita, T., Otsu, N., Hirata, K.: A sketch retrieval method for full color image database-query by visual example. Pattern Recognit. (1992)
13. Niblack, C.W., Barber, R., Equitz, W., Flickner, M.D., Taubin, E.H.G.D.P., Yanker, P., Faloutsos, C., Taubin, G.: QBIC project: querying images by content, using color, texture, and shape. In: Storage and Retrieval for Image and Video Databases, San Jose, USA (1993)
14. Santini, S., Jain, R.: Beyond query by example. In: ACM Multimedia, Bristol, UK (1998)
15. Cox, I.J., Miller, M.L., Minka, T.P., Papathomas, T.V., Yianilos, P.N.: The Bayesian Image Retrieval System, PicHunter: Theory, Implementation, and Psychophysical Experiments. IEEE Trans. Image Process. (2000)
16. Slaney, M., White, W.: Similarity Based on Rating Data. In: 8th Int. Conf. on Music Information Retrieval (ISMIR), Vienna, Austria (2007)
17. Babas, K., Chalkiadakis, G., Tripolitakis, E.: You Are What You Consume: A Bayesian Method for Personalized Recommendations. In: RecSys 2013, Hong Kong, China (2013)
18. Jeckmans, A.J.P., Beye, M., Erkin, Z., Hartel, P., Lagendijk, R.L., Tang, Q.: Privacy in Recommender Systems. Social Media Retrieval, 263–281 (2013)
19. Polat, H., Du, W.: SVD-based collaborative filtering with privacy. In: ACM Symp. on Applied Computing (SAC), Santa Fe, USA (2005)
20. McSherry, F., Mironov, I.: Differentially Private Recommender Systems: Building Privacy into the Netflix Prize Contenders. In: ACM SIGKDD Int. Conf. on Knowledge Discovery and Data Mining (KDD), Paris, France (2009)
21. Selke, J., Balke, W.T.: Extracting Features from Ratings: The Role of Factor Models. In: Workshop on Advances in Preference Handling, Lisbon, Portugal (2011)
22. Khoshneshin, M., Street, W.: Collaborative filtering via euclidean embedding. In: 4th ACM Conf. on Recommender Systems (RecSys), Chicago, Illinois, USA (2010)
23. Lofi, C., Nieke, C., Balke, W.-T.: Mobile Product Browsing Using Bayesian Retrieval. In: IEEE Conf. Commerce and Enterprise Comp. (CEC), Shanghai, China (2010)

Identifying Explicit Features
for Sentiment Analysis in Consumer Reviews

Nienke de Boer, Marijtje van Leeuwen, Ruud van Luijk,
Kim Schouten, Flavius Frasincar, and Damir Vandic

Erasmus University Rotterdam
P.O. Box 1738, 3000 DR, Rotterdam, The Netherlands
{nien_de_boer,marijtje93}@hotmail.com,
ruudvanluijk91@gmail.com,
{schouten,frasincar,vandic}@ese.eur.nl

Abstract. With the number of reviews growing every day, it has become more important for both consumers and producers to gather the
information that these reviews contain in an effective way. For this, a
well performing feature extraction method is needed. In this paper we
focus on detecting explicit features. For this purpose, we use grammatical relations between words in combination with baseline statistics of
words as found in the review text. Compared to three investigated existing methods for explicit feature detection, our method significantly
improves the F_1-measure on three publicly available data sets.

1 Introduction

In the last decade, the World Wide Web has changed enormously. E-commerce
is expanding at a rapid pace as more and more people have access to the Internet
nowadays. Because of this, the amount of reviews given on products and services
is also increasing. Some products and services have hundreds of reviews, scattered
over many websites. These reviews are a valuable source of information for both
consumers [4] and producers [15]. Since the amount of reviews is large, it is hard,
if not impossible, to read all of them and to keep track of all expressed opinions
on the different features of the product or service. Selecting a few reviews to read
may give a false impression of the product or service, so it is clear that more
advanced and automated methods for processing and summarizing reviews are
needed [10].

Reviews contain characteristics of products or services, so-called features. The
literature reports several works on extracting features from texts. Some of these
methods concentrate on finding explicit features while others focus on extracting
implicit features. In this paper, we propose a new method to extract explicit
features on which reviewers have expressed their opinions, by employing and
adapting various techniques that proved to be useful in previous works. Where
existing approaches use either frequency counts or grammatical relations, we
propose to use both. First, we use grammatical relations between words to find

B. Benatallah et al. (Eds.): WISE 2014, Part I, LNCS 8786, pp. 357–371, 2014.
© Springer International Publishing Switzerland 2014

feature candidates. Then, we check whether these feature candidates occur more often than expected with respect to a general corpus, before annotating them as actual features. We hypothesize that these two techniques complement each other in such a way that combining them will yield higher precision and recall.

The paper is organized as follows. We start by discussing some of the related work in Sect. 2. In Sect. 3, we present the proposed method. In Sect. 4, the data that is used to test the method is discussed and then in Sect. 5 the evaluation results are presented. Last, our conclusions and suggestions for further research are given in Sect. 6.

2   Related Work

In this section we discuss some of the related work that has been done on finding features in customer reviews. Due to space limitations, we investigate only three existing methods which are representative for their classes: a frequency-based approach, a statistical approach, and a (grammatical) relation-based approach.

One of the most well known methods to find explicit features is proposed by Hu and Liu in [7]. This method first extracts the features that are frequently mentioned in the review corpus. Since it is assumed here that explicit features are most likely to be nouns or noun phrases, these are extracted from all sentences and are included in a transaction file. In order to find features that people are most interested in, Association Rule Mining [1] is used to find all frequent item sets. In this context, an item set is a set of words or phrases that occur together. When the final list of frequent features is known, the method extracts all opinion words that are nearby the frequent features. To that end, it exploits the fact that opinion words are most likely to be adjectives. The found opinion words are used to find infrequent features, based on the idea that people often use the same opinion words for both frequent and infrequent features. If a sentence does not contain a frequent feature but does contain one or more opinion words, the proposed method extracts the noun nearest to that opinion word and this noun is stored in the feature set as an infrequent feature. A disadvantage of this method is that nouns or noun phrases that are more used in general will also be annotated as feature, favoring false positives. In the next paragraph we present a method which alleviates this issue.

In the paper by Scaffidi et al. [17], a method is proposed that uses baseline statistics of words in English and probability-based heuristics to identify features. The main idea is that nouns that occur more frequently in the review corpus than in a random section of English text are more likely to be features. Since reviewers focus on a specific topic, the relevant words for that topic will occur more in the review than in a normal English text. The probability that a certain noun or noun phrase (a maximal length of 2 is used for a noun phrase) occurs in a normal text as much as it does in the review is calculated. If this probability is small, it is more probable that the noun or noun phrase is a feature. This method has a high precision but the recall is low. For this method, it is important that the English text that is used for determining the baseline is in the same language and roughly the same style as the reviews. Using baseline statistics thus addresses the discussed disadvantage of the previous method.

A major shortcoming, shared by both [7] and [17] is that only the number of times a word occurs in all the reviews taken together matters. The methods do not take into account the grammatical structures that are within the review sentences, which could be useful to find infrequent features that appear seldom in reviews as well as in general text. In [6], Hai et al. propose a method that focuses on implicit feature identification in Chinese written reviews. However, since explicit features are used to find implicit ones, an algorithm for finding explicit features is also presented. In this method, all nouns and noun phrases that are in certain grammatical dependency relations are added as explicit features. The relations that are used here are the nominal subject relation, the root relation, the direct object relation, and object of a preposition relation. In this method, it is assumed that the sentences which contain an explicit feature are already known. The method first checks whether the sentence contains an explicit feature before it investigates the dependency relations. This makes feature extraction relatively easy and it allows for a higher recall and precision. However, in general it is not known in advance which sentences contain features and which sentences do not, so that precision is generally low when using this method. Also, this method does not use information about how often nouns and noun phrases occur at all, while this certainly is valuable information.

3 Method

In this section we propose a new method for finding explicit features by addressing the shortcomings identified in the previous work. For this purpose, we reuse and adapt techniques described in [6] and [17]. Since in the first method, precision is low but recall is relatively high, and in the second the recall is rather low but precision is high, an ensemble method that combines these two methods seems a logical next step. Therefore, we use the dependency relations of [6] to obtain a high recall and the baseline statistics of [17] to obtain a high precision.

First, the proposed method analyzes dependency relations in the review sentences to find feature candidates. This can be seen as the first step in [6]. To find feature candidates, we check the grammatical structures in each of the review sentences. If a word is in one of the used dependency relations, we first check if it is a noun or part of a noun group, before we add it to the list of feature candidates. This is because earlier research has shown that features which are explicitly mentioned are most likely to be nouns [14]. We use the Stanford Parser [11] to find the grammatical structures. Besides the dependencies mentioned by Hai et al., we also use some additional dependencies. First the dependencies which are used in [6] are explained. One of these dependencies is the 'direct object' (dobj) dependency. We explain this dependency by the following example sentence:

"He tried to clear a table for six."

If a word is the object of a verb, for instance the word "table" in the above sentence, we add this word to the feature group. Also, we add the dependent word of a 'preposition-object' (pobj) relationship. This relationship is often combined with the 'prepositional dependency' (prep). For instance, take the next sentence, in which one of the features is "brunch":

"Great for groups, great for a date, great for early brunch or a nightcap."

The dependencies prep(great, for) and pobj(for, brunch) are combined into prep_for(great, brunch). Therefore, we use both the dependencies 'pobj' and all the combined dependencies such as prep_for. Two other dependencies that are used in [6] are the 'root' and the 'nominal subject' relationships. We explain these dependencies with the help of another sentence example:

"The food is very good too but for the most part, it's just regular food."

The dependencies between the words of this sentence are shown in Fig. 1. Additional information regarding the various dependencies mentioned in this figure can be found in [12]. We add nouns to the feature group which are the 'root' (root) of the sentence, the word the whole sentence relates to. In the example, the root is "good", which is not a noun, so in this case we do not add this word to the feature group. We also add a noun to the feature group if it is the dependent word in the 'nominal subject' (nsubj) relationship. In our example sentence, this means we add the second word of the sentence, the word "food", to the feature group.

The dependencies we discussed so far are the original dependencies that were used in the method which was proposed in [6], but that method is based on Chinese written reviews. Since the data sets we use contain English reviews, it is useful to add extra dependencies. One of the extra dependencies we add is the dependency 'conjunction' (conj). This is a relation between two words which are connected by words like "and" or "or". In the above example sentence, the words "good" and "food" are connected by the word "but". Because of this, "food" is added to the list of feature candidates. Furthermore, we examine the 'noun compound modifier' (nn). This is a dependency between two nouns, in which first noun modifies the meaning of the second noun. The next sentence shows an example of this dependency:

"A wonderful jazz brunch with great live jazz."

Here, the noun "jazz" modifies the meaning of the noun "brunch", so "brunch" is added to the list of feature candidates. The last dependency we use is the 'appositional modifier' (appos): a noun that serves to define or modify the meaning of another noun, which is located at the left of the first noun. We explain this with the help of another sentence:

"A gentleman, maybe the manager, came to our table, and without so much as a smile or greeting asked for our order."

Here the noun "manager" defines the meaning of the noun at the left of it, the noun "gentleman". In this case, "manager" is added to the list of feature candidates. The pseudocode of the total process of retrieving the feature candidates is shown in Algorithm 1.

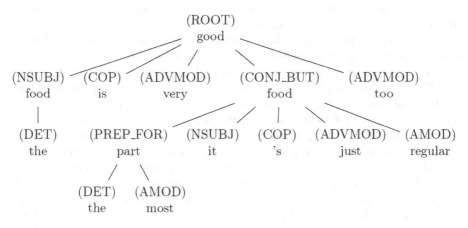

Fig. 1. Grammatical dependency relations

Now that we have a list with feature candidates based on grammatical structures, we want to improve precision. The above method is likely to find a lot of features, but it will also find a lot of non-features. If a possible feature is mentioned in many reviews, it is more likely to be an actual feature. At this point, it is important to note that some words are just more common to use in English text than others. For example, the word "lens" might not be used a lot in a normal English text, but when it is used in a review about a digital camera, it might be mentioned a lot more. Therefore, it could very well be a feature about which the writer is expressing his opinion. For this reason, we check for each of the feature candidates that we found using Algorithm 1 whether it occurs more than would be expected based on a general English text. This is the main idea that is used in [17] as well. When the word is in a list of stop words, we do not add it as a possible feature. Also, if a found noun is more likely to be used as a non-noun, it is probably not a feature, so those will not be added as features as well. For example, the word "count" can be used as a noun, but it is more probable that it is used as a verb. We check this aspect for all nouns using the frequency counts that are provided by WordNet [13]. However, for some words, the frequency counts are not given in WordNet or the word is not in the Word-Net database. In such cases, we assume that the word is more likely to be used as a noun.

To explain the idea of using baseline probabilities in more detail, we start by focusing on single noun features only. To check whether the noun occurs more often than expected, one needs to check whether the number of times it occurs in the reviews is larger than the expected amount in a generic English text of the same length. If it does occur more often, the probability that the noun occurs exactly that often is calculated. In this case, we do not look at the word itself, but we take the dictionary form of the word, the lemma, and count how many times the lemma occurs. We denote by n_x the number of times that lemma x occurs in the review corpus. Thus we calculate the probability $P(n_x)$ that lemma x occurs n_x times in a general English text with N words, where N is the total number of words in the review corpus.

To calculate this probability, the probability p_x that the lemma of a randomly selected word in a generic English text equals x needs to be known first. We use the word frequency list on conversational English from Leech et al. [9] for this. With the provided word counts, we calculate the baseline probability by dividing the count by the total amount of words in the general English text. When a review word does not occur in the general English text, we assign the average probability of words in the general English text to this word.

We can now estimate the probability that lemma x occurs n_x times in the general English text with a Poisson distribution, as we want to determine the probability of a certain count in an interval, in our case, the entire length of the review corpus [3]. However, to avoid numerical underflow, it is convenient to take the logarithm of the distribution. For this, we need to apply Stirling's approximation to estimate $\ln(n_x!)$ [16]. As a result, Eq. 1 is used to calculate the probabilities. The last term, namely $ln(\sqrt{2\pi})$ is just a constant, and thus not necessary when comparing probabilities.

$$ln(P(n_x)) \simeq (n_x - p_x N) - n_x ln(\frac{n_x}{p_x N}) - \frac{ln(n_x)}{2} - ln(\sqrt{2\pi}) \qquad (1)$$

We can use a similar approach for calculating probabilities of noun phrases. In order to do this, we do need to make some assumptions. First, we assume that the occurrence of lemma x in position i is independent of whether lemma x occurs in any other position j in a sentence. We also assume that the occurrence of lemma x in position i is independent of i. Although these assumptions are generally not true, it helps to simplify the problem while they have no serious consequences for the results [19]. Because of these assumptions, the probability p_b that the bi-gram b occurs in generic English is simply p_x times p_y. In this context, a bi-gram is a noun phrase with two nouns, the probability of occurrence of the first word in the bi-gram is p_x and the probability of occurrence of the second one is p_y. We can now use Eq. 1 where n_x represents the number of occurrences of the bi-gram. To obtain the probability of a tri-gram, one must first multiply the probabilities of the single words to get p_x and use Eq. 1 again after that.

Now that the probabilities of a noun or bi-gram occurring with a certain frequency are known, we select the ones with the smallest probabilities to be features. It is important to determine how many features are added and for this a threshold value is required. If the probability of a noun or noun phrase is smaller than the threshold value, the noun or noun phrase is added to the list of features. The threshold value may differ among different data sets so this value should be trained per data set. In [17], there was only one threshold value for single-nouns and n-grams, but there might be different optimal values for single-noun features, bi-grams, tri-grams, et cetera, so the values should be allowed to be different. This gives rise to a new problem, namely that there are not many feature candidates that consist of three or more words (as will be discussed in Sect. 4). Because of this, it is hard to train the threshold value for tri-grams and higher n-grams. Therefore, it is better to apply the approach of baseline probabilities comparison only on single-noun features and bi-grams. Algorithm 2 shows how the probability of a noun or bi-gram occurring a given number of times is computed.

In order to determine the best threshold value, we use training data only. As mentioned before, the threshold values of single-nouns and bi-grams are allowed to be different. We choose the combination of threshold values for which the F_1-value, the values that we use for evaluation, is the highest. For this, we cannot use a gradient ascent method, since the function of the F_1-values is not concave (there are some local maxima points). Therefore, a linear search method is used instead of the gradient ascent method. First, the probabilities computed with Algorithm 2 are sorted. These probabilities are the possible threshold values. We iterate over the possible threshold values by selecting the percentage of single nouns or bi-grams that we want to add as features. The step size we use is 0.01.

Now that we have all probabilities and threshold values, we iterate over each possible feature found with Algorithm 1 and check whether the corresponding probability is smaller than the threshold value. If this is the case, we add the feature to the final list of features. If the feature is a tri-gram or a n-gram of higher degree, we add this feature to the final list without considering the probabilities.

Algorithm 1. Generating a group of possible explicit features

---

Input: review sentences in the corpus
Output: a list of possible explicit features F
for each sentence $s \in$ corpus **do**
    for each word or wordgroup $w \in s$ **do**
        if w is in grammatical relationship of specified types **then**
            if each POS tag of w is noun **then**
                add w to F
            end if
        end if
    end for
end for

---

4   Data Analysis

In this section a brief overview of the used data sets is presented. Since the performance of different algorithms depends a lot on the used data set, three different data sets are used to train and evaluate the algorithms. The first two data sets are from the SemEval competition [2]. The first is a set containing

---

Algorithm 2. Calculating probabilities of single nouns and bi-grams

---

Input: D : a list of all words in the corpus. N : a count of all words in the corpus.
O : a list of all unique nouns in the corpus. B : a list of all bi-grams in the corpus.
p_x : a list of probabilities how often a word appears in an English text.
Output: P_S : a list with the logarithm of the probability that a single noun occurs exactly n_x times in the review corpus with x denoting the lemma of the noun. P_B : a list with the logarithm of the probability that a bi-gram occurs exactly n_b times in the review corpus with b denoting the bi-gram.

for each word $x \in D$ **do**
    if x is a noun **then**
        $count(x) + +$
        if $nextWord(x)$ is a noun **then**
            $b = concat(x, nextWord(x))$
            $countB(b) + +$
        end if
    end if
end for
for each word $x \in O$ **do**
    find baseline probability p_x
    calculate the probability that the word occurs n_x times, with $n_x = count(x)$:
    $P_S(x) = (n_x - p_x * N) - n_x * ln(\frac{n_x}{p_x * N}) - \frac{ln(n_x)}{2} - ln(\sqrt{2\pi})$
end for
for each bi-gram $b \in B$ **do**
    calculate $p_b = p_x \times p_{nextWord(x)}$
    calculate the probability that the bi-gram occurs n_b times, with $n_b = countB(b)$:
    $P_B(b) = (n_b - p_b * N) - n_b * ln(\frac{n_b}{p_b * N}) - \frac{ln(n_b)}{2} - ln(\sqrt{2\pi})$
end for

---

reviews about restaurants [5]. The second data set contains reviews about laptops. The third data set contains a collection of reviews of a set of products [8]. These products include a camera, a printer, a DVD-player, a phone and an mp3. It turns out that the differences between these data sets are substantial. The characteristics of these sets will now be presented in more detail.

4.1   Restaurant Data Set

The restaurant data set contains 3041 sentences. In these sentences, one can find a total of 1096 unique explicit features. As can be seen in Fig. 2a, about one third of the sentences does not contain a feature, which means that two thirds of the sentences contain at least one feature. Since most sentences contain a feature, it is relatively easy to find features. The features in this data set are mostly single word features (75.44%) and bi-grams (16.84%), but 7.72% of the features consist of 3 or more words. This is illustrated in Fig. 2b. The largest feature is a composite of nineteen words, but this only appears one single time.

Most algorithms extract single-noun features and bi-gram features rather well. The longer features are harder to find as the difficulty of finding a feature increases with the amount of words it consists of. This results in most algorithms performing well on this data set.

4.2   Laptop Data Set

The laptop data set has only four more sentences than the restaurant data set, but there are 231 less unique features to be found. This means that the data set contains only 865 unique features. As can be seen in Fig. 2a, more than half of the sentences does not contain any feature. This is about 20%-point more than in the restaurant data set. The distribution of the sentences that contain one or more features is similar as in the restaurant set. The ratio between sentences with one feature and sentences with two, three or more features is about the same. Fig. 2b shows that there are relatively less features that consist of one single word, but there are more bi-grams and tri-grams in this data set in comparison with the restaurant data set.

Although some of the characteristics of this data set seem to be quite similar to the characteristics of the restaurant data set, it is harder to get a good performance on this data set. This is mainly caused by the fact that there are more sentences that do not contain any features. The fact that there are more features that consist of more than one word makes it also harder to get a good performance.

4.3   Product Data Set

The product data set has very different characteristics in comparison with the restaurant data set and the laptop data set. There are 904 more sentences in this data set than in the restaurant data set, which gives a total of 3945 review sentences. However, there are less features to be found. The number of unique features in all these sentences is only 231. There are 2850 sentences without any feature and only 1095 of the sentences contain one or more features. A visual representation of this can be found in Fig. 2a. In this data set, most features contain only one or two words. The corresponding percentages are respectively 63.20% and 35.43%. The largest feature in this data set has a size of four words. In Fig. 2b the differences between the data sets are illustrated. While the product data set has more bi-grams, there are almost no n-grams with n higher than two. Thus an algorithm that extracts bi-grams rather well is needed here. Since the amount of features is small and the amount of sentences without features is rather large, it is more difficult to get a good performance on this data set. A lower F_1 can thus be expected.

Because of the different characteristics, the use of these three sets is ideal for developing and testing an algorithm. If an algorithm performs very well on only one of the three sets, it might not be very useful in general. Performing good on all three data sets is a good indicator that it might perform well on other data sets as well.

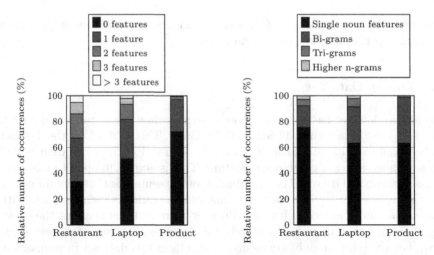

(a) The distribution of number of features per sentence

(b) The distribution of n-grams in the data sets

Fig. 2. Characteristics of the data sets

5   Evaluation

For the evaluation of the proposed method, a 10-fold cross-validation is performed. For this, the available data is randomly divided in ten equally sized groups. Nine groups are used to train the threshold values. The sentences that are in the tenth group are used as test data. We repeat this approach ten times, with each time another group as test data. In the end, all ten groups are used as test data once, and nine times as training data. For the evaluation of the method, and to determine the best threshold values, we used the F_1-value, which is the harmonic mean of precision and recall.

Since the proposed method uses some of the ideas that were expressed in the papers [17] and [6], the performance of the proposed method is compared with the performance of these methods. Furthermore, we use the method proposed in [7] to compare the results of the new method with. The results of the earlier proposed methods on the different data sets can be found in Table 1. To show the effects of the various components of the proposed method, a stepwise evaluation is performed. The first step is to improve the two original methods of [17] and [6].

In [17], a single threshold for both single-noun features and bi-grams is used. As argued before, it can be better to allow different threshold values for single noun features and bi-gram features respectively. The performance of the method as proposed in the paper, having only one threshold value, and the performance of the extended method that allows for threshold values to be different are shown in Table 2. It can be seen that for the laptop data set and for the product data set, the F_1-score improves by more than one percentage point when the threshold values are allowed to be different. For the laptop data set, the increase

Table 1. The performance of the methods proposed in the papers by Hu and Liu [7], Scaffidi et al. [17] and Hai et al. [6]

	Restaurant data set			Laptop data set			Product data set		
Method	Hu	Scaffidi	Hai	Hu	Scaffidi	Hai	Hu	Scaffidi	Hai
Precision	0.371	0.432	0.384	0.137	0.188	0.200	0.056	0.126	0.060
Recall	0.627	0.575	0.715	0.405	0.451	0.591	0.480	0.356	0.417
F_1	0.467	0.493	0.500	0.204	0.266	0.299	0.100	0.186	0.105

is caused by an increase in precision. For the product data set, both precision and recall have increased. Results for the restaurant data set are slightly different. Although recall has increased in this case, the F_1-score is the same for both methods. Combining the results on the different data sets, we conclude that it is better to allow the threshold values to be different, since the F_1-scores of that method are the same or higher than the scores of the method that uses a single threshold.

Table 2. Performance of the method proposed by Scaffidi et al. [17] with one threshold value and the performance of the extended method that uses two threshold values on the different data sets

	Restaurant data set		Laptop data set		Product data set	
Threshold values	One	Two	One	Two	One	Two
Precision	0.432	0.412	0.188	0.226	0.126	0.133
Recall	0.575	0.614	0.451	0.380	0.356	0.380
F_1	0.493	0.493	0.266	0.283	0.186	0.197
Diff. in F_1	**+0.000**		**+0.017**		**+0.011**	

In [6], the used grammatical relations are the nominal-subject, the root, the direct object, and the object of a preposition relations. This method was designed for Chinese written reviews and for English reviews it is better to use more types of relations. Adding the relation types 'conjunction', 'appositional modifier' and 'noun compound modifier' boosts the recall, but the precision declines. Since we want to get a high recall using the grammatical dependencies and we expect precision to grow using the baseline probabilities, we use the relation types 'conjunction', 'appositional modifier' and 'noun compound modifier' in addition to the relation types that were used in [6]. In Table 3 the performance of the method proposed in [6] with the original dependencies and the performance of the same method but with the proposed dependencies are shown.

Table 3. Performance of the method proposed by Hai et al. [6] with the original dependencies and the performance of the method with the proposed dependencies on the different data sets

	Restaurant data set		Laptop data set		Product data set	
Dependencies	Original	Proposed	Original	Proposed	Original	Proposed
Precision	0.384	0.380	0.200	0.197	0.060	0.054
Recall	0.715	0.771	0.591	0.646	0.417	0.477
F_1	0.500	0.509	0.299	0.303	0.105	0.098
Diff. in Recall	**+0.056**		**+0.055**		**+0.060**	

Now, the performance of the proposed method is discussed. We evaluate the performance of the method using only the grammatical structures used in [6] and a single threshold value first. The results are shown in Table 4, with 'original' referring to the fact that only the grammatical structures of [6] are used. As discussed in the previous part of this section, adding more grammatical structures may improve performance and allowing for different threshold values causes an increase in performance as well. Therefore, we also tested the proposed method with the additional grammatical structures and with different threshold values. The results of the method with proposed grammatical structures are shown in Table 4 under the heading 'proposed'. The bottom row of each subtable shows the difference between the F_1-value of the used method in that column and the F_1-value of the proposed method that uses only one threshold value and the original dependencies.

In Table 4 it is shown that for all data sets, the F_1-measure of the method in which more grammatical relations and different threshold values are used is the highest. When comparing the F_1-values of this method with the performance of the existing methods as shown in Table 1, we find that our method improves the F_1-value of the methods proposed in [7], [6] and [17] on each of the used data sets. For the restaurant data set, the F_1-value of the proposed method is 8.1%-point higher than the method proposed in [7], 5.5%-point higher than the method proposed in [17] and 4.8%-point higher than the method proposed in [6]. For the laptop data set, these values are 11.0%-point, 4.8%-point and 1.5%-point, respectively, and for the product data set the values are 9.1%-point, 0.5%-point and 8.6%-point, respectively.

To test whether the found differences are statistically significant, we perform a t-test. We also test whether the differences between the proposed method and the extended versions of the existing methods proposed in [17] and [6] are statistically significant. In order to perform the tests, it is necessary to have multiple evaluations. Therefore, we construct 30 bootstrap samples [18] for each of the used data sets. Every bootstrap sample is expected to contain about 63.2% of the unique sentences of the original data set. By using this sampling technique, we have obtained 30 new data sets for each of the original data sets.

Table 4. The performance of the proposed method for the different combinations on the different data sets

Restaurant data set				
Threshold values	One threshold		Two thresholds	
Dependencies	Original	Proposed	Original	Proposed
Precision	0.523	0.518	0.521	0.516
Recall	0.542	0.580	0.547	0.585
F_1	0.532	0.547	0.534	0.548
Diff. in F_1		**+0.015**	**+0.001**	**+0.016**

Laptop data set				
Threshold values	One threshold		Two thresholds	
Dependencies	Original	Proposed	Original	Proposed
Precision	0.242	0.235	0.279	0.269
Recall	0.415	0.442	0.354	0.377
F_1	0.306	0.307	0.312	0.314
Diff. in F_1		**+0.001**	**+0.006**	**+0.008**

Product data set				
Threshold values	One threshold		Two thresholds	
Dependencies	Original	Proposed	Original	Proposed
Precision	0.132	0.131	0.140	0.139
Recall	0.271	0.285	0.290	0.305
F_1	0.177	0.180	0.189	0.191
Diff. in F_1		**+0.003**	**+0.012**	**+0.014**

These new sets will be used for method evaluation. We use the data sets as input for the existing methods, the extended versions of the existing methods and for the proposed method. A one-tailed paired t-test is performed to test whether the differences between the F_1-value of the proposed method and the F_1-values of the (extended) existing methods is significantly larger than zero. The results of the test are shown in Table 5. It shows that the F_1-measure of the proposed method is significantly higher than the F_1-values of the existing methods proposed in [7], [17] and [6] for each of the three used data sets at a 1.0% significance level. Furthermore, the proposed method performs significantly better in terms of the F_1-measure than the extended versions of the existing methods proposed in [17] and [6] for the restaurant and for the laptop data set. For the product data set, only the extended version of the method proposed in [17] performs slightly better than the proposed method.

Table 5. Results of the t-test on whether the F_1-value of the proposed method is significantly higher than the F_1-values of the (extended) existing methods

Statistics	Restaurant data set		Laptop data set		Product data set	
	Mean	P-value	Mean	P-value	Mean	P-value
Proposed method	0.547		0.313		0.191	
Hu et al.	0.470	0.000	0.207	0.000	0.104	0.000
Scaffidi et al.	0.493	0.000	0.269	0.000	0.187	0.001
Scaffidi et al. with double threshold	0.494	0.000	0.281	0.000	0.198	1.000
Hai et al.	0.501	0.000	0.300	0.000	0.099	0.000
Hai et al. with extra dependencies	0.509	0.000	0.304	0.000	0.098	0.000

6　Conclusion

In this paper, we proposed a new method that extracts explicit features from consumer reviews. We employed and adapted various techniques that were used in two existing methods. The F_1-value of the proposed method is higher than the F_1-values of the separate methods on three different data sets. The differences are statistically significant at a 1.0% significance level for a restaurant, a laptop and a product data set. For the restaurant and for the laptop data sets, the proposed method also performs significantly better than the extended versions of the existing methods. For the product data set, the difference in F_1-value is statistically significant with respect to one of the extended versions of the earlier proposed methods but not the other.

Possible future work includes using domain knowledge, for example by employing ontologies, to find features, instead of purely statistical information. Also of interest are implicit features which were out of scope for this research. However, when performing aspect-level sentiment analysis, it is definitely beneficial to detect all features, not just the explicitly mentioned ones. Last, the method now assigns the average probability for words in a review text which are not in the general text. However, given the highly skewed, zipfian distribution of word frequencies, this could be improved upon. For instance, if the general text corpus is large enough, it stands to reason that all regular words are included, and that therefore words which do not appear in the text corpus, should have a very low probability associated to them, instead of the average probability.

Acknowledgment. The authors are partially supported by the Dutch national program COMMIT.

References

1. Agrawal, R., Srikant, R.: Fast Algorithms for Mining Association Rules in Large Databases. In: Proceedings of the 20th International Conference on Very Large Databases (VLDB 1994), vol. 1215, pp. 487–499. Morgan Kaufmann (1994)
2. Androutsopoulos, I., Galanis, D., Manandhar, S., Papageorgiou, H., Pavlopoulos, J., Pontiki, M.: SemEval-2014 Task 4 (March 2014), http://alt.qcri.org/semeval2014/task4/
3. Bain, L.J., Engelhardt, M.: Introduction to Probability and Mathematical Statistics, 2nd edn. Duxbury Press (2000)
4. Bickart, B., Schindler, R.M.: Internet Forums as Influential Sources of Consumer Information. Journal of Interactive Marketing 15(3), 31–40 (2001)
5. Ganu, G., Elhadad, N., Marian, A.: Beyond the Stars: Improving Rating Predictions using Review Content. In: Proceedings of the 12th International Workshop on the Web and Databases (WebDB 2009) (2009)
6. Hai, Z., Chang, K., Kim, J.-J.: Implicit Feature Identification via Co-occurrence Association Rule Mining. In: Gelbukh, A.F. (ed.) CICLing 2011, Part I. LNCS, vol. 6608, pp. 393–404. Springer, Heidelberg (2011)
7. Hu, M., Liu, B.: Mining Opinion Features in Customer Reviews. In: Proceedings of the 19th National Conference on Artifical Intelligence (AAAI 2004), pp. 755–760. AAAI (2004)
8. Hu, M., Liu, B.: Mining and Summarizing Customer Reviews. In: Proceedings of 10th ACM SIGKDD International Conference on Knowledge Discovery and Data Mining (KDD 2004), pp. 168–177. ACM (2004)
9. Leech, G., Rayson, P., Wilson, A.: Word Frequencies in Written and Spoken English: Based on the British National Corpus. Longman (2001)
10. Liu, B.: Sentiment Analysis and Opinion Mining. Morgan & Claypool (2012)
11. Marneffe, M.C.D., MacCartney, B., Manning, C.D.: Generating Typed Dependency Parses from Phrase Structure Parses. In: Proceedings of International Conference on Language Resources and Evaluation (LREC 2006), vol. 6, pp. 449–454 (2006)
12. Marneffe, M.C.D., Manning, C.D.: Stanford Typed Dependencies Manual (September 2008), http://nlp.stanford.edu/downloads/lex-parser.shtml
13. Miller, G., Beckwith, R., Felbaum, C., Gross, D., Miller, K.: Introduction to Word-Net: An On-Line Lexical Database. International Journal of Lexicography 3(4), 235–312 (1990)
14. Nakagawa, H., Mori, T.: A Simple but Powerful Automatic Term Extraction Method. In: Proceedings of the 19th International Conference on Computational Linguistics (AAAI 2004), pp. 29–35. Morgan Kaufmann Press (2002)
15. Pang, B., Lee, L.: Opinion Mining and Sentiment Analysis. Foundations and Trends in Information Retrieval 2(1-2), 1–135 (2008)
16. Ross, S.M.: Introduction to Probability Models, 10th edn. Academic Press (2010)
17. Scaffidi, C., Bierhoff, K., Chang, E., Felker, M., Ng, H., Jin, C.: Red Opal: Product-Feature Scoring from Reviews. In: Proceedings of the 8th ACM Conference on Electronic Commerce (EC 2007), pp. 182–191. ACM (2007)
18. Tan, P.N., Steinbach, M., Kumar, V.: Introduction to Data Mining. Addison-Wesley (2005)
19. Wu, H., Salton, G.: A Comparison of Search Term Weighting: Term Relevance vs. Inverse Document Frequency. In: Proceedings of the 4th Annual International ACM Conference on Information Storage and Retrieval, pp. 30–39 (1981)

Facet Tree for Personalized
Web Documents Organization

Róbert Móro, Mária Bieliková, and Roman Burger

Slovak University of Technology in Bratislava
Faculty of Informatics and Information Technologies
Ilkovičova 2, 842 16 Bratislava, Slovakia
name.surname@stuba.sk

Abstract. Vast amount information and resources in the digital libraries and in general on the Web demands effective methods of archiving and organization. However, most of the existing solutions support only very specific use case scenarios, or are not flexible enough to accommodate to the changes in the document collections over time. We propose a method for web documents organization based on a facet view of the personal information structure. Facet chaining in a tree can create any depth of the structure and thus specify any context of resources. We enhanced this method by clustering similar resources and by using a special Search facet that allows users to specify arbitrary keyword queries as an input for collection's categorization. In order to evaluate the proposed approach, we carried out a user study in the bookmarking system Annota.

Keywords: digital library, personal information management, facet tree, web document clustering, user study, empirical evaluation, Annota.

1 Introduction

When browsing the Web, we encounter daily tens or even hundreds of web pages. For the purpose of their later retrieval and reference we bookmark those that are of any value, using either browser built-in bookmarking capability, or any of the available services, such as Delicious[1], Readability[2] or Pocket[3].

Similarly, with the continual shift of traditional libraries to the digital ones, we have now whole libraries within the reach of our hands (and mouse clicks). The researchers have to work with many resources when e.g. writing a thesis or a new paper; therefore, it becomes very important to them to be able to properly archive, maintain and retrieve all this information.

Based on [14], we can identify three basic operations as a part of the (personal information) organization process:

---

[1] https://delicious.com/
[2] https://www.readability.com/
[3] https://getpocket.com/

B. Benatallah et al. (Eds.): WISE 2014, Part I, LNCS 8786, pp. 372–387, 2014.
© Springer International Publishing Switzerland 2014

— *archiving a new resource* – it is a process of expanding personal collection with new resources. Input to this process is the resource itself and metadata describing it, giving it context. It is up to the user how specific the context is. Output of the process is a resource archived in the user's document collection.

— *retrieving an archived resource* – it is a process of searching and retrieving the resource. Retrieving can be either destructive (resource is removed from the collection) or preserving (resource is kept in the collection). Resource query is the input to the process. Output is usually a set of best matching resources to the query.

— *editing an archived resource* – it is a process of updating resource information, usually the metadata and relationships between resources. This process can be actually carried out as a series of destructive retrieval and archiving with new information.

The interface for personal documents organization and management should support all these operations minimizing the time and effort that the users have to spend in the process. Another important aspect to consider is, how the interface supports refactoring (restructuring) of the collection.

In addition, the user needs may vary greatly between individuals, which is a fact often overlooked and ignored by typical frameworks and solutions for organizational tasks. Even users in the same domain can have radically different information management strategies as observed in [4]. These strategies are mostly based on personal preferences of individuals, but can also be influenced by various tasks or events (such as preparing a paper for the upcoming conference).

In this paper, we explore the problem of personal information management of the web resources. We propose a new method of organizing and archiving web resources in an effective, easy to use and user friendly manner based on a concept of facet trees. We provide an empirical evaluation of the proposed method; we carried out a qualitative user study comparing the facet tree organization with folders commonly employed in many tools for personal documents management, such as Mendeley[4].

2   Related Work

There has been an extensive research in identifying main strategies commonly used in personal information management, based on which we can identify three basic strategies (or roles) that most users can be mapped to [3]:

1. piling strategy,
2. filing strategy,
3. structuring strategy.

Piling strategy is on the context-free side of the organizational spectrum with the users archiving the resources in an unstructured pile (or a stack), while structuring strategy is on the context-full side. Filing strategy is somewhere in the middle. However,

---

[4] http://www.mendeley.com/

it is not about using average amount of context to describe resources, but it is rather more of a combination. Some parts of the personal library are in context-free zone, having stacks or piles of resources that user wants to dig in later (or never). Other parts of the library are reasonably structured, giving the user option to file new resources that are in great importance to the user.

Typical task with personal libraries is recollecting and re-finding the archived or previously visited resources [1]. Semantic maps were used for this purpose in [13] in the domain of web search history.

However, two of the most prevalent approaches of organizing web resources are bookmarks and tags services. Bookmarks usually utilize folder structure so they are suited for structuring strategy. Problem with maintaining huge structured libraries was tried to be solved using information retrieval algorithms such as clustering and classification. Authors in [5] used n-grams in documents to find clusters of similar documents. In [10] authors used incremental clustering to simulate more typical user scenarios. Hierarchical clustering based on the documents' metadata and zoom-based navigation in the personal document collections have been utilized also in [9].

Tags are keywords assigned to a web resource that have special meaning to the user and are usually visualized by a means of a tag cloud [11, 6]. The resource can be easily retrieved by the keyword-resource association. Tags represent one type of (in this case user-added) metadata. However, there can be other domain-specific metadata types.

Approach that uses them to navigate, search and explore the document collection is called faceted classification or search [8]. Facet browsers have been very successful in recent years; the extensive survey can be found in [18]. There are several problems associated, e.g. how to personalize the faceted interface to the user needs [15, 16].

We have identified several limitations of the existing approaches:

— *low adaptability and limited support of multiple organization strategies* – existing approaches usually support only one strategy for which they were designed, thus ignoring different personal organizational preferences of the users as well as their habits.
— *manual filing of a new resource* – every resource has to be filed into a predefined structure manually, either into a folder or by assigning a tag. In addition, resources can be often assigned into only one category (folder). This is not true for tags, but on the other hand, they do not offer stable transparent library structure required for users using structuring strategy.
— *limited support for reorganization of the collection* – as the collection grows, the originally designed organizational structure can become too limited or simply no longer sufficient for maintaining the desired level of transparency and navigability. Existing approaches require manual reorganization which can be very demanding with respect to time and effort.
— *limited support for resources cleaning* – it is common for filing strategy, however users usually need to manually edit each unsorted resource and file it into the right location in the library.

Facets can address many of these limitations with their ability to automatically classify the collections of resources and to construct ad-hoc views. But because of their dynamic nature, they are rarely used for organization, in which the users usually rely on static personal structures, the state of which does not change until explicitly updated, thus allowing them to re-find the documents in the collection.

3   Method for Personalized Web Resources Organization

We propose our organization method based on the faceted search paradigm. The main advantage of faceted search is, that it allows users to construct arbitrary views on the underlying collection of documents, in our case the user's personal resources library.

Each resource in the collection is described by its associated metadata. In the domain of digital libraries of research papers, these would be e.g. authors, title, publication year, publication name, pages, etc. They represent non-overlapping (orthogonal) categories, i.e. facets, each describing particular aspect of a resource. Since they do not overlap, it is possible to combine them to better specify the given resource.

In order to find a balance between a static organization structure and a dynamic nature of facets, we utilize a concept of a *facet tree* based on the facet folders originally proposed in [17], which we enrich by the special *Search* facet and by the clustering of the documents with co-occurring facet values.

3.1   Facet Tree

Facet tree allows the users to define hierarchies of the selected facets, thus automatically organizing their collection of documents. An example of such a hierarchy can be seen in Fig. 1; the user selected *Keywords* facet representing the keywords added to the documents by their authors at the first (root) level and *Year* facet representing the publication year at the second level.

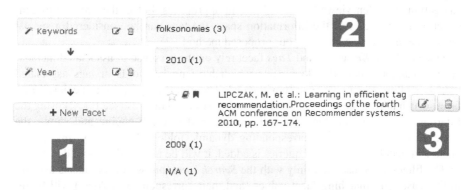

Fig. 1. Example of a facet tree hierarchy, with *Keywords* and *Year* facets selected (1). Folder with keyword *folksonomies* (2) is expanded showing three subfolders containing also *NA* value. User can edit the resource and its metadata or remove it from the collection (3).

Thus, the collection is automatically divided into dynamic folders, where each folder represents one facet value (one author-specified keyword in our example, such as *collaborative tagging*). Each first-level folder is further divided into folders based on the publication year. Only non-empty folders are visualized to the users. In case of a missing facet's value, the document is assigned into the special *NA* (not available) folder at its corresponding level. If the new document is added into collection, it will now be automatically added into existing structure based on its metadata. If the document matches more than one facet value (e.g. when it has more than one associated keyword), it will be added into each of the corresponding facet tree branches. This eliminates the problem the users often face when using the classical folders, which allow the documents to be added only at one place in the folder hierarchy.

Important is, that the facet tree maintains its state (structure defined by the user). On the other hand, it can be easily adjusted if necessary. Individual facets in chained facet tree can be removed or added creating context views on demand (thus allowing to be used as a search tool as well).

3.2   Proposed Facets

Because we focus on the domain of digital libraries and specifically on the collections of research papers, we use metadata usually associated with these documents as our facets, namely *author*, *publication year*, *publication name* and *publication type* (proceedings, journal, book etc.). We also use date when the document was *added* to the personal library and the date, when it was *last accessed* by the user.

These types of metadata are more categorical and less descriptive, i.e. they do not convey much information regarding the papers' content. For this purpose, we use already mentioned *Keywords* facet (representing the keywords specified by the papers' authors). It is an example of a narrow folksonomy; however, as shown in [2] a broad folksonomy, i.e. created collaboratively by the users, is better for navigational purposes, as it provides more paths to the resources and utilizes directly the vocabulary of the users. Therefore, we use *Tags* facet as well. Faceted search and tag navigation are often viewed as two different approaches, but in their ability to construct arbitrary views of the information space the tags can be considered a special facet – the fact that we utilize in our proposed method.

Since both the *Keywords* and *Tags* facet rely on the presence of associated metadata, they cannot handle the documents which have no keywords or tags associated (which is often the case especially with tags). Therefore, we propose a special *Search* facet, which allows the user to specify arbitrary keyword queries (see Fig. 2) using the filtering feature. It runs the specified queries in the search engine and shows the retrieved documents in the corresponding dynamic folders. Now every time a new document matching one of the queries is added, it will be retrieved in its folder.

The filter can be used not only with the *Search* facet, but with other facets as well; in that case only matching facet values (and their corresponding folders) will be retrieved (e.g. when the user is interested only in certain authors or only in papers from certain year range).

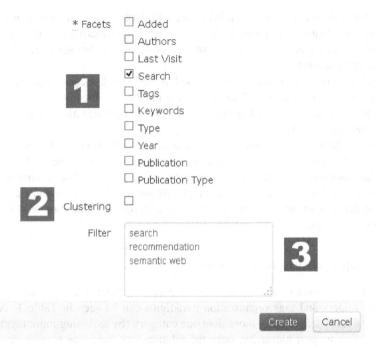

Fig. 2. Interface for specifying the new facet in the tree hierarchy (1) with optional clustering (2) and filtering (3). More than one filter value can be added (one per line).

3.3   Identification of the Clusters of Related Documents

Keywords as well as *Tags* facets are useful for decomposing the collection by documents' topics, but they can lead to many relatively small folders, when only a few documents share the same keyword or tag. Therefore, we provide the users with a possibility to cluster related documents with the co-occurring metadata values. This can be used with other facets as well, namely with *Authors* facet; in that case the authors are clustered based on the co-authorship relationships.

The clustering algorithm works as follows:

```
foreach folder (facet value) F do {
  foreach candidate folder (facet value) CF do {
    if |F ∩ CF| >= |F|/2 {
      parent_cluster(F) <- parent_cluster(CF)
      foreach CC in child_clusters(F) do {
        parent_cluster(CC) <- parent_cluster(CF)
}}}}
```

It is a simple algorithm that merges two folders (based on their facet's value), if they have at least half of the documents in common. Any given folder can be merged multiple times, thus allowing to find different clusters (cluster combinations) in the collection. In addition, it is possible that two folders with no intersection will end up together in a cluster, if there is a third folder with large enough intersection with both of them. This feature can prove to be useful, when considering e.g. the authors; we can find this way communities that have a common author, but are themselves different (not collaborating).

Other clustering algorithms could be used as well, e.g. k-means with k set based on the collections' size and the desired average number of documents in a cluster, or hierarchical clustering algorithm that would allow to optimize also the number of the clusters presented to the user.

Overall, clustering provides a better view of the collection and helps the users to discover hidden, or not so obvious relationship between the documents. Its power lies in further combination of facets. An interesting example would be to have *Keywords* facet at the first level, which clusters topically similar documents and then *Authors* facet on the second, which helps to visualize different communities working on the same research topics.

3.4 Support of Organization Strategies

Comparison of some of the features of the proposed facet tree organization method with the folders and tags organization paradigms can be seen in Table 1. Although tags allow resources to be in more than one category (by assigning more tags), it does not (usually) support hierarchy, only the relationships based on tags co-occurrence. As to the arbitrary views construction, it can be partially achieved also with folders, but it will always be a static structure, i.e. no ad-hoc views will be possible. Lastly, folders as well as tags require user action in order to categorize resource (inserting it to the selected folder or assigning tags).

Table 1. Feature comparison of the facet tree with most common organization paradigms

	Resource in more categories	Supports hierarchy	Provides arbitrary and ad-hoc views	Categorize resource automatically
Folders	✗	✓	✗	✗
Tags	✓	✗	✓	✗
Facet tree	✓	✓	✓	✓

In addition, we have designed our approach to meet most of the common user needs and use cases in personal information management (PIM) strategies. Table 2 shows, how each of the user actions identified in [3] is mapped to the actions of our proposed method and reflects, how the result can be achieved using our proposed method.

Archiving a new resource is simple, because it is done automatically based on the resource's metadata. If the metadata is missing, the resource is added to the special *NA* folder. Users are thus motivated to clean their collection by adding the missing values as it directly influences the filing of the resource in the facet tree structure. Resources can be searched by either navigating in the facet tree structure or by creating an ad-hoc structure using the *Search* facet with the search query as its filter value. Lastly, the structure can be easily reorganized by changing the defined facet hierarchy.

Table 2. Mapping of PIM strategies to the corresponding actions in our proposed approach

User action	Proposed approach
Piling strategy	
Simple resource archiving	Only one click is needed to archive resource.
Resources listing based on archiving time	Select *Added* facet.
Resources search	Select facet and input search criteria for its values. Can use special *Search* facet.
Filing strategy	
Archive resource by filing into structure	Adding new resource automatically extracts metadata and the resource is filed into the facet tree structure.
Personal library browsing	Chain facets as required.
Cleaning up large collection	Users see resources with missing metadata in the *NA* folder. They can edit the resource, so that it correctly files within the structure.
Structuring strategy	
Upfront structure creating	Chain facets as required; can be done at any time.
Archive resource	Adding new resource automatically extracts metadata.
Navigation to the specific resource	Chain facets as required.
Resources search	Select facet and input search criteria for its values. Can use special *Search* facet.

4   Evaluation

We evaluated our proposed approach in a bookmarking system *Annota*[5] [12], which is being developed at our university as a part of a research project TraDiCe [7]. It allows users to bookmark any resource on the Web using the browser extension and manage their personal libraries using the web interface. Currently, it is used by 185 users, mostly consisting of the students and staff of our faculty.

---

[5] http://annota.fiit.stuba.sk/

Annota provides special support for resources from digital libraries; it can automatically download the associated metadata, such as authors, title, year, etc. Users can annotate resources in their library with various annotation types (tags, comments or highlights). Important feature is sharing of the bookmarks within groups, thus supporting collaboration of researchers, or of students and their supervisors.

Because Annota supports different kinds of personal resources management, namely folders and tags, we used it to compare our proposed approach with these traditional paradigms.

4.1   Analysis of Folders Usage

We wanted to find out, how the users usually work with their document collections, how many documents these collections typically contain, or if users use folders to organize them. Therefore, we analyzed real usage data from Mendeley for 31 users who synchronized their libraries with Annota (mainly master or doctoral students).

We summarize the data in the Table 3. We found out that the analyzed users have in average 233 documents in their personal libraries. This number is, however, influenced by a small number of users with more than 1000 documents. Average user has typically a lot less, with median being 95 documents per collection.

From the analyzed users 77.4% uses folders to organize their documents, while 22.6% of the users have no folders. When they do have folders, they have typically no more than four and only at first level with no subfolders. Interestingly, even users, who use folders, have a quite large part of their collection unorganized with about 30% of documents, which do not belong to any folder. Only 19% of users have every document filed in an appropriate folder. This can suggest problems with folders' usage; they can either not know, where they should file the document or simply do not have time to do it. Our data also show us that the ideal number of documents in a folder for the users is about 11, although there are some extremes as well.

Table 3. Summary of folder usage analysis

	Mean	Median	Min	Max
No. of documents in collection	233.10	95	3	1336
No. of folders	10.32	4	0	44
No. of documents not in any folder	42%	28%	0%	100%
No. of documents in a folder	11.23	10.81	0	167
Maximal depth	1.29	1	0	4

Based on these data, we can conclude that about 20% of users use piling strategy, while probably relying heavily on the provided search, 20% of users use structuring strategy and the largest group – about 60% – prefer filing strategy. There is also a difference in the folders' purpose (based on the names of folders); 56% of users use folders predominantly for task-oriented organization, 24% organizes documents into folders based on their topics and 20% mixes both strategies.

4.2   User Study

In order to evaluate our proposed approach, we carried out a qualitative experiment – a user study with six participants. The participant were all male between 23 and 30 years with strong background in computer science and informatics (one master student, two doctoral students and three post-docs).

Our hypothesis was that the users can more easily (as measured by time and effort) manage their personal library using the facet tree as compared to the traditional folder approach. We also assumed that the facet tree structure will be robust enough to support various tasks without the need to change it.

Participants of the Study. Before the experiment we interviewed the participants to assess their information management habits and preferred strategies; results are summarized in Table 4. All the participants except one had previous experience with using Annota, although only two were using it actively to bookmark resources on the Web and in the digital libraries. Mendeley was stated as the preferred personal library management tool in five cases. One participant claimed not to use any available tool; instead, he uses folders provided by the operational system.

Five participants use folders, but their use is different. Only one participant prefers structuring strategy; two participants use folders based on their tasks (such as thesis, research paper etc.), two combine topic and task folders and one uses solely topic folders. One user claimed not use folders; he piles all the resources and uses search to retrieve the documents. Interestingly, the users tend not to change their defined structure; mostly because it suits their needs or because it would be too demanding.

Table 4. Results of pre-experiment questionnaire

	Experience with Annota or Mendeley	Uses folders	Retrieves documents in the structure	Reorganizes the structure	Preferred inform. strategy
Participant #1	✗	✓	✓	✓	filing
Participant #2	✓	✓	✓	✗	filing
Participant #3	✓	✓	✓	✓	structuring
Participant #4	✓	✓	✗	✗	filing
Participant #5	✓	✗	✗	-	piling
Participant #6	✓	✓	✗	✗	filing

We can say that the distribution of preferences and strategies in the selected group of participants corresponds with the distribution discovered during the analysis of the folders usage described in previous section.

Experimental Setting. Each participant was supposed to solve four tasks during the experiment. First task was to organize the given document collection using the folders and facet tree (half of the participants used first folders and half facet tree). We considered the task to be successful, if the users identified the underlying topics in the dataset and organized it accordingly.

The second task was to archive a new source in the digital library, file it into folder structure constructed in the first task and try to locate the resource in the facet tree structure. Third task emulated conditions, when the users want to locate the resource they know they have in the library, but their information is incomplete (they do not remember all the necessary information); in our case we provided them the name of the first author, main topic and the conference, at which it was published. And lastly, the fourth task was to reorganize the collection using folders as well as facet tree.

All the participants were provided with the same collection of 125 documents that was created as a subset of the Annota library dataset, which is also publicly available[6] for research purposes. Overall, the dataset consists of approximately 16,000 documents which were collected as the users of Annota travelled in the information space of digital libraries, such as ACM DL[7] or IEEE Xplore[8]. It thus contains groups of related documents which reflect research interests of the users. Beside the documents themselves, it contains extracted information on other entities, such as authors (currently 230,000) or author-added keywords (about 130,000).

The collection provided to the participants covered different topics, namely search, recommendation, tags and folksonomies, semantics, web and summarization with some random documents to simulate "noise".

After each task, participants were asked to fill in the corresponding questions in the prepared questionnaire. Similarly, at the end of the experiment, we asked them to globally evaluate the proposed facet tree interface.

Task 1. The participants were confronted with a new interface. At first, they had problems with understanding the facets' meaning, but after a quick explanation and a few trials with different facets, they were able to use it. For the organization of the collection, they almost exclusively used *Keywords* or *Search* facet at the first level. Typical sequence was to try the *Keywords* facet with clustering option first; this helped them to discover the underlying topics in the collection. For better results, they then proceeded to use the *Search* facet, where they could specify their own search terms.

A few participants suggested that they would want to use combination of the *Tags*, *Keywords* and *Search* facets, as they are all concerned with the documents' topics. At the second level (if they used it at all), participants used *Authors* or *Year* facet. *Added* and *Last access* facets were also mentioned as helpful, but they would prefer it as a sorting criterion instead of the separate facets.

---

[6] http://annota.fiit.stuba.sk/dataset
[7] https://dl.acm.org/
[8] http://ieeexplore.ieee.org

When the participants used first the facet tree, it helped them to quickly design the folder structure (more than a half of the participants agreed that it helped them with the domain overview). Otherwise the participants went through the documents in a sequence and created the corresponding folders. On the other hand, when creating the folders first, it also helped them to more quickly understand the collection; in that case, the participants usually applied directly the *Search* facet with the terms discovered during the folders creation. The chart in the Fig. 3 shows, how the users perceived the effort when creating the organizational structure, which is in favor of the facet tree.

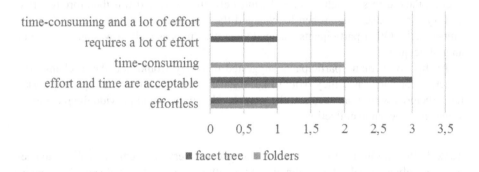

Fig. 3. Effort required for organizing the collection of documents (values are missing, if no participant selected that answer, as is the case e.g. with "time-consuming" for the facet tree)

We also enquired about the satisfaction with the resulting structure (see Fig. 4). On the scale 1 to 5, the four was the most frequent with the facet tree. Folders were rated variously, but on average worse. As a main disadvantage of folders was perceived the possibility of assigning the resource only to a single folder in the hierarchy.

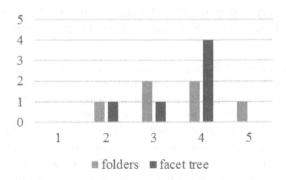

Fig. 4. Satisfaction with the resulting structure

Task 2. Finding the newly archived resource and filing it into appropriate folder was trivial, because it was first on the list (resource were sorted by the date of last access). However, two participants reported, that they hesitated, into which folder to file the resource.

Locating the resource in the facet tree was for most of the participants easy as well. The metadata from the added resource were extracted correctly and thus the participants did not have to locate it in the *NA* folder. However, they appreciated that the facet tree approach motivates them to clean the data (i.e. to add missing metadata values) in order to remove the resources from the *NA* folder to the appropriate one.

One participant had a *Keyword* facet at the first level, which resulted into going over all the clusters, which was not the most effective way and was therefore reported as very time-consuming. When he changed the structure to a better one, he found it immediately. Other participants changed their structures as well; it was an example of an ad-hoc query.

At this point, many participants suggested that they would not like to change the structure for each task; they would like to have more facet trees for repeating tasks, i.e. structures with more facets at the same level, which would provide them different views over the same collection.

Task 3. Re-finding a resource in a collection is a very frequent task. We often remember only a partial information, e.g. only author, or part of the title etc. This can be problematic to do with folders, because they usually provide only one view of the collection. In addition, if the users hesitate between two folders when archiving the resource and then decide for one of them, they will probably check both of them in the future, because they will not remember, where they filed it; a problem reported by one of our participants. Also, if we forget to file the resource (as is the case in 30% of the users' documents according to our analysis), the structure becomes useless and the user has to go over the resources sequentially or use a search.

Re-finding with facet tree depends on the used structure and the available metadata. If the current structure does not allow for easy location of the resource, users can easily change it; this happened in 67% of cases during our experiment.

Task 4. We also discussed with the participants the task of reorganizing their personal libraries. According to our pre-experimental questionnaire, this does not seem a very frequent task. The participants tend to add new folders to their library, but do not change the existing ones. One of the reasons is probably the effort that it requires – 60% of participants rated merging as well as partitioning of the folders as difficult (rating 4 and 5 on the 5-point scale). It largely depends on the provided folder interface, but it usually means going over the whole collection manually and moving the resources or folders to the new ones, which can be very time consuming.

Here, the facet tree clearly prevails, as was proved in the three experimental tasks. Cost of changing the tree is very low, allowing the users to experiment and use it also to explore new views of the collection or create an arbitrary ad-hoc view for the task at hand.

5   Discussion and Conclusions

In this paper, we proposed a facet tree approach as a method for organization of Web resources with a special focus on the digital libraries. Our main contribution is in providing the clustering of the facets' values in order to create more meaningful grouping of documents and the special *Search* facet which enables to organize documents based on user-specified search queries.

We evaluated our proposed approach in bookmarking system Annota by carrying out a user study. Our results indicate that the facet tree is an efficient tool for organization of the personal libraries of the web resources:

— It provides users with an overview of the domain and helps them to uncover not so obvious or even hidden relationships using a combination of facets and clustering of the related documents based on their facet' values.
— It supports creation of stable organizational structures, as well as arbitrary ad-hoc queries (views) for the task at hand.
— It automatically files the resources into a predefined structure.
— It can help retrieve documents with only a partial information about it.
— It supports various personal information management strategies (piling, filing and structuring) and different folder usage styles (topic-based, task-oriented). Task-oriented folders can be easily achieved with an added effort of tagging the resources according to the task and then using the *Tags* facet.

As to our hypothesis, we proved that the facet tree outperforms the folders approach in terms of time and effort when organizing the collection for the first time or in the process of its reorganization as well as when locating (re-finding) the resource in the collection. It is also easier to add new resources to the structure, since this happens automatically.

On the other hand, our second hypothesis that the structure will be robust to changes proved to be false. As it turned out, most of the participants reorganized structure for each task. As a solution, we propose the concept of *facet forest* of individual trees that provides different views at the collection at the same time and thus eliminates the need to create a new facet tree structure again and again for the repeating tasks.

We can conclude that the participants of our user study appreciated the clustering option and used it frequently with the *Keywords* facet. On the other hand, there were too many small clusters (of one or two documents) that could have been clustered together, as the ideal number of first level clusters seems 7 ± 2 with 10-15 documents in each cluster according to our analysis. If there are more documents, than this structure could contain, the second level should be added following the same rule of thumb. For most of the users' collections the second level would be enough, since it could contain $7\times7\times10 = 490$ documents and as we found out, users have typically far less documents than that (with median of 95).

In addition, the participants found the *Search* facet very useful, as it allowed them to formulate their own search queries and organize resources accordingly. This gives them a powerful tool, especially if it is further combined with *Keywords* and *Tags* facet as was suggested by numerous participants during the experimental evaluation.

The reaction and feedback of the participants was overall positive. The average rating of the global satisfaction with the interface at its current state was 3.3 (on the five-point scale).

Positive is that the participants did not feel constrained by the fact that they could not directly change the filing of the resource within the automatically generated clusters. In fact, they spoke against it, saying, that then they could easily forget, that they did such changes and it would make it harder to re-find the resources in the future. On the other hand, hiding the unwanted resources from the search facet that match the queries would be appreciated by some of the participants.

Most of the participants (5 out of 6) were content with the provided functionality and would give even higher global rating if their suggestions to the functionality would be incorporated. From these the most important that we plan to add to the next version of the interface are the following:

— Enable to create more independent facet trees (i.e. a facet forest) in order to provide different views at the collection at once and to lower the need to create new views for each new task, i.e. to make the structure more stable with respect to the future changes.
— Combine *Tags*, *Keywords* and *Search* facet into one *Topic* facet. Also enable to assign names to specified facet values, which can be more or less complicated user queries. Because we log search queries performed by the users in the digital libraries, we could extend this functionality to actually suggest the users the queries for the combined *Topic* facet based on their search history.
— Add the possibility to set granularity of the clustering algorithm and avoid creating too small clusters (or too big for that matter).
— Enable sorting of the resources in the dynamic folders based on the document metadata values, as well as their easier filtering.
— Add examples of facets or previews to make them easier to understand and use by the novice users.

In addition, there is a potential to use the proposed approach not only for the organization of the personal document collections. Other viable scenarios include exploratory search or finding of new resources in the public library that match the criteria given by the facet tree structure.

Acknowledgement. This work was partially supported by the Scientific Grant Agency of the Slovak Republic, grant No. VG1/0675/11 and VG1/0971/11 and by the Slovak Research and Development Agency under the contract No. APVV-0208-10.

References

1. Elsweiler, D., Baillie, M., Ruthven, I.A.N.: On Understanding the Relationship between Recollection and Refinding. Journal of Digital Information 10(5), 1–31 (2009)

2. Helic, D., Körner, C., Granitzer, M., Strohmaier, M., Trattner, C.: Navigational Efficiency of Broad vs. Narrow Folksonomies. In: HT 2012: Proc. of the 23rd ACM Conf. on Hypertext and Social Media, pp. 63–72. ACM Press, New York (2012)
3. Henderson, S., Srinivasan, A.: Filing, Piling & Structuring: Strategies for Personal Document Management. In: IEEE Proc. of 44th Hawaii Int. Conf. on System Sciences, pp. 1530–1605. IEEE Press (2011)
4. Kelly, D.: Evaluating Personal Information Management Behaviors and Tools. Communications of the ACM 49(1), 84–86 (2006)
5. Miao, Y., Kešelj, V., Milios, E.: Document Clustering Using Character N-grams. In: CIKM 2005: Proc. of the 14th ACM Int. Conf. on Information and Knowledge Management, pp. 357–358. ACM Press, New York (2005)
6. Molnár, S., Móro, R., Bieliková, M.: Trending Words in Digital Library for Term Cloud-based Navigation. In: SMAP 2013: Proc. of the 8th Int. Workshop on Semantic and Social Media Adaptation and Personalization, pp. 53–58. IEEE CS, Washington, DC (2013)
7. Návrat, P.: Cognitive Traveling in Digital Space: From Keyword Search through Exploratory Information Seeking. Central European J. of Comp. Science 2, 170–182 (2012)
8. Perugini, S.: Supporting Multiple Paths to Objects in Information Hierarchies: Faceted Classification, Faceted Search, and Symbolic Links. Information Processing & Management 46(1), 22–43 (2010)
9. Rástočný, K., Tvarožek, M., Bieliková, M.: Web Search Results Exploration via Cluster-Based Views and Zoom-Based Navigation. J. of Universal Computer Science 19(15), 2320–2346 (2013)
10. Sahoo, N., Callan, J., Krishnan, R., Duncan, G., Padman, R.: Incremental Hierarchical Clustering of Text Documents. In: CIKM 2006: Proc. of the 15th ACM Int. Conf. on Information and Knowledge Management, pp. 357–366. ACM Press, New York (2006)
11. Skoutas, D., Alrifai, M.: Tag Clouds Revisited. In: CIKM 2011: Proc. of the 20th ACM Int. Conf. on Information and Knowledge Management, pp. 221–230. ACM Press, New York (2011)
12. Ševcech, J., Móro, R., Holub, M., Bieliková, M.: User Annotations as a Context for Related Document Search on the Web and Digital Libraries. Informatica 38(1), 21–30 (2014)
13. Šimko, J., Tvarožek, M., Bieliková, M.: Semantic History Map: Graphs Aiding Web Revisitation Support. In: DEXA 2010: Workshop on Database and Expert Systems Applications, pp. 206–210. IEEE Press (2010)
14. Teevan, J., Jones, W., Capra, R.: Personal Information Management (PIM) 2008. ACM SIGIR Forum 42(2), 96–103 (2008)
15. Tvarožek, M., Bieliková, M.: Personalized Faceted Navigation in the Semantic Web. In: Baresi, L., Fraternali, P., Houben, G.-J. (eds.) ICWE 2007. LNCS, vol. 4607, pp. 511–515. Springer, Heidelberg (2007)
16. Tvarožek, M.: Exploratory Search in the Adaptive Social Semantic Web. Information Sciences and Technologies Bulletin of the ACM Slovakia 3(1), 42–51 (2011)
17. Weiland, M., Dachselt, R.: Facet Folders: Flexible Filter Hierarchies with Faceted Metadata. In: CHI 2008: Proc. of the 26th Annual Conf. on Human Factors in Computing Systems, pp. 3735–3740. ACM Press, New York (2008)
18. Wei, B., Liu, J., Zheng, Q., Zhang, W., Fu, X., Feng, B.: A survey of Faceted Search. Journal of Web Engineering 12(1&2), 41–64 (2013)

Mobile Web User Behavior Modeling

Bozhi Yuan[1,2], Bin Xu[1,2], Chao Wu[1,2], and Yuanchao Ma[1,2]

[1] Department of Computer Science and Technology, Tsinghua University, China
[2] Tsinghua National Laboratory for Information Science and Technology, China
lawby1229@163.com, xubin@tsinghua.edu.cn, ariesnix93@gmail.com,
myccs@outlook.com

Abstract. Models of mobile web user behavior have broad applicability in fields such as mobile network optimization, mobile web content recommendation, collective behavior analysis, and human dynamics. This paper proposes and evaluates URI model, a novel approach to analyze user mobile Web usage behavior, which combines user interest modeling with location analysis. The URI model takes as input mobile user web logs associated with coarse-grained location drawn from real data, such as Event Detail Records(EDRs) from a cellular telephone network. We use probabilistic topic modeling to discover latent *user interest* from user mobile Web usage log. We validated the URI model against billions of mobile web logs for millions of cellular phones in Beijing metropolitan areas. Experiments show that the URI model achieves a good performance, and offers significantly high fidelity.

Keywords: Behavior modeling, Location mining, Behavior pattern analysis, Web Behavior, Latent Web Interest.

1 Introduction

Mobile web user behavior reflects human mobility, and has broad uses in mobile computing research and other fields of study. Models of mobile web user behavior can help answer questions in area like web user content recommendation, mobile network optimization, collective behavior analysis, and human dynamics.

Our work aims to produce accurate model of how people access mobile web and move in a city. To achieve this general aim, we define a number of more specific goals. The first goal is to discover the geographical region when users access mobile web, such as living, work, business, and way. Different users have the habits to access mobile web in different locations. The second goal is to discover user's interests on mobile web, such as news, sports, and entertainment etc. Different user likes access different kinds of website, which reflects his/her interests on mobile web. We use probability distribution to represent user's interest. The third goal is to discover the correlation among user, region and interest. Through analyzing users' regions and interests on accessing mobile web, we can find the behavior pattern of users.

The contributions of the paper are: *1)* We propose a novel probabilistic model to analyze mobile web user behavior, only based on the usage history of mobile

B. Benatallah et al. (Eds.): WISE 2014, Part I, LNCS 8786, pp. 388–397, 2014.
© Springer International Publishing Switzerland 2014

Web. The model proves its effectiveness on large scale of EDRs from real cellular operator. The model can be used to do city-level collective behavior analysis, as well as mobile Web usage prediction/service recommendation. *2)* A new approach is presented to discover regions by leveraging geographical feature and Web usage traffic history. Raw web log like EDRs can only give coarse information about user location, while our approach can give semantic information of user location such as living/work/business/way, which is important to make the model applicable to different scenarios.

The rest of the paper is organized as follows. In Section 2, we describe recent related work on Web user behavior analysis and location analysis. In Section 3, we give an overview of the data we use, and observe some important characteristics of the data in several aspects. Section 4 describes the discovery of *Region* from raw HTTP log. Section 5 formally describes the probabilistic topic model we use for user behavior modeling in detail. Section 6 describes the experimental results applying the framework on our dataset, and gives detailed analysis on the result in several different angels. Finally, we conclude the paper in Section 7.

2   Related Work

User interest and behavior mining based on Web log data has been a hot topic[1][2]. Some researchers use clustering methods to extract types of users [3][4]. They either do clustering on the users' perspective and cluster user into different types, or on the websites' perspective and make URL groups. *Nasraoui, et, al.* [5] study user behavior of a particular website based on tracking user profiles and their evolving.

As the mobile Web takes more and more proportions of people's total Web usage, study of mobile Web user behavior also gains a great attention recently. *Cui and Roto* [6] describe how people use the web on mobile devices by contextual inquiries, and analyze contextual factor as well as user activity patterns. *Tseng and Lin* [7] mine user behavior patterns in mobile web systems based on location trace, and do experiments using simulation. *Phatak and Mulvaney* [8] propose a fuzzy clustering method on URLs and users based on a distance matrix, and further do user profiling and recommendation. *Do and Gatica-Perez*[9] mine user pattern using mobile phone app usage, including mobile Web usage on mobile phone. *Verkasalo* [10] analyzes contextual patterns in mobile service usage statistically, using handset-based data, which includes location and Web usage data. Most of the studies use mobile phone collected data, which can hardly scale up due to the deployment limitation of their applications. Our method mine user behavior patterns in mobile Web usage from the mobile network service provider's perspective, which is both comprehensive and large scale.

3   Data Description

The data we use is the mobile Web usage log (HTTP request log) of the cellular network (including 2G and 3G) of a mobile operator. The dataset covers the geographical range of Beijing, the capital of China. The time range is from Oct. 24 to Nov. 14, 2012. One line in the dataset corresponds to a HTTP request/response pair occurred using cellular network. The structure of the data is shown in Table 1.

Table 1. Field details of the dataset

Name	Data Type	Description
User Id	String	Imsi Id, which is unique identifier of a SIM card
Latitude	Float	Latitude of the cellular tower
Longitude	Float	Longitude of the cellular tower
Request Time	DateTime	Time when the request occurs
Host	String	The domain name of the host
Content Type	String	The ContentType in HTTP

Overview. There are totally **578,134,225** complete records in the dataset. We clean the raw data by record with meaningless field or empty identification. After cleaning, there are **66.823% (386,332,325)** records left.

Users. There are **3,524,929** distinct users appears in the dataset. **40%** of all users make less than 10 requests in a whole week; and over **95%** of all users make less than 1,000 requests in a week. The average number of requests per user is **137.27**.

Hosts. There are totally **363,841** hosts that appear in the dataset. However, only **40.12%** of these hosts are visited by more than 1 user. Top 990 hosts take 97.877% of all records, and top 10 hosts take 58.1% of all records. It can be observed that the websites / services people use on mobile Web is much more concentrated than traditional desktop Web.

Locations. 856 distinct cellulars appear in the dataset, represented by the latitude and longitude of cellular towers. Raw cellulars are less meaningful for behavior modeling, so we cluster them into regions (See Section 4).

4   Geographical Region Discovery

Definition 1. *Region*. *A function region of the city is a minimal geographical region that serves a particular set of functions. The function set of the region is the same for most of the citizens.*

As a function region of a city is usually larger and contains several cellular towers, we have to cluster cellular towers into larger function regions, and find characteristics of each region.

4.1   Clustering Algorithm

We use an improved DBSCAN algorithm for cluster cellular towers. The definition of a cluster in DBSCAN is based on the notion of density reachability.

There are two parameters in DBSCAN: ϵ and MinPts. The $\epsilon - neighborhood$ of point p, denoted by $N_\epsilon(p)$, is defined by $N_\epsilon(p) = \{q \in D | dist(p, q) \leq \epsilon\}$. A native approach could require for each point in a cluster that there are at least a minimum number $(MinPts)$ of points in a $\epsilon - neighborhood$ of that point. A cluster, which is a subset of the points of the database, satisfies two properties: all points within the cluster are mutually density-connected, if a point is density-connected to any point of the cluster, it is part of the cluster as well[11].

We change the definition of $\epsilon - neighborhood$. In our algorithm, the definition of $\epsilon - neighborhood$ contains three part of restrictions: (A and B as two cellular towers)

- Record similarity sim_r, which represents the similarity of A and B on record amount. We believe that cellulars in the same region would have similar record distributions over different time of day.
- Record migration similarity sim_m. If there is a record of user u at A, followed by another record of u at B in a short time, we call this a *migrating record*.
- Geographical distance dg. $dg_{(A,B)}$ donates the geographical distance (in meters) between A and B.

4.2 Clustering Result

Using our algorithm and the parameter we set, 717 cellular towers are clustered into 126 regions, with the rest 139 as single-cellular regions marked by different colors. In total we get 265 regions. The regions are shown in Figure 1.

Fig. 1. Regions on map marked by cellular towers in the region

5   Model for User Interest Discovery

First of all, we give the definition of "user interest".

Definition 2. *User Interest. Interest of a user is a specific type of service / material that the user is interested in. A user may have several different interests, and each interest is associated with a weight, standing for the degree of fondness. On the other hand, user interest types are also website style types, and a website can also serves different user interest types.*

In order to extract latent user interests from their Web usage log, we propose a probabilistic topic modeling method based on LDA (Latent Dirichlet Allocation). In our model, the extracted latent layer represents a *User Interest* defined in Definition 2.

Topic Model is commonly used in text mining for discovering abstract "topics" in a set of documents. LDA (Latent Dirichlet Allocation)[12] is a commonly used topic model currently, and it has also been applied in discovering user behavior patterns[13]. LDA is a unsupervised, generative model, which models the generation of a document into a two-step process: choosing a topic based on topics distribution over a document; and choosing a word based on words distribution over a topic.

We use the bag-of-website representation of a user as a document, and propose a probabilistic topic model for user behavior modeling.

5.1   Website Discovery

Firstly, we will discuss the details about how we extract website from raw HTTP log of mobile Web. There are some formal definitions of the concepts used.

Definition 3. *User: A user if uniquely identified by a UserId in the dataset. It corresponds to a real person (Imsi number of the SIM card) using the mobile Web, regardless of what devices are used.*

Definition 4. *Host: A host is the domain name of the HTTP request. It may or may not be the address that is directly requested by the user / app.*

Definition 5. *Website. A website stands for a unit that provides material/ services to users as a independent entity, using several different hosts.*

Websites and hosts do not have strong corresponding relationships. We use a simple strategy to cluster hosts into websites. We treat each host as a vector $H_i :< C_{h_i,u_1}, C_{h_i,u_2}, ..., C_{h_i,u_n} >$, where C_{h_i,u_j} is the number of requests user j makes to host i. Then we calculate pairwise cosine-similarities between all hosts pairs using the vectors. Then we merge the websites with higher similarity.

5.2   Bag-of-Website Representation of User

We treat a user as a document, and websites as words. Each occurrence of a website in a user is associated with a region, corresponding to the location where the user visits the website.

Since raw HTTP log may not truly reflect user behavior, we do some transformation for forming the bag-of-websites of a user. we firstly divide a day into 48 half-hour time slot and treat each slot as a period of user behavior[13]. Then for each user u and each time slot t, the top 3 most visited websites are treated as been "visited". A user is then represented as $U = \{< w_{ui}, r_{ui}, c_{u,w_{ui},r_{ui}} >\}$, where $c_{u,w_{ui},r_{ui}}$ is the number of times that $< w_{ui}, r_{ui} >$ is visited by u.

5.3   User-Region-Interest (URI) Model

We combine users, geographical regions, user interests and websites in unified generative models. Distinguished by the practical interpretation of the "interest" layer and the strategies to engage *region*, we propose two different generative processes and two corresponding URI models.

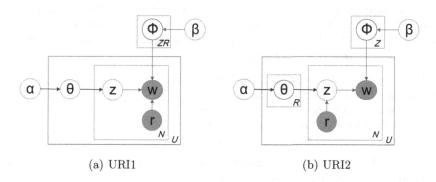

(a) URI1   (b) URI2

Fig. 2. Plate representation of User-Region-Interest (URI) model

URI Model 1. In the first model (URI1), the latent "interest" layer is treated as frequent user Web-using patterns, which is more related with users. Each user has a "interest" distribution, regardless of location; and within each "interest", there is a website distribution specific to each region. The generative process of URI1 is as follows (Figure 2(a)):

1. For each user u, draw θ_u from Dirichlet prior α;
2. For each interest z, draw $\phi_{(z,r)}$ for all $r \in R$ from Dirichlet priors β_z;
3. For each website appearance w_{ui} in user u:
   - draw a topic z_{ui} from a multinomial distribution θ_u;
   - draw a term for w_{ui} from multinomial distribution $\phi_{(z_{ui}, r_{ui})}$.

We use Gibbs sampling to estimate the model parameters, following [14]. For simplicity, we take fixed values for hyperparameters α and β (i.e. $\alpha = 50/T$, $\beta = 0.01$). We use Gibbs sampling to estimate the posterior distribution on w, z and r, then use the result to estimate θ and ϕ. The posterior probability can be calculated by :

$$P(z_{ui}|\mathbf{z}_{-ui}, \mathbf{w}, \mathbf{r}, \alpha, \beta) \propto \frac{m_{u,z_{ui}}^{-ui} + \alpha_{z_{ui}}}{\sum_z (m_{u,z}^{-ui} + \alpha_z)} \frac{n_{z_{ui},r_{ui},w_{ui}}^{-ui} + \beta_{w_{ui}}}{\sum_w (n_{z_{ui},r_{ui},w}^{-ui} + \beta_w)} \tag{1}$$

$m_{u,z}$ denotes the times that interest z is assigned to user u, and $n_{z,r,w}$ denotes the times that interest z is assigned to user website w in region r. Superscript $-ui$ means that the quantity is counted excluding the current instance.

After the sampling convergences, the multinominal distribution parameters θ and β can be estimated as follows:

$$\theta_{u,z} = \frac{m_{u,z} + \alpha_z}{\sum_{z'} (m_{u,z'} + \alpha_{z'})} \tag{2}$$

$$\phi_{z,r,w} = \frac{n_{z,r,w} + \beta_w}{\sum_{w'} (n_{z,r,w'} + \beta_{w'})} \tag{3}$$

URI Model 2. In the second model (URI2), the "interest" layer is treated as commonly appeared website clusters, which is more related with websites. Each "interest" has a unified website distribution regardless of location, while each user has a "interest" distribution specific to each region. The generative process of URI2 is as follows (Figure 2(b)):

1. For each user u, draw $\theta_{(u,r)}$ from Dirichlet prior α_u for all $r \in R$;
2. For each interest z, draw ϕ_z from Dirichlet priors β;
3. For each website appearance w_{ui} in user u:
   - draw a topic z_{ui} from a multinomial distribution $\theta_{(u,r_{ui})}$;
   - draw a term for w_{ui} from multinomial distribution $\phi_{z_{ui}}$.

Like in URI1, we use Gibbs sampling to estimate the model parameters, and the posterior probability can be calculated by :

$$P(z_{ui}|\mathbf{z}_{-ui}, \mathbf{w}, \mathbf{r}, \alpha, \beta) \propto \frac{m_{u,r_{ui},z_{ui}}^{-ui} + \alpha_{z_{ui}}}{\sum_z (m_{u,r_{ui},z}^{-ui} + \alpha_z)} \frac{n_{z_{ui},w_{ui}}^{-ui} + \beta_{w_{ui}}}{\sum_w (n_{z_{ui},w}^{-ui} + \beta_w)} \tag{4}$$

θ and β can be estimated as follows:

$$\theta_{u,r,z} = \frac{m_{u,r,z} + \alpha_z}{\sum_{z'} (m_{u,r,z'} + \alpha_{z'})} \tag{5}$$

$$\phi_{z,w} = \frac{n_{z,w} + \beta_w}{\sum_{w'} (n_{z,w'} + \beta_{w'})} \tag{6}$$

6 Experiments and Discussion

In this section, we will demonstrate the experiment results of applying the framework on our dataset, and give analysis based on the experiment results.

6.1   User Interest Discovery

We use $U = 469,297$ users, $W = 924$ websites, and $R = 265$ regions to evaluate the models. We have run both URI model and LDA model (using traditional LDA generative process) on same datasets. For comparison, we keep track of the corresponding region of a word when it is assigned a interest during the LDA sampling process, and use Equation 3 to calculate a $K \times R \times W$ Φ parameter matrix, and follow Equation 5 to calculate a $M \times R \times K$ Θ matrix, where M is number of users in the training set. The original Φ and Θ in LDA is called $\Phi 0$ and $\Theta 0$ here. We use LDA0 to donate the LDA model using $\Phi 0$ and $\Theta 0$, and LDA1 using Φ and $\Theta 0$, LDA2 donates the model using $\Phi 0$ and Θ.

Metrics. We use average Jensen Shannon divergence among topics to evaluate the quality of latent "interests". Jensen Shannon divergence (JSD) is commonly used for measuring similarity between probabilistic distribution describing the same random variable. Using different Φ and Θ matrix as parameters, we can calculate divergence for each model. For LDA1 and URI, in which website distribution within "interest" is specific to regions, JSD is calculated as:

$$JSD = \frac{\sum_{i=0}^{T} \sum_{j=i+1}^{T} \sum_{r=1}^{R} jsd(\Phi_{i,r}, \Phi_{j,r})}{\frac{T(T+1)}{2} \times R} \qquad (7)$$

For the rest of the models (LDA0, LDA2 and URI2), JSD is calculated as:

$$JSD = \frac{\sum_{i=0}^{T} \sum_{j=i+1}^{T} jsd(\Phi 0_i, \Phi 0_j)}{\frac{T(T+1)}{2}} \qquad (8)$$

The estimated distribution of websites for users can be used to represent their (potential) preference about websites, which can be used for understanding their behavior pattern, and for recommendation. We use perplexity to evaluate the performance of these proposed model probability distributions. Perplexity is commonly used how well a proposed distribution can predict samples from the target distribution. It is defined as

$$perp(\widetilde{\mathbf{p}}, \mathbf{q}) = 2^{H(\widetilde{\mathbf{p}}, \mathbf{q})},$$

where the exponent $H(\widetilde{p}, q)$ is the cross-entropy:

$$H(\widetilde{\mathbf{p}}, \mathbf{q}) = -\sum_{x} \widetilde{p}(x) log_2^{q(x)}$$

Smaller perplexity means better prediction. The averaged perplexity over all users is defined as the perplexity of the proposed distribution.

Topic number. We have tried different numbers of user interests (topics) K. Generally, reproduction performance improves(perplexity drops) when increase K, and stays stable after K reaches a certain value. For topic divergence, JSD decreases with K. The experiment result is shown in Figure 3(a) and 3(b).

Since , Considering both generalization and reproduction ability of the model, JSD and *Perplexity* are our major concern, we set $K = 50$ in our model.

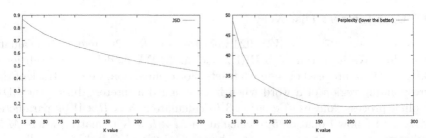

(a) JSD with different K value (URI model) (b) Perplexity of **p** with different K value (URI model)

Fig. 3. The experiment results

Table 2. Performance of LDA and URI models

Model Type	D0		D1		D2	
	JSD	Perplexity	JSD	Perplexity	JSD	Perplexity
LDA0	0.040	203.805	0.079	183.021	0.064	305.282
LDA1	0.211	181.824	0.307	89.910	0.247	158.881
LDA2	0.211	181.824	**0.990**	69.853	**0.992**	64.621
URI1	0.925	**98.229**	0.751	**34.423**	0.818	**54.079**
URI2	**0.979**	103.498	0.981	73.668	0.974	121.978

The performance of different methods on the the whole dataset and two random-generated subsets are shown in Table 2.

It can be seen that, generally, the URI model out-performs LDA significantly. URI Model 2 and LDA 2 discover better latent "interest", reflecting website clusters, while URI Model 1 can better estimate Web using behavior for each user.

7   Conclusion

In this paper, we have presented a framework for mining user behavior pattern in mobile Web usage. A method is proposed to cluster cellular towers into regions according to their mobile Web traffic log. We have also presented a probabilistic topic modeling method to extract latent "user interests" from mobile Web usage log with location. We have applied the proposed framework to a real-world large scale dataset from a Beijing, capital of China, covering more than 3 million users.

Acknowledgement. This work is supported by China National Science Foundation under grant No.61170212, China National High-Tech Project (863) under grant No.SS2013AA010307, and Ministry of Education-China Mobile Research Fund under grant No.MCM20130381. Beijing Key Lab of Networked Multimedia also supports our research work.

References

1. Kosala, R., Blockeel, H.: Web mining research: a survey. SIGKDD Explor. Newsl. 2(1), 1–15 (2000)
2. Srivastava, J., Cooley, R., Deshpande, M., Tan, P.N.: Web usage mining: discovery and applications of usage patterns from web data. SIGKDD Explor. Newsl. 1(2), 12–23 (2000)
3. Xu, J., Liu, H.: Web user clustering analysis based on kmeans algorithm. In: 2010 International Conference on Information Networking and Automation (ICINA), vol. 2, pp. V2-6–V2-9 (October 2010)
4. Mobasher, B., Cooley, R., Srivastava, J.: Creating adaptive web sites through usage-based clustering of urls. In: Proceedings of the 1999 Workshop on Knowledge and Data Engineering Exchange (KDEX 1999), pp. 19–25 (1999)
5. Nasraoui, O., Soliman, M., Saka, E., Badia, A., Germain, R.: A web usage mining framework for mining evolving user profiles in dynamic web sites. IEEE Transactions on Knowledge and Data Engineering 20(2), 202–215 (2008)
6. Cui, Y., Roto, V.: How people use the web on mobile devices. In: Proceedings of the 17th International Conference on World Wide Web, WWW 2008, pp. 905–914. ACM, New York (2008)
7. Tseng, V.S., Lin, K.W.: Efficient mining and prediction of user behavior patterns in mobile web systems. Information and Software Technology 48(6), 357–369 (2006), WAMIS 2005 Workshop
8. Phatak, D., Mulvaney, R.: Clustering for personalized mobile web usage. In: Proceedings of the 2002 IEEE International Conference on Fuzzy Systems, FUZZ-IEEE 2002, vol. 1, pp. 705–710 (2002)
9. Do, T.M.T., Gatica-Perez, D.: By their apps you shall understand them: mining large-scale patterns of mobile phone usage. In: Proceedings of the 9th International Conference on Mobile and Ubiquitous Multimedia, MUM 2010, pp. 1–27. ACM, New York (2010)
10. Verkasalo, H.: Contextual patterns in mobile service usage. Personal and Ubiquitous Computing 13, 331–342 (2009)
11. Ester, M., Kriegel, H., Sander, J., Xu, X.: A density-based algorithm for discovering clusters in large spatial databases with noise. In: Proceedings of the 2nd International Conference on Knowledge Discovery and Data Mining, pp. 226–231. AAAI Press (1996)
12. Blei, D.M., Ng, A.Y., Jordan, M.I.: Latent dirichlet allocation. J. Mach. Learn. Res. 3, 993–1022 (2003)
13. Farrahi, K., Gatica-Perez, D.: Discovering routines from large-scale human locations using probabilistic topic models. ACM Trans. Intell. Syst. Technol. 2(1), 3:1–3:27 (2011)
14. Griffiths, T., Steyvers, M.: Finding scientific topics. Proceedings of the National Academy of Sciences of the United States of America 101(suppl. 1), 5228–5235 (2004)

Effect of Mood, Social Connectivity and Age in Online Depression Community via Topic and Linguistic Analysis

Bo Dao, Thin Nguyen, Dinh Phung, and Svetha Venkatesh

Deakin University, Victoria 3216, Australia
{dbdao,thin.nguyen,dinh.phung,svetha.venkatesh}@deakin.edu.au

Abstract. Depression afflicts one in four people during their lives. Several studies have shown that for the isolated and mentally ill, the Web and social media provide effective platforms for supports and treatments as well as to acquire scientific, clinical understanding of this mental condition. More and more individuals affected by depression join online communities to seek for information, express themselves, share their concerns and look for supports [12]. For the first time, we collect and study a large online depression community of more than 12,000 active members from Live Journal. We examine the effect of mood, social connectivity and age on the online messages authored by members in an online depression community. The posts are considered in two aspects: what is written (topic) and how it is written (language style). We use statistical and machine learning methods to discriminate the posts made by bloggers in low versus high valence mood, in different age categories and in different degrees of social connectivity. Using statistical tests, language styles are found to be significantly different between low and high valence cohorts, whilst topics are significantly different between people whose different degrees of social connectivity. High performance is achieved for low versus high valence post classification using writing style as features. The finding suggests the potential of using social media in depression screening, especially in online setting.

Keywords: weblog, linguistic, mental health, depression communities.

1 Introduction

Depression is a mental disorder that is widespread in the world. It affects more than 350 million people and is believed to be responsible for almost 1 million suicides each year [19]. The demand for controlling depression is on the rise, requiring a continuing support from various sources - relatives, friends and neighbors. The task is challenging since people with mental disorder often face a decline of physical contact and social cohesion. In this situation, social media become an ideal communication channel for people that need support. It provides them a convenient venue to make friend and share experience to fight depression [12].

B. Benatallah et al. (Eds.): WISE 2014, Part I, LNCS 8786, pp. 398–407, 2014.
© Springer International Publishing Switzerland 2014

The rise of using social media as a place to vent about mental problems makes social media a new venue for health studies. Indeed, recent studies have investigated into depression in online setting, including Twitter [3,14] and Facebook [7]. Still, these studies have not considered the community context in communication within this online channel.

In this study, we collect data from *depression.livejournal.com*, the largest community in Live Journal interested in "depression". We investigate into the impact of mood, age and social connectivity on the content (topic) that members of the community are interested in and the way they express arguments in their blog posts (linguistic). We first use statistical tests to examine if topics and language styles are discriminative between different cohorts defined by mood, age and social connectivity. We then further validate the discrimination by using topics and language styles as features for classification of the cohorts.

The main contribution of this work is to provide of depression community on their topics of interest and language styles. Another contribution is a set of potentially powerful predictors for depression, which possibly can be used as a base for depression screening, especially in online setting.

This paper is organized as follows. Related work is discussed in Section 2. Method and experimental setup are described in Section 3. Section 4 presents the results of hypothesis testing and classification of different cohorts by age, mood and social connectivity using topics and language styles as features. Section 5 concludes the paper and remarks on potential implications.

2   Related Work

The link between depression and age has been analyzed in both social and epidemiological sciences [5,21]. These studies show that depression declines in early adulthood and raises in late life with the highest at 80 years old or older. In a study of depression and mood, mental disorders were found to be linked with mood swings [20]. Mood swings and other social aspects have also been shown to be connected to mental social capital [16].

Researches on major depression issues in online setting have been proposed in the literature. For Twitter, depressive mood of bloggers was found to be portrayed in tweets [13]; linguistic style and diurnal patterns of tweeting were used as predictors of affective disorders [3]. For Facebook, status updates were found to be strong indicators of major depressive episodes [6,7]; suicide notes were also found in the social network [17].

To characterize an online community, two popular ways are via topics of interest and language style – the way people express the topics. For example, for tweets, topics and linguistic style were used as features to predict the life satisfaction of US states [18]. For web-blogs, latent topics were found to have greater predictive power than linguistic features when predicting if a blog post is made by an autism community [11]; the variation of these two features in general Live Journal communities were found to be associated with age, sentiment and social connectivity [8,9]; and the features were used to determine if a post is

Fig. 1. Cloud visualization of mood tags used in the community

made in depression community [10]. However, using linguistic style and topics to differentiate cohorts defined by mood, social connectivity and age within a depression community has not been investigated.

3 Method and Experiment Setup

3.1 Dataset

In this paper, data for experiments was crawled from *depression.livejournal.com*, the largest community in Live Journal interested in "depression", describing itself as "*a safe and open community for those experiencing depression*". Live Journal data was chosen since it enables community setting, allowing people to join communities of interest. Also, Live Journal offers a list of 132 predefined moods for bloggers to tag to their posts. It is suitable for a sentiment analysis of the online communities.

The crawled dataset spans over 15 years from 2000 to 2014. More than 12,000 active members – who have posted and/or commented to the community – have made more than 40,000 posts and approximately 180,000 comments within the period. Of the posts, about 15,000 were tagged with one of the predefined moods. A cloud visualization of the mood tags in the community is shown in Figure 1. As expected, the mood depressed is dominantly tagged in the community. We construct three sub-corpus from the data, based on mood valence, age and social connectivity of users, as follows.

Age based corpus: The first corpus is on age difference. In the depression community, about 2,500 bloggers reveal birth date in their profile, and the age distribution in the community is plotted in Figure 2. The corpus is created by selecting users at the extremes of the age spectrum. To avoid noise, we set the lower bound to the number of the 2.5th percentile of data points and the upper bound is the 97.5th percentile. This results in a set of 500 *young* bloggers aged from 22 (at the 2.5th percentile) to 26. Similarly, a set of *old* bloggers is formed by adding those aged 51 (at the 97.5th percentile) and less, to a cut-off of 500 bloggers.

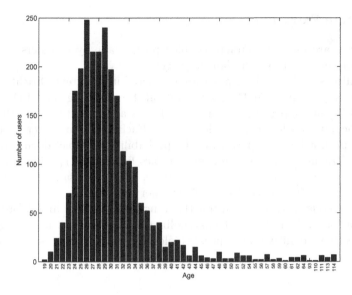

Fig. 2. Age distribution in the depression community

Valence based corpus: This corpus consists of 400 posts tagged with high valence moods (≥ 7.0) and 400 posts tagged with low valence mood (e.g., *depressed* mood with valence of 1.83). The valence value of mood is measured using Affective Norms for English Words (ANEW) [2] – a sentiment bearing lexicon. Words in this lexicon are rated in terms of valence, ranging from 1 (*very unpleasant*) to 9 (*very pleasant*).

Social connectivity based corpus: In term of social connectivity, Live Journal offers two types of connecting: making friends and joining communities. On making friends, for a user, incoming links are considered as *followers*, and outgoing links are *friends*. Then three corpora are built based on the extreme numbers of followers, friends, and community membership. To avoid noise, again, the 2.5th and 97.5th percentiles of data points are used as the lower and upper bounds, respectively.

- *Based on number of friends:* The high category consists of 1,000 bloggers having from 51 to 162 friends. The low category consists of 1,000 bloggers having from two to six friends.

- *Based on number of followers:* The high category includes 1,000 bloggers having from 40 to 170 followers. The low category includes 1,000 users having from one to three followers.

- *Based on number of community memberships:* The high category consists of 1,000 users having joined 167 communities and less. The low category consists of 1,000 users having joined from one to three communities.

3.2   Feature Extraction

Two features are used to characterize blog posts made by members of the depression community: topic and language style.

Topics discussed in the blog posts are extracted using latent Dirichlet allocation (LDA) [1] – a popular Bayesian probabilistic modeling tool. LDA assigns each word in a blog post text to a topic that is a distribution over the set of distinct vocabularies found in all documents. Each post can be represented by a mixture proportion of topics using the probability p (topic| document). The topic model requires that the number of topics be set to a specific number. In this paper, we set the number of topics to 50, run Gibbs sampling with 5,000 samples, and use the last Gibbs sample to infer the results.

Language style is captured using the Linguistic Inquiry and Word Count (LIWC) package [15]. It returns 68 psycholinguistic categories, such as linguistic, social, affective, cognitive, perceptual, biological, relativity, personal concerns, and spoken.

3.3   Hypothesis Testing

We investigate how mood, age and the degree of social connectivity influence the text of blog posts authored by members of *depression.livejournal.com* community. Here the text is considered in two aspects: (1) the topics are discussed and (2) the language style is conveyed. In particular, we examine three hypotheses:

(1) There is a significant difference in the use of topics and language style between young and old bloggers.

(2) There is a significant difference in the use of topics and language style among posts tagged with low and high valence moods.

(3) There is a significant difference in the use of topics and language style among bloggers having low and high level of social connectivity.

We perform nonparametric Wilcoxon rank sum test on the hypothesis of equal medians in the use of topics and language style between two categories in each corpus. Each of 50 topics and 68 LIWC categories is tested. The null hypothesis is rejected at the significance level of 5% ($p \leq 0.05$). A feature set is considered to be significantly different in the use between different cohorts defined by age, mood and social connectivity if a majority of features in the set are found significantly different.

3.4   Classification

Whilst the statistic test is conducted on each topic and LIWC category, a classifier takes all of them as a feature set to classify a post or a blogger into the cohort it likely belongs to. A classifier whose feature set has strong predictive power of the outcome will gain a high accuracy in the classification. In this work, we examine if topics or language style are predictive in classification of the cohorts stratified by age, sentiment orientation and the level of social connectivity.

Denote by B the corpus of all N blog posts. Given a blog post $d \in B$, based on features extracted from the post d, denoted as $x^{(d)} = [..., x_i^{(d)}, ...]$, where i indexes the features. If topics are used as features, $x_i^{(d)}$ is the probability of topic i in blog post d. When LIWC are the features, $x_i^{(d)}$ is the value of psycholinguistic process i in blog post d. Furthermore, let $U = (u_1, ..., u_N)$ be a corpus of all N users. When mixture proportion of topics and LIWC categories are chosen to be features, u_n denotes an average vector from all blog posts in B made by user n.

We perform the least absolute shrinkage and selection operator (Lasso) [4] – a regularized regression model – to do logistic regression and select features. Lasso assigns positive and negative weight to features, thus by examining these weights we can find out the importance of each feature in the prediction.

The regularization parameter (λ) in the regression model chosen is the largest number such that the accuracy of the model is still within one standard error of the optimum (1se model). This rule may prevent over-fitting in the model while the accuracy of classification is still guaranteed. Ten-fold cross validations are conducted on the training data to estimate the prediction model and accuracy is used to evaluate the performance of the classification.

Table 1. The numbers of topics and linguistic features are rejected from the hypothesis testing of equal median in the use of different cohorts classified by mood, age and social connectivity

Properties	Topic	LIWC
Mood valence	2	38
Age	9	11
Number of friends	42	3
Number of followers	20	8
Number of community	23	18

4 Experiment Results

4.1 Hypothesis Testing

In this section, we present the results of the hypothesis testing using Wilcoxon rank sum test. The null hypothesis H_0 of equal medians in the use of a feature – a topic or an LIWC category – is considered to be rejected at $p \leq 0.05$. When the null hypothesis is rejected, its alternative hypothesis is accepted, meaning the use of the feature is significantly different between different respective cohorts.

As shown in Table 1, in topic usage, out of 50 topics, a majority are significantly different in the use between people whose low and high social connectivity. It indicates that there is a statistically significant difference in the use of topics of interest between bloggers whose few and many social connectivity in the depression community. On the other hand, there is no strong evidence about the difference in the use of topics of interest between young and old bloggers, as well as even between people in low and high valence mood when blogging.

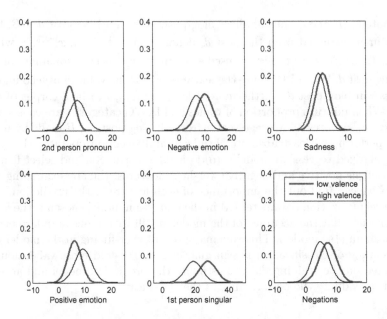

Fig. 3. Examples of LIWC categories different in the use between posts tagged with low and high valence mood

In the use of psycholinguistic processes, out of 68 LIWC categories, 38 are found significantly different between blog posts made when people are in low and high valence. In contrast, only a minority of LIWC categories are found different between different cohorts defined by age and social connectivity. Figure 3 shows examples of LIWC features that have significant difference between high and low valence cohorts. While the high valence cohort prefers using *positive emotion* and *2nd personal pronoun* words in the posts, the low valence cohort uses *negative emotion, 1st personal singular, sadness* and *negations* words more frequently.

4.2   Classification

In Lasso [4], the features unpredictive of the independent variable receive small weight (zero or closely to zero), whereas the predictive ones receive large weight, providing interpretable predictors. Figure 4 illustrates an example of feature selection by Lasso, which builds a regression model to predict whether a post is tagged a high or low valence mood, using topics or LIWC categories as features. Initially, all 50 topics or 68 LIWC categories are included into the model. Then, for the *1se* model, Lasso assigns nonzero weight to 8 features in both models.

Classification results are presented in Figure 5. The best is with the classification of blog posts into low and high valence mood using LIWC categories as predictors, gaining an accuracy of 78.32%. The model to predict posts tagged with high valence mood, learned by Lasso, is shown in Table 2. We observe that *positive emotion*, as expected, is a positive predictor of high valence posts. On

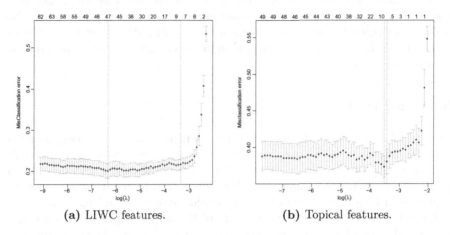

(a) LIWC features.   (b) Topical features.

Fig. 4. Features selected by Lasso to predict posts in high valence mood using topics and LIWC features as predictors

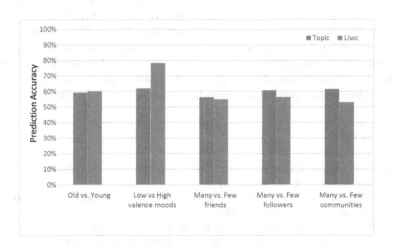

Fig. 5. Prediction performance

Table 2. Examples of LIWC features selected by Lasso to predict posts in high valence

LIWC	Coef.	Example	LIWC	Coef.	Example
Positive emotion	12.02	Nice, love	Negative emotion	-2.64	Hurt, ugly
2nd personal pronoun	4.57	you, your	1st personal singular	-1.00	I, me, mine
Negations	-5.18	No, never	Home	-0.57	Kitchen
Swear words	-2.83	Swear, damn	Sadness	-0.19	Sad, crying

the other hand, *negative emotions* and *sadness* words, are negative predictors of the outcome. Interestingly, *2nd personal pronoun* is positively associated with high valence posts, whereas *1st personal singular* is linked with low valence posts.

The result in other classifications is quite modest, achieving an accuracy of approximately 60%.

5   Conclusions

The need for research in depression and other mental health disorders is on the rise. The physically isolated and mentally ill is over-represented in social media communities partially due to limited independence, difficulties with face-to-face communication, and the scarcity of fellow suffers. The topics and language styles expressed in blog posts may indicate depression symptoms in bloggers.

From the posts authored by members of the online depression community, topics and psycholinguistic processes were extracted to construct classifiers to discriminate young versus old, high versus low valence mood, and high versus low social capital cohorts. Language styles were found to be significantly different between low and high valence cohorts, whilst topics were significantly different between people whose different degrees of social connectivity. Psycholinguistic features were the best features for classification of low and high valence mood.

The finding provides a comprehensive understanding on the aspects of topics of interest and linguistics in several cohorts of bloggers within the depression community. It also suggests a set of powerful predictors of depression in different cohorts. The results indicate the potential of social media as a platform for capturing depression, providing a foundation for online early warning systems.

References

1. Blei, D.M., Ng, A.Y., Jordan, M.I.: Latent Dirichlet allocation. Journal of Machine Learning Research 3, 993–1022 (2003)
2. Bradley, M.M., Lang, P.J.: Affective norms for English words (ANEW): Instruction manual and affective ratings (1999)
3. De Choudhury, M., Gamon, M., Counts, S., Horvitz, E.: Predicting depression via social media. In: Proceedings of the AAAI International Conference on Weblogs and Social Media, ICWSM (2013)
4. Friedman, J., Hastie, T., Tibshirani, R.: Regularization paths for generalized linear models via coordinate descent. Journal of Statistical Software 33(1), 1 (2010)
5. Mirowsky, J., Ross, C.E.: Age and depression. Journal of Health and Social Behavior (1992)
6. Moreno, M.A., Christakis, D.A., Egan, K.G., Jelenchick, L.A., Cox, E., Young, H., Villiard, H., Becker, T.: A pilot evaluation of associations between displayed depression references on facebook and self-reported depression using a clinical scale. The Journal of Behavioral Health Services & Research 39(3), 295–304 (2012)
7. Moreno, M.A., Jelenchick, L.A., Egan, K.G., Cox, E., Young, H., Gannon, K.E., Becker, T.: Feeling bad on facebook: Depression disclosures by college students on a social networking site. Depression and Anxiety 28(6), 447–455 (2011)

8. Nguyen, T., Phung, D., Adams, B., Venkatesh, S.: Prediction of age, sentiment, and connectivity from social media text. In: Bouguettaya, A., Hauswirth, M., Liu, L. (eds.) WISE 2011. LNCS, vol. 6997, pp. 227–240. Springer, Heidelberg (2011)

9. Nguyen, T., Phung, D., Adams, B., Venkatesh, S.: Towards discovery of influence and personality traits through social link prediction. In: Proceedings of the International AAAI Conference on Weblogs and Social Media, pp. 566–569 (2011)

10. Nguyen, T., Phung, D., Dao, B., Venkatesh, S., Berk, M.: Affective and content analysis of online depression communities. IEEE Transactions on Affective Computing PP(99), 1 (2014)

11. Nguyen, T., Phung, D., Venkatesh, S.: Analysis of psycholinguistic processes and topics in online autism communities. In: Proceedings of the IEEE International Conference on Multimedia and Expo (ICME), San Jose, USA (July 2013)

12. Nimrod, G.: From knowledge to hope: Online depression communities. International Journal on Disability and Human Development 11(1), 23–30 (2012)

13. Park, M., Cha, C., Cha, M.: Depressive moods of users portrayed in Twitter. In: Proceedings of the ACM SIGKDD Workshop on Healthcare Informatics (2012)

14. Park, M., McDonald, D.W., Cha, M.: Perception differences between the depressed and non-depressed users in Twitter. In: Proceedings of the AAAI International Conference on Weblogs and Social Media, ICWSM (2013)

15. Pennebaker, J.W., Francis, M.E., Booth, R.J.: Linguistic Inquiry and Word Count (Computer Software). LIWC Inc. (2007)

16. Phung, D., Gupta, S.K., Nguyen, T., Venkatesh, S.: Connectivity, online social capital and mood: A Bayesian nonparametric analysis. IEEE Transactions on Multimedia 15, 1316–1325 (2013)

17. Ruder, T.D., Hatch, G.M., Ampanozi, G., Thali, M.J., Fischer, N.: Suicide announcement on facebook. Crisis: The Journal of Crisis Intervention and Suicide Prevention 32(5), 280–282 (2011)

18. Andrew Schwartz, H., Eichstaedt, J.C., Kern, M.L., Dziurzynski, L., Agrawal, M., Park, G.J., Lakshmikanth, S.K., Jha, S., Seligman, M.E.P., Ungar, L., Lucas, R.E.: Characterizing geographic variation in well-being using tweets. In: Proceedings of the AAAI International Conference on Weblogs and Social Media, ICWSM (2013)

19. WHO. Depression: A global public health concern (2012)

20. Wirz-Justice, A.: Diurnal variation of depressive symptoms. Dialogues in Clinical Neuroscience 10(3), 337 (2008)

21. Yang, Y.: Is old age depressing? growth trajectories and cohort variations in late-life depression. Journal of Health and Social Behavior 48(1), 16–32 (2007)

A Review Selection Method
Using Product Feature Taxonomy

Nan Tian, Yue Xu, and Yuefeng Li

Faculty of Science and Engineering, Queensland University of Technology,
Brisbane, Australia
{n.tian,yue.xu,y2.li}@qut.edu.au

Abstract. As of today, online reviews have become more and more important in decision making process. In recent years, the problem of identifying useful reviews for users has attracted significant attentions. For instance, in order to select reviews that focus on a particular feature, researchers proposed a method which extracts all associated words of this feature as the relevant information to evaluate and find appropriate reviews. However, the extraction of associated words is not that accurate due to the noise in free review text, and this affects the overall performance negatively. In this paper, we propose a method to select reviews according to a given feature by using a review model generated based upon a domain ontology called product feature taxonomy. The proposed review model provides relevant information about the hierarchical relationships of the features in the review which captures the review characteristics accurately. Our experiment results based on real world review dataset show that our approach is able to improve the review selection performance according to the given criteria effectively.

Keywords: Review Selection, Review Quality, Review Model, Ontology, Product Feature Taxonomy.

1   Introduction

The advent of Web 2.0 has promoted huge amount of user generated information which contains rich personal opinions such as user reviews. As of today, online reviews have become increasingly important in decision-making process for the users. Since the number of online reviews has been increasing significantly at commercial websites, more and more researchers attempt to find an effective way to find helpful reviews for the users [1–6]. The early approaches tried to analyse review quality by examining a number of features related to writing quality such as the length of the review. Then, the focus turns to determine the helpfulness of the review based upon its content such as the features discussed by the reviewer. In particular, [2] present an approach which aims to select reviews that comprehensively discuss a certain feature. In detail, the method finds all words which are relevant to a given feature based upon semantic meaning and co-occurrence from the review collection in order to determine if this feature

B. Benatallah et al. (Eds.): WISE 2014, Part I, LNCS 8786, pp. 408–417, 2014.
© Springer International Publishing Switzerland 2014

has been comprehensively discussed. However, due to the characteristics of free text written by online users, the reviews contain a lot of noises and unrelated information, the identified relevant words are very often not accurate, which affects the performance negatively.

On the other hand, ontology learning has attracted significant attention in recent years. Researchers made a lot of efforts to find the relationship between different terms or concepts more effectively and accurately. By making use of various techniques such as text mining and ontology learning, people now are able to generate product ontology or taxonomy about product features and relationships between features from data about products or even from user generated information such as tags and review text [8–10]. In this paper, we introduce a review selection method called RMS (Review Model based review selection). Instead of analysing writing quality or finding relevant information from review text for review quality prediction, we make use of a hierarchical product profile called product feature taxonomy to capture the characteristics of a review, which helps improving the performance of review selection.

2   Related Work

In recent years, the explosion of user generated information such as online reviews provides a lot obstacles for people to find and utilize useful information. The rapid development of data mining especially text mining has made analysing and utilizing review data a reality. However, a user may still prefer to read vivid and complete reviews to make purchase decision. The overwhelming volume of review data makes it extremely difficult and time-consuming to find the useful ones. As a result, the research on review quality prediction and review selection has attracted significant attentions recently. A number of research works have been proposed to make use of textual and social features for review helpfulness estimation. For instance, [1] proposed a method that uses radial basis functions to determine review helpfulness rating based upon three factors (reviewer expertise, writing style and timeliness). Similarly, [6] presented a non-personalized classifier to predict the helpfulness based on writing style and the expertise of the reviewer. One significant drawback of these methods is that some required information is not widely available (e.g., reviewer's expertise information). Therefore, some researchers attempted to extract useful reviews purely based upon the review content. Specifically, [4] make use of Greedy algorithm to extract a small set of highly rated reviews that cover maximum product features and users' opinions buried in the whole review collection. Since the user may be interested in a particular product feature when he/she looks for helpful reviews, [2] proposed a review selection approach which identifies those reviews that focus on a certain feature. In detail, by utilizing the idea of Kolmogorov complexity, all relevant words for a feature (e.g., words that have similar semantic meaning and opinion words that have been used to modify this feature) are extracted to calculate the information distance of each review. The reviews that obtain minimum information distance are considered most specialized reviews on this feature.

Meanwhile, product classification or taxonomy is often available, provided by product manufacture organizations or companies for promotion or marketing purposes. Moreover, ontology learning has been a wide studied area. Recent years, some researchers seek to create a hierarchical structure about products or items from user generated content. [8] proposed a method which exploits a probabilistic model to identify the relations between tags. Based upon the relations, a hierarchical structure between tags is constructed. [9] proposed to construct tag ontology from folksonomy based on WordNet and also personalized the tag ontology based on user clusters. [10] presented an approach to construct a hierarchical product profile which contains product features and relationships between them. Specifically, association rules and sentiment words shared between features are used to identify the product feature relationships. [4] make use of a pre-identified set of product features to identify useful reviews, and [2] try to find a set of associated words with the concerned feature to determine the helpfulness of reviews. However, these existing works did not consider the relationships between product features. In this paper, we propose an approach to assess the quality of product reviews based on the product feature ontology, especially to make use of the hierarchical relationships between features.

3   The Proposed Approach

In order to select reviews for the user, we first generate a review model for each of the reviews, then rank the reviews based on the quality of the reviews characterized in their review model. The input includes a given product feature taxonomy generated from the reviews or given by domain experts, and a collection of reviews. The output is a number of highly-ranked reviews according to the user specified criteria (e.g., the concerned feature provided by the user).

3.1   Product Feature Taxonomy

Reviews may vary in terms of coverage and focuses by considering different product features. For instance, some users may prefer reviews which are talking about a number of unrelated features (e.g., "battery life", "picture quality", and "size", each of which indicates a different attribute of the camera); while some other users may prefer reviews which focus on one feature only by analysing it from different angles. We believe that a review's quality can be better predicted on how it covers the product features than its writing style or the writer's reputation. In order to identify the aforementioned characteristics based on discussed features in a review, we need a structural product profile which provides the relationships between different features. It could be a standard ontology provided by domain experts or an ontology automatically generated from domain data such as reviews by using ontology learning methods. In this paper, we make use of the product profile called product feature taxonomy proposed in [10], defined below, for assisting the review analysis. Figure 1 shows part of a feature taxonomy for a product (i.e., camera) generated from a collection of reviews. As shown, it is a tree structure describing the relationships between product features.

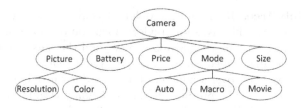

Fig. 1. Product Feature Taxonomy

Definition 1 (Feature Taxonomy): A feature taxonomy consists of a set of features and their relationships, denoted as $FT = \{F, L\}$, F is a set of features where $F = \{f_1, f_2, ..., f_n\}$ and L is a set of relations. The feature taxonomy has the following constraints:

(1) The relationship between a pair of features is the sub-feature relationship. For $f_i, f_j \in F$, if f_j is a sub-feature of f_i, then (f_i, f_j) is a link in the taxonomy and $(f_i, f_j) \in L$, which indicates that f_j is more specific than f_i. f_i is called the parent feature of f_j denoted as $P(f_j)$.

(2) Except for the root, each feature has only one parent feature. This means that the taxonomy is structured as a tree.

(3) The root of the taxonomy represents the product itself.

3.2 Review Models

In this step, we aim to utilize the information of the given feature taxonomy to generate the review model for each individual review based on the processed review text. According to the product feature taxonomy FT, we first identify all discussed features in a review r: $F_r = \{f | \forall f \in F, f \in r\}$.

Review Model. In this section, we present an approach to represent a review in terms of its diversity and comprehension in order to facilitate the review quality prediction and selection task. Based upon observations, from users' perspective, people prefer those reviews that not only cover more relevant features, but also describe more detailed aspects for a particular feature that they are interested. For instance, if a review does not only discusses *vivid color* and *high resolution* of the captured pictures, but also mentions how decent the *movie mode* is, it is actually quite helpful for those users who concern imaging system. Driven by this motivation, we attempt to formalize a review by capturing such characteristics in order to determine the review quality.

Definition 2 (Review Model): A review model consists of a set of features and corresponding characteristic information which is used to determine the review quality, denoted as $RM_r = \{F_r, Q_r\}$, F_r is identified set of product features mentioned in review r and Q_r is a set of 3-tuple $q_{r,f} = (f, div_{r,f}, comp_{r,f})$ where $f \in F_r$, $div_{r,f}$ and $comp_{r,f}$ indicate the diversity and comprehension value of f, respectively.

Maximum Sub Tree. Reviews have different characteristics in terms of the discussed features. Specifically, some reviews focus on one or a number of particular features (e.g., the user talks about how good the *lens* of a camera is by describing each detailed aspect such as *autofocus* and *image stabilizer*). In contrast, some reviews may cover a number of unrelated features but do not discuss each of them in depth. The product feature taxonomy is able to facilitate the analysis in this regard. First of all, we need to find the maximum coverage of each feature in terms of its sub features in the product feature taxonomy. In detail, we attempt to generate all maximum sub trees for a review according its identified features. For easy understanding, the maximum sub tree is defined as follows:

Definition 3 (Maximum Sub Tree): Let $FT = \{F, L\}$ be a product feature taxonomy, $F_r \subseteq F$ be a set of features identified from review r and $f \in F_r$ be a specified feature. The maximum sub tree rooted at f is defined as $MST_{r,f} = \{SF_f, SL_f\}$ which satisfies the following constraints:

- f is the root of the sub tree, $f \in SF_f$
- $SF_f \subseteq F_r$, $SL_f \subseteq L$
- $\forall g \in SF_f$, there must be a path $< f, f_1, f_2, ..., f_n, g >$ between f and g in the feature taxonomy FT, $f_i \in SF_f$, $i = 1, ...n$, and $(f, f_1), ..., (f_i, f_{i+1})$, $(f_n, g) \in SL_f$, $j = 1, ..., n-1$
- $\forall g \in F_r$ and $g \notin SF_f$, there is no path $< f, f_1, f_2, ..., f_n, g >$ between f and g in the feature taxonomy FT and $f_i \in SF_f$, $i = 1, ..., n$.

These sub trees should be exclusive with each other, i.e., there is no any overlap between any two sub trees. The features in one sub tree are considered related in terms of F_r since they are linked in the sub tree, while the features from different sub trees are not considered related since there is no path or link between these features. Let $MST_r = \{MST_{r,f_1}, ..., MST_{r,f_m}\}$ be a set of m maximum sub trees generated for review r, $f_1, ..., f_m \in F_r$ are the root for the sub trees respectively, then $SF_{f_1} \cap SF_{f_2} \cap ... \cap SF_{f_m} = \emptyset$, that is, $\{SF_{f_1}, SF_{f_2}, ..., SF_{f_m}\}$ is a partition to F_r.

Review Characteristics. In this section, we present methods to calculate review characteristics according to the structural relationships in the feature taxonomy. As aforementioned, reviews are different from each other in terms of the coverage of features as well as the depth of abstract level discussed for a certain feature. The features discussed in a review may be scattered over the whole feature taxonomy or concentrated in a certain area, or both. In order to describe such characteristics, we propose two measures: *diversity* and *comprehension*.

Diversity: The diversity is a measure of the distance between the features in a review based upon their positions in the product feature taxonomy. In this paper, we utilize two different aspects proposed in [7] to calculate the diversity: Hierarchical Relationship Distance (HRD) and Concept Level Distance (CLD).

For easy understanding, we first propose the following terms and formula to be used in this paper:

- *ca*: (common ancestor) is the closet taxonomy feature which is the parent feature of two features.
- *TaxonomyHeight*: is the maximum number of features on a path from the root to a feature located at the lowest hierarchical level.
- Hierarchy level of a feature: the hierarchy level of the root is 1, denoted as $HL(root)=1$. The hierarchy level of a feature in the feature taxonomy is larger than the level of its parent feature by 1.
- Number of levels difference: let f_1 and f_2 be two features in FT, the number of hierarchy levels difference between two features:

$$NLD(f_1, f_2) =\mid HL(f_1) - HL(f_2) \mid \tag{1}$$

Hierarchical Relationship Distance: The HRD between two features examine how close two features are in terms of a hierarchical relationship from a common ancestor. Specifically, the basic idea is that the greater the number of hierarchy levels difference between two features and their common ancestor, the more hierarchical distance between two features, which make the review more diverse. Therefore, the HRD can be calculated as follows:

$$HRD(f_1, f_2) = \frac{NLD(f_1, ca) + NLD(f_2, ca)}{2 \times TaxonomyHeight} \tag{2}$$

Concept Level Distance: The CLD is defined based on the difference between the hierarchy levels of two features. The higher the CLD is, the more concepts are between the two features. And this makes the review more diverse as well. Thus, the CLD is calculated as follows:

$$CLD(f_1, f_2) = \frac{NLD(f_1, f_2)}{(TaxonomyHeight - 1)} \tag{3}$$

Based on the above two aspects, we can calculate the diversity value for a certain feature. We measure the diversity of a feature by considering all discussed features in a review that have similar semantic meaning with it. For instance, a review may mention both *picture* and *movie*; they are similar but locate in different positions of the feature taxonomy. Therefore, we define semantic related features as follows:

For a given feature f, its semantic related feature is a feature in the feature taxonomy which has the similar semantic meaning with f. We can make use of semantic similarity tools such as WordNet to assist identifying semantic related features of f. Let $RF_{r,f} = \{rf_{r,1}, rf_{r,2}, ..., rf_{r,n}\}$ be a set of semantic related features of f including f itself in review r; then the feature diversity of f in review r is defined as:

$$div_{r,f} = \alpha \frac{\sum_{i=1}^{n-1} \sum_{j=i+1}^{n} HRD(rf_{r,i}, rf_{r,j})}{n(n-1)} + (1-\alpha) \frac{\sum_{i=1}^{n-1} \sum_{j=i+1}^{n} CLD(rf_{r,i}, rf_{r,j})}{n(n-1)} \tag{4}$$

Where $0 < \alpha < 1$. The value of α is set to 0.5. We calculate the diversity of the review based upon the generated maximum sub trees. In detail, let $MST_r = \{MST_{r,f_1}, ..., MST_{r,f_m}\}$ be a set of maximum sub trees generated from review r, the diversity of review r is calculated as:

$$DIV_r = \beta \frac{\sum_{i=1}^{m-1} \sum_{j=i+1}^{m} HRD(MST_{r,f_i}, MST_{r,f_j})}{m(m-1)}$$
$$+(1-\beta) \frac{\sum_{i=1}^{m-1} \sum_{j=i+1}^{m} CLD(MST_{r,f_i}, MST_{r,f_j})}{m(m-1)} \tag{5}$$

Where $0 < \beta < 1$. The value of β is set to 0.5 in the experiment. The HRD and CLD between two maximum sub trees are defined as the average HRD and CLD between two features in two maximum sub trees as follows:

$$HRD(MST_{r,f_i}, MST_{r,f_j}) = \frac{\sum_{f_x \in SF_{f_i}} \sum_{f_y \in SF_{f_j}} HRD(f_x, f_y)}{|SF_{f_i}| \times |SF_{f_j}|} \tag{6}$$

$$CLD(MST_{r,f_i}, MST_{r,f_j}) = \frac{\sum_{f_x \in SF_{f_i}} \sum_{f_y \in SF_{f_j}} CLD(f_x, f_y)}{|SF_{f_i}| \times |SF_{f_j}|} \tag{7}$$

Comprehension: We define comprehension to indicate how comprehensively a review discusses one or a number of particular features. The sub-feature relationships of the feature taxonomy can be a good indicator for this measurement. The more sub features of a feature appear in a review, the more comprehensive this feature is. As a result, we calculate the ratio between the number of the feature's sub features appearing in the review and the total number of sub features in the feature taxonomy based upon the generated maximum sub trees. Let $MST_{FT,f} = \{SF_f, SL_f\}$ be a maximum sub tree in which feature f is the root, the comprehension of f can be derived by the following equation:

$$comp_{r,f} = 1 + \frac{|SF_f \cap F_r|}{|SF_f|} \tag{8}$$

It is easy to identify the difference between the features of a review based upon comprehension. In addition, we calculate the average feature comprehension based on the maximum sub trees $MST_r = \{MST_{r,f_1}, ..., MST_{r,f_m}\}$ generated from the review r to represent the comprehension value of the review:

$$COMP_r = \frac{\sum_{MST_{r,f_i} \in MST_r} comp_{r,f_i}}{|MST_r|} \tag{9}$$

After generating the review model for each review, we use them to estimate the quality of the review in terms of a particular feature for review selection.

3.3   Review Ranking and Selection

Users are usually interested in a particular feature instead of all features [2]. To tackle this problem, we make use of the proposed review model to determine the review quality according to the user-specified feature. Specifically, we aim to rank the reviews based upon the diversity and comprehension of a certain feature. In this paper, we define review feature relatedness to indicate how a certain review is related to a specified feature.

Definition 4 (Review Feature Relatedness): Let $q_{r,f} = (f, div_{r,f}, comp_{r,f})$ be a tuple given in the review model of review r and a user-specified feature f, the review feature relatedness for review r to the feature f is defined below:

$$RFR_{r,f} = \gamma div_{r,f} + \delta comp_{r,f} \qquad (10)$$

$0 < \gamma, \delta < 1$. The value of γ and δ is set to 0.8 and 0.2, respectively, in the experiments. The review feature relatedness value of each review is calculated based on its review model. Those reviews that obtain the highest RFR value are considered the best reviews.

4 Experiment and Evaluation

In this section we present a set of experimental results to evaluate the effectiveness of our proposed approach. Our experiment is carried out using real data collected from one of the most popular e-commerce websites: Amazon (www.amazon.com). We choose digital camera review data for testing in this paper. In Amazon, users are able to rate each review to indicate if it is helpful from their perspective. Thus, we use the ratio between the number of positive rating votes and the total number of votes as the gold standard of review quality. For instance, if 8 out of 10 users like a review, the rating score of this review is *0.8*. We collected all online users' reviews for a digital camera and kept those that have received at least two votes (e,g., *like* or *dislike*) for further review selection.

The method proposed in [2] is chosen as a baseline for comparison. In detail, Long's method is to extract all relevant words (e.g., opinion words and words of similar semantic meaning) for a given feature from review text. These generated words are used for determine how much relevant information for this feature has been covered in each review. It has been proven effective in identifying reviews that focus on a certain feature. Both the baseline and our proposed approach are run to rank reviews according to a user-specified feature and select a number of top-ranked reviews as the result. The evaluations are twofold: evaluation on the quality of the selected reviews and evaluation on the comprehension of the specified feature in the selected reviews.

4.1 Review Quality Evaluation

First of all, we evaluate the performance of our approach by comparing the average review rating score of the Top N selected reviews generated by our proposed method and the baseline. Three camera features are chosen in the experiment. They are: Feature 1: *picture*, Feature 2: *mode*, and Feature 3: *lens*. The experimental results are given in Tables 1, 2, and 3 below.

Table 1. Feature 1 Average Rating Score Comparison

	Top 10	Top 20	Top 30	Average
Baseline	0.870	0.892	0.885	0.882
RMS	0.904	0.907	0.889	0.9

Table 2. Feature 2 Average Rating Score Comparison

	Top 10	Top 20	Top 30	Average
Baseline	0.884	0.862	0.877	0.874
RMS	0.906	0.885	0.883	0.891

Table 3. Feature 3 Average Rating Score Comparison

	Top 10	Top 20	Top 30	Average
Baseline	0.902	0.866	0.876	0.881
RMS	0.895	0.874	0.884	0.884

Table 1, 2 and 3 illustrate the average rating score of top 10, 20 and 30 selected reviews generated by our approach and the baseline, respectively. From the results, we can see that the average rating scores of the selected reviews generated by our method are better than that of the baseline in most cases. Therefore, we can believe that the review characteristics captured in the proposed review model improve the performance of review selection.

4.2   Specified Feature Comprehension Evaluation

We also undertake an experiment to compare the performance of both methods in terms of the coverage of the user specified feature in the selected reviews. In detail, we utilize the WordNet to generate a list of equivalent words that are similar to the specified feature (e.g., *image* and *movie* for feature *"picture"*). We calculate the ratio between the number of sentences that contain these equivalent words and the total number of sentences in the review. This ratio is called feature occurrence ratio, which indicates how popular this feature is.

Table 4. Feature 1 Average Occurrence Ratio Comparison

	Top 10	Top 20	Top 30	Average
Baseline	0.017	0.021	0.023	0.020
RMS	0.030	0.031	0.030	0.030

Table 5. Feature 2 Average Occurrence Ratio Comparison

	Top 10	Top 20	Top 30	Average
Baseline	0.011	0.011	0.011	0.011
RMS	0.013	0.015	0.015	0.014

Table 4, 5 and 6 illustrates the evaluation results of the average feature occurrence ratio value of the top 10, 20, and 30 selected reviews for our approach and the baseline, respectively. According to the comparison, the specified feature and its equivalent words appear more frequently in the reviews selected by our method. By using the product feature taxonomy, our method is able to find the relevant information about the specified feature more accurately, which helps selecting more appropriate reviews.

Table 6. Feature 3 Average Occurrence Ratio Comparison

	Top 10	Top 20	Top 30	Average
Baseline	0.033	0.031	0.026	0.030
RMS	0.036	0.060	0.052	0.049

5 Conclusion

In this paper, we proposed a method for selecting reviews according to a certain feature based on a product feature ontology which contains both features and relationships. The objective is to capture the review characteristics (e.g., diversity and comprehension) to find most helpful reviews. Our experiments show that the proposed approach is promising in review selection task. In the future, we plan to improve our techniques, and use them to select reviews based on multiple features.

References

1. Liu, Y., Huang, X., An, A., Yu, X.: Modeling and Predicting the Helpfulness of Online Reviews. In: Eighth IEEE International Conference on Data Mining, pp. 443–452 (2008)
2. Long, C., Zhang, J., Huang, M., Zhu, X., Li, M., Ma, B.: Estimating Feature Ratings through an Effective Review Selection Approach. Knowledge and Information Systems 38, 419–446 (2012)
3. Moghaddam, S., Jamali, M., Ester, M.: ETF: Extended Tensor Factorization Model for Personalizing Prediction of Review Helpfulness. In: Proceedings of the fifth ACM International Conference on Web Search and Data Mining, pp. 163–172 (2012)
4. Tsaparas, P., Ntoulas, A., Terzi, E.: Selecting a Comprehensive Set of Reviews. In: Proceedings of the 17th ACM SIGKDD International Conference on Knowledge Discovery and Data Mining, pp. 168–176 (2011)
5. Tseng, Y.-D., Chen, C.C.: Using an Information Quality Framework to Evaluate the Quality of Product Reviews. In: Lee, G.G., Song, D., Lin, C.-Y., Aizawa, A., Kuriyama, K., Yoshioka, M., Sakai, T. (eds.) AIRS 2009. LNCS, vol. 5839, pp. 100–111. Springer, Heidelberg (2009)
6. O'Mahony, M.P., Smyth, B.: Learning to Recommend Helpful Hotel Reviews. In: Proceedings of the Third ACM Conference on Recommender Systems, pp. 305–308 (2009)
7. Shaw, G., Xu, Y., Geva, S.: Interestingness Measures for Multi-Level Association Rules. In: Proceedings of the 14th Australasian Document Computing Symposium, pp. 27–34 (2009)
8. Tang, J., Leung, H.-F., Luo, Q., Chen, D., Gong, J.: Towards Ontology Learning from Folksonomies. In: Proceedings of the 21st International Jont Conference on Artifical Intelligence (2009)
9. Djuana, E., Xu, Y., Li, Y., Cox, C.: Personalization in tag ontology learning for recommendation making. In: Proceedings of the 14th International Conference on Information Integration and Web-Based Applications and Services, pp. 368–377 (2012)
10. Tian, N., Xu, Y., Li, Y., Abdel-Hafez, A., Josang, A.: Product Feature Taxonomy Learning based on User Reviews. In: WEBIST 2014 10th International Conference on Web Information Systems and Technologies (2014)

A Genetic Programming Approach
for Learning Semantic Information Extraction
Rules from News

Wouter IJntema, Frederik Hogenboom, Flavius Frasincar, and Damir Vandic

Erasmus University Rotterdam
P.O. Box 1738, 3000 DR, Rotterdam, The Netherlands
wouterijntema@gmail.com, {fhogenboom,frasincar,vandic}@ese.eur.nl

Abstract. Due to the increasing amount of data provided by news sources and the user specific information needs, recently, many news personalization systems have been proposed. Often, these systems process news data automatically into information, while relying on underlying knowledge bases, containing concepts and their relations for specific domains. For this, information extraction rules are frequently used, yet they are usually manually constructed. As it is difficult to efficiently maintain a balance between precision and recall, while using a manual approach, we present a genetic programming-based approach for automatically learning semantic information extraction rules from (financial) news that extract events. Our evaluation results show that compared to information extraction rules constructed by expert users, our rules yield a 27% higher F_1-measure after the same amount of rules construction time.

1 Introduction

The immense growth of the World Wide Web in the past decades resulted in enormous amounts of data that are readily available to the average user. Many researchers have hence developed ways to convert these vast amounts of data into valuable information. The Semantic Web aims to organize the currently unstructured Web. While to a human reader, the meaning of text is easily interpretable, machines interpret an average document currently found on the Web as a random collection of characters, without properly associating meaning to it. The Semantic Web is organized by using numerous languages and technologies to convey meaning. A structural means for representing Web information is provided by ontologies.

An ontology is formally defined as a specification of a conceptualization and can be used to store domain-specific knowledge in the form of concepts with various relations between them. Utilizing ontologies within Web information systems allows one to perform searches based on these concepts and relations. Often, ontologies are used as a knowledge base to support the information-intensive

B. Benatallah et al. (Eds.): WISE 2014, Part I, LNCS 8786, pp. 418–432, 2014.
© Springer International Publishing Switzerland 2014

operations. An example of a finance-oriented Web information system is the Hermes framework [10]. Its ontology consists of lexicalized concepts that exist in the financial domain and is used for the classification of Web news items, as well as querying.

A major problem in such news processing frameworks is that in order to turn news item data into information, a knowledge engineer has to keep up with the incoming data and process it in order to determine the value of the data. The ontology needs to be kept up-to-date in an efficient and timely manner. For this, user-defined patterns can be used that extract information that is needed for ontology updating. Usually, lexico-syntactic patterns are employed [11], yet the problem with these type of patterns is that they are based on syntactical elements, such as Part-of-Speech (POS) tags, and thus do not make use of the available domain semantics. Their application is hence often limited to hypernym (generalization), hyponym (specialization), meronym (part of), and holonym (the whole) relations.

In order to overcome the aforementioned problems with common information extraction patterns, in earlier work, we have proposed the lexico-semantic Hermes Information Extraction Language (HIEL) [12], which makes use of lexical and syntactical elements as well as semantic elements. While the use of information extraction rules allows for semi-automatic information extraction, the construction of the patterns remains a non-trivial, tedious, and time-consuming process, because a trade-off needs to be made between the rules' precision and recall. Therefore, in this paper, we propose a method that assists the construction of information extraction rules. Additionally, the method is implemented and evaluated on a data set containing news, while employing the learned rules for fact extraction (relations between concepts) within the financial domain. In this research, we consider facts to represent (financial) events like acquisitions, profit announcements, CEO changes, etc., which are captured by triples consisting of a subject, a predicate, and an object.

The remainder of this paper is organized as follows. First, we discuss related work in Sect. 2. Next, we introduce our information extraction pattern language and our rule learning framework in Sects. 3 and 4, respectively. Subsequently, our implementation is discussed in Sect. 5 and Sect. 6 presents the performance evaluation of our algorithm. Last, we draw conclusions and discuss some directions for future work in Sect. 7.

2 Related Work

In the mature field of pattern learning, there is a lot of work previously done. Hence, we discuss only work that is closely related to ours, i.e., learning patterns for information extraction. This section elaborates on a small, yet representative, part of the current body of knowledge, ranging from work as old as 1992 to more recent work from 2010.

In the early 1990's, Hearst [11] showed that simple lexico-syntactic patterns can be used to extract hyponyms from text. While the patterns of Hearst generated high precision, they did not perform well recall-wise, hence, driving the development of new pattern languages that were (loosely) based on the patterns of Hearst. For example, in 2002, the authors of [7] proposed JAPE rules for information extraction which are nowadays widely employed, e.g., in the work of Maynard et al. [14]. Other applications that use patterns for general purpose information extraction are presented in for example [2], while [8] uses patterns specifically for news processing, analogous to Hermes, our news processing framework [10]. Even though some of the aforementioned pattern languages incorporate some semantics, they could easily result in verbose rules and they could become rather complex [12]. Hence, we defined our own information extraction language which we incorporate in the Hermes news processing framework, i.e., the Hermes Information Extraction Language (HIEL) [12].

Since the composition of information extraction rules is a tedious process which requires a domain expert to invest a lot of time, a vast amount of effort has been put into automation of this process. We distinguish between supervised and unsupervised learning, where in the former method a model is learned from data of which the correct outcomes (classifications) are known, while the latter method does not rely on any prior knowledge. Due to the fact that on free text supervised methods generally perform better compared to unsupervised methods [6], we aim to employ a supervised learning technique. A problem many supervised approaches have to deal with, is the sparse amount of training examples, for which bootstrapping has proven to be an effective solution [4].

In [16] hypernym relations are learned from text using a supervised learning technique. The authors collect noun pairs from a corpus in order to identify new hypernym pairs, and for each of these pairs sentences are gathered in which both nouns occur. New hypernym classifiers are trained based on patterns extracted from the gathered sentences, using classifiers like Naïve Bayes, genetic algorithms, and logistic regression. When such methods are applied in rule learning processes, rules are generated randomly during initialization and are altered in such a way that the built rules perform better in terms of a predefined metric, which is often a combination of precision and recall. With respect to rule generalization and specialization, the authors of [16] distinguish between top-down and bottom-up approaches. The first type starts with a general rule and then aims to specialize it, while the latter starts with a specialized rule which is then generalized. Our approach goes beyond the one from [16] by allowing domain concepts and relationships to be extracted from text.

WHISK [17] employs a supervised top-down rule learning method. The rules learned in this system are based on regular expressions, which is similar to our approach. In addition to simple literals, syntactic and semantic tags are used to generalize the rules. In the learning process, these tags are determined by means of heuristics. While classes are allowed, no is-a hierarchy or other relationships are employed, while at the heart of our system is a domain ontology with both concepts and relations, which is used to create generic lexico-semantic patterns.

KNOWITALL [9] uses an unsupervised bottom-up approach in order to extract named-entities from the Web. The system employs patterns that incorporate POS tags to extract new information. Pattern learning is based on Web searches, where for each occurrence of an instance, a prefix of a specific amount of words and a suffix of a number of words is added to the pattern. The learned patterns consist only of an entity surrounded by words, unlike our approach, which employs a larger amount of linguistic information like orthographical categories and ontology elements, and not only POS tags. Furthermore, the expressiveness of the learned patterns appears to be limited, since repetition and logical operators are not allowed. KNOWITALL focuses on learning named entity extraction patterns rather than on the extraction of new relationships between entities, which is something we pursue.

While many of the above methods have proven to be effective when using lexico-syntactic rules, Genetic Algorithms (GA) are suitable for rule learning as well, since the input is often a bit string. One can encode a pattern such that every bit represents a token or its corresponding features. By employing different genetic operators, such as inheritance, selection, mutation, and cross-over, the optimal information extraction rule can be determined. A similar method is applied in [5] for learning lexico-syntactic patterns, that only incorporate POS tags, in order to extract entities, which is different from our approach, since we aim to generate lexico-semantic patterns to extract concepts, relationships, and events (complex concepts) from text.

A branch of Genetic Algorithms is Genetic Programming (GP), where generally each problem is represented as a tree instead of a bit string. This makes it easier to encode the problem. Each node either represents a sequence, a logical operator – e.g., conjunction, disjunction, and negation – or a repetition operator. Similarly, terminal tree nodes are suitable to represent a literal, syntactic category, orthographical category, or a concept. Genetic algorithms often converge fast to a good solution when compared to other meta-heuristics, such as simulated annealing [18]. By performing the default genetic operators, trees can evolve until the desired performance is achieved. In a similar manner [3] employs trees to represent rules, containing POS tags, that are used in genetic programming to discriminate between definitions and non-definitions in text. Because of the identified advantages of Genetic Programming approaches over other approaches, in our research, we use a Genetic Programming approach to pattern learning.

3   HIEL: The Hermes Information Extraction Language

The Hermes Information Extraction Language (HIEL) has been extensively described and evaluated in earlier work [12]. This section provides an overview of the basic constructs of HIEL, which are captured by Fig. 1. The latter figure shows an example rule that links CEOs to their companies. Lexical and syntactic elements are indicated by white labels, whereas semantic elements (which make use of a domain ontology) are indicated by shaded labels.

In HIEL, a rule typically consists of a left-hand side (LHS) and a right-hand side (RHS). Once the pattern on the RHS has been matched, it is used in the LHS, consisting of three components, i.e., a subject, predicate, and an object,

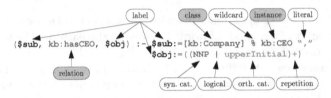

Fig. 1. Example HIEL rule

where the predicate describes the relation between the subject and the object (in this case `kb:hasCEO`). The RHS supports sequences of many different features, as explained below.

First, labels (preceded by `$`) on the RHS associate sequences using assignment (`:=`) to the correct entities specified on the LHS. Second, syntactic categories (e.g., nouns, verbs, etc.) and orthographical categories (i.e., token capitalization) can be employed. Next, HIEL supports the basic logical operators *and* (`&`), *or* (`|`), and *not* (`!`), and additionally allows for repetition (regular expression operators, i.e., `*`, `+`, `?`, and `{...}`). Moreover, wildcards are also supported, allowing for ≥ 0 tokens (`%`) or exactly 1 token (`_`) to be skipped.

Of paramount importance is the support for semantic elements through the use of ontological classes, which are defined as groups of individuals that share the same properties, i.e., the instances of a class. A concept (class or instance) or relationship may consist of several lexical representations that are stored using the synonym property in the (lexicalized) domain ontology. The hierarchical structure of the ontology allows the user to make rules either more specific or generic, depending on the needs at hand.

4   Rule Learning

In order to assist domain experts with rule creation, we propose to employ a genetic programming approach to rule learning. Our information extraction language, HIEL, which can intuitively be implemented using tree structures, fits the required tree structure of the genetic programming operators. Additionally, a genetic programming approach offers transparency in the sense that it gives the user insight into how information extraction rules are learned. Also, a genetic approach – as opposed to other meta-heuristics such as simulated annealing – often converges to a good solution in a relatively small amount of time [18].

4.1   Rule Learning Process

Figure 2 depicts the basic steps of our genetic programming approach to rule learning. First rules are initialized, followed by the evaluation of the fitness of each of these rules. Rule evolution is done by applying a genetic operator on the rules, such as elitism, cross-over, and mutation. Based on a selection procedure which takes into account the fitness of individuals we determine the

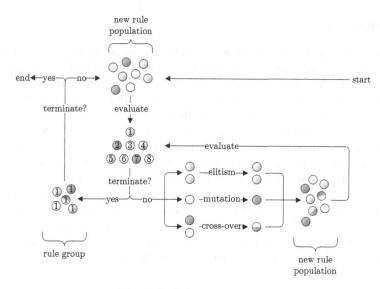

Fig. 2. Rule learning process

rules on which these operators are applied. This process continues until one of the termination criteria (see Subsect. 4.7) is fulfilled, after which the rule with the highest fitness is collected in a rule group. The latter is a group of rules, in which each rule aims to extract the same type of information, albeit covering different situations. Since a single rule is not likely to achieve a high recall, because of the many different sentence structures, a collection of rules could achieve this goal. After the best rule, i.e., the rule with the highest fitness, has been selected and the rule group does not yet meet its termination requirements, a new population is initialized and another iteration is performed.

4.2   Representation

While in genetic algorithms, individuals are generally encoded in the form of an array of bits, in genetic programming individuals are specified as trees. In our representation, a tree consists of functions and terminals. Functions have functions and terminals as children, whereas terminals cannot have child nodes. We differentiate between five functions, i.e., a sequence, a conjunction operator, a disjunction operator, a negation operator, and a repetition. Also, we distinguish four terminals, i.e., a syntactic category, an orthographic category, a concept, and a wildcard.

Each information extraction rule can be represented by a tree. As HIEL requires labels to be placed on separate elements on the first level of the tree and each label should be bound to different tokens in the text, the root of each tree is a sequence, which can have one or more function or terminal child nodes.

4.3   Initialization

The first phase in the genetic programming process is the initialization of the rules. During the initialization, a population of N individuals is created. For initialization, each node needs to be created such that it is syntactically correct. In addition, a maximum number of nodes per tree and a maximum tree depth helps constraining the rule size and complexity.

Generally, in genetic programming, information extraction rules are generated randomly at initialization phase. A commonly used method is ramped-half-and-half, which is proven to produce a wide variety of trees of various sizes and shapes. The ramped-half-and-half initialization procedure consists of two methods, i.e., *full* and *grow*. The full method generates trees for which the leaves (terminal nodes) are all at the same level (i.e., *maxdepth*), while the grow method generates more variously shaped trees. Because neither of the methods provide a wide variety of individuals, half of the population is constructed using the full method and half of the population is constructed using the grow method.

4.4   Fitness Evaluation

Each individual in the population is evaluated for each generation in order to determine its fitness. We compare the extracted information with manually annotated information by evaluating the F_1-measure and the number of nodes within a tree. The F_1-measure is defined as the harmonic mean of precision (the correctly found items) and recall (the correctly found items with respect to the should-be-found items). We calculate the number of nodes within a tree in order to control the amount of bloat (i.e., uncontrolled growth of information extraction rules during the evolutionary process) in the population. Both measures are combined into one fitness measure that determines how well an individual performs compared to others, where F_1-scores of longer rules are penalized more than those of rules showing less bloat.

A common problem in genetic programming is tree size explosion. Often, rules are learned that have the same fitness, but that are slightly different. In order to overcome the problem of learning rules consisting of unnecessary nodes, we introduce some parsimony pressure by including a small penalty in the overall fitness measure for the total number of nodes of the rule. Let α denote the amount of bloat (optimized later on) and R represent a rule, then the fitness of a rule (when taking into account both F_1 and rule length l) is determined as:

$$Fitness(R) = \begin{cases} 0 & \text{if } F_1(R) = 0 \\ F_1(R) - \alpha \cdot l(R) & \text{if } F_1(R) > 0 \,. \end{cases} \tag{1}$$

4.5   Selection

For each genetic operator one or more individuals from the population need to be selected. According to the Darwinian principles, the strongest individuals survive, therefore it is better to select individuals based on their fitness. A common selection method is tournament selection. One of the advantages of this method is that the selection pressure, which determines the degree to which it favors fit individuals over less fit individuals, remains constant. In tournament selection, ts (tournament size) individuals are selected randomly from the population, and the individual with the highest fitness is selected. By adjusting the tournament size, the selection pressure can be adapted.

4.6   Genetic Operations

After the rules have been initialized, the actual process of evolving can be initiated. During the evolution, several genetic operators are applied, i.e., elitist selection, cross-over, and mutation.

Elitist Selection. The first operation, elitist selection, resembles the survival of the fittest principle from Darwin. After the fitness of each individual in the population has been determined, the best r performing individuals are selected and copied to the next generation. The user may alter the portion of the population that is allocated for selection. Generally r is set to a value between 5% and 10%, in order for the algorithm to keep just a small set of the best performing individuals.

An advantage of applying the selection operator is that it helps the process to remember the best performing individuals until a better one is found. If the operator is omitted, these well performing rules might disappear from the population due to the cross-over and mutation operators.

Cross-over. During the cross-over operation two parents are selected from the population to produce either one or two offsprings. The former method randomly selects a cross-over point in both parents and interchanges the selected nodes, producing two children. The latter also randomly selects a cross-over point in both parents, but generates one child by combining the selected parts from both parents. Each parent is chosen based on its fitness using tournament selection and could be selected more than once in each generation, making it possible to use the same individual for multiple cross-over operations.

The selection of the cross-over points is generally not done with uniform probability, since the majority of the nodes will be terminal nodes. In order to overcome this problem, we select 90% of the time a function and 10% of the time a terminal node. While individuals are selected based on their fitness, the nodes interchanged during cross-over are selected in a random manner. This can result in offspring that do(es) not necessarily perform well, while the originating trees can have a relatively good performance. This is the case if a node (including its child nodes), also called a subpattern, is almost never discovered in the text.

Mutation. The mutation operator aims to introduce more variety into the population. Several approaches are identified in mutation for genetic programming. The first is subtree mutation, also known as headless chicken cross-over, where a random point in the tree is replaced by a randomly generated subtree. A second approach is point mutation, where only the randomly selected point is replaced by a function or terminal. If no replacement is possible (i.e., if the randomly generated node is not allowed within the selected parent node), the mutation is not performed. We implement the headless chicken cross-over method, because of its reported good performance with respect to the other approaches [1,13].

4.7 Termination Criteria

A genetic programming run terminates when one of the termination criteria is satisfied. In our system we have implemented two termination criteria, one for a run and one for a rule group. Each run generates a certain maximum number of generations, which can be specified by the user. Because of the wide variety in sentence structures it is not plausible that one rule would be able to achieve high recall and precision values, yet a group of rules might be able to achieve this goal for a particular event. Once a termination criterion has been fulfilled, the rule with the highest fitness is saved to the assembled rule group. This group is a set of rules that intend to extract the same information (i.e., triple type). For example, it is likely that one needs several rules to extract all instances of the CEO relationship that has been mentioned in examples earlier. If the triple to extract is defined by *Company hasCEO Person*, at least two rules are needed to extract both the instance in *"Apple's chief executive, Steven P. Jobs"* and *"Steve Ballmer, Microsoft's chief executive"* as the order of the company and the CEO is different in these two cases.

Once a rule is learned and added to the rule group, the information extracted by this rule is excluded while learning additional rules. If a rule does match a previous annotation, it is not taken into account for its fitness, and hence each rule will extract different information. After the termination criterion for the current population fires, the rule with the highest fitness is only collected in the rule group if it causes the rule group to achieve a higher overall fitness value. In case it lowers the fitness of the entire group, it is omitted. The entire rule learning process, i.e., assembling the rule group, terminates when T iterations of updates have passed in a sequence, which did not manage to produce rules that increased the fitness of the rule group, meaning the algorithm is stuck in a (local, possibly sub-optimal) solution.

5 Implementation

We implemented our information extraction language and rule learning approach in the Hermes framework [10], which can be found at http://people.few.eur. nl/fhogenboom/hermes.html. At the core of the Hermes framework lies a lexicalized financial domain ontology which specifies domain concepts and their relationships. The implementation, the Hermes News Portal (HNP), provides components for the import and classification of the news articles extracted from various RSS news feeds. During this process the classifier adds annotations, such as syntactical categories, orthographical categories, and concepts, to the text which can subsequently be used for creating and matching information extraction rules. The details of this process can be found in our previous work [10].

In our rule learning environment, the user is able to keep track of the current generation, the learned rules, and their fitness. Several controls are put in place for managing the rule learning process. Additionally, current generations and learned rules are displayed. Last, the user is able to fine-tune the algorithm parameters.

6   Evaluation

To evaluate the performance of our information extraction language and the genetic programming approach to automatic rule learning, we have selected 500 news articles from the Web with an average length of 700 words from the financial and technology domain originating from various sources, including New York Times, Reuters, Washington Post, and Businessweek. Each news item has been processed using Hermes, with at the back-end a knowledge base containing over 1,200 concepts, including companies, persons, products, financial terms, etc. The learned rules are employed for fact extraction (relations between concepts, i.e., triples that denote an event) within the financial domain.

Three domain experts have annotated the documents, while distinguishing between ten different financial relations, such as profits, products, CEOs, and competitors of companies. In order to decrease the amount of subjectivity we have used a democratic voting principle for the selection of annotations, meaning two out of three annotators should have proposed the annotation to consider it valid. As displayed in Table 1, this resulted in an average Inter-Annotator Agreement (IAA) of 71% for 1,153 unique annotations among all the relations. The table shows that there is a clear difference between the different relations. For instance, the *Competitor* relation is often subjective and therefore hard to determine whether a clear competitor relationship is stated in the text. The same can be argued for the *Partner* relation, which indicates a partnership between two companies. This is in contrast to, for instance, the *CEO* relation, which is often indicated by words like "CEO", "chief", or "chief executive".

Furthermore the table shows the number of annotations per relation found by the annotators in the set of 500 news items. While the knowledge experts have selected subjects and objects that appeared in separate sentences, which is shown in the second column of Table 1, we have made a selection of annotations for which the subject and object appeared in the same sentence, displayed in the third column. The reason for doing this, is that restricting it to finding relations in a single sentence speeds up the algorithm significantly, while losing only a small portion of the annotations. In future work we intend to experiment with matching a rule onto several sentences, instead of just one. This may also increase the recall, because it often occurs that the subject and the object lie within a certain range from each other, while such an approach still takes less computation time compared to matching the full news item.

Table 1. Inter-Annotator Agreement (IAA) for each of the 10 considered relationships

Relation	Articles	Sentences	IAA
Competitor	157	126	0.62
Loss	56	31	0.67
Partner	61	59	0.63
Subsidiary	115	97	0.63
CEO	161	135	0.83
President	64	58	0.68
Product	344	300	0.73
Profit	68	46	0.72
Sales	45	20	0.78
ShareValue	82	77	0.78
Total	1153	949	0.71

Using a hill-climbing procedure, we optimized our algorithm parameters. When learning rules using the genetic programming algorithm with ramped-half-and-half initialization, tournament selection (with a tournament size of 0.25), and a population size of 100, a tree depth of 3 and a maximum amount of children of 7 yielded the best results. Here, the mutation rate and elitism rate are 0.3 and 0.05, respectively, whereas the bloat parameter α equals 0.001, making it only effective for situations where F_1-measures are approximately the same. The group size equals 10, and in our optimal configuration, we only allow for $T = 50$ generations with the same fitness values. Also, during rule learning, we put an emphasis on precision scores with $\beta = 0.3$ for F_β, i.e., an increase in precision is considered to be more important than an increase in recall.

The results of the evaluation are presented in Table 2, which underlines that, when compared to a full manual approach to rule creation, the use of genetic programming for rule learning can be useful for the considered relations within our evaluated financial domain. The learned rules are used for extracting relations between subjects and objects (facts), i.e., both subject and object have to be correctly identified, as well as the other components used in the rules. Small errors in classification of individual tokens (words) easily disrupt relation detection. Correct classification of relations thus is less trivial than regular named entity recognition, leading to lower results than one would initially expect [10].

For automatic rule learning, the *CEO* relation performs best with a precision, recall, and F_1-measure of 90%. In a similar manner rules are learned for the *President* and *Product* relations. For the latter relation we obtain a rule group with a precision and recall of 79%, yielding a 79% F_1-measure. For the *President* relation, we measure a precision and recall of 82% and 79%, respectively, resulting in a slightly higher F_1-measure of 80%. The *President* relation hence performs slightly worse than the *CEO* relation, even though the structure of text is somewhat similar. This may be caused by the lower number of annotations for the *President* relation. In addition, we have shown in Table 1 that the IAA for this relation is slightly lower compared to the *CEO* relation.

Table 2. Precision, recall, and F_1 scores for all 10 financial relations (rule groups) after 5 hours of automatic rule learning (left) and manual creation (right)

Relation	Automatic Learning			Manual Creation			$\Delta\%$
	Precision	Recall	F_1-measure	Precision	Recall	F_1-measure	
Competitor	0.667	0.508	0.577	0.875	0.280	0.424	36.0%
Loss	0.905	0.613	0.731	0.818	0.333	0.474	54.3%
Partner	0.808	0.356	0.494	0.450	0.391	0.419	18.0%
Subsidiary	0.698	0.309	0.429	0.611	0.239	0.344	24.8%
CEO	0.904	0.904	0.904	0.824	0.700	0.757	19.5%
President	0.821	0.793	0.807	0.833	0.455	0.588	37.2%
Product	0.788	0.793	0.791	0.862	0.596	0.704	12.3%
Profit	0.960	0.522	0.676	1.000	0.273	0.429	57.7%
Sales	0.900	0.450	0.600	0.455	0.455	0.455	32.0%
ShareValue	0.939	0.805	0.867	0.530	0.778	0.631	37.5%
Total	0.839	0.605	0.703	0.726	0.450	0.555	26.6%

For the *Competitor, Subsidiary,* and *Partner* relations, the precision, recall, and F_1-measure are lower in comparison with the aforementioned relations, approximately ranging between 40% and 60%. This could be caused by the fact that both the subject and the object of these relations are expected to be of type *Company*, while for other types of relation – e.g., *Product* and *CEO* – the subject and object are of different types, increasing the importance of finding contextual concepts that specifically describe the relation at hand. Additionally, in retrospect, the structure of the sentences in our data describing such relations is more complex than for other relations. In order to find more suitable patterns, the patterns need to be more complex by, for instance, adding more *and* and *not* operators, with the risk of overfitting. Future work should therefore focus on determining how patterns can be learned from more complex sentences, by for instance pre-analyzing the rules for often returning concepts and increasing the probability of appearance for these concepts during initialization and mutation.

The remaining relations, i.e., *Loss, Profit, Sales,* and *ShareValue* are all data properties, meaning they do not require a concept for the object of the relation. Examples of the data property values are "10.5 million euros", "\$12", or "53 thousand yen". In order to match those values one may need a complex pattern, and hence we decided to use the classification component of Hermes to annotate currency values as a single token. For example, the string "10.5 million euros" is annotated with a single annotation, e.g., *CurrencyValue*, which can be used in the information extraction rules. This allows us to treat these data properties in a similar manner as the object properties.

Last, the results for automatic rule generation depicted in Table 2 show that among the data properties *ShareValue* achieved the highest F_1-value, i.e., 87%, followed by the *Loss* relation, which measured an F_1-value of 73%. The *Profit* and *Sales* relations performed slightly worse, resulting in F_1-measures between 60% and 70%.

Our experiments show that the used fitness function – defined in (1) – is expensive because the F_1-measure has to be calculated for each rule in each generation of a population, and is heavily dependent on available computing power. On our machine, using an Intel® 2.66 GHz Core^{TM} i7 920 processor with 6 GB of RAM, jobs finished within 5 hours each. On average, the generation of a rule group representing a relation takes approximately 4 and a half hours. The largest amount of time needed for one rule group was 5 hours, whereas the smallest amount of time required was 3 and a half hours.

We also let a domain expert create rules manually for 5 hours per rule group on the same machine to ensure a fair comparison of our automatic system with the manual creation of rules. Again, most time is consumed by evaluating rules, yet a manual approach is less efficient. Where the genetic programming approach generates precision, recall, and F_1-values of 84%, 61%, and 70%, respectively, on average, the manually created rule groups show lower performances. For manual rule creation, the resulting F_1-values are on average about 27% lower (displayed under $\Delta\%$ in the rightmost column of Table 2). Hence, within the same amount of time (i.e., 5 hours per rule group), a domain expert manually writing rules would end up with worse performing rules than an automated genetic programming-based approach. We do not question the potential quality of the rules manually created by the experts when allowing for more time, yet within the limited amount of time advantages of automatic generation are clearly shown. We do, however, observe similar performance patterns as have been described above.

The largest improvements (up to 58%) we observe for relations that involve data properties that deal with more complex constructions (e.g., using datatype variants), which are cumbersome for human experts to include in their rules, hence leading to lower recall. For example, *Loss* and *Profit* involve complex sentences with currencies, which have many different variants in our data set. On the other hand, rule groups that cover many structurally homogeneous examples for which the subject and object are concepts having different types, e.g., *Product* and *CEO*, show improvements as low as 12%, as these are straightforward to implement for domain experts, thus diminishing the need for automation.

For the domain expert, the actual writing took up a few percent of the total time (5 to 10 minutes). A considerable amount of time was used for reading news messages, analyzing matched patterns, verifying results, etc. Additionally, perfecting rules took up increasingly more time, as one needs to abstract away from the given examples in the training set. When increasing the training set size, it would become virtually impossible for domain experts to keep up with a genetic programming-based approach, underlining the added value for automatic rule generation for detecting complex semantic relations in large data sets.

7   Conclusions

Answering to the need for ontology update languages, in this paper we have introduced the Hermes Information Extraction Language (HIEL). The language supports many of the features that exist in regular expressions, such as sequences, literals, logical operators, repetition, and wildcards. In addition to this, syntactic and orthographic categories are supported. In order to allow the user to create generic information extraction rules for a domain we made use of semantic entities in rules by employing ontological elements, such as classes and instances.

Information extraction rules are often used in automatic information extraction, yet they are usually manually constructed. We have presented a genetic programming-based approach for automatically learning these rules from financial news. Genetic programming approaches provide rules expressed in a user-understandable way, and usually find adequate solutions within a reasonable amount of time. In general, our system performs good in terms of recall and precision, and hence also yields good F_1-values of 70% across all considered financial relations. Our experiments show that compared to information extraction rules constructed by expert users, we are able to find rules that yield a higher F_1-value (i.e., 27% higher on average) after the same amount of time (i.e., 5 hours). A frequently encountered problem for the genetic programming approach is that the quality of the initial population is too low, because the probability that the right concepts are initially chosen becomes smaller as the total number of concepts in the knowledge base increases.

As future work, we aim to investigate solutions to the aforementioned problem, e.g., by implementing heuristics and bootstrapping our algorithms. We hypothesize that frequently appearing concepts in a certain domain can be given a higher probability during initialization to increase the quality of the initial population. Moreover, manually derived rules can be useful when deployed in the initial population. Also, we plan to extend our evaluation to also include single rule matching on multiple sentences. Additional directions are extracting other types of information (from different domains than the financial domain, such as the political, medical, and weather domains), as well as connecting automatic rule learning with (semi-)automated ontology updating mechanisms [15] that process the extracted information and update ontologies accordingly.

Acknowledgment. The authors are partially supported by the NWO Physical Sciences Free Competition project 612.001.009: Financial Events Recognition in News for Algorithmic Trading (FERNAT), the Dutch national program COMMIT, and the NWO Mozaiek scholarship project 017.007.142: Semantic Web Enhanced Product Search (SWEPS).

References

1. Angeline, P.J.: Subtree Crossover: Building Block Engine or Macromutation? In: 2nd Ann. Conf. on Genetic Programming (GP 1997), pp. 9–17. Morgan Kaufmann (1997)
2. Black, W.J., Mc Naught, J., Vasilakopoulos, A., Zervanou, K., Theodoulidis, B., Rinaldi, F.: CAFETIERE: Conceptual Annotations for Facts, Events, Terms, Individual Entities, and RElations. Technical Report TR–U4.3.1, UMIST (2005)

3. Borg, C., Rosner, M., Pace, G.J.: Automatic Grammar Rule Extraction and Ranking for Definitions. In: 7th Int. Conf. of Language Resources and Evaluation (LREC 2010). European Language Resources Association (2010)
4. Carlson, A., Betteridge, J., Wang, R.C., Hruschka Jr., E.R., Mitchell, T.M.: Coupled Semi-Supervised Learning for Information Extraction. In: 3rd Int. Conf. on Web Search and Data Mining (WSDM 2010), pp. 101–110. ACM (2010)
5. Castellanos, M., Gupta, C., Wang, S., Dayal, U.: Leveraging Web Streams for Contractual Situational Awareness in Operational BI. In: Int. Workshop on Business intelligencE and the WEB (BEWEB 2010) in Conjunction with EDBT/ICDT 2010 Joint Conf., pp. 1–8. ACM (2010)
6. Chang, C.H., Kayed, M., Girgis, M.R., Shaalan, K.: A Survey of Web Information Extraction Systems. IEEE Transactions on Knowledge and Data Engineering 18(10), 1411–1428 (2006)
7. Cunningham, H., Maynard, D., Bontcheva, K., Tablan, V.: GATE: A Framework and Graphical Development Environment for Robust NLP Tools and Applications. In: 40th Anniversary Meeting of the Association for Computational Linguistics (ACL 2002), pp. 168–175. Association for Computational Linguistics (2002)
8. Domingue, J., Motta, E.: PlanetOnto: From News Publishing to Integrated Knowledge Management Support. IEEE Intelligent Systems 15(3), 26–32 (2000)
9. Etzioni, O., Cafarella, M., Downey, D., Popescu, A., Shaked, T., Soderland, S., Weld, D.S., Yates, A.: Unsupervised Named-Entity Extraction From The Web: An Experimental Study. Artificial Intelligence 165(1), 91–134 (2005)
10. Frasincar, F., Borsje, J., Hogenboom, F.: Personalizing News Services Using Semantic Web Technologies. In: E-Business Applications for Product Development and Competitive Growth: Emerging Technologies, pp. 261–289. IGI Global (2011)
11. Hearst, M.A.: Automatic Acquisition of Hyponyms from Large Text Corpora. In: 14th Conf. on Computational Linguistics (COLING 1992), vol. 2, pp. 539–545 (1992)
12. IJntema, W., Sangers, J., Hogenboom, F., Frasincar, F.: A Lexico-Semantic Pattern Language for Learning Ontology Instances from Text. J. of Web Semantics: Science, Services and Agents on the World Wide Web 15(1), 37–50 (2012)
13. Jones, T.: Crossover Macromutation and Population-based Search. In: 6th Int. Conf. on Genetic Algorithms (ICGA 1995), pp. 73–80. Morgan Kaufmann (1995)
14. Maynard, D., Saggion, H., Yankova, M., Bontcheva, K., Peters, W.: Natural Language Technology for Information Integration in Business Intelligence. In: Abramowicz, W. (ed.) BIS 2007. LNCS, vol. 4439, pp. 366–380. Springer, Heidelberg (2007)
15. Sangers, J., Hogenboom, F., Frasincar, F.: Event-Driven Ontology Updating. In: Wang, X.S., Cruz, I., Delis, A., Huang, G. (eds.) WISE 2012. LNCS, vol. 7651, pp. 44–57. Springer, Heidelberg (2012)
16. Snow, R., Jurafsky, D., Ng, A.Y.: Learning Syntactic Patterns for Automatic Hypernym Discovery. In: 18th Ann. Conf. on Neural Information Processing Systems (NIPS 2004). Advances in Neural Information Processing Systems, vol. 17, pp. 1297–1304. MIT Press (2004)
17. Soderland, S.: Learning Information Extraction Rules for Semi-Structured and Free Text. Machine Learning 34(1-3), 233–272 (1999)
18. Thompson, D.R., Bilbro, G.L.: Comparison of a Genetic Algorithm with a Simulated Annealing Algorithm for the Design of an ATM Network. IEEE Communications Letters 4(8), 267–269 (2000)

Ontology-Based Management
of Conflicting Products in Pixel Advertising

Ferry Boon, Sabri Bouzidi, Raymond Vermaas,
Damir Vandic, and Flavius Frasincar

Erasmus University Rotterdam
Erasmus School of Economics
P.O. Box 1738, 3000 DR, Rotterdam, The Netherlands
ferry.boon@gmail.com, sabribouzidi@outlook.com,
research@raymondvermaas.nl, {vandic,frasincar}@ese.eur.nl

Abstract. Pixel advertising represents the placement of multiple pixel blocks on a banner for the purpose of advertising companies and their products. In this paper, we investigate how one can avoid product conflicts in the placement of pixel advertisements on a Web banner, while maximizing the overall banner revenue. Our solution for this problem is based on a product ontology that defines products and their relationships. We evaluate three heuristic algorithms for generating allocation patterns, i.e., the left justified algorithm, the orthogonal algorithm, and the GRASP constructive algorithm. The results show that the left justified algorithm and the orthogonal algorithm are most effective in terms of profit per pixel, while the GRASP constructive algorithm is identified as most efficient in terms of computational time.

1 Introduction

Web advertising is a billion dollar business in which most large Web companies have found their main stream of revenue. The total advertising revenues from Google were 37.9 billion USD in 2011, 46.1 billion USD in 2012, and 55.5 billion USD in 2013 [5]. Companies such as Yahoo!, Facebook, Microsoft, and AOL also report advertising revenues in the billions, making online advertisement revenues larger than the revenues obtained in printed media [7].

Pixel advertising is a form of display advertising on the Web in which the cost of each advertisement is calculated based on the number of pixels it occupies. The general idea is to have a banner that consists of several small advertisements (i.e., a multi-picture banner), instead of just one advertisement occupying the whole banner. Pixel advertising was invented in 2005 by the English student Alex Tew, who created the "Million Dollar Homepage" [12]. This Web page holds a 1000 by 1000 pixel grid from which blocks of 10 by 10 pixels can be bought for 1 dollar per pixel. Buyers can place an image on their pixels and have the image link to their Website.

Although the "Million Dollar Homepage" has been incredibly successful, pixel advertising has not been yet widely adopted. There are a number of issues that need to be addressed in order to make the concept more appealing

B. Benatallah et al. (Eds.): WISE 2014, Part I, LNCS 8786, pp. 433–447, 2014.
© Springer International Publishing Switzerland 2014

to advertising companies. One of these issues is that there is no motivation for consumers to return to a pixel advertisement banner. Furthermore, when advertisements are placed on a banner, the content of the advertisement is often not taken into account, making it possible to place conflicting advertisements of competing products and/or brands in one Web banner. In this paper, we do take the content of an advertisement into account and tackle the problem of placing conflicting advertisements. Customers often have to choose between products from various domains. Marketing differentiated products, i.e., products that serve the same need of the customer and thus need to be pushed to the consumer via advertisements, frequently develop and compete on the basis of brands or labels. The Coca-Cola Company vs. Pepsi is a typical example of such a competition.

Placing advertisements on a banner is often done by employing heuristics, as the problem of finding an optimally constructed banner for a set of advertisements is NP-hard [15]. In the literature, there are studies that focus on finding heuristics for the purpose of optimally placing rectangular advertisements on a rectangular banner, in such a way that the revenue generated by the banner is maximized [3, 9, 15]. Unfortunately, none of these approaches takes the content of an advertisement into account.

In this paper, we propose an approach that avoids conflicts when placing pixel advertisements on a banner, while maximizing banner revenue. Using an ontology-based solution, our current research extends the heuristics discussed in [3, 9] by avoiding placing advertisements of conflicting brands on a banner. Conflicts can be identified by categorizing products using a domain ontology. In order to avoid product conflicts, we restrict the number of advertisements per category that is placed on the banner to one. We investigate which heuristic copes best with this new constraint and discuss the implementation of such an approach in a Web application.

The paper is structured as follows. In Section 2 we discuss related work. In Section 3 we give a formal problem definition, after which we present heuristics that deal with the conflict constraint banner placement problem. Section 4 explains the experimental setup and presents the obtained results. The concluding remarks and proposed future work are given in Section 5.

2 Related Work

Even though pixel advertising has not been successful so far, it is still interesting to study this topic, as it has great potential for new forms of advertising campaigns. In [14] the success of the "Million Dollar Homepage" and the failure of the many copycats that spawned from the original success is discussed. The authors argue that visitors do not return to the "Million Dollar Homepage", they only visit it once to check it out. The paper proposes some improvements to the concept of pixel advertising in Web pages. In [15], the authors extend the idea of pixel advertising by placing small advertisements in banners.

The work presented in [3] is the most related research to this paper and goes one step further than [15]. The authors propose a Web application that can automatically fill the banner with provided advertisements. Several heuristics are explored that optimize the building time and the revenue generated by the banner. More specific, heuristics for optimally placing rectangular advertisements on a rectangular banner in such a way that the revenue generated by the banner is maximized are discussed. The authors show that the orthogonal heuristic is the most suitable for a Web application. Furthermore, the authors experimented with the left justified algorithm, the GRASP constructive algorithm, and the greedy stripping algorithm. However, when placing the advertisements on a banner, the content is not considered, with the possible result of two competitive advertisements on the same banner. The pixel advertising and the multiple advertising allocation approach is related to other problems which are further discussed in [3], including the ad placement problem, knapsack problem, MINSPACE, and MAXSPACE problem, which are known NP-hard problems, hence the need for heuristics.

As previously mentioned, when placing advertisements on the banner, the advertisement content, and its associated semantics, is not taken into account. Especially with a small set of possible advertisements, it is highly probable that advertisements with conflicting messages or competing products are placed on the same banner. When we are looking at the nature of the conflicts that arise, they can be traced back to the same problem brands have in stores. Customers only have a limited amount of time and money to spend on products and services, and often have to choose between products. Marketing differentiated products frequently develop and compete on the basis of brands or labels. Several examples of this inter-brand competition would be Coca Cola vs. Pepsi-Cola, Levi vs. Pall Mall Jeans, and Pizza Hut vs. Dominos.

Another possible conflict would be the intra-manufacturer conflict, where a manufacturer owns, produces, and/or sells different brands that (in)directly compete with one another. A good example would be The Coca-Cola Company owns several soft drink brands such as Coca Cola, Fanta, and Sprite. These products obviously compete with each other as they are all carbonated soft drinks, but according to [11] we can disregard this form of competition as being a conflict. This form of brand extension is the so-called substitute brand extension, where a substitute product is branded differently than the original product. In the case of the Coca-Cola Company, all products are marketed separately with each brand having its own management and thus each brand has its own goals. Nevertheless, one can also argue that the intra-manufacturer conflicts are negligible as often in printed media brands from the same manufacturer are promoted near or even next to each other.

In order to avoid product conflicts, we extend the heuristics presented in [3]. Furthermore, we aim to identify which heuristic in the new setup is the most effective in terms of profit and which one is the most efficient in terms of speed. Speed is important because we plan to use the heuristics in a Web application, and according to a psychology research Web users are not willing to wait for more than 15 seconds without feedback from the system that it is working [10]. Therefore our constraint on time is 15 seconds. Because the problem that we consider in this paper is NP-hard, heuristics are needed to obtain good solutions for large inputs in a timely manner. To identify conflicts we use an ontology-based approach. The type of conflict that this study uses is called Class Assertion Conflict. Such conflicts occur when constraints placed on classes are violated, as shown in [4].

3 Optimal Advertisement Allocation

In this section, we discuss in more detail the problem and solutions of optimally allocating advertisements, under the constraint that no conflicting advertisements should be placed and the revenue should be maximized. In Section 3.1 we give a formal problem definition that results in an integer programming model. Section 3.2 presents the considered heuristics to solve the presented problem. Last, in Section 3.3 we discuss a Web implementation of our approach.

3.1 Problem Definition

The formal definition of the problem to be solved is similar and based on the problem definition given in [3]. We have a set $\mathcal{A}$ with a fixed number of advertisements $|\mathcal{A}|$ to allocate in a banner. We assume we have more advertisements in $\mathcal{A}$ would fit on the banner, thus not every advertisement is placed. Each advertisement $a_i \in \mathcal{A}$ has a width w_i, height h_i, and a price per pixel pp_i, with $i \in \{1, ..., |\mathcal{A}|\}$. The banner has width $\mathcal{W}$ and height $\mathcal{H}$. The advertisements from $\mathcal{A}$ should be allocated on the banner such that the total value of the set of allocated advertisements $\mathcal{A}'$ (subset of $\mathcal{A}$) is maximized. Each advertisement a_i in $\mathcal{A}'$ has its top-left corner at position (x, y) on the banner, starting from $(0, 0)$ which represents the top left corner of the banner. The value of an allocated advertisements $\mathcal{A}'$ is defined by v_i, where $v_i = pp_i \times w_i \times h_i$. Our objective is to maximize the total value of allocated advertisements in $\mathcal{A}'$.

We can formulate the problem as a 0-1 integer programming problem, which is a simplification of the problem formulation from [6], since our problem assumes that every advertisement can be allocated only once. In order to make sure that advertisements do not overlap on the banner we have reused a constraint from [2]. Let

$$\mathcal{X}_i = \{x | 0 \le x \le \mathcal{W} - w_i\}, \forall i \in \{1, ..., |\mathcal{A}|\}. \tag{1}$$

be the set of all possible points along the width of the banner such that an advertisement a_i from $\mathcal{A}$ can be placed on the banner with its top-left corner at these x-axis positions. Similarly we define

$$\mathcal{Y}_i = \{x | 0 \le y \le \mathcal{H} - h_i\}, \forall i \in \{1, ..., |\mathcal{A}|\}. \tag{2}$$

as the set of all possible allocation points along the height of the banner. We define

$$x_{ip} = \begin{cases} 1 & \text{if } a_i \text{ is placed with top-left corner at x-position } p, \text{ where } p \in \mathcal{X}_i \\ 0 & \text{otherwise.} \end{cases} \tag{3}$$

$$y_{ip} = \begin{cases} 1 & \text{if } a_i \text{ is placed with top-left corner at y-position } q \text{ where } q \in \mathcal{Y}_i \\ 0 & \text{otherwise.} \end{cases} \quad (4)$$

and let

$$b_{ipqrs} = \begin{cases} 1 & \text{if advertisement } i, \text{ placed with top-left corner at } (p,q), \\ & \text{cuts out point } (r,s) \text{ of the banner} \\ 0 & \text{otherwise.} \end{cases} \quad (5)$$

which can be restated as

$$b_{ipqrs} = \begin{cases} 1 & \text{if } 0 \le p \le r \le p + w_i - 1 \le W - 1 \\ & \text{and } 0 \le q \le s \le q + h_i - 1 \le \mathcal{H} - 1 \\ 0 & \text{otherwise.} \end{cases} \quad (6)$$

Figure 1 visualizes x_{ip}, y_{iq}, and b_{ipqrs}. Now the integer programming formulation can be stated as follows:

$$max \sum_{i=1}^{|\mathcal{A}|} v_i \sum_{p \in \mathcal{X}_i} x_{ip} \quad (7)$$

subject to

$$\sum_{i=1}^{|\mathcal{A}|} \sum_{p \in \mathcal{X}_i} \sum_{q \in \mathcal{Y}_i} b_{ipqrs} \cdot x_{ip} \cdot y_{iq} \le 1, \forall r \in \{0, ..., W-1\}, \forall s \in \{0, ..., \mathcal{H}-1\} \quad (8)$$

$$\sum_{p \in \mathcal{X}_i} x_{ip} \le 1, \forall i \in \{1, ..., |\mathcal{A}|\} \quad (9)$$

$$\sum_{p \in \mathcal{X}_i} x_{ip} = \sum_{q \in \mathcal{Y}_i} y_{iq}, \forall i \in \{1, ..., |\mathcal{A}|\} \quad (10)$$

$$x_{ip}, y_{ip} \in \{0,1\}, \forall i \in \{1, ..., |\mathcal{A}|\}, \forall p \in \mathcal{X}_i, \forall q \in \mathcal{Y}_i \quad (11)$$

In order to avoid conflicting advertisements, a conflict matrix $\mathcal{C}$ is introduced where c_{ij} is the conflict between advertisement i and advertisement j and is defined as:

$$c_{ij} = \begin{cases} 1 & \text{if advertisement } i \text{ and } j \text{ are conflicting} \\ 0 & \text{otherwise.} \end{cases} \quad (12)$$

A new constraint can be added so that no conflicts are allowed between the allocated advertisements on the banner:

$$\sum_{p \in \mathcal{X}_i} x_{ip} = 1 - \sum_{j \ne i} c_{ij} \sum_{p \in \mathcal{X}_j} x_{jp}, \forall i \in \{1, ..., |\mathcal{A}|\} \quad (13)$$

In Eq. 7 the objective function maximizes the total value of the allocated advertisements. Constraint 8 ensures that any banner point is used by at most one advertisement. Constraints 9 and 10 ensure that any advertisement is allocated at most once on the whole banner. The ranges of p and q, i.e., $\mathcal{X}_i$ and $\mathcal{Y}_i$, respectively, ensure that advertisements are always placed inside the banner. Constraint 11 is the integrality constraint. Constraint 13 is added so that no conflicts are allowed in the allocated advertisements on the banner. The model can be linearized by replacing variables x_{ip} and y_{ip} with a variable that represents both x_{ip} and y_{ip} (e.g., z_{ipq}), as shown in [2].

3.2   Heuristics

Following the typology presented in [13], we can characterize the problem that we presented in the previous section as a *two-dimensional, single, orthogonal, knapsack problem* with conflict restrictions. The knapsack problem is a combinatorial optimization problem where strongly heterogeneous assortment of small items has to be allocated to one or more larger objects. The limitation here is that there is not enough space for all items, which means that a choice has to be made. The term *single* indicates that we only deal with one large object in which the smaller items (i.e., the advertisements) have to be placed. The term *orthogonal* refers to the fact that the edges of the smaller items are orthogonal to the edges of the larger object(s) and that rotation is not allowed.

In order to deal with the fact that the previously identified problem is NP-hard, one often employs optimized search algorithms to find solutions. These algorithms work by searching through the solution space in order to find the optimal solution. Such approaches can be classified as uninformed and informed algorithms. An uninformed algorithm tests all the possible solutions whereas an informed algorithm uses knowledge of the search space to find the solution. The downside of finding the optimal solution is that it takes very long to find it using any of the current approaches. This is an important limitation because we aim to have a Web-service that allows users to create an optimal banner, which requires a relatively short execution time.

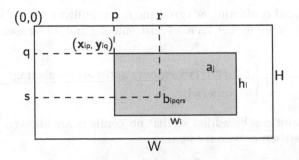

Fig. 1. Visualization of x_{ip}, y_{iq}, and b_{ipqrs}

Because optimal solutions are hard to find, we focus on finding, in a relatively short amount of time, solutions that are close to the optimal one. For this purpose, we analyze and adapt three heuristic algorithms that can solve the problem presented in the previous section. For all three considered heuristics, which will be discussed shortly, we introduce an extra condition that is meant to avoid placing conflicting advertisements, i.e., we only place an advertisement on the banner if there is no other banner placed from the same product category (obtained from the product ontology).

Sorting. The heuristic approaches we use rely on the order the advertisements are processed by the algorithm. For this reason we apply a sorting procedure as an initial step. Besides considering the possibility of randomly sorting the advertisements, we define a set of properties of the advertisement on which we can sort. These are price per pixel (pp), width (w), height (h), total area ($w \times h$), flatness (w/h), and proportionality ($|\log w/h|$). The flatness property indicates whether an advertisement is flat ($w > h$) or tall ($h > w$). The proportionality property refers to what extent the advertisement dimensions resembles a square, where a value of 0 represents an exact square advertisement. Using these properties, we apply a two-way sort, i.e., a primary and a secondary sort. This means that we first apply the primary sort on a particular criterion and for advertisements that are ranked the same we apply the secondary sort (on one of the remaining criteria).

Left justified heuristic. The first heuristic that we consider iterates through the list of available advertisements. For each advertisement, it scans the columns of the banner from top to bottom. If the end of a column is reached the algorithm continues to the next column on the first row. This process is then repeated for each advertisement. When an empty location has been found and the advertisement size fits this location, the advertisement is placed on the banner with the top left corner at the current position. If the advertisement goes horizontally out of bounds, we are unable to get it allocated and continue with the next advertisement. In contrast, if the advertisement goes vertically out of bounds, we move the 'current location' to the first row of the next column. The algorithm stops once we iterated through all available advertisements.

Orthogonal heuristic. In the orthogonal algorithm we iterate through the list of available advertisements and place advertisements as close as possible to the top left corner. The algorithm looks for empty locations for the current advertisement by moving along the diagonal from the top left corner of the banner. At each step, the algorithm first searches, on the right and the bottom of the current location, for empty locations where the advertisement can be allocated. After that, we compare the two found free locations (one on the right of the current location and one below the current location) with respect to the sum of the distances to the top and to the left. We allocate the currently considered advertisement on the position that yields the smallest sum. When there is a tie we choose the one on the vertical search path. After the advertisement is allocated, we start again in the top-left corner of the banner, trying to allocate the next advertisement.

If we fail to allocate an advertisement for a certain location, we continue to walk along the diagonal. When the final row is reached, and there are still columns left, we deviate from or walk on the diagonal and move to the next column. When the final column is reached, and there are still rows left, we move to the next row. This means that after we start walking diagonally, we will eventually switch to walking either right or down, except for the situation in which the banner is a square. When the final row and column are reached and we have failed to allocated the currently considered advertisement, we start again in the top left corner of the banner and try to allocate the next available advertisement. The algorithm stops once we iterated through all available advertisements.

GRASP constructive heuristic. The GRASP constructive algorithm is based on the greedy randomized adaptive search procedure (GRASP) for the constrained two-dimensional non-guillotine cutting problem [1]. It has a different approach than the algorithms discussed previously. Instead of searching the banner for free space to place the currently considered advertisement, it considers the rectangles of free space and finds matching advertisements that fit into these free spaces.

Initially, the full banner is considered as one large free space. First, we take the smallest rectangle of free space in which an advertisement that has not yet been placed can fit, after which we place the corresponding advertisement in this smallest rectangle of free space. Whenever an advertisement is placed in a rectangle, new free rectangles are formed and added to set of free spaces, while the original rectangle is marked as non-free. We always place the advertisement in a corner of the rectangle which is closest to a corner of the banner, and cut the free space left in such a way that it yields optimal new free rectangles. In order to obtain the optimal new free rectangles we choose to merge rectangles in such a way that the largest rectangle can accommodate the next advertisement. If there is a tie, we choose the merge which yields a new free rectangle with the largest area. We mark an empty location as used whenever we fail to allocate an advertisement to it, otherwise the algorithm could fall into an endless loop. After we have processed a rectangle we continue with the next smallest rectangle that has not been used before. When there are no free rectangles left (i.e., the whole banner is allocated) or no more available advertisements that fit any of the remaining rectangles, the algorithm stops.

We propose some modifications to the original GRASP approach that aim to decrease the total execution time. Differently from the original approach, we merge immediately adjacent free rectangles after we obtain new free rectangles. This is done in order to increase the probability of allocating advertisements to free rectangles.

3.3 Web Implementation

For the Web implementation we need a different approach than presented in the previous section, since we have to deal with an additional number of constraints. First, the Web implementation has to be fast and accurate, since users on the Web expect good results and do not want to wait too long. Second, it must be easy for users to submit their advertisement data and conflicting products so that this information can be used by the algorithm.

The Web application is available on `http://pixmax.damirvandic.com`. The software is implemented in Java. For the trade-off between the execution time and the quality of the solution, we set the maximum processing time to 15 seconds. This setting seems to give good results for a relatively low amount of processing time. Figure 2 shows the result for a 728×90 banner, with the orthogonal algorithm and sorting on price per pixel and proportionality. As we can see in this example, there are no conflicting items on the banner.

Users are also able to upload a zip file with images, a configuration file, and a domain ontology containing different product groups. The Web application provides the user with an example ontology that covers the beverages domain. The relationships between objects in the ontology specify how ontology individuals are related to each other. The relations used in the beverages ontology is the subsumption relation. By the use of the subsumption relation a taxonomy (a tree-like structure) of products is created. It is possible to create conflicts on all levels in the tree except the root.

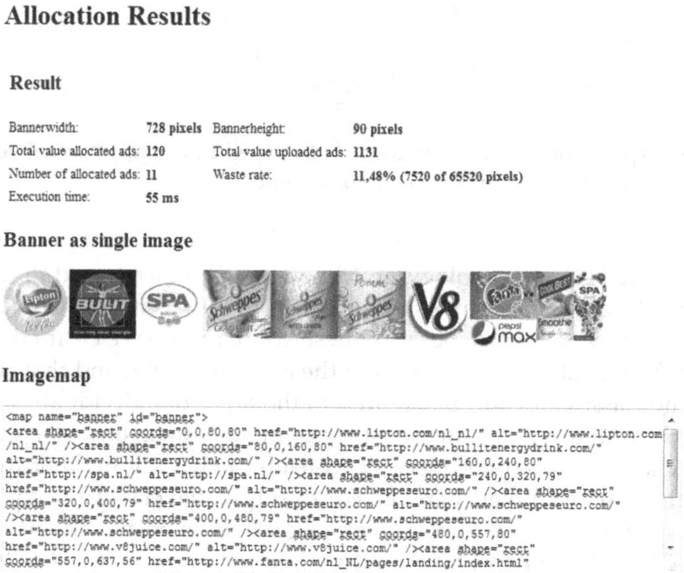

Allocation Results

Result

Bannerwidth:	728 pixels	Bannerheight:	90 pixels
Total value allocated ads:	120	Total value uploaded ads:	1131
Number of allocated ads:	11	Waste rate:	11,48% (7520 of 65520 pixels)
Execution time:	55 ms		

Banner as single image

Imagemap

```
<map name="banner" id="banner">
<area shape="rect" coords="0,0,80,80" href="http://www.lipton.com/nl_nl/" alt="http://www.lipton.com
/nl_nl/" /><area shape="rect" coords="80,0,160,80" href="http://www.bullitenergydrink.com/"
alt="http://www.bullitenergydrink.com/" /><area shape="rect" coords="160,0,240,80"
href="http://spa.nl/" alt="http://spa.nl/" /><area shape="rect" coords="240,0,320,79"
href="http://www.schweppeseuro.com/" alt="http://www.schweppeseuro.com/" /><area shape="rect"
coords="320,0,400,79" href="http://www.schweppeseuro.com/" alt="http://www.schweppeseuro.com/"
/><area shape="rect" coords="400,0,480,79" href="http://www.schweppeseuro.com/"
alt="http://www.schweppeseuro.com/" /><area shape="rect" coords="480,0,557,80"
href="http://www.v8juice.com/" alt="http://www.v8juice.com/" /><area shape="rect"
coords="557,0,637,56" href="http://www.fanta.com/nl_NL/pages/landing/index.html"
```

Fig. 2. A screenshot of our Web application, where users can generate optimally allocated banner advertisements

4   Evaluation

In this section, the results of our experiments are discussed. First, we give more insight in the dataset we have used and our experimental setup, after which we discuss the performance of the considered heuristics.

4.1 Dataset

In order to test our implementation, we built a dataset that contains 113 adver-tisements of different sizes for non-alcoholic beverages. The price per pixel of each advertisement is drawn from a uniform distribution between 9 and 11. We chose for non-alcoholic beverages, because it offers a large variety of products. The ontology contains 34 different product groups, with 23 products groups as leaf nodes that contain product instances. An excerpt of the ontology and its structure is shown in Figure 3.

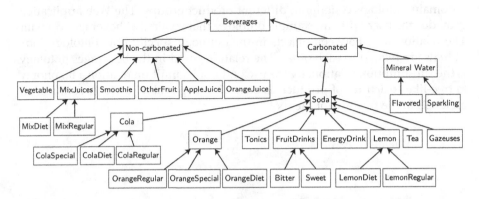

Fig. 3. An excerpt of the product ontology that is used in our approach. Every arrow represents an 'is-a' relationship between two concepts.

The individuals in our ontology, also known as instances, used in this study are concrete objects. For example, beverages from the Coca-Cola Company and PepsiCo Inc. are represented, because these companies have clear competitive products, which could cause conflicts in the advertisement, and that should be avoided. The classes, or concepts, represent the types of individuals. Each class has a number of individuals. In the advertisement banner, only one individual from each class can be represented. We consider here only the first two levels, i.e., the leaf classes and their parents. For example, the class Cola contains the individuals Coca-Cola and Pepsi, when Coca-Cola is on the advertisement it is not possible to have Pepsi on the advertisement banner as well. This is one example of conflicting advertisements, which is given here for illustrative pur-poses. Other types of conflicts can be defined and incorporated in the domain ontology and our application.

4.2 Experimental Setup

For each of the three algorithms, we tested all combinations between the 121 sorting combinations and the 9 banner sizes. The 121 sorting combinations fol-low from the fact that we have 6 fields (price per pixel, width, height, total area, flatness, and proportionality), two directions (ascending and descending), and one random order, which yield $\frac{(6 \cdot 2)!}{10!} - 12 + 1 = 121$ combinations. This can be considered as choosing a pair from 12 options (6 fields $\times$ 2 directions) where the order matters and no duplicates are allowed. Our experimental setup resulted in 3267 simulation cycles (121 sorting combinations $\times$ 9 banner sizes $\times$ 3 algorithms), resulting in a total run time of 5 hours on a AMD Phenom II X4 810 with 4GB of memory.

We have chosen to use 5 banner sizes that are commonly used in Web advertising [8]. The five banner sizes (expressed in $w \times h$) are 728 $\times$ 90 (leader board), 234 $\times$ 60 (half banner), 125 $\times$ 125 (square button), 120 $\times$ 600 (skyscraper), and 336 $\times$ 280 (large rectangle). Because we also consider inverting every banner size, and one banner size is square, we have a total of $2 \times (5 - 1) + 1 = 9$ banner sizes.

4.3 Results

For overall performance, we consider the mean and five point statistics summary for the profit per pixel and the execution of all simulations, without taking the banner size or sorting into account. For this purpose, we take an average over all banner sizes and sorting combinations for each heuristic. Table 1 shows an overview of these results. As we can see, the orthogonal algorithm performs best on account of profit per pixel, closely followed by the left justified algorithm. The profit per pixel for the GRASP constructive algorithm is significantly lower than that of the first two.

Table 1. Mean and five point summary of the profit per pixel for each algorithm

Algorithm	Minimum	Q_1	Median	Mean	Q_3	Maximum
Left justified	3.620	5.130	6.130	6.347	7.720	9.590
Orthogonal	3.620	5.200	6.250	6.439	7.720	9.590
GRASP con.	0.520	1.180	2.060	2.949	4.590	9.240

Table 2. Mean and five point summary of the execution time in seconds for each algorithm

Algorithm	Minimum	Q_1	Median	Mean	Q_3	Maximum
Left justified	0.090	2.480	4.210	6.454	6.890	38.200
Orthogonal	1.891	4.384	7.188	9.602	12.620	35.700
GRASP con.	0.003866	0.009210	0.017000	0.019610	0.024560	0.125300

Table 2 shows the average execution time for each of the algorithms. The GRASP constructive algorithm performs best on execution time, since all the values of the five point summary and the mean are below 1 second. The execution time for the other two algorithms are significantly higher, where the left justified algorithm scores better in almost all of the cases. The orthogonal algorithm scores only better on the maximum value which is 35.7 seconds compared to 38.2 seconds with the left justified algorithm.

From Tables 1 and 2 we can conclude that the GRASP algorithm is fast but performs poorly. The left justified and orthogonal algorithms have the same performance with respect to profit per pixel, but the left justified algorithm is in general faster. Furthermore, the interquartile range (IQR), defined as $Q_3 - Q_1$, is 4.41 and 8.24 for the left justified and orthogonal algorithm, respectively. The lower IQR of the left justified algorithm indicates that the variance of the execution time is lower than the orthogonal algorithm, which is an advantage because it allows for more precise predictions of the execution time.

Sorting Criteria. The heuristics are influenced significantly by the sorting of the incoming advertisements. Table 3 shows the profit per pixel and the execution time, averaged over all banner sizes and secondary sorting combinations, for combinations between the 13 primary sorting combinations and the three algorithms. The table shows only combinations that have a price per pixel higher than 9. The results show again that the orthogonal algorithm and the left justified perform much better than the GRASP constructive algorithm. We observe that sorting descending on the price per pixel (PPP) property gives the best results for all three algorithms. We also observe that for most of the sorting combinations the descending order performs better than the ascending order. This can be explained by the fact that for most of the values we sort on, the highest values add the most value to the banner.

We also notice that dependency on the sorting criteria for the GRASP algorithm differs from the patterns we encounter for the left justified and orthogonal algorithms. The GRASP algorithm seems to be less dependent on the sorting criteria, as only sorting on price per pixel yields a total price per pixel of 9.24, while sorting on other properties yields the same price per pixel of 9.11. This might indicate that the GRASP algorithm, while having a lower overall average price per pixel, is more robust than the left justified and orthogonal algorithms.

Banner Sizes. Table 4 shows for each considered banner size and for each algorithm, the price per pixel and the total execution time. For each banner size, the rows are sorted on the average price per pixel. The results indicate that for most of the banner sizes both the orthogonal and the left justified algorithms give the same results on profit per pixel, although

the left justified algorithm has a lower or equal execution time in all cases (which was also clear when we considered the Q_1 and Q_3 of the execution time). The GRASP constructive algorithm is the fastest, while the price per pixel is lower than or equal to the price per pixel of the other two algorithms (for all banner sizes, except for the 60×234 banner).

In order to get a better sense of the difference in effectiveness (i.e., profit) between the left justified and the orthogonal algorithm, we compute the total profit if all the banners that are displayed in Table 4 would be sold to

Table 3. The evaluation results based on the primary sorting and considered algorithms

Algorithm	Primary sort	Exec. time (s)	P_{pixel}
Orthogonal	PPP Desc.	11.143	9.59
Left-Justified	PPP Desc.	4.200	9.59
Orthogonal	Width Desc.	20.931	9.48
Left-Justified	Width Desc.	17.350	9.48
Orthogonal	Total Size Desc.	22.932	9.32
Left-Justified	Total Size Desc.	15.240	9.32
GRASP	PPP Desc.	0.007	9.240
Orthogonal	Height Desc.	17.084	9.11
Left-Justified	Proportional Asc.	11.00	9.11
Left-Justified	Height Desc.	11.450	9.11
GRASP	Proportional Asc.	0.006	9.11
GRASP	Total Size Desc.	0.006	9.11
GRASP	Height Desc.	0.006	9.11
GRASP	Width Desc.	0.006	9.11
Orthogonal	Proportional Asc.	16.626	9.11

Table 4. The evaluation results for each of the banner sizes

Size	Algorithm	Sorting	Exec. time (s)	P_{pixel}
728×90	Orthogonal	PPP Desc. & Proportional Asc.	14.106	9.59
	Left-Justified	PPP Desc. & Proportional Asc.	11.620	9.59
	GRASP	PPP Desc. & Width Desc.	0.034	5.26
600×120	Orthogonal	Width Desc. & PPP Desc.	12.341	9.02
	Left-Justified	Width Desc. & PPP Desc.	11.140	9.02
	GRASP	PPP Asc. & Height Asc.	0.030	3.47
336×280	Left-Justified	Total Size Desc. & Height Desc.	13.280	7.85
	Orthogonal	Total Size Desc. & Height Desc.	13.970	7.85
	GRASP	PPP Desc. & Proportional Asc.	0.038	7.02
280×336	Left-Justified	Total Size Desc. & Height Desc.	13.050	7.85
	Orthogonal	Total Size Desc. & Height Desc.	13.763	7.85
	GRASP	PPP Desc. & Proportional Asc.	0.041	7.17
234×60	Orthogonal	PPP Desc. & Height Asc.	7.112	7.07
	Left-Justified	PPP Desc. & Height Asc.	1.370	7.07
	GRASP	PPP Desc. & Width Asc.	0.009	7.02
125×125	Left-Justified	PPP Desc. & Proportional Asc.	10.860	9.24
	Orthogonal	PPP Desc. & Proportional Asc.	15.778	9.24
	GRASP	PPP Desc. & Proportional Asc.	0.007	9.24
120×600	Orthogonal	Width Desc. & PPP Desc.	21.718	8.51
	Left-Justified	Width Desc. & PPP Desc.	19.580	8.51
	GRASP	Width Desc. & PPP Asc.	0.0553	7.37
90×728	Left-Justified	PPP Desc. & Proportional Asc.	4.200	9.59
	Orthogonal	PPP Desc. & Proportional Asc.	11.143	9.59
	GRASP	PPP Asc. & Width Asc.	0.049	7.88
60×234	GRASP	Height Asc. & PPP Desc.	0.008	6.96
	Orthogonal	Height Asc. & PPP Desc.	6.573	6.81
	Left-Justified	Height Asc. & PPP Desc.	1.310	6.81

advertisers. As a result, the left justified algorithm would bring in €4,335,139 and the orthogonal algorithm €4,337,246. As we can see, the difference is negligible on such a high total amount. These results support our previous claim that with respect to profit, the left justified and orthogonal algorithms do not differ much. Because of this, and the fact that the left justified algorithm has a IQR lower execution time, we prefer the left justified algorithm over the orthogonal algorithm.

5   Conclusions

This paper presents an extension to the pixel advertising concept. We focused on avoiding conflicting advertisements on the same banner, using an ontology and heuristic algorithms adopted from [3], including the left justified algorithm, the orthogonal algorithm, and the GRASP constructive algorithm. The results of the experiments indicate that the left justified and orthogonal algorithm are most effective, which means that these algorithms give the highest price per pixel. The most efficient is the GRASP constructive algorithm, however the price per pixel is significantly lower for this algorithm. Furthermore, the left justified algorithm has on average a lower execution time than the orthogonal algorithm. Therefore, the left justified algorithm is considered to be the best choice when it comes to avoiding conflicts in placing pixel advertisements on a banner while maximizing banner revenue.

The key contribution of this study is that we use a domain ontology to manage conflicts when placing advertisements in a banner. Furthermore, we incorporate this approach in three existing heuristics, which we then compare and evaluate. We can conclude that, from the currently available algorithms, the best performing are the left justified and orthogonal algorithms.

As future work, the use of pixel advertising can be made more attractive by coupling the content of the banner to the content of the Web page it is shown on. In this way, the price per pixel can be increased as the advertisements are more useful when they are custom tailored to specific target groups. Another technical approach would be the use of degrees of conflict, where products that are not direct substitutes of each other have a degree of conflict lower than one and are allowed on the same banner given that they are separated by some predefined minimum distance. This might allow for better coverage of the advertisement banner.

Acknowledgment. Damir Vandic is sponsored by the NWO Mosaic project 017.007.142: Semantic Web Enhanced Product Search (SWEPS).

References

[1] Alvarez-Valdes, R., Parreño, F., Tamarit, J.M.: A GRASP algorithm for constrained two-dimensional non-guillotine cutting problems. The Journal of the Operational Research Society 56(4), 414–425 (2005)

[2] Beasley, J.: An exact two-dimensional non-guillotine cutting tree search procedure. Operations Research 33(1), 49–64 (1985)

[3] Boskamp, V., Knoops, A., Frasincar, F., Gabor, A.: Maximizing revenue with allocation of multiple advertisements on a web banner. Computers & Operations Research 38, 1412–1424 (2011)

[4] Budak Arpinar, I., Karthikeyan Giriloganathan, B.A.M.: Ontology quality by detection of conflicts in metadata. In: 4th International Workshop on Evaluation of Ontologies for the Web (EON 2006) (2006)

[5] Google: Financial tables 2014 (2014),
     http://investor.google.com/financial/tables.html

[6] Hadjiconstantinou, E., Christofides, N.: An exact algorithm for general, orthogonal, two-dimensional knapsack problems. European Journal of Operational Research 83(1), 39–56 (1995)

[7] Hof, R.: Online ad revenues to pass print in 2012,
     http://www.forbes.com/sites/roberthof/2012/01/19/
     online-ad-revenues-to-pass-print-in-2012/

[8] Interactive Advertising Bureau: Ad unit guidelines,
     http://www.iab.net/iab_products_and_industry_services/
     1421/1443/1452

[9] Knoops, A., Boskamp, V., Wojciechowski, A., Frasincar, F.: Single pattern generating heuristics for pixel advertisements. In: Vossen, G., Long, D.D.E., Yu, J.X. (eds.) WISE 2009. LNCS, vol. 5802, pp. 415–428. Springer, Heidelberg (2009)

[10] Nah, F.F.H.: A study on tolerable waiting time: how long are web users willing to wait? Behaviour & Information Technology 23(3), 153–163 (2004)

[11] OECD: Competition: Economic issues, http://www.oecd.org/

[12] Tew, A.: Million dollar homepage,
     http://www.milliondollarhomepage.com/

[13] Wäscher, G., Haussner, H., Schumann, H.: An improved typology of cutting and packing problems. European Journal of Operational Research 183(3), 1109–1130 (2007)

[14] Wojciechowski, A.: An improved web system for pixel advertising. In: Bauknecht, K., Pröll, B., Werthner, H. (eds.) EC-Web 2006. LNCS, vol. 4082, pp. 232–241. Springer, Heidelberg (2006)

[15] Wojciechowski, A., Kapral, D.: Allocation of multiple advertisement on limited space: Heuristic approach. In: Mauthe, A., Zeadally, S., Cerqueira, E., Curado, M. (eds.) FMN 2009. LNCS, vol. 5630, pp. 230–235. Springer, Heidelberg (2009)

Exploiting Semantic Result Clustering to Support Keyword Search on Linked Data

Ananya Dass[1], Cem Aksoy[1], Aggeliki Dimitriou[2], and Dimitri Theodoratos[1]

[1] New Jersey Institute of Technology, USA
[2] National Technical University of Athens, Greece

Abstract. Keyword search is by far the most popular technique for searching linked data on the web. The simplicity of keyword search on data graphs comes with at least two drawbacks: difficulty in identifying results relevant to the user intent among an overwhelming number of candidates and performance scalability problems. In this paper, we claim that result ranking and top-k processing which adapt schema unaware IR-based techniques to loosely structured data are not sufficient to address these drawbacks and efficiently produce answers of high quality. We present an alternative solution which hierarchically clusters the results based on a semantic interpretation of the keyword instances and takes advantage of relevance feedback from the user. Our clustering hierarchy exploits graph patterns which are structured queries clustering together result graphs of the same structure and represent possible interpretations for the keyword query. We present an algorithm which computes r-radius Steiner patterns graphs using exclusively the structural summary of the data graph. The user selects relevant pattern graphs by exploring only a small portion of the hierarchy supported by a ranking of the hierarchy components.Our experimental results show the feasibility of our system by demonstrating short reach times and efficient computation of the relevant results.

1 Introduction

In recent years, the proliferation of semistructured data (e.g., XML and RDF data) has sparked a lot of interest on developing techniques for effectively and efficiently querying tree and graph data. Keyword search is, by far, the most popular technique for querying data with loose structure on the web. Its success comes from the flexibility it provides to the user to retrieve information from a data source without mastering a complex query language (e.g., XQuery, SPARQL) and without knowing the structure of the data source. The advantages of keyword search on semistructured data come with a number of disadvantages. Keyword queries are ambiguous in determining both: the user intent and the form of the results. For this reason, keyword search on tree and graph data faces three major challenges:

(a) **Determining the form of the results:** In contrast to the IR domain where the answer of a keyword query is a set of flat documents, in the domain of tree and graph databases, each result in the keyword query answer is a

B. Benatallah et al. (Eds.): WISE 2014, Part I, LNCS 8786, pp. 448–463, 2014.
© Springer International Publishing Switzerland 2014

database substructure (e.g., node, subtree, subgraph). This not only multiplies enormously the number of candidate results and, therefore, makes the evaluation more complex, but it also raises the issue of determining an appropriate form for the results. The goal is to return results (substructures) which are meaningful to the user. Different approaches on tree data define the results as LCA nodes [24,4], minimum connecting trees [12], instance trees [2], etc. In the context of graph data multiple approaches return trees [23,8,13] usually constraint by the adopted search algorithms. Indeed, traditional keyword search algorithms on graphs associate keywords only to vertices and, by construction, compute and return minimum spanning trees [3,14,11,9]. However, tree structures do not appropriately capture the semantics of queries on graph data which should naturally return graph structures. Further, in RDF data, semantics are assigned to graph elements through their association to schema elements. This information should be taken into account in the search process and integrated in the query results in order to help disambiguating the queries and their results. In this direction, some approaches attempt to exploit predicates [22,7,13].

(b) **Identifying the relevant results.** Because of the ambiguity of keyword queries there is usually an overwhelming number of results matching the query keywords (candidate results) of which only a tiny portion is relevant to the user intent. Multiple approaches assign semantics to keyword queries by exploiting structural and semantic features of the data in order to automatically filter out irrelevant results [20]. Although filtering approaches are intuitively reasonable, they are sufficiently ad-hoc and they are frequently violated in practice resulting in low precision and/or recall. A better technique followed by some other approaches ranks the candidate results in descending order of their estimated relevance [10,20]. Given that users are typically interested in a small number of query results, some of these approaches combine ranking with top-k algorithms for keyword search [23,22,8]. Keyword search over graph data returns a multitude of candidate results and of extended diversity. Therefore, current algorithms compute candidate results in an approximate way by considering only those which maintain the keyword instances in close proximity [3,14,5,11,9,18,21,15]. The ranking and top-k processing of the filtered results usually employ IR-style metrics for flat documents (e.g., tf*idf or PageRank) [10,23,22,7,20] adapted to the structural characteristics of the data. However, the occurrence statistics alone cannot capture effectively the diversity of the results represented in a large graph dataset neither identify the intent of the user. As a consequence the produced rankings are, in general, of low quality.

(c) **Coping with the performance scalability issue.** As mentioned above, the number of candidate results can be very large. Computing all the results of a certain form is intractable. For instance, the problem of finding the Steiner trees for a set of keywords in a data graph is NP-complete [15]. The current algorithms which compute all the results of a certain form restricted so that their size is below a certain threshold are still of high complexity. Consequently, these algorithms do not scale satisfactorily when the size of the data graph and the number of query keywords increases.

Our Approach. In this paper, we present a novel approach for keyword search on RDF graph data. Our approach utilizes a semantic two-level hierarchical clustering of the keyword query results. The first—fine-grained—level of clustering, partitions the results based on pattern graphs. These pattern graphs, when used as queries on the RDF graph, compute the results of the corresponding cluster. The second—coarser—level of clustering, partitions the patterns (and their results) based on the different types of construct (e.g., class, property, value) each query keyword matches. However, our approach does not exhaustively generate and cluster all the results and the hierarchy components. Instead, it benefits from relevance feedback at different levels of granularity to identify the pattern graphs which are relevant to the user intent. The hierarchy components are presented to the user ranked to facilitate their selection. The candidate pattern graphs are generated using the structural summary of the data graph (a concept analogous to that of 1-index in tree databases [16]) without accessing the data graph. Only the selected pattern graphs are evaluated on the data returning all and only the relevant results. This way, our system addresses efficiently the problems of relevant result identification and performance scalability.

Contribution. The main contributions of the paper are the following:
- We define query results as meaningful subgraphs of the data graph that appropriately connect together keyword *matching constructs* (elementary subgraphs representing semantic interpretations of the keyword instances). This addresses the problem (a) of determining the form of query result (Section 2.)
- We use the concept of patterns graph of a keyword query to cluster together result graphs that share the same structure. The pattern graphs involve all the matching constructs of the query keywords on the structural summary and express the possible interpretations of the keyword query on the graph data (Section 3.1).
- We design an algorithm which takes as input the matching constructs of the query keywords on the structural summary and computes patterns forming r-radius Steiner graphs that involve these matching constructs. Our algorithm computes the patterns on the structural summary without accessing the data (Section 3.2).
- We design a clustering hierarchy for the results which is based on pattern graphs and keyword matching constructs. Our clustering hierarchy allows the user to disambiguate the query and compute the relevant results while examining only a small portion of the hierarchy components. To shorten the user interaction we devise ranking techniques for the hierarchy components that take into account structural and semantic information and occurrence frequency statistics (Section 4).
- We implemented and experimentally evaluated our approach. Our results on measuring reach time show that the user can find the relevant patterns in short time supported effectively by our ranking of the hierarchy components, and that the system efficiently computes the required hierarchy components on the structural summary and evaluates the relevant pattern graph on the data (Section 5).

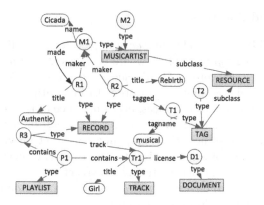

Fig. 1. An RDF graph

2   Data Model and Query Language Semantics

2.1   Data Model

Resource Description Framework (RDF) provides a framework for representing information about web resources in a graph form. The RDF vocabulary includes elements, that can be broadly classified into Classes, Properties, Entities and Relationships. All the elements are resources. RDF has a special class, called *Resource* class and all the resources that are defined in an RDF graph belong to the *Resource* class. Our data model is an RDF graph defined as follows:

Definition 1 (RDF Graph). An *RDF graph* is a quadruple $G = (V, E, L, l)$ where:

V is a finite set of vertices, which is the union of three disjoint sets: V_E (representing entities), V_C (representing classes) and V_V (representing values).

E is a finite set of directed edges, which is the union of four disjoint sets: E_R (inter-entity edges called *relationship* edges which represent entity relationships), E_P (entity to value edges called *property* edges which represent property assignments), E_T (entity to class edges called *type* edges which represent entity to class membership) and E_S (class to class edges called *subclass* edges which represent class-subclass relationship).

L is a finite set of labels that includes the labels "type", "subclass" and "resource".

l is a function from $V_C \cup V_V \cup E_R \cup E_P$ to L. That is, l assigns labels to class and values vertices and to relationship and property edges.

Every entity belongs to a class. Figure 1 shows an example RDF graph (a subgraph of the Jamendo Dataset[1]).

---

[1] http://dbtune.org/jamendo/

2.2   Queries and Answers

A *query* is a set of keywords. The *answer* of a query Q on an RDF graph G is a set of subgraphs (*result graphs*) of G, where each result graph involves at least one instance of every keyword in Q. A *keyword instance* of a keyword k in Q is a vertex or edge label containing k. In order to facilitate the interpretation of the semantics of the keyword instances, every instance of keyword in a query is matched against a small subgraph (*matching construct*) of the graph G which involves this keyword instance. Each matching construct provides a deeper insight about the context of a keyword instance in terms of classes, entities and relationship edges. We link one matching construct for every keyword in the query Q through edges (*inter-construct connections*) and common vertices into a connected component to form a *result graph*.

Definition 2 (Matching Construct). Given a keyword k of a query and an RDF graph G, for every instance of k in G, we define a *matching construct* as a small subgraph of G. If the instance i of k in G is:

- the label of a class vertex $v_c \in V_C$, the matching construct of i is the vertex v_c (*class matching construct*).
- the label of a value vertex $v_v \in V_V$, the matching construct of i comprises the value vertex v_v, the corresponding entity vertex, and its class vertices along with the property and type edges between them (*value matching construct*).
- the label of relationship edge $e_r \in E_R$, the matching construct of i comprises the relationship edge e_r, its entity vertices and their class vertices along with the type edges between them (*relationship matching construct*).
- the label of property edge $e_p \in E_P$, the matching construct of i comprises the property edge e_p, its value and the entity vertices, and the class vertices of the entity vertex along with the type edges between them (*property matching construct*).

Figures 2(a), (b), (c) and (d) show a class, value, relationship and property matching construct, respectively, for different keyword instances in the RDF graph of Figure 1. Underlined labels in a matching construct denote the keyword instances on which the matching construct is defined (called *active keyword instances* of the matching construct).

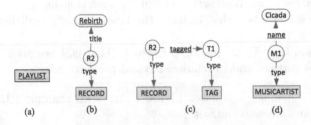

Fig. 2. Matching Constructs for (a) class, (b) value, (c) relationship and (d) property

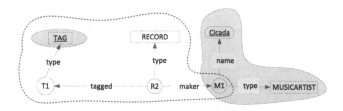

Fig. 3. Inter-construct connection

Definition 3 (Query Signature). Given a query Q and an RDF graph G, a *signature* of Q is a function from the keywords of Q that matches every keyword k to a matching construct of k in G.

Figures 2(a), (b), (c) and (d) show a query signature for the query {*Playlist, Rebirth, tagged, name*}. Note that a signature of a query Q can have less matching constructs than the keywords in Q, since one matching construct can have more than one active keyword instance.

Definition 4 (Inter-construct Connection). Given a query signature S, an *inter-construct connection* between two distinct matching constructs C_1 and C_2 in S is a simple path augmented with the class vertices of the intermediate entity vertices in the path (if not already in the path) such that: (a) one of the terminal vertices in the path belongs to C_1 and the other belongs to C_2, and (b) no vertex in the connection except the terminal vertices belong to a construct in S.

Figure 3 shows an inter-construct connection between the matching constructs for keywords *Tag* and *Cicada* in the RDF graph of Figure 1. The matching constructs are shaded and the inter-construct connection is circumscribed with a dotted line.

In order to define result graphs, we need the concept of acyclic subgraph with respect to a query signature. Let G_s be a subgraph of the RDF graph that comprises all the constructs in the signature of a query. We construct an undirected graph G_c as follows: there is exactly one vertex in G_C for every matching construct and for every vertex not in a matching construct in G_s. Further: (a) if v_1 and v_2 are non-construct vertices in G_c, there is an edge between v_1 and v_2 in G_c iff there is an edge between the corresponding vertices in G_s, (b) If v_1 is a construct vertex and v_2 is a non-construct vertex in G_c, there is an edge between v_1 and v_2 in G_c iff there is an edge between a vertex of the construct corresponding to v_1 and the vertex corresponding to v_2 in G_s, and (c) if v_1 and v_2 are two construct vertices, there is an edge between them in G_c iff there exists in G_s an edge between a vertex of the construct corresponding to v_1 and a vertex of the construct corresponding to v_2 and that edge does not occur in any one of the constructs. Graph G_s is said to be *connection acyclic* if there is no cycle in G_c.

Consider the query $Q = \{$ *Cicada, musical, Playlist* $\}$ on the RDF graph G of Figure 1. Figure 4 shows two subgraphs of G which comprise a signature of

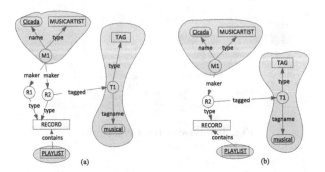

Fig. 4. (a) Invalid result graph (b) Valid result graphs

Q on G. The active keyword instances are underlined and the corresponding matching constructs are shaded. One can see that the subgraph in Figure 4(a) is connection cyclic while the other subgraph Figure 4(b) is connection acyclic.

Definition 5 (Result Graph). Given a signature S for a query Q over an RDF graph G, a *result graph* of Q for S is a connected connection acyclic subgraph G_R of G which contains only the matching constructs in S and possibly inter-construct connections between them.

Therefore, a result graph of a query contains all the matching constructs of a signature of the query and guarantees that they are linked with inter-construct connections into a connected whole. Note that a result graph might not contain any inter-construct connection (this can happen if every matching construct in the query signature overlaps with some other matching construct). However, if inter-construct connections are used within the result graph, no redundant (cycle creating) inter-construct connections are introduced.

We now define the *answer* of a query Q on an RDF graph G as the set of result graphs of Q on G.

3   Computing Pattern Graphs on the Structural Summary

We formally introduce in this section the structural summary of a data graph and query pattern graphs. Then, we present an algorithm for computing pattern graphs on a structural summary.

3.1   The Structural Summary and Pattern Graphs

In order to construct pattern graphs we use the structural summary of the RDF graph. Intuitively, the structural summary of an RDF graph G is a special type of graph which summarizes the data graph showing vertices and edges corresponding to the class vertices and property, relationship and subclass edges in G.

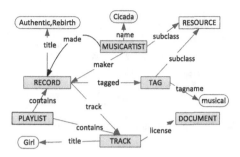

Fig. 5. Structural Summary

Definition 6 (Structural Summary). The *structural summary* of an RDF graph $G(V, E, L, l)$ is a vertex and edge labeled graph $G'(V', E', L', l')$ where

- $V' = V'_C \cup V'_v$ where:
  (a) V'_C is a set of class vertices which has a one to one mapping f onto V_C,
  (b) V'_v is a set of value vertices which contains a vertex for every distinct pair (c, l_p) such that there exists an entity of class c in the RDF graph G having a property labeled by l_p.
  A class vertex c in V_C is labeled by the label of the corresponding class vertex $f(c)$ in G.
- $E' = E'_p \cup E'_r \cup E'_s$ where:
  (a) E'_p is a set of edges from vertices in V'_c to vertices in V'_v such that there is an edge $(c, v) \in E'_p$ labeled by l_p iff there is an entity of class $f(c)$ in G which has a property edge labeled by l_p,
  (b) E'_r is a set of edges from a vertex in V'_C to another vertex in V'_C such that there is an edge $(c_1, c_2) \in E'_r$ labeled by l_r iff there is an edge from an entity of $f(c_1)$ to an entity of $f(c_2)$ in G labeled by l_r.
  (c) E'_s is a set of edges from a vertex in V'_C to another vertex in V'_C such that there is an edge $(c_1, c_2) \in E'_s$ labeled by *subclass* iff there is an edge from $f(c_1)$ to $f(c_2)$ in G labeled by *subclass*.
- L' is the set of labels of vertices in V' and edges in E'.
- l' is a function assigning labels to vertices and edges in G' as already described above.

Figure 5 shows the structural summary for the RDF graph G of Figure 1. Similarly to matching constructs on the data graph, we define matching constructs on the structural summary and we refer to them as MC. The pattern graphs comprise MCs for every keyword in the query and the connections between them.

The next objective is to compute subgraphs of the structural summary, strictly consisting of one matching construct for every keyword in the query Q and the connections between them without forming any cycle. These subgraphs are called *result pattern graphs*.

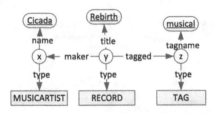

Fig. 6. Query Pattern Graph

Definition 7 (Pattern Graph). A *(result) pattern graph* for a keyword query Q is a graph similar to a result graph for Q, and with the same keyword instances, but with the following two exceptions

(a) the labels of the entity vertices in the result graph, if any, are replaced by distinct variables in the pattern graph. These variables are called *entity variables* and they range over entity labels.

(b) The labels of the value vertices are replaced by distinct variables whenever these labels are not the keyword instances in the result graph. These variables are called *value variables* and they range over value labels in the RDF graph.

Figure 6 shows an example of a pattern graph, for the keyword query $Q = \{Cicada,\ Rebirth,\ musical\}$. X, Y and Z are entity variables.

3.2   An Algorithm for Computing Query Pattern Graphs

Our algorithm computes r-radius Steiner pattern graphs. The r-radius Steiner graph computation is inspired by the algorithm of [18]. However, unlike that algorithm, our algorithm allows the keywords to match the edges labels of the graph.

Our algorithm is shown in algorithm 1. It takes as input the structural summary G' of a data graph and a signature S of a query Q on G' (the set of MCs for the keywords of Q on G'). It produces as output the r-radius Steiner query pattern graphs on G' that contain S whose radius r is minimum.

Every MC in S is associated with one or two class vertices. Set $\mathcal{K}$ is initialized with all the class vertices in S. The adjacency matrix A is a square matrix of order $n+1$, where n is the number of class vertices in G'. The first row and column of A record the class identifiers. $\mathcal{A}(i, j)$ holds information about the relationship edges connecting the classes $\mathcal{A}(i, 0)$ and $\mathcal{A}(0, j)$. Initially the algorithm tries to find Steiner pattern graphs with radius $r = 1$. If no graph is found, r is gradually increased. Sets $\mathcal{P}$ and $\mathcal{C}$ represent the data structures recording the pattern graphs and their corresponding connecting vertices, respectively. Procedure *GeneratePattern* generates the pattern graphs using the information recorded about the relationship edges between the connecting vertex and the class vertices in $\mathcal{K}$ and the MCs in S.

Algorithm 1. Query Pattern Graphs Computation

Input: S: signature, G': structural summary.
Output: $\mathcal{P}$: set of pattern graphs.
1: $\mathcal{K} \leftarrow$ extract distinct class vertices from S;
2: $\mathcal{A} \leftarrow$ adjacency matrix encoding class vertices V'_C and rel. edges E'_R in G';
3: $r \leftarrow 1$;   ▷ radius of the Steiner graph
4: $\mathcal{P} \leftarrow \emptyset$;
5: $\mathcal{C} \leftarrow \emptyset$;   ▷ set of connecting vertices
6: **while** $\mathcal{C} = \emptyset$ **do**
7:     **for all** rows $i \in \mathcal{A}$ **do**
8:         $conn \leftarrow \mathcal{A}(i,0)$;   ▷ adding class id to set conn
9:         **for all** cols $j \in \mathcal{A}$ **do**
10:             **if** $\mathcal{A}(i,j) \neq \emptyset$ **then**
11:                 $conn = conn \cup (\mathcal{A}(0,j))$;   ▷ adding class id to set conn
12:         **if** $\mathcal{K} \subseteq conn$ **then**
13:             $\mathcal{C} \leftarrow \mathcal{A}(i,0)$;   ▷ a connecting node is found
14:             $P \leftarrow$ GeneratePattern$(G', r, \mathcal{S}, \mathcal{A}(i,0))$;
15:             **if** $P \notin \mathcal{P}$ **then**
16:                 $\mathcal{P} \leftarrow \mathcal{P} \cup P$;
17:         $conn \leftarrow \emptyset$;
18:     **if** $\mathcal{C} = \emptyset$ **then**
19:         r=r+1;
20:         $\mathcal{A}' \leftarrow$ modify adjacency matrix $\mathcal{A}$ to represent how the class vertices are connected to each other at a distance less than or equal to r;
21:         $\mathcal{A} \leftarrow \mathcal{A}'$

4   Semantic Hierarchical Clustering and Ranking

We now describe our result clustering hierarchy, how the user navigates through the hierarchy and graph ranking.

Semantic Hierarchical Clustering. The hierarchy has two levels on top of the result graph layer. The pattern graphs of a query Q on an RDF graph G define a partition of the result graphs of Q on G. The pattern graphs constitute the *first level* of the clustering hierarchy. Multiple pattern graphs can share the same signature. The signatures determine a partition of the pattern graphs of Q on G. They, in turn, define a partition of the results which is coarser than that of the pattern graphs. The signatures constitute the *second (top) level* of the hierarchy.

Hierarchy Navigation. In order to navigate through the hierarchy after issuing a query the user starts from the top level. The top level may have numerous signatures. However, the user does not have to examine all the signatures. Instead, she is presented with the MC list for one of the query keywords. We describe below how this list of MCs is ranked. As mentioned earlier, the MCs of a keyword provide all the possible interpretations for this keyword in the data. The user selects the MC that she considers relevant to her intent. Subsequently, she is presented with the MC lists of the other query keywords, though some of the MC lists can be skipped. This can happen if the user selects an MC which involves more than one keyword instances that she wants to see combined together in one MC. Once MCs for all keywords have been selected, that is, a query signature has been determined, the system presents a ranked list of all the pattern graphs that comply with the signature. The user chooses the pattern graph of her preference which is evaluated by the system. The result graphs are returned to the user.

Ranking. The MCs for a keyword are ranked in an MC list based on the following rules: (a) MCs that involve more than one active keyword instances are ranked first in order of the number of active keyword instances they contain, (b) class MCs, relationship MCs and property MCs are ranked next in that order, (c) value MCs follow next and are ranked in descending order of the frequency of their value. The *value frequency* f_m^v of a value MC m with property p, class c and value (keyword) v is the number $n_{p,c}^v$ of occurrences of the value v in matching constructs involving p and c in the data graph divided by the number $n_{p,c}$ of occurrences of property matching constructs in the data graph involving p and c. That is,

$$f_m^v = n_{p,c}^v / n_{p,c}$$

This ranking of the MCs favors MCs with multiple keyword instances based on the assumption that keywords that occur in close proximity are more relevant to the user's intent. Further, it favors MCs whose active keyword matches a schema element (class, relationship or property), favoring most class MCs which have unique occurrences in the data graph. Finally, value MCs are ranked at the end since they are more specific. The value frequency of a value MC reflects the popularity of this MC in the data. Therefore, value MCs with high value frequency are ranked higher than value MCs for the same value with low value frequency.

The pattern graphs the system ranks share the same signature. Thus, they are r-radius graphs with the same r. In almost all the cases they have the same number of edges and they differ only in the relationship edges which are not part of any MC. For this reason, the pattern graphs are ranked in descending order of their connecting edge frequency defined next. Given a pattern P, its *connecting edge frequency*, $f_c(P)$, is the sum of the number n_e occurrences in the data graph of the relationship edges e in P that do not occur in an MC in P divided by the total number $|E_R|$ of relationship edges in the data graph. That is, if E_c is the set of these relationship edges in P,

$$f_c(P) = \sum_{e \in E_c} n_e / |E_R|$$

In order to rank MCs and pattern graphs our system needs statistics about value MCs and their property edges and about connecting relationship edges in pattern graphs. This information is precomputed and stored with the structural summary when this one is constructed. Therefore, no access to the data graph is needed.

Table 1. Queries used in the experiments and their statistics

Query	keywords	#I	#MC	#S
1	Track Obsession Divergence format title mp32	699	29	2,646
2	biography guitarist track lemonade	633	18	216
3	Knees Cicada recorded_as	59	7	9
4	sweet recorded_as Signal onTimeLine 104734	177	17	48
5	Track Nuts chillout ACExpress	618	21	252
6	Mako La deux date love time	2846	36	24,300
7	Fantasie recorded_as factor published_as format date title	145	25	588
8	Fantasie recorded_as Performance Paure	68	8	8
9	Briareus Vampires Infirmary Cool	154	18	288
10	Fantasie text Paure Document	144	13	42

5 Experimental Evaluation

We implemented our approach and run experiments to evaluate our system. The goal of our experiment is to assess: (a) the effectiveness of our clustering approach, (b) the efficiency of our techniques in providing suggestions to the user and in obtaining results from the selected pattern graphs in real time. It is not meaningful to run experiments to measure precision and recall since our approach exploits relevance feedback and returns all and only the results which are relevant to the user intent (perfect precision and recall).

Dataset and Queries. We use Jamendo, a large repository of Creative Commons licensed music. Jamendo is a dataset of 1.1M triples and of 85MB size containing information about musicians, music tracks, records, licenses of the tracks, music categories, track lyrics and many other details related to them. Its structural summary was extracted and stored in a relational database. Experiments are conducted on a standalone machine with an Intel i5-3210M @ 2.5GHz processors and 8GB memory.

Users provided different queries on the Jamendo dataset and navigated through our hierarchical clustering system to select a relevant MC (for every keyword) and a relevant pattern graph (when more than one were proposed by the system for the selected MCs). We report on 10 of them. The queries cover a broad range of cases. They involve from 3 to 7 keywords. Table 1 shows the keyword queries and statistics about them. For every query it shows the total number of keyword instances in the data graph (#I), the number of MCs (for all the keywords) in the structural summary (#MC), and the total number of signatures (#S). As we can see, a query can have many pattern graphs (their number is greater than or equal to that of the signatures). However, thanks to our hierarchical clustering approach, the user has to examine only one or at most two of them. Further, exploiting our ranking of the MCs, the user has to examine only a fraction of the MCs for every query.

Effectiveness of Hierarchical Clustering. In order to measure the retrieval effectiveness of our hierarchical clustering, we adapted the *reach time* metric used in [17,19]. The reach time sets forth to quantify the time spent by a user to locate the relevant results. In our case, the relevant results are represented by the relevant pattern graph. The relevant results in terms of subgraphs of the data graph can then be retrieved by evaluating the pattern graph against the database. For simplicity, we assume that the user always selects one relevant pattern graph. We employ different versions of reach time in two different settings: when the components (MCs and pattern graphs) are ranked, and when they are not. rt_{avg} and rt_{max} apply to the case the components are not ranked. rt_{avg} (resp. rt_{max}) denotes the average (resp. maximum) number of components the user needs to examine in order to retrieve the relevant pattern. rt_{rank} denotes the number of components the user examines in order to retrieve the relevant pattern when the components are ranked. For instance, if a query has k keywords and the user needs to examine m_i of the ranked MCs for keyword i and p of the ranked pattern graphs for the selected MCs,

$$rt_{rank} = \sum_{i=1}^{k} m_i + p$$

Figure 7 shows the reach times for the queries of Table 1.

As one can see, the user has to examine on the average a small number of components even for queries with many keywords. Further, when the components are ranked, rt_{rank} is smaller than rt_{avg} in most cases and never comparable to rt_{max}. This demonstrates the feasibility of our hierarchical clustering system and the quality of the component ranking process.

Efficiency of the System. In order to asses the efficiency of out system, we measured the time needed to compute the ranked list of MCs for the query keywords and the time needed to evaluate the selected pattern graphs.

Figure 8(a) shows the total time (*totalMC*) needed to compute and rank the MCs for the keyword queries. It also shows the shortest(*minMC*) and the longest (*maxMC*) time needed to compute the rank list of MCs of a keyword in a given

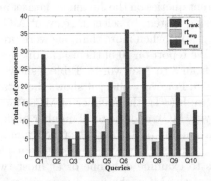

Fig. 7. Reach times of the two clustering approaches

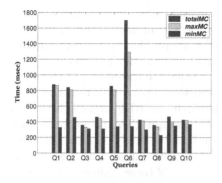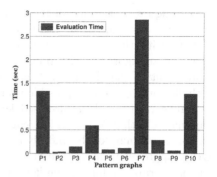

Fig. 8. (a) Time to compute the MCs for the query keywords (b) Evaluation time for the selected pattern graphs

query. The list of MCs for the first keyword in the query is presented to the user for selection as soon as it is computed. The plot displays interactive time for the all the queries which do not delay the selection process. The time needed to compute the pattern graphs on the structural summary after the MCs for the query keywords are selected by the user are insignificant and are not displayed here. This is expected since the pattern graphs are computed using exclusively the structural summary whose size is very small compared to the size of the data.

Figure 8(b) displays the time needed to evaluate the selected pattern graphs on the data graph. This diagram again shows interactive times even though it is a prototype system and no optimizations have been applied.

6 Related Work

Several algorithms explore data graphs to support keyword search. In [3] a backward search algorithm BANK is presented for finding Steiner trees. The Steiner tree problem is NP complete. Hence different techniques are used to work around NP completeness. In [5], a dynamic programming approach applicable to only few keywords and having an exponential time complexity is employed. In [9] a polynomial delay algorithm is introduced. The algorithm in [14] produced trees rooted at distinct vertices which was supplemented by BLINK [11] with an efficient indexing structure. Tree-based methods produce succinct answers but answers from graph-based methods are more informative. Recent graph based approaches like [18] compute all possible r-radius Steiner graphs and index them. This method is prone to produce redundant results since it is possible that a highly ranked r-radius Steiner graph is included in another Steiner graph having a larger radius. [21] finds multi-centered subgraphs called communities containing all the keywords, such that there exists at least one path of distance less than or equal to R_{max} between every keyword instance and a center vertex. Later in

[15], r-cliques containing all the keywords are found such that the distance between any two pair of keyword matching vertices is no more than r. Finding
r-clique with the minimum weight is an NP-hard problem, hence the authors
provided an algorithm with polynomial delay to find the top-k r-cliques where r
is an input to the algorithm. Predicting an optimal r for producing r-cliques is
a challenge because it is possible that there exists no clique with that r or less.

Unlike traditional graph-based keyword search, on RDF graphs, keywords
can match both a vertex and an edge label of the graph [22,6]. Some approaches
exploit a summary of the underlying RDF graph to find query patterns [23,22,8].
Other approaches directly search for connections between keyword matches on
the data graph [6] for a given keyword query. None of these approaches exploits
hierarchical clustering and relevance feedback. User interaction for predicate
selection is explored in [13] but without leveraging pattern graphs and structural
summaries. Hierarchical clustering mechanisms and user interaction at multiple
levels are discussed in [19,1]. However, these studies focus on tree-structured
XML data and do not consider graphs.

7   Conclusion

We have presented a novel approach to address the problems related to keyword
search on large RDF data. Our approach hierarchically clusters the result graphs
and leverages relevance feedback from the user. In order to form the clustering
hierarchy, we use matching constructs and pattern graphs which are subgraphs
representing semantic interpretations for keywords and queries, respectively. We
presented an algorithm to efficiently compute r-radius Steiner pattern graphs. All
hierarchy components are computed efficiently on the structural summary without accessing the (much larger) data graph. We presented a technique that ranks
the hierarchy components based on their structural and semantic features and occurrence frequencies. Our approach allows the user to explore only a tiny portion
of the clustering hierarchy in selecting the relevant graph patterns supported by
the component ranking. The experimental evaluation of our approach shows its
feasibility by demonstrating short reach times to the relevant graph patterns and
efficient computation of the relevant result graphs on the data graph.

As a future work we are planning to study relaxation techniques for the selected relevant pattern graphs in order to enable the extraction of loosely connected result graphs which are still of interest to the user.

References

1. Aksoy, C., Dass, A., Theodoratos, D., Wu, X.: Clustering query results to support
   keyword search on tree data. In: Li, F., Li, G., Hwang, S.-w., Yao, B., Zhang, Z.
   (eds.) WAIM 2014. LNCS, vol. 8485, pp. 213–224. Springer, Heidelberg (2014)
2. Aksoy, C., Dimitriou, A., Theodoratos, D., Wu, X.: xReason: A semantic approach
   that reasons with patterns to answer XML keyword queries. In: Meng, W., Feng,
   L., Bressan, S., Winiwarter, W., Song, W. (eds.) DASFAA 2013, Part I. LNCS,
   vol. 7825, pp. 299–314. Springer, Heidelberg (2013)

3. Bhalotia, G., Hulgeri, A., Nakhe, C., Chakrabarti, S., Sudarshan, S.: Keyword searching and browsing in databases using BANKS. In: ICDE, pp. 431–440 (2002)
4. Dimitriou, A., Theodoratos, D.: Efficient keyword search on large tree structured datasets. In: KEYS, pp. 63–74 (2012)
5. Ding, B., Yu, J.X., Wang, S., Qin, L., Zhang, X., Lin, X.: Finding top-k min-cost connected trees in databases. In: ICDE, pp. 836–845 (2007)
6. Elbassuoni, S., Blanco, R.: Keyword search over RDF graphs. In: CIKM, pp. 237–242 (2011)
7. Elbassuoni, S., Ramanath, M., Schenkel, R., Weikum, G.: Searching RDF graphs with sparql and keywords. IEEE Data Eng. Bull., 16–24 (2010)
8. Fu, H., Gao, S., Anyanwu, K.: Disambiguating keyword queries on RDF databases using "Deep" segmentation. In: ICSC, pp. 236–243 (2010)
9. Golenberg, K., Kimelfeld, B., Sagiv, Y.: Keyword proximity search in complex data graphs. In: SIGMOD, pp. 927–940 (2008)
10. Guo, L., Shao, F., Botev, C., Shanmugasundaram, J.: Xrank: ranked keyword search over XML documents. In: SIGMOD, pp. 16–27 (2003)
11. He, H., Wang, H., Yang, J., Yu, P.S.: Blinks: ranked keyword searches on graphs. In: SIGMOD, pp. 305–316 (2007)
12. Hristidis, V., Koudas, N., Papakonstantinou, Y., Srivastava, D.: Keyword proximity search in XML trees. IEEE Trans. Knowl. Data Eng., 525–539 (2006)
13. Jiang, M., Chen, Y., Chen, J., Du, X.: Interactive predicate suggestion for keyword search on RDF graphs. In: Tang, J., King, I., Chen, L., Wang, J. (eds.) ADMA 2011, Part II. LNCS, vol. 7121, pp. 96–109. Springer, Heidelberg (2011)
14. Kacholia, V., Pandit, S., Chakrabarti, S., Sudarshan, S., Desai, R., Karambelkar, H.: Bidirectional expansion for keyword search on graph databases. In: VLDB, pp. 505–516 (2005)
15. Kargar, M., An, A.: Keyword search in graphs: Finding r-cliques. In: VLDB, pp. 681–692 (2011)
16. Kaushik, R., Bohannon, P., Naughton, J.F., Korth, H.F.: Covering indexes for branching path queries. In: SIGMOD Conference, pp. 133–144 (2002)
17. Kummamuru, K., Lotlikar, R., Roy, S., Singal, K., Krishnapuram, R.: A hierarchical monothetic document clustering algorithm for summarization and browsing search results. In: WWW, pp. 658–665 (2004)
18. Li, G., Ooi, B.C., Feng, J., Wang, J., Zhou, L.: Ease: an effective 3-in-1 keyword search method for unstructured, semi-structured and structured data. In: SIGMOD, pp. 903–914 (2008)
19. Liu, X., Wan, C., Chen, L.: Returning clustered results for keyword search on XML documents. IEEE Trans. Knowl. Data Eng., 1811–1825 (2011)
20. Liu, Z., Chen, Y.: Processing keyword search on XML: a survey. World Wide Web, 671–707 (2011)
21. Qin, L., Yu, J.X., Chang, L., Tao, Y.: Querying communities in relational databases. In: ICDE, pp. 724–735 (2009)
22. Tran, T., Wang, H., Rudolph, S., Cimiano, P.: Top-k exploration of query candidates for efficient keyword search on graph-shaped (RDF) data. In: ICDE, pp. 405–416 (2009)
23. Wang, H., Zhang, K., Liu, Q., Tran, T., Yu, Y.: Q2semantic: A lightweight keyword interface to semantic search. In: Bechhofer, S., Hauswirth, M., Hoffmann, J., Koubarakis, M. (eds.) ESWC 2008. LNCS, vol. 5021, pp. 584–598. Springer, Heidelberg (2008)
24. Xu, Y., Papakonstantinou, Y.: Efficient LCA based keyword search in XML data. In: EDBT, pp. 535–546 (2008)

Discovering Semantic Mobility Pattern
from Check-in Data

Ji Yuan, Xudong Liu, Richong Zhang, Hailong Sun,
Xiaohui Guo, and Yanghao Wang

School of Computer Science and Engineering, Beihang University,
Beijing, 100191 China
{yuanji,liuxd,zhangrc,sunhl,guoxh,wangyh}@act.buaa.edu.cn

Abstract. The wealth of check-in data offers new opportunities for better understanding user movement patterns. Existing studies have been focusing on mining explicit frequent sequential patterns. However, the sparseness of check-in data makes it difficult that all explicit patterns be precisely discovered. In addition, due to the weakness in expressing semantic knowledge of explicit patterns, the need for discovering semantic pattern rises. In this paper, we propose the Topical User Transition Model (TUTM) to discover the semantic mobility patterns by analyzing topical transitions. Via this model, we discover semantic transition properties and predict the user movement preferences. Furthermore, Expectation-Maximization (EM) algorithm incorporating with Forward-Backward algorithm is provided for estimating the model parameters. To demonstrate the performance of TUTM model, experimental studies are carried out and the results show that our model can not only reasonably explain user mobility patterns, but also effectively improve the prediction accuracy in comparison with traditional approaches.

Keywords: Mobility pattern, Spatial semantics, Check-in data.

1 Introduction

The location-based data is growing exponentially with the pervasive usage of GPS-enabled mobile devices and mobile applications. These location data contains a large quantity of mobility patterns, which represent the common regulations that people follow when they move among locations in daily life. In practice, user mobility patterns can be explained through many forms, such as destination preference [1], frequent transitions [2], periodic regulations [3], etc. These existing studies mainly focus on interpreting the mobility patterns as explicit frequent sequential movements, such as frequent visiting regulations [4] and trajectory patterns [5]. However, the explicit patterns does not carry any semantic descriptions. In addition, the sparseness, incompleteness and randomness of the user-generated data makes it impossible that all transitional patterns be precisely characterized by statistically analyzing the explicit data. Therefore, the requirement for discovering semantic mobility pattern naturally rises. As a

B. Benatallah et al. (Eds.): WISE 2014, Part I, LNCS 8786, pp. 464–479, 2014.
© Springer International Publishing Switzerland 2014

simple example, the semantic pattern of *Office - Food* can be abstracted from the explicit sequential movement *"Office Building A - Restaurant B"*. Obviously, this semantic pattern generalizes the human explicit sequential movements and overcomes the data uncertainties.

To model the semantic patterns, researches propose probabilistic approaches [6] [7] by exploiting latent factors for discovering the user hidden mobility preferences. Yet, the transition patterns between locations are not captured. There are some researches that combine the mobility transition and user preferences. For instance, Kustrama [1] *et al.* combine topic model and Markov model for recommending possible destinations. Still, the semantic transitional pattern is not modeled by this hybrid approach, which simply considers the explicit transition as a calculation factor.

In order to discover the semantic mobility pattern from location data, we assume that there are latent topics corresponding to each location, which represents the implicit semantic attributes of location itself and when people transiting from one place to another, semantic transitions between latent topics are generated accordingly. Our objective is to find these common semantic transitions. As the previous example "many people would go to restaurants after work" indicates, such phenomena represents a coarse-gained transition among the latent semantic topics in adjacent locations. These transitions reflect the latent spatial topical semantic pattern.

Based on the previous analysis, in this study we propose the Topical User Transition Model (TUTM): a probabilistic model which leverages the topical transitions among gathered location sequences to model the user mobility patterns in daily life. Specifically, we synthesize the strong dependencies between the consecutive locations and exploit the transition between locations in building this model. Since transitions between locations obviously reveal knowledge about users' latent mobility regulation, the transitional relations are established between consecutive latent topics rather than explicit locations. Generally, evolving from LDA, our model takes into consideration the dependencies between topics, which will be specified in Section 3.1. Parameters of this model are estimated through the widely-used Expectation-Maximization algorithm.

Experimental studies are carried out on a real-world dataset shared by Cheng *et al.* [8] and the results confirm the effectiveness of our model on inferring the semantic mobility patterns by analyzing the topical transitions of the model. Moreover, the applicability of our model on predicting the user destinations is also demonstrated through the comparison with existing methods.

The rest of the paper is organized as follows. We indicate the existing studies related to the problems of mobility patterns modeling firstly in Section 2. Through augmenting the first-order Markov property to the locations' topic assignments, we formulate the topical location transition problem as a probabilistic generative process in section 3. The EM-styled inference method and the next-place predictive probability follows up. Section 4 contains the experiments carried out for verifying model performance. Finally, we conclude our contribution and the future work.

2   Related Work

Mining the knowledge from the location based services and social networks is an appealing direction in the research domain. Cheng *et al.* [8] and Daniel *et al.* [4] provide some large scale statistical quantitative analysis of human mobility pattern in check-in data, considering the temporal-spatial influence and even economic factors. Besides, the work of GPS trajectory pattern mining which is considered by Zheng *et al* [9], develop a framework to mine interesting locations and travel sequences from GPS trajectories. Travel patterns are discovered by Zheng *et al.* [2] based on geo-tagged photo sharing community. They leverage the geo-tags and textual features annotated on the photos to detect the travel patterns at local city level using a markov chain model. Similarly, the frequent trip patterns are also mined in [10]. In the work by Bayir *et al* [11], the spatio-temporal mobility patterns and profiles of mobile phone users are captured using cell phone log data. In the urban computing, the taxicab traces are also utilized to inferring mobility patterns in [12]. Obviously, a variety of location datasets are analyzed in those researches. But they pay less attention on the semantic aspect of the location transition and deterministically mine the frequent patterns, which usually suffer from highly-cost computation and performance degradation due to the sparse and large scale data.

Other than frequent pattern mining, the semantics knowledge mining of human mobility close to our target are also concerned by academic community. In [13], the social-spatial properties in social network are discussed. They study the connections between spatial properties and users' relations. Eagle *et al.* [14] infer the friendships and the spatio-temporal patterns in their physical proximity and calling patterns based on the observational device mobility data, and compare the results with the self-reported data. Xiao *et al.* [15] estimate the similarity between users according to their GPS trajectories. They map every location in the trajectory to a certain category, and measure the similarity between trajectories. Similar research has been conducted on Facebook data in [16] to analyze the correlations between the distance and social relations, and further utilize it to predict users' current location. These works are closely relevant to our work of analyzing the mobility semantics. However, they mostly focus on discovering the impact of the location based services on the social relations, rather than the human transition behaviors semantics itself.

Actually, the transition pattern mining and semantics analysis serve the purpose of facilitating the diversified realistic applications, such as, convenient routes recommendation, next-place prediction, and city governance. In [17], frequent routes are generated for travel recommendation based on trajectory patterns discovering. Monreale *et al.* [5] presents a model called *WhereNext*, which aims at predicting the next location also by trajectory pattern mining. These approaches based on frequent patterns have limitations on discovering semantic transition patterns for modeling mobility. As for semantics based methods, the prediction algorithms considering several semantic temporal-spatial factors are also used for modeling user mobility. Noulas *et al.* [18] have examined plenty of methods to predict the next location, which are statistical models considering various temporal or spatial factors. Baumann *et al.* [19] also study the influence of temporal and spatial features on prediction algorithms. As the method concerning temporal factors, Wang *et al.* [3] propose a periodicity based prediction model to detect the periods for the next place recommendation problem.

There are also several probabilistic models for the location recommendation problem emerged. In the work [20], a Mobility Markov Chain model is established by generalizing the Hidden Markov Model, in which the hidden states do not correspond only to a single location, but rather a sequence of the N previous visited places. Yu *et al.* [6] use the conventional LDA for location preference analysis and recommend locations for mobile users. A hybrid and approximative approach combining Markov and topic models in [1], named Photographer Behavior Model, is presented to handle the trip planning problem by Kurashima *et al.* It seems to coincide with our modeling intentions; however, we coherently build our model using the consistent language of probabilistic graphical models.

All in all, the mobility pattern and semantics analysis for location data are indeed of great practical significance. Nevertheless, seldom works focus on the topical transition semantic analysis of the location based service data in a probabilistic way.

3   Topical User Transition Model

In this section, we represent Topical User Transition Model as a probabilistic generative model. As we analyzed in previous sections, there are two facts that should be highlighted in modeling the focused problem. One is that user generated trajectories collaboratively and statistically exhibit latent spatial topical semantics. The other is that the sequential venues shift might be governed by some coarse-grained transitions among the latent topics.

3.1   Model Specification

Suppose we have the trajectory (i.e. location sequences without ambiguity) set S with the cardinal $|S| = N$. Each sequence $s \in S$ is composed by N_s successive locations, $l_1, l_2, \cdots, l_{N_s}$, and $l_i \in L$; $i = 1, \cdots, N_s$. Here L is the total involved locations or venues set with the cardinal $|L| = M$. When it comes to the latent topic recognizing issue, the probabilistic hierarchical generative model LDA, shown in Fig.1(a), is competent to a large extent. For t^{th} observed location l_t of some s, one first samples l_t's topic assignment $z_t \in \{1, \cdots, K\}$ from the corresponding multinominal distribution probability vector θ_s of the trajectory s. For the trajectory level topic distributions θ_s, LDA imposes the common Dirichlet prior parameterized by α on them. Then, given the topic z_t, the location l_t could be drawn from the multinomial distribution over locations, denoted by ϕ_{z_t}, which is also regulated by a conjugate Dirichlet prior sharing the common concentration hyperparameter vector β. We pile all the ϕs into a matrix Φ of size $K \times M$.

With the simple "bag of words" assumption, i.e. the topic assignments are independent in a specific "bag", LDA is effective in many domains. But in the

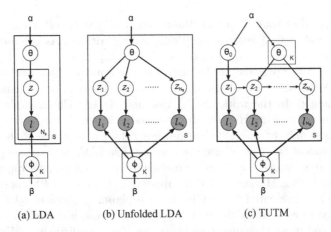

(a) LDA   (b) Unfolded LDA   (c) TUTM

Fig. 1. Graphical Notation Representations of the Models

user mobility application, the topic of current "word", i.e. location, is usually strongly correlated and dependent on the previous locations in the trajectory "bag", owing to the geographical restriction or users' intentions. For the simplicity and effectiveness, the first-order markov property is employed to address those conditional location transitions probabilistically.

The word layer plate unfolded version of LDA is given in Fig.1(b), where the latent topics' independence assumption is explicitly expressed. Then a Markov chain is introduced to manifest the conditional transitions among the latent topic assignments of locations, as shown in Fig.1(c). Furthermore, we lift the topic multinomial distributions θs from the trajectory layer to the top layer. Meanwhile, $\theta_0 \sim Dirichlet(\alpha)$ is used to initialize the hidden Markov chains' starting states, and $\theta_{1,\cdots,k} \sim Dirichlet(\alpha)$, are stacked by rows to form the $K \times K$ Markov state transition probability matrix Θ. Specifically, each matrix entry θ_{ij} governs the probability of location's topic assignment i shifting to j $(i, j = 1, \cdots, K)$. Thus for two successive locations l_t and l_{t+1} with topic assignments z_{l_t} and $z_{l_{t+1}}$, the corresponding transition probability can be directly determined by $p(z_{l_{t+1}} = j | z_{l_t} = i) = \Theta_{ij}$.

Totally, the proposed Topical User Transition Model (TUTM) can be formulated as following generative process.

1. For each topic $z = 1, \cdots, K$
     draw $\phi_z \sim Dir(\beta)$
2. For the initial topics and the Markov topic transition matrix
     (a) draw $\theta_0 \sim Dir(\alpha)$
     (b) draw $\theta_k \sim Dir(\alpha)$, $k = 1, \cdots, K$
3. For each trajectory $s \in S$
     (a) For the first location
           i. draw its topic assignment $z_1 \sim Mul(\theta_0)$
           ii. draw $l_1 \sim Mul(\phi_{z_1})$
     (b) For the subsequent locations l_t, $t = 2, \cdots, N_s$

     i. draw its topic assignment $z_t \sim Mul(\Theta_{z_{t-1}})$
     ii. draw $l_t \sim Mul(\phi_{z_t})$

Collaboratively, the above generative process is actually encoding following joint probability distribution over latent variables, $Z = \cup_{s=1}^{N}\{z_1^s, z_2^s, \cdots, z_{N_s}^s\}$, and observed variables, $L = \cup_{s=1}^{N}\{l_1^s, l_2^s, \cdots, l_{N_s}^s\}$, and the given model parameters, $\Lambda = \{\theta_0, \Theta, \Phi\}$, For simplicity, the hyperparameters α and β are omitted and assumed to be explicitly dependent.

$$p(Z, L|\Lambda) = \prod_{s=1}^{N} p(z_0^s|\theta_0)\Big[\prod_{t=2}^{N_s} p(z_t^s|z_{t-1}^s, \Theta)\Big] \prod_{r=1}^{N_s} p(l_r^s|z_r^s, \Phi) \tag{1}$$

Compared with traditional LDA, TUTM can obviously and felicitously characterize the human daily mobility or migration, through introducing the Markov property to express the transitional dependency of the successive location's latent topics.

3.2 Inference

The posteriors of the probabilistic models, embedding the LDA as a building block, are usually intractable, and inferred by variational methods, Markov chain Monte Carlo, and expectation propagation, etc. Instead, we employ the widely used Expectation-Maximization (EM) algorithm to estimate our TUTM's model parameters, for the fact that we can incorporate the endowments of the off-the-shelf Forward-Backward algorithm to treat the Markov chain's estimation at the expectation step. Specifically, the Expectation and Maximization steps are formulated as follows:

E-step: We take the initial or previously estimated parameters Λ^{old} values to calculate the posterior of the latent variables $p(Z|X, \Lambda^{old})$. We then use this posterior distribution to evaluate the expectation of the logarithm of the complete-data likelihood function parameterized by Λ of the current iterative round, by summing up all the configuration of the latent variables Z as follows,

$$Q(\Lambda, \Lambda^{old}) = \sum_Z p(Z|L, \Lambda^{old}) \ln p(L, Z|\Lambda) \tag{2}$$

We subtitute the joint distribution Eq.1 into Eq.2, then yields,

$$Q(\Lambda, \Lambda^{old}) = \sum_{s=1}^{N}\Big\{\sum_{k=1}^{K} p(z_1^s = k|L, \Lambda^{old}) \ln \theta_0 + \sum_{t=1}^{N_s}\sum_{k=1}^{K} p(z_t^s = k|L, \Lambda^{old}) \ln p(l_t^s|\phi_k)$$
$$+ \sum_{t=2}^{N_s}\sum_{j=1}^{K}\sum_{k=1}^{K} p(z_t^s = k, z_{t-1}^s = j|L, \Lambda^{old}) \ln \Theta_{jk}\Big\} \tag{3}$$

Utilizing the Forward-Backward algorithm, we achieve the posteriors of latent variables, $p(z_t^s|L, \Lambda^{old})$, and their co-occurrences, $p(z_t^s = k, z_{t-1}^s = j|L, \Lambda^{old})$, in Eq.3. Firstly, the two posterior are factorized as follows:

$$p(z_t^s|L, \Lambda^{old}) \propto \alpha(z_t^s)\beta(z_t^s); \quad p(z_t^s = k, z_{t-1}^s = j|L, \Lambda^{old}) \propto \alpha(z_{t-1}^s)\phi_k\Theta_{jk}\beta(z_t^s)$$

Where $\alpha(z_t^s) = p(l_1, \cdots, l_t^s, z_t^s)$ atd $\beta(z_t^s) = p(l_{t+1}^s, \cdots, l_{N_s} | z_t^s)$ are the forward factors and the backward factors respectively. Then, the algorithm uses a two-stage message passing procedure to exactly calculate the factors through these recursion relations.

$$\alpha(z_t^s) = \sum_{z_{t-1}^s} \alpha(z_{t-1}^s) \Theta_{z_{t-1}^s, z_t^s}^{old} \Phi_{z_t^s, l_t^s}^{old}; \quad \beta(z_t^s) = \sum_{z_{t+1}^s} \beta(z_{t+1}^s) \Theta_{z_t^s, z_{t+1}^s}^{old} \Phi_{z_{t+1}^s, l_{t+1}^s}^{old}$$

Thus, the Eq.3 can be easily evaluated, so as to be used to feed the maximization step. During the Forward-Backward algorithm, we also collect the expectation number of the topic transition pairs (i, j): $i, j = 1, \cdots, K$, and topic-location pairs (p, q): $p = 1, \cdots, K$, and $q = 1, \cdots, M$, i.e.

$$C_{0k} = \mathbb{E}(z_1 = k) = \sum_{s=1}^{N} p(z_1^s = k | L, \Lambda^{old});$$

$$C_{ij} = \mathbb{E}(z_t = i, z_{t+1} = j) = \sum_{s=1}^{N} \sum_{t=1}^{N_s-1} p(z_t^s = i, z_{t+1}^s = j | L, \Lambda^{old});$$

$$C_{pq} = \mathbb{E}(z_t = p, l_t = q) = \sum_{s=1}^{N} \sum_{t=1}^{N_s} p(z_t^s = p, l_t^s = q | L, \Lambda^{old}).$$

M-step: We maximize the evaluated Eq.3 with respect to the nondeterministic parameters $\Lambda = \{\theta_0, \Theta, \Phi\}$. Lagrange Multiplier methods is employed to determine the new parameters θ_0, Θ, and Φ.

$$\theta_{0k} \propto C_{0k} + \alpha; \quad \Theta_{ij} \propto C_{ij} + \alpha; \quad \Phi_{pq} \propto C_{pq} + \beta$$

Thereby, the EM algorithm starts with some initial selection for the model parameters, Λ^{init}, and then alternates between E and M steps until some convergence criterion is satisfied, for instance when the change in the complete-data likelihood function is below some threshold. Besides in the later experiment sections, we empirically assume that the α and β are fixed and heuristically set $\alpha = 50/K$, $\beta = 0.01$.

3.3   Prediction

Due to the TUTM's powerful ability of modeling user mobility, it is suitable for us to utilize the model for the destination recommendation or prediction. We first define the next-place prediction problem. When predicting the next place, we aim at computing the most possible location conditioned user's visiting history. Therefore, a set of training sequences S and a set of locations L are defined as input of the focused problem. There is a check-in set C_s corresponding to each $s \in S$, which is composed by a list of check-ins $c_{l,t}$, where $l \in L$ and t represents the timestamp. The order of location list $l_1, l_2, \cdots, l_{N_s}$ is indicated by the timestamp. The prediction problem is formalized as: Given a sequence

$l_1, l_2, \cdots, l_t$, how to rank the locations in L based on $P(l_{t+1}|l_1, l_2, ..., l_t)$ in order that the actual visited location is top-ranked.

The prediction probability of the next place could be calculated as follows:

$$p(l_{t+1}|l_1, \cdots, l_t) \propto \sum_{z_{t+1}} \Phi_{z_{t+1}, l_{t+1}} \sum_{z_t} \Theta_{z_t, z_{t+1}} \alpha(z_t)$$

4 Experiments

In order to demonstrate the performance of TUTM, the following two experiments are carried out. Firstly, as the main concern of this work, the topical transition probabilities are inferred to discover the semantic mobility patterns of the check-in data. The Foursquare categorical label is elaborately employed to show the conformance between the transition semantics and common sense. Secondly, we assess the next-place prediction accuracy of TUTM, and compare it with some relevant benchmark models to exhibit the feasibility and effectiveness of our model.

4.1 Dataset Description and Analysis

The dataset used in this paper is contributed by Cheng [8], which contains check-in data crawled from the location sharing status on Twitter . The original dataset is composed by 22 million check-ins by over 220 thousand users, each of which is represented as { *UserId, CheckInId, Longitude, Latitude, Timestamp, TwitterContent*}.

We only extract those check-ins occurred in New York City from the above mentioned dataset for our experiments. The total number of check-ins samples is already 840,046, with 12,623 users and over 50,000 venues involved. This data

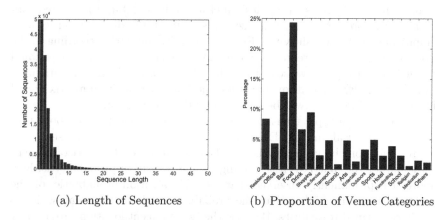

(a) Length of Sequences   (b) Proportion of Venue Categories

Fig. 2. Preliminary Analysis on Dataset

subset, without loss of generality, should be sufficient to cover most of the transition patterns in urban area. Through querying the Foursquare API , we firstly map the check-in's geo-coordinates and twitter contents onto unique pre-defined Foursquare Venue Entity, and meanwhile download the corresponding category labels of every venue for the later semantic analysis, which categorizes the venues into 19 pre-defined classes. Last but not least, we reorganize the user and theirs check-in points set into consecutive venue sequences according to the timestamps. Specifically, we group every user's check-ins per a given idle time span threshold.

As shown in Fig.2, we conduct a preliminary analysis on the dataset. The histogram of sequence length shown in Fig.2(a) prominently reflects that the distribution of sequence length apparently conforms the long tail effect. Thereby, the number of check-ins included in each sequence is highly heterogeneous: few sequences have large number of check-ins, and more than 50% of sequences have less than 5 check-ins, which causes the average length of sequence reaching down to 5. The proportion of venue categories shown in Fig.2(b) illustrates that the use frequency of location-based services varies in different type of venues. We can clearly observe that check-ins occur more in the places of restaurants or bars, while seldom people update the location status in religious places. This observation is quite consistent with common sense, therefore, we can consider the category distribution as the empirical distribution over topics, which is helpful for topical semantic analysis.

4.2 Topical Transition Analysis

In order to quantitatively demonstrate the TUTM's ability of identifying the semantics transition patterns, we resort to the prepared venues categories information and align them to the Markov transition matrix parameters Θ , so as to manifest our results as intuitively as possible. To this end, firstly, the topic number of TUTM is set to 19, the same as the number of venues' categories. Secondly, we assume that the venues categories' label naturally exhibits the desired topical semantics, and hence, we initialize the starting topic distribution θ_0 with the empirical distribution shown by Fig.2(b). Thirdly, similarly according to the category information, we create a topic-location concurrence binary matrix and normalize it as per topic rows, so as to use it as the prior for the model parameter Φ. Thus after the EM algorithm converges, the Markov transition matrix Θ and final venues distribution Φ will be fitted under these configurations.

Fig.3 visualize the topic transition matrix, i.e. Θ learned by the above procedure, in a gray-scale image. Every transition in the matrix is presented as the source being on the vertical axis and the destination on the horizontal axis. And, the darker the matrix cell is, the higher the transition probability is. Furthermore, it is prominently observed that the self-transition probabilities on the diagonal are generally higher than the probabilities off the diagonal, which is obviously conformed to realities. Because the two adjacent locations are usually generated under the same user intention. We will detail the inferred mobility

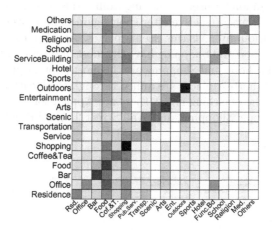

Fig. 3. Topical Transition Matrix

Table 1. Lowest and Highest Self-Transitions

Topic	Probability		Topic	Probability
Hotel	0.0828		Shopping	0.3952
Religion	0.0901		Outdoors	0.3770
PublicService	0.1187		Transportation	0.3399
FunctionalBuilding	0.1338		School	0.3296
Office	0.1678		Arts	0.3164
Medication	0.1792		Bar	0.3127
Residence	0.1836		Food	0.3119

patterns displayed by this matrix in two aspects: 1) self-transitions; 2) inter-topic transitions.

The self-transitions on the diagonal of topic transition matrix, represent that one transits between two locations of the same topic. As shown in Table.1, the self-transition of the Hotels and Religious venues have the lower probabilities. Because one or a few venues from these topics could be enough to accomplish people's intended tasks. For instance, the user who has checked in to a hotel will not visit another hotel next. While the self-transitions of Shopping, Outdoors and Transportation topics are with the higher probabilities. Multiple venues of these topics are combined to reach the users' aims. This fact happens frequently in real life. Taking the Shopping topic as an example, people usually go to several stores to achieve the more opportunities to compare the commodities and make purchase decisions. Moreover, as for the transportation topics, it also make sense that people regularly transform among several stops to reach their destination.

We then analyze the inter-topic transition probabilities which are shown in Fig.4. In this graph, the transitions with probability less than 0.15 are discarded in order to clearly discover significant patterns. In addition, isolated nodes are discarded too. Notes are attached to each topic showing the top 5 locations with highest probability according to the location distribution of that topic.

As observed, it is common for users transiting from residences or offices to restaurants, and bars are always attractive after sports games. Moreover, tourists usually take transportation after visiting scenic places. Another fact can be easily observed is that the Food and Shopping have the most in-degrees, indicating that the places under topic Food and Shopping are the most popular places to be transited to. Besides, the hottest places are mined through the location probabilities Φ. As we can see in the attaching notes, there are Apple Store, Rockefeller Center, Topshop under topic Shopping, and Wall Street Journal Building, Goldman Sachs under Office.

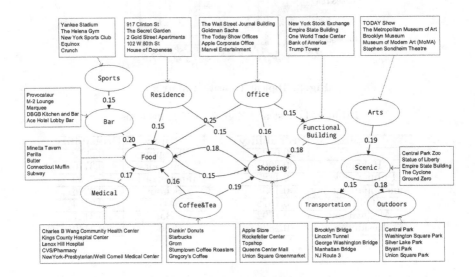

Fig. 4. Inter-Topic Transitions

The previous analysis demonstrates that our model can reasonably explain the user mobility patterns. Due to the expressiveness of this model, we are convinced that TUTM can better understand user mobility features for destination recommendation and prediction. To further evaluate the usefulness of TUTM, we put the model into practice by predicting the next place.

4.3   Next-Place Prediction

Experiments on next-place prediction are carried out based on the aforementioned predicting problem in Section 3.3. We divide the dataset into training and testing set: for each sequence $s \in S$, the former $N_s - 1$ locations $\{l_1, \cdots, l_{N_s-1}\}$ become training sequence, and the last location l_{N_s} to be predicted composes the testing set. Due to the large quantity of the potential target locations in L, the ranking performance cannot be well reflected by the measurement of exact prediction accuracy. In this case, we assess the performance with prediction metric *Accuracy@N* used in [18],

$$Accuracy@N = \frac{|Hit\ Sequences@N|}{|S|} \qquad (4)$$

where N denotes the number of candidates chosen with the highest ranking. $Hit\ Sequence@N$ represents the number of sequences with successful prediction, in which the test venue is ranked within top N.

Before training, model parameters are initialized as random probabilities. In order to achieve the best performance, parameters are tuned based on the prediction accuracy. Thus, we first try to discover how performance trends with increasing topic number and iteration round. Fig.5(a) shows that the performance improves overall during the topic number grows until it gets 210. As the influence of iteration round shown in Fig.5(b), the EM algorithm converges approximately at round 100, and the performance will not improve any longer.

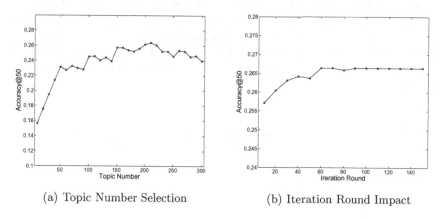

(a) Topic Number Selection   (b) Iteration Round Impact

Fig. 5. Parameter Selection

Moreover, in order to manifest the capability of TUTM on predicting or recommending next locations, we compare our model with the following baseline models:

- **Popularity** (PopRec): The popularity recommendation method assumes the popular places will be more likely to be visited, thus it utilizes the frequency of location occurrence for predicting, i.e. $p(l_{N+1}|l_1, \cdots, l_N) \propto Count(l_{N+1})$. This approach is commonly used in recommendation field.
- **Markov Model** (MM): Markov Model considers the most likely transitions from current place for prediction. Thus locations are ranked based on the transition probability, i.e. $p(l_{N+1}|l_1, \cdots, l_N) \propto Count(l_N, l_{N+1})$.
- **Latent Dirichlet Allocation** (LDA): As a topic model, LDA extracts the topics to model user preferences to locations. LDA is applied for prediction using user's preference extracting, which is presented as $p(l_{N+1}|l_1, \cdots, l_N) \propto \sum_{z_{N+1}} p(l_{N+1}|z_{N+1})p(z_{N+1}|U)$.

- **Hidden Markov Model** (HMM): HMM has been used for next-place prediction in [20]. HMM predicts the next location by inferring the most likely
  hidden states of z_{N+1} and sum up the products with the emission probability
  $p(l_{N+1}|l_1, \cdots, l_N) \propto \sum_{z_{N+1}} p(l_{N+1}|z_{N+1}) p(z_{N+1}|l_1, \cdots, l_N)$

PopRec and MM are statistical methods to deterministically calculate the
occurrence. As for probabilistic models LDA and HMM, we tune the parameters
by selecting their best performance respectively, which are 50 topics for LDA,
and 150 hidden states for HMM.

Our model is compared with the baseline models in terms of prediction accuracy specified in Eq.4. As shown in Fig.6, the curves of different approaches
demonstrates that our model outperforms the others, and the details are presented in Table.2.

We can observe from the result that the Popularity method performs worst,
because it has badly reduced the inter-user diversity. Despite of the high score
on top-10, MarkovModel suffers slow growth on accuracy during the increase of
candidate set size. It reveals the limits of statistical transition patterns, since only
those explicit transitions are recorded. However, TUTM can model the topical
transition patterns, which are capable to mine probabilistic relations between
locations.

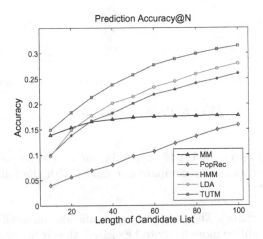

Fig. 6. Accuracy Curves of Different Models

Table 2. Prediction Accuracy

	Acc@10	Acc@50	Acc@100
Pop.Rec	0.0393	0.0973	0.1586
MM	0.1389	0.1734	0.1772
HMM	0.1001	0.2013	0.2598
LDA	0.0979	0.2144	0.2795
TUTM	**0.1493**	**0.2636**	**0.3183**

As for the probabilistic models, HMM and LDA employ the latent variables in modeling user mobility behavior. Similar to our model, HMM could also model the transition between hidden states. However, the transitions are independent to each other. This deficiency is addressed in TUTM by introducing an unified prior Dirichlet distribution to each transition probabilities. Therefore, as we can observe from the results, an improvement of 24% over HMM has been attained by TUTM on $Acc@50$. The accuracy score of LDA is the closest to our model, since it considers the user's preferences to locations which is widely utilized for recommendation. TUTM improve 20% over LDA in terms of $Acc@50$. In general, TUTM achieves the highest prediction accuracy owing to the great expressiveness in modeling the user mobility pattern.

5 Conclusion and Future Work

In this paper, we proposed a probabilistic model TUTM, which utilizes topical transition for semantic mobility pattern analysis. This model can be used for topical transition mining and next-place prediction. Evolving from LDA, TUTM is established by adding transitional dependencies between adjacent latent variables. The Expectation Maximization algorithm incorporating with Forward Backward algorithm is also delivered for estimating the model parameters. We conduct two experiments on real-world check-in dataset to demonstrate the effectiveness of TUTM. Specifically, the topical transition probabilities are analyzed to discover the semantic patterns, which demonstrate that TUTM can reasonably explain the user mobility patterns. Also, in comparison with existing approaches, the next-place prediction performance is confirmed.

As for the future work, we are going to interpret the meaning of the discovered semantics. Since the latent variables corresponding to each location always seem inexplicable, the meaning of them are artificially speculated. In that case, currently we can only temporarily regard them as the attributes or categories of locations. Therefore, reasonably explaining the spatial semantics under the topical transitions is our primary target recently. Moreover, we are planning to further study the quality of next-place prediction in our model. Currently, the predictive probabilities of our model generated only from the spatial information, without considering the detailed temporal factors. Therefore, we are going to incorporating more context information, such as the distance, time, and even hybrid factors. A more complex model will be expected with more context factors considered. Meanwhile, the social factors are also supposed to be taken into consideration as many existing studies support.

Acknowledgments. This work was supported partly by National Natural Science Foundation of China (No. 61300070, No. 61103031), partly by China 973 program (No. 2014CB340305), China 863 program (No. 2012AA011203), partly by the State Key Lab for Software Development Environment (SKLSDE-2013ZX-16), partly by A Foundation for the Author of National Excellent Doctoral Dissertation of PR China(No. 201159) and partly by Program for New Century Excellent Talents in University.

References

1. Kurashima, T., Iwata, T., Irie, G., Fujimura, K.: Travel route recommendation using geotags in photo sharing sites. In: Proceedings of the 19th ACM International Conference on Information and Knowledge Management, CIKM 2010, pp. 579–588. ACM, New York (2010)
2. Zheng, Y.T., Zha, Z.J., Chua, T.S.: Mining travel patterns from geotagged photos. ACM Trans. Intell. Syst. Technol. 3(3), 56:1–56:18 (2012)
3. Li, Z., Wang, J., Han, J.: Mining event periodicity from incomplete observations. In: Proceedings of the 18th ACM SIGKDD International Conference on Knowledge Discovery and Data Mining, KDD 2012, pp. 444–452. ACM, New York (2012)
4. Preotiuc-Pietro, D., Cohn, T.: Mining user behaviours: a study of check-in patterns in location based social networks. In: Davis, H.C., Halpin, H., Pentland, A., Bernstein, M., Adamic, L.A. (eds.) WebSci, pp. 306–315. ACM (2013)
5. Monreale, A., Pinelli, F., Trasarti, R., Giannotti, F.: Wherenext: A location predictor on trajectory pattern mining. In: Proceedings of the 15th ACM SIGKDD International Conference on Knowledge Discovery and Data Mining, KDD 2009, pp. 637–646. ACM, New York (2009)
6. Yu, K., Zhang, B., Zhu, H., Cao, H., Tian, J.: Towards personalized context-aware recommendation by mining context logs through topic models. In: Tan, P.-N., Chawla, S., Ho, C.K., Bailey, J. (eds.) PAKDD 2012, Part I. LNCS, vol. 7301, pp. 431–443. Springer, Heidelberg (2012)
7. Liu, Q., Ge, Y., Li, Z., Chen, E., Xiong, H.: Personalized travel package recommendation. In: Proceedings of the 2011 IEEE 11th International Conference on Data Mining, ICDM 2011, pp. 407–416. IEEE Computer Society, Washington, DC (2011)
8. Cheng, Z., Caverlee, J., Lee, K., Sui, D.Z.: Exploring millions of footprints in location sharing services. In: Adamic, L.A., Baeza-Yates, R.A., Counts, S. (eds.) ICWSM. The AAAI Press (2011)
9. Zheng, Y., Zhang, L., Xie, X., Ma, W.Y.: Mining interesting locations and travel sequences from gps trajectories. In: Proceedings of the 18th International Conference on World Wide Web, WWW 2009, pp. 791–800. ACM, New York (2009)
10. Arase, Y., Xie, X., Hara, T., Nishio, S.: Mining people's trips from large scale geotagged photos. In: Proceedings of the International Conference on Multimedia, MM 2010, pp. 133–142. ACM, New York (2010)
11. Bayir, M.A., Eagle, N., Demirbas, M.: Discovering spatiotemporal mobility profiles of cellphone users. In: Proceedings of the 10th IEEE International Symposium on a World of Wireless, Mobile and Multimedia Networks (WoWMoM 2009), pp. 1–9 (2009)
12. Ganti, R., Srivatsa, M., Ranganathan, A., Han, J.: Inferring human mobility patterns from taxicab location traces. In: Proceedings of the 2013 ACM International Joint Conference on Pervasive and Ubiquitous Computing, UbiComp 2013, pp. 459–468. ACM, New York (2013)
13. Scellato, S., Noulas, A., Lambiotte, R., Mascolo, C.: Socio-spatial properties of online location-based social networks. In: Adamic, L.A., Baeza-Yates, R.A., Counts, S. (eds.) ICWSM. AAAI Press, Menlo Park (2011)
14. Eagle, N., Pentland, A.S., Lazer, D.: Inferring friendship network structure by using mobile phone data. Proceedings of the National Academy of Sciences 106(36), 15274–15278 (2009)

15. Xiao, X., Zheng, Y., Luo, Q., Xie, X.: Finding similar users using category-based location history. In: Proceedings of the 18th SIGSPATIAL International Conference on Advances in Geographic Information Systems, GIS 2010, pp. 442–445. ACM, New York (2010)
16. Backstrom, L., Sun, E., Marlow, C.: Find me if you can: Improving geographical prediction with social and spatial proximity. In: Proceedings of the 19th International Conference on World Wide Web, WWW 2010, pp. 61–70. ACM, New York (2010)
17. Lu, X., Wang, C., Yang, J.M., Pang, Y., Zhang, L.: Photo2trip: Generating travel routes from geo-tagged photos for trip planning. In: Proceedings of the International Conference on Multimedia, MM 2010, pp. 143–152. ACM, New York (2010)
18. Noulas, A., Scellato, S., Lathia, N., Mascolo, C.: Mining user mobility features for next place prediction in location-based services. In: ICDM, pp. 1038–1043 (2012)
19. Baumann, P., Kleiminger, W., Santini, S.: The influence of temporal and spatial features on the performance of next-place prediction algorithms. In: Proceedings of the 2013 ACM International Joint Conference on Pervasive and Ubiquitous Computing, UbiComp 2013, pp. 449–458. ACM, New York (2013)
20. Gambs, S., Killijian, M.O., del Prado Cortez, M.N.: n.: Next place prediction using mobility markov chains. In: Proceedings of the First Workshop on Measurement, Privacy, and Mobility, MPM 2012, pp. 3:1–3:6. ACM, New York (2012)

An Offline Optimal SPARQL Query Planning Approach to Evaluate Online Heuristic Planners

Achille Fokoue, Mihaela Bornea, Julian Dolby,
Anastasios Kementsietsidis, and Kavitha Srinivas

IBM T.J. Watson Research
P.O. Box 704, Yorktown Heights, NY 10598, USA

Abstract. In graph databases, a given graph query can be executed in a large variety of semantically equivalent ways. Each such execution plan produces the same results, but at different computation costs. The query planning problem consists of finding, for a given query, an execution plan with the minimum cost. The traditional greedy or heuristic cost-based approaches addressing the query planning problem do not guarantee by design the optimality of the chosen execution plan. In this paper, we present a principled framework to solve the query planning problem by casting it into an Integer Linear Programming problem, and discuss its applications to testing and improving heuristic-based query planners.

Keywords: RDF, SPARQL, Query Optimization, Query Planning, ILP.

1 Introduction

Obtaining good performance for declarative query languages requires an optimized total system, with an efficient data layout, good data statistics, and careful query optimization (e.g. [2]). One key piece of such systems is a query planner that translates a declarative query into a concrete execution plan with minimal cost. This problem has been extensively studied - in particular, in the relational database literature [19,7,4,5]. The traditional solution builds a cost-model that, based on data statistics [15,17], is able to estimate the cost of a given query execution plan. However, since the number of execution plans can be extremely large, only a small subset of all valid plans are constructed (using heuristics and/or greedy approaches that consider plans likely to have a low cost) [10,11]. The cost of those selected candidate plans are then estimated using the cost-model, and the cheapest plan is selected for execution. The chosen plan is a local optimal and not guaranteed to be a global optimal. Even with sub-optimal plans, the performance of an optimizer is still considered satisfactory, if it performs better (in terms of evaluation times) when compared to other competing optimizers. Yet, there is an alternative metric to measure how well the optimizer performs: how far its local optimal plans are from global optimal plans. However, finding a global optimal is challenging and is one of the reasons why the heuristic planners were devised in the first place. To the best of our knowledge, there is

B. Benatallah et al. (Eds.): WISE 2014, Part I, LNCS 8786, pp. 480–495, 2014.
© Springer International Publishing Switzerland 2014

no practical mechanism for assessing how good these planners are, i.e. whether they produce optimal plans given the data layout and statistics available.

In this paper, we describe an efficient offline technique to find optimal plans for the graph query planning problem. The problem is NP-hard, as shown in [9,1]. Our approach to find optimal plans works by casting it as an integer programming (or ILP) problem. An ILP problem consists of an objective function (in our case, the cost of executing the query) that needs to be minimized under a set of inequality constraints (in our case, these constraints encode the semantics of the input graph query and of valid execution plans). Both the objective function and the set of constraints are expressed as linear expressions over a set variables - some of which are restricted to take only integer values. Although ILP is also known to be NP-hard in the worst case, in practice, highly optimized solvers exist to efficiently solve our formulation of the query planning problem (see section 4). Furthermore, we show that our ILP formulation can be used to evaluate the effectiveness of any greedy/heuristic planning solution. In fact, the two approaches can be potentially combined, with the ILP formulation being used to precompile specific queries that may occur frequently within a workload, or test the heuristic solution to find how far away it is from optimal.

Our main contributions in the paper are as follows:

1. We present an abstract general formulation of the planning problem which decouples it from the actual resolution of the problem.
2. We show how to translate that abstract formulation into an ILP problem.
3. Using an implementation of this approach for SPARQL 1.0, we show how it can be concretely applied to test and improve a state of the art online heuristic-based planner in DB2RDF [2]. This formal and principled evaluation approach helps uncover new opportunities for further optimization in a mature system which outperforms existing systems (Virtuoso, Jena, Sesame and RDF3X) across four benchmark and real datasets [2].

The rest of this paper is organized as follows. Section 2 presents an algebraic representation of queries, and uses it to introduce the universe of alternative query plans for an input query q. In Section 3, we cast the planning problem as an ILP problem. In particular, our approach is inspired by electronic circuit design and we intuitively construct a single concise circuit board that captures the whole universe of plans. Then, we introduce appropriate constraints and cost functions and use an ILP solver to identify an optimal sub-circuit in the board that connects all the circuit components (i.e. all the input query sub-patterns) which corresponds to the optimal query plan. Section 4 empirically demonstrates that our approach is a practical formalization of the optimization problem, for testing and improving query planners and offline query optimization.

2   The SPARQL Query Planning Problem

2.1   Planning Problem Input

There are three inputs to the process of SPARQL planning:

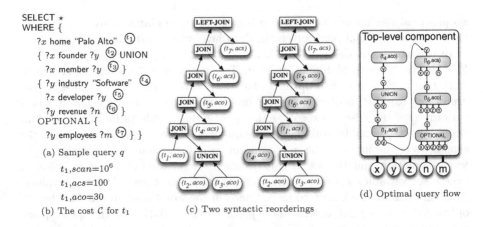

Fig. 1. Sample input, alternative plans, and optimal flow

1. The query q : The SPARQL query conforms to the SPARQL standard. Therefore, each query q is composed of a set of hierarchically nested graph patterns $\mathcal{P}$, with each graph pattern $p \in \mathcal{P}$ being either a simple triple pattern[1] or more complex patterns such as BGP, UNION, or OPTIONAL.

2. The access methods $\mathcal{M}$: Access methods provide alternative ways to evaluate a pattern $P \in \mathcal{P}$. The methods are system-specific, and dependent on existing indexes in the store. For example, a store might have subject indexes to access a triple by subject (*access-by-subject* (acs)), or by object (*access-by-object* (aco)), by a scan (*access-by-scan* ($scan$)).

3. The cost $\mathcal{C}$: Each access method for a pattern P is annotated with a cost, based on system specific notions of how expensive it is to execute the method. The $\mathcal{C}$ may be derived from statistics maintained by the system about the characteristics of a particular dataset, or well known costs for a particular access method (e.g, scans are more expensive than index based access).

Figure 1 shows a sample input where query q retrieves the people that founded or are board members of companies in the software industry, and live in "Palo Alto". For each such company, the query retrieves the products that were developed by it, its revenue, and optionally its number of employees. Three different access methods are assumed in $\mathcal{M}$, one that performs a data scan (*scan*), one that retrieves all the triples given a subject (*acs*), and one that retrieves all the triples given an object (*aco*). The cost $\mathcal{C}$ for accessing a specific pattern p, given an access method, is the third input, an example of which is shown for triple t_1 in Figure 1(b). This query will form our running example and the rest of the figure will be explained in the following sections.

To simplify the planning process, we introduce a function **flat**(q) to eliminate unnecessary syntactic nesting that might occur in the query. Specifically, since

---

[1] We reuse the same notation as the SPARQL algebra (section 18 of the spec), and assume, w.l.o.g, that every triple appears in singleton Basic Graph Pattern (BGP).

each query q is composed of a set of hierarchically nested graph patterns $\mathcal{P}$, for each graph pattern $p \in \mathcal{P}_{\text{BGP}}$ (i.e., the set of BGP patterns in q), we flatten nested BGP patterns because they do not reflect any change in the semantics of the SPARQL query. Note that when we flatten the query q, we ensure that any OPTIONAL pattern associated with a nested BGP pattern stays scoped with the BGP pattern to make it equivalent to the query.

2.2   Universe of Valid Plans

Given a query q, the SPARQL specification (section 18) defines a transformation of q into an algebraic expression, denoted $\mathbf{algebra}(q)$, that corresponds to a valid evaluation of the query. The tree on the left in Figure 1(c) shows $\mathbf{algebra}(\mathbf{flat}(q))$ for our example query. Due to the guaranteed correctness of the transformation from a query to a SPARQL algebraic expression, the SPARQL Algebra is clearly a good starting point to define our notion of a valid execution plan of a query. However, the algebraic expression generated for a query q suffers from two key limitations. First, it is underspecified: it implies an execution order, but, for example, it does not specify the access method to use to access a given triple pattern. Second, the implied execution order only mirrors the order in which patterns appear in the original query (no join order optimization). Thus, the evaluation order entailed by the generated expression is likely to be suboptimal.

In this section, we first formally define a valid plan as an annotated SPARQL algebraic expression. Annotations make an algebraic expression fully specified and executable. Annotations indicate, for example, the precise access method used to access a triple pattern, or, for a JOIN node, whether it is a PRODUCT (i.e. when the two operands of the JOIN node have no variables in common). Then, we present a generalization of the transformation from a SPARQL query to a SPARQL algebraic expression that, for a given query q, generates a very large universe $\mathcal{U}_q$ of valid plans of q. Plans in $\mathcal{U}_q$ are obtained by considering all permutations of elements in all BGP patterns of the flattened query $\mathbf{flat}(q)$ and all valid annotations of all algebraic nodes. Finally, the query planning problem is defined as finding a plan in $\mathcal{U}_q$ with the lowest cost.

Annotated SPARQL Algebra. The access method annotation function, denoted $\mathbf{am}$, maps a Basic Graph Pattern (BGP) containing a single triple t in $\mathbf{algebra}(\mathbf{flat}(q))$ to an access method $m \in \mathcal{M}$ to use to evaluate t. But patterns are not the only part of the algebra that requires an annotation. Indeed, if left without annotation, the JOIN operator is *ambiguous*. A JOIN(e_1, e_2) operation can stand for one of many types of join implementations, each with a different cost. So, a join can be implemented like a Nested Loop Join, with a cost that might be quadratic to the size of the join inputs, or it can be implemented like a Hash Join, with a cost that is only linear to its input. And like the access methods for patterns, the available join implementations are system-specific. For a query plan to be costed appropriately though, it is important to identify which join implementation will be used for each join operation. Therefore, a second annotation function is used, called join annotation function and denoted $\mathbf{jan}$, which maps a join operation to one element of a set like $\mathcal{J} = \{\text{PRODUCT}, JoinImp1,$ $JoinImp2, \ldots\}$ to indicate the precise nature of the join to be performed. Notice that we consider a cartesian product as a special form of a join. Beyond joins, the same annotation function **jan** will be used in plans to maps left-outer-join operations to their respective implementations (left-outer-joins operations are needed for the OPTIONAL operator in SPARQL).

A few additional functions are necessary, in addition to our annotation functions. Given an access method annotation function **am** and a join annotation function **jan**, we define the required variables function, denoted **required[am, jan]** (or simply **required** when there is no ambiguity), and the available variables function, denoted **available[am, jan]** (or simply **available**). For an algebraic sub-expression e in **algebra(flat(q))**, **required(e)** is the set of all variables required to evaluate e, and **available(e)** is the set of all variables available after the evaluation of e. The intuition for these additional functions is illustrated through an example. Assume that **am(t_1)** = acs, that is, the access method for pattern t_1 in Figure 1(a) is acs. To evaluate t_1 with this access method, the subject variable $?x$ must be provided. Therefore, **required(t_1)** = $\{?x\}$. Similarly, it is easy to see that if **am(t_5)** = aco, then **available(t_5)** = $\{?z, ?y\}$.

Definition 1. *An annotated SPARQL algebraic expression is a tuple $(e, \mathbf{am}, \mathbf{jan})$ such that : (1) e is an SPARQL algebraic expression whose BGP sub-expressions consist of a single triple, (2) $\mathbf{am}$ is a function that maps each BGP sub-expression of e to an access method $m \in \mathcal{M}$, and (3) $\mathbf{jan}$ is a function that maps each JOIN or LEFTJOIN sub-expression of e to an available implementation.*

We can now formally define a query plan as an annotated SPARQL algebraic expression that does not require any variable

Definition 2. *A query plan is an annotated SPARQL algebraic expression $(e, \mathbf{am}, \mathbf{jan})$ such that $\mathbf{required}(e) = \emptyset$. $\mathcal{PL}$ denotes the set of all plans.*

Universe of Valid Plans Considered. A query q_1 is said to be a syntactic reordering of q_2, denoted $q_1 \sim q_2$, when q_1 and q_2 are syntactically identical after reordering elements in BGP patterns of q_1. For a given query q, we can now define a set of equivalent queries $\mathcal{EQ}_q$ as follows: $\mathcal{EQ}_q = \{q' | q' \sim \mathbf{flat}(q)\}$[2].

Finally, the universe of plans considered, $\mathcal{U}_q$, is defined as :

$$\mathcal{U}_q = \{p = (e, \mathbf{am}, \mathbf{jan}) \in \mathcal{PL} | e = \mathbf{algebra}(q') \wedge q' \in \mathcal{EQ}_q\}$$

If q consists of a single BGP group with n triple patterns, and, for each triple pattern, there are k possible access methods, the cardinality of $\mathcal{U}_q$ can be as large as $n! k^n$ (assuming only one implementation for joins other than PRODUCT).

2.3 The Planning Problem

The planning problem consists in finding a minimal cost plan $p \in \mathcal{U}_q$ for a query q. Plans in $\mathcal{U}_q$ are obtained by considering all permutations of elements in all

---

[2] Due to space limitation, the special treatment of optional patterns is discussed only in details in the technical report [1].

BGP patterns of the flattened query **flat**(q). The planning problem is NP hard since choosing an ordering in a single BGP is NP hard: the classic join order optimization is showed to be NP-hard in [9], and our formulation is shown to be NP-hard through reduction from TSP in [1].

3   Integer Linear Programming Approach

For a query q, the set $\mathcal{U}_q$ of plans defined in section 2 is too large for an exhaustive search of an element with the lowest cost. For q with 15 triple patterns and 3 access methods for each triple, assuming enough compute power to generate and cost a billion plans per second, 594 years are needed for the $15! \times 3^{15}$ plans!

In this section, we present an efficient, principled, and general approach to solve an arbitrary complex query planning problem by casting it into an Integer Linear Programming (ILP) problem. It consists in the following key steps:

• **Control-aware Data Flow Construction:** The access methods applicable to a given triple pattern depend on the variables that are available (in-scope variables) when it is evaluated. Since patterns typically share variables, the evaluation of one is often dependent on the evaluation of another. For example, in Figure 1, the triple pattern t_1 shares the variable $?x$ with triple patterns t_2 and t_3 appearing in the union. Hence, there is an inter-dependency between t_1 and the union pattern containing t_2 and t_3 as, depending on the execution methods used and the order of execution of t_1 and $\text{UNION}(t_2, t_3)$, the variable $?x$ may "flow" from t_1 to $\text{UNION}(t_2, t_3)$ or in the reverse direction. The Data Flow Construction step builds a data structure that captures all potentially valid ways in which variables can "flow" between various parts of the query. This data flow structure is control aware because it explicitly rules out variable flows that would clearly violate the semantics of control statements in the query. For example, the $?y$ variable shared between t_2 and t_3 cannot be produced by one and used by the other because it would violate the semantics of a UNION pattern.

• **Constraint Generation:** To ensure completeness (i.e., all plans in $\mathcal{U}_q$ are considered), the Control-aware Data Flow has to capture all potentially valid flows and execution orders. Unfortunately, it also contains many invalid flows and execution orders that cannot be ruled out a-priori. For example, it encodes a cyclic flow of variable $?x$ from t_1 to $\text{UNION}(t_2, t_3)$ and from $\text{UNION}(t_2, t_3)$ to t_1. To ensure soundness (i.e., all solutions of the ILP problem can be converted into a valid plan in $\mathcal{U}_q$), the constraint generation step generates, from the Control-aware Data Flow structure, constraints that "dynamically" rule out all invalid flows and execution orders (e.g., constraints ruling out cyclic data flows). These constraints constitute the linear constraints of the ILP problem formulation.

• **Cost Function Formulation:** The cost function is expressed as a linear expression of the various elements of the Control-aware Data Flow structure. It is such that, in an optimal plan, *cheaper* patterns are evaluated first before feeding their bindings to more *expensive* patterns.

• **Solving the Resulting ILP Problem:** Using an optimized ILP solver (e.g., IBM ILOG CPLEX), we solve the ILP problem of minimizing the cost function under the generated set of constraints.

In the remainder of this section, we provide more details of these key steps.

3.1 Control-Aware Data Flow Construction

Our approach to build a Control-aware Data Flow for any arbitrary complex graph pattern is inspired from electronic circuit design. A Control-aware Data Flow consists of set of hierarchically nested *components*, analog to the hierarchically nested patterns in the SPARQL query. A *component* c is responsible for the evaluation of an arbitrary complex graph or triple pattern p. p is the key of component c (denoted **key**(c)). Multiple components may be assigned to the same key p as they represent alternative ways of evaluating p (e.g., multiple access methods for the same triple pattern).

A component can be viewed from the outside (i.e., its external view) as a black box connected to a set of input pins (one for each variable it may be required to perform its function), and a set of output pins (one for each variable that may become available to other components as a result of its evaluation). Each pin can be in one of two states: activated or deactivated. An activated input pin indicates that its corresponding variable is indeed available to use inside the black box. An activated output pin indicates that its corresponding variable is available to other components after the evaluation of its black box. Likewise, the black box to which input and output pins are connected can be activated (i.e., enabled and performing its function) or deactivated (disabled). The top level external component of a query, an example of which is presented in Figure 2, has no input pins.

While the external view defines how the component interacts with the rest of the data flow, the internal view determines how its sub-components interact with each other and captures all possible connections among all the sub-components. The component/sub-component relation in the data flow corresponds to a pattern/sub-pattern relation in the SPARQL query. In the internal view of a component the sub-components are represented through their external view since the component level planning is not concerned with the internal details of the sub-components.

Figure 2 shows all the internal sub-components of the top level component responsible for the evaluation of the main pattern of our query. There are three sub-components associated with each triple pattern. These sub-components have the same key (i.e, the triple pattern to evaluate) and each component corresponds to a different access method for the triple (e.g., the sub-components (t_1, aco), $(t_1, scan)$, and (t_1, acs) represent different access methods to evaluate the triple t_1). The UNION (resp. OPTIONAL) component in the figure is associated with the pattern UNION(t_2, t_3) (resp. OPTIONAL(t_7)). In addition to sub-components responsible for the evaluation of each sub-pattern of the main pattern, Figure 2 shows two special components: one *product* component and five *join* components. A *product* component represents a product operation performed on the two components connected to its input pins, whereas a *join* component corresponds to a regular join performed on the two components connected to its input pins. Figure 2 shows all potential connections to the x input pin of the

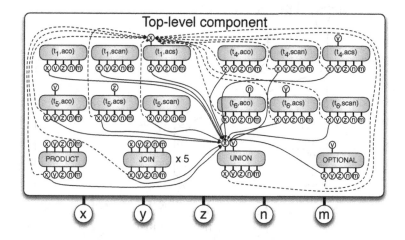

Fig. 2. Top Level Component for q

component (t_1, acs) (dotted lines) and all potential connections to the x input pin of the union component (continuous lines). In the internal view of a component associated with an BGP pattern all sub-components are connected as follows: for each variable x, all output pins corresponding to x are connected to all input pins corresponding to x. As illustrated in Figure 2 , in general, the main operations inside the internal component are the following:

1. Wiring external inputs to inputs of internal components and wiring of outputs of internal components to inputs of other internal components based on the semantics of the graph pattern type.
2. Conservatively enumerating all potentially valid wirings since the optimal data flow inside a component is not known a-priori. However, in a given plan (solution to our problem), only a subset of wires will be activated.

Figure 3 shows the internal view of the union component, whose external view is present in the internal view of the top level component in Figure 2. The internal component keeps track of its two inputs x and y and their origin. The two sub-components (BGP (t_2) and BGP (t_3)), called *child* components, are responsible for the evaluation of the two sub-patterns of the union pattern. Finally, Figure 3 shows all connections between the external view of all sub-components of the union component (continuous lines for x flows and dotted lines for y flows). As opposed to the internal view for an BGP component, the connections in the internal view of a *union* component are limited to connections from input pins to *child* components. There are no connections between sub-components of a UNION component as they would violate the semantics of a UNION pattern: for example, the $?y$ variable shared between t_2 and t_3 cannot be produced by one and used by the other. In the output pins of a component c a variable is available or active after the execution of c if it is either an in-scope variable of the graph

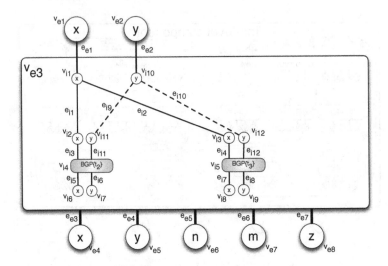

Fig. 3. Union Component in q

pattern associated with c or it is a variable provided by the parents of c (i.e, components whose output pins are connected to input pins of c)[3].

A component is formally defined as follows:

Definition 3. *Let V be an infinite set of variables. Let T be a finite set of types. A component C is a triple (EV, IV, t). $t \in T$ is the type of the component C. EV, called the external view, is defined as a pair (G^e, var) consisting of:*

- *a directed graph $G^e = (V^e = IP \cup \{bb\} \cup OP, E^e)$ whose set of vertices V^e is a partition of 3 disjoint sets (the singleton $\{bb\}$ with the back box bb, the set IP of input pins, and the set OP of output pins), and whose set of edges $E^e = \{(p, bb)|p \in IP\} \cup \{(bb, p)|p \in OP\}$, and*
- *a function var, called variable function, that maps an element of $IP \cup OP$ to a variable in V such that if p_1 and p_2 are in IP (or in OP), then $var(p_1) \neq var(p_2)$.*

IV, called the internal view of C, is defined by a pair (SC, G^i) consisting of:

- *A finite set of components $C_k \in SC$ of C.*
- *A graph $G^i = (V^i, E^i)$, called internal graph of C and representing all potentially valid data flows inside C.*

The function *blackbox* maps a component to its black box bb.

For the UNION component in Figure 3 the external view is represented as $G^e = (V^e = IP \cup \{bb\} \cup OP, E^e)$ where $IP = \{v_{e1}, v_{e2}\}$, $bb = v_{e3}$, $OP = \{v_{ek}|4 \leq k \leq 8\}$ and $E^e = \{e_{ek}|1 \leq k \leq 7\}$. The variable function var associated with

---

[3] The details are discussed in the technical report [1] when addressing the output pin constraints.

the external view is $var = \{v_{e1} \rightarrow x, v_{e2} \rightarrow y, v_{e4} \rightarrow x, v_{e5} \rightarrow y, v_{e6} \rightarrow n, v_{e7} \rightarrow m, v_{e8} \rightarrow z\}$. The internal view of the UNION component is represented by $G^i = (V^i, E^i)$ where $V^i = \{v_{ik}|1 \leq k \leq 12\}$, $E^i = \{e_{ik}|1 \leq k \leq 12\}$.

3.2   Constraint Generation

In this section, we introduce constraints to rule out invalid flows.

Decision Variables and Candidate Solutions. Assume a given set of components $\mathcal{C}$ responsible for the evaluation of a pattern GP. Constraints are expressed in terms of decision variables and they are defined for each component and sub-components $C \in \mathcal{C}$. The function α maps each vertex n (resp. edge (n_1, n_2)) in the graphs G^e and G^i associated with the external and internal views of C, to a unique boolean variable $\alpha(n)$ (resp. $\alpha((n_1, n_2))$), called decision variable associated with n (resp. (n_1, n_2)). The decision variable indicates whether n (resp. (n_1, n_2)) is activated. The range of α, denoted $range(\alpha)$, is the set of all the decision variables associated with all components and sub-components of $\mathcal{C}$. A candidate solution for GP is a function δ from the set of decision variables (i.e., $range(\alpha)$) to $\{0, 1\}$. It assigns an activation state (0 or 1) to each vertex and edge of components directly or indirectly contained in $\mathcal{C}$.

For the UNION component in Figure 3, we present below the decision variables for two possible candidate solutions δ_x^{UNION} and δ_y^{UNION} corresponding to the UNION nodes in the two syntactic reorderings in Figure 1(c). δ_x^{UNION}, is a the restriction to the UNION component of a candidate solution when the only activated input pin is x (i.e., the solution corresponding to the left tree in Figure 1(c)); whereas δ_y^{UNION} is the restriction to the UNION component of a candidate solution when y is the only actived input pin (i.e., the solution corresponding to the right tree in Figure 1(c)); .

$\delta_x^{UNION} = \{\alpha(v_{ik}) \rightarrow 1|1 \leq k \leq 9\} \cup \{\alpha(v_{ik}) \rightarrow 0|10 \leq k \leq 12\} \cup \{\alpha(e_{ik}) \rightarrow 1|1 \leq k \leq 8\} \cup \{\alpha(e_{ik}) \rightarrow 0|9 \leq k \leq 12\} \cup \{\alpha(v_{ek}) \rightarrow 1|k \in \{1, 3, 4, 5\}\} \cup \{\alpha(v_{ek}) \rightarrow 0|k \in \{2, 6, 7, 8\}\} \cup \{\alpha(e_{ek}) \rightarrow 1|k \in \{1, 3, 4\}\} \cup \{\alpha(e_{ek}) \rightarrow 0|k \in \{2, 5, 6, 7\}\}$

$\delta_y^{UNION} = \{\alpha(v_{ik}) \rightarrow 0|1 \leq k \leq 3\} \cup \{\alpha(v_{ik}) \rightarrow 1|4 \leq k \leq 12\} \cup \{\alpha(e_{il}) \rightarrow 0|1 \leq l \leq 4\} \cup \{\alpha(e_{il}) \rightarrow 1|5 \leq l \leq 12\} \cup \{\alpha(v_{ek}) \rightarrow 1|k \in \{2, 3, 4, 5\}\} \cup \{\alpha(v_{ek}) \rightarrow 0|k \in \{1, 6, 7, 8\}\} \cup \{\alpha(e_{ek}) \rightarrow 1|k \in \{2, 3, 4\}\} \cup \{\alpha(e_{ek}) \rightarrow 0|k \in \{1, 5, 6, 7\}\}$

Figure 1(d) shows the internal view of the top level component for an optimum solution δ_y for the top level component of the query (note that the restriction of δ_y to the UNION component is δ_y^{UNION} shown above) . This solution corresponds to the right hand plan shown in Figure 1(c). Please node that for clarity only the active components are shown in the optimum flow. The remaining components and their connections that appear in Figure 2 are not active in the optimum solution.

Constraint Definition and Classification. We introduce constraints to rule out invalid candidate solutions. A constraint is a logical expression written as a function of decision variables that expresses a relation that must hold for all valid candidate solutions. We express a constraint as a linear inequality of the form: $a_0 \times x_0 + \ldots + a_k \times x_k \geq b$ or $a_0 \times x_0 + \ldots + a_k \times x_k \leq b$, where k is a positive integer, and, for $0 \leq i \leq k$, a_i and b are real number constants and x_i are decision variables. Constraints fall in one of the following categories: generic component constraints, generic graph constraints, predecessor constraints, output pin constraints, and component-type specific constraints. Due to space limitation, we present only 7 of the 20 constraints. The set of all 20 constraints is available in the technical report [1].

Generic Component Constraints. These constraints, shown below, are applicable to the the external view of every component and they define valid ways in which the component is connected to the rest of the data flow. Constraints $(C1)$ to $(C4)$ trace the binding of query variables to values. In particular, when a variable x in the SPARQL query is required by a component, the planning mechanism should make sure that there is an active component that already extracted the matching values of x from the database. Failure to enforce this constraint results in faulty candidate solutions that do not produce correct query results. In the example in Figure 3, $(C1)$ prevents candidate solutions with $\alpha(v_{e4}) \to 1$ and $\alpha(v_{e3}) \to 0$ which makes variable x available without executing any component. In the same way $(C2)$ prevents candidate solutions with $\alpha(e_{e4}) \to 0$ and $\alpha(v_{e4}) \to 1$. For $(C3)$ the focus is the internal view of the UNION component. $\{\alpha(v_{i2}) \to 1, \alpha(e_{i1}) \to 0\}$ is an example of decision variable assignment that violates the constraint $(C3)$ and allows the use of x without any information about its provenance (x is produced ex nihilo). Finally, an example of violation for $(C4)$ is the following decision variable assignments: $\alpha(e_{i1}) \to 1$ and $\alpha(v_{i1}) \to 0$.

(C1) If a black box bb is not activated (i.e., $\alpha(bb) = 0$), then each of its input or output pin p is also deactivated (i.e., $\alpha(p) = 0$): $\alpha(p) \leq \alpha(bb)$

(C2) A pin p is connected to its black box bb iff. it is activated:
   (C2-a) For p an input pin: $\alpha((p, bb)) = \alpha(p)$
   (C2-b) For p an output pin: $\alpha((bb, p)) = \alpha(p)$

(C3) In the internal view of a component c, whose internal graph is $G = (V, E)$, if an input pin ip of a sub-component sc of c is activated, then ip must have at least one activated incoming edge: $\sum_{(op,ip) \in E} \alpha((op, ip)) \geq \alpha(ip)$

(C4) If an edge (n, m) is activated, then nodes n and m must also be activated: $\alpha(n) + \alpha(m) \geq 2 \times \alpha((n, m))$

(C5) Each key k (query fragment) must be executed exactly once: $\sum_{c \ s.t. \ key(c)=k} \alpha(blackbox(c)) = 1$. Components of types JOIN and PRODUCT do not have any key, so this constraint is not applied to them.

Generic Graph Constraints. A valid plan is an acyclic graph.

(C6) The internal graph $G = (V, E)$ of a component c must be acyclic. For each vertice $v \in V$, we map v to a new integer decision variable representing its position, denoted $pos(v)$ and such that $0 \leq pos(v) \leq |V| - 1$ (where $|V|$ denotes the cardinality of the set V). The position associated to each vertice introduces an implicit ordering that we use to informally express the acyclicity constraint as follows: if an edge (n, m) is activated, then $pos(n) + 1 \leq pos(m)$ (i.e. $pos(n) < pos(m)$). The formal ILP acyclicity constraint is expressed as follows for an edge $(n, m) \in E$:

$$pos(n) + 1 + (|V| \times (\alpha((n, m)) - 1)) \leq pos(m)$$

Note that if (n, m) is activated (i.e., $\alpha((n, m)) = 1$), the previous constraint becomes what we wanted (i.e., $pos(n) < pos(m)$); otherwise, it is always satisfied as $pos(n) + 1 - |V| \leq 0$ (by definition of $pos(n)$) and $pos(m) \geq 0$.

3.3   Cost Function Formulation

For each component c, we associate a new positive real number variable, denoted $cost(c)$, for its cost. The cost structure of a component c is defined as:

$$cost(c) =$$

$$\lambda_0 \times \alpha(blackbox(c)) + \sum_{sc \in subcomp(c)} \lambda_{sc} \times cost(sc) + \sum_{sc \in parent(c)} \lambda'_{sc} \times cost(sc)$$

where λ_0, λ_{sc}, and λ'_{sc} are positive real number constants, parent(c) is the set of parents of c (i.e, components with at least one output pin connected to at least one input pin of c), and $\alpha(blackbox(c))$ indicates whether the component c is activated. We include a fraction of parent cost (i.e., $\lambda'_{sc} \times cost(sc)$ for each parent sc) to reflect properties of a component inputs such as its cardinality. λ_0 is a fixed cost defined for each component type. For example, for a component c corresponding to the evaluation of a triple pattern t with the access method $acm \in \mathcal{M}$, λ_0 is the cost of evaluating t using access method acm. In our current implementation, λ_{sc} is set to 1 (all sub-components have the same weight).

3.4   Soundness and Completeness

Before presenting soundness and completeness results, we briefly introduce the notations. Let q be a query whose main graph pattern is GP. Let $\mathcal{C}$ be the set of components responsible for the evaluation of GP. The set Φ_q denotes the set of constraints generated for $\mathcal{C}$ and presented in section 3.2. The set of candidate solutions satisfying all constraints in Φ_q is denoted $\mathcal{ILPS}_q$. For $\delta \in \mathcal{ILPS}_q$, $cost(\delta)$ is defined as the $cost(\delta) = \sum_{c \in \mathcal{C}} \delta(cost(c))$

Finally in a plan $p \in \mathcal{U}_q$, for some operators (e.g. product, and union) the order of evaluation of their operands does not affect the total estimated cost. We say that two plans p_1 and p_2 are cost equivalent, denoted $p_1 \approx p_2$, iff. one can be transformed into the other by a sequence of applications of commutative operator com and associative operator $asso$ on commutative and associative algebraic operators (e.g., PRODUCT and UNION). For a commutative and associative algebraic operator op, $com(op(e_1, e_2)) = op(e_2, e_1)$ and $asso(op(e_1, op(e_2, e_3))) = op(op(e_1, e_2), e_3)$. The soundness and completeness of the ILP approach is established by the following Theorem (proof in the technical report [1] with concrete implementations of σ and β) :

Theorem 1. *Let q be a query whose main graph pattern is GP. There exists a pair of functions (β, σ) such that β is a function from $\mathcal{U}_q$ to $\mathcal{ILPS}_q$ and σ is a function from $\mathcal{ILPS}_q$ to $\mathcal{U}_q$ such that: 1) if $p \in \mathcal{U}_q$, then $\sigma(\beta(p)) \approx p$ and 2) if $\delta \in \mathcal{ILPS}_q$, then $cost(\beta(\sigma(\delta))) = cost(\delta)$*

4   Evaluation

To examine the effectiveness of the ILP based planner as a testing framework, we conducted experiments with 5 different benchmarks: LUBM [6], SP2Bench [18], DBpedia [14], UOBM [12] and a private benchmark PRBench used in earlier work [2]. Our focus in this paper was to determine whether the ILP based planner could be used to test the greedy approach outlined in the DB2RDF system [2], given that this is one relatively mature implementation of a greedy approach to SPARQL planning. Our evaluation of the ILP testing framework had two goals: (1) to demonstrate that the framework can actually compute optimal plans for a wide variety of queries, (2) to determine if the framework could be used to uncover optimization opportunities in a mature planner.

We describe each of the benchmarks briefly:

• **LUBM:** The LUBM benchmark queries consist of the 12 queries, and an ontology that was modified to OWL QL expressivity.

• **UOBM:** The UOBM benchmark queries consist of the 14 queries, and an ontology that was modified to OWL QL expressivity. OWL QL query expansion is applied to LUBM and UOBM queries.

• **SP2Bench:** SP2Bench is an extract of DBLP data with corresponding SPARQL queries. We used this benchmark as is, with no modifications.

• **DBpedia:** The DBpedia SPARQL benchmark is a set of query templates derived from actual query logs against the public DBpedia SPARQL endpoint [14]. We used these templates with the DBpedia 3.7 dataset, and obtained 20 queries that had non-empty result sets.

• **PRBench:** The private benchmark reflects data from a tool integration scenario where specific information about the same software artifacts are generated by different tools, and RDF data provides an integrated view on these artifacts across tools. The benchmark has fairly complex queries, and is therefore a good test for the ILP planner because the search space is large.

Experiments were conducted on a machine with two 2.3 GHz processors, each with 6 cores, and 64 GB of RAM (the max. java heap size allocated was 5G) running 64-bit Linux. The ILP solver used was IBM ILOG CPLEX Version 12.5.

For each query, we first computed an optimal plan op using our ILP approach. Then, we translated the plan returned by the greedy planner into a solution s to the ILP planning problem. Finally, we compared the cost of s with the cost of the optimal solution *op* to check the optimality of the greedy plan. Table 1 shows a summary of the ILP results on the 5 benchmarks. As shown in the table, the average time for ILP query planning of all 91 queries indicates that the ILP approach is very practical for testing SPARQL planners. As shown in the table, the average time for queries ranged from 0.45-27.2 s on all the benchmarks, which is impressive if one considers the size of the search space for many of the 91 queries that were tested. A significant number of queries (66/91) ran under 1 second, as shown in the table. The slowest planning problem took 11 minutes for the largest query in PRBench (a 1005 line query).

Further, as shown in the Figure, it helped identify 7 cases where the greedy plans were suboptimal. For at least one of those cases, the ILP planner's optimal plan helped us identify obvious opportunities for improving the greedy algorithm. Specifically, the greedy planner in DB2RDF missed opportunities for exploiting star queries (i.e., queries on the same entity for which DB2RDF [2] data layout is designed to provide a very efficient evaluation without any join) due to heuristics that did not adequately reflect the performance gain from stars. Once the optimal plans highlighted the problem, we were able to tune the greedy planner with better heuristics and verify that these new heuristics made that plan optimal with negligible added overhead. In the other 6 cases, it was quite clear that any greedy approach would arrive at a suboptimal plan.

Table 1. Query optimality results

Dataset	#Queries	Avg Time(s)	StDev (s)	Min - Max (s)	# Queries < 1 s	#Suboptimal
LUBM	12	0.4	0.5	0.02 - 1.3	9	1
UOBM	14	2	6	0.02 - 24	11	2
SP2Bench	17	1.7	3.2	0.01 - 10	8	2
DBpedia	20	1	2.2	0.02 - 6.5	15	2
PRBench	28	27.2	127	0.06 - 673	23	None

5   Related Work and Conclusion

Query optimization has been researched in the context relational databases. Greedy and heuristic based algorithms have been introduced to avoid the exponential cost of producing all possible plans, at the expense of query performance. At the base of relational optimizers is the System-R optimization framework [19] which was subsequently extended to Starburst [7] and Volcano/Cascade [4,5]. Numerous techniques have been proposed to improve query performance in relational databases and are referenced in [10,11,3]. There is currently a large body of research in SPARQL query optimization based on graph-specific greedy and heuristic algorithms. Typical approaches perform bottom-up SPARQL query optimization, i.e., individual triples [20] or conjunctive SPARQL patterns [8,16] are independently optimized, and then the optimizer orders and merges these individual plans into one global plan. These approaches rely on statistics [13] to assign costs to query plans. Optimization here is restricted to conjunctive patterns.

The work in [21] contrasts with these approaches and adopts a heuristic based optimization mechanism where the statistics are ignored. [2] focusses on important characteristics of SPARQL queries, often with deep, nested sub-queries whose inter-relationships are lost when optimizations are limited by the scope of single triple or individual conjunctive patterns (as in prior work). The query optimizer captures the inherent inter-relationships due to the sharing of common variables or constants of different query components. These inter-relationships often span the boundaries of simple conjuncts and are often across the different levels of nesting of a query, i.e., they are not visible to existing bottom-up optimizers.

In this paper, we investigated the optimal SPARQL query planning problem, in the context of offline query optimization and planner testing. We formally introduced the universe of alternative query plans for an input query q . To efficiently solve the planning problem, we devised an approach that casts our planning problem as an ILP problem. We experimented with well-known datasets and large numbers of queries and illustrated that our approach consistently finds optimal plans in reasonable amount of time (in a few minutes in the worst case).

Acknowledgment. Research was sponsored by the U.S. Army Research Laboratory and the U.K. Ministry of Defence and was accomplished under Agreement Number W911NF-06-3-0001. The views and conclusions contained in this document are those of the author(s) and should not be interpreted as representing the official policies, either expressed or implied, of the U.S. Army Research Laboratory, the U.S. Government, the U.K. Ministry of Defence or the U.K. Government. The U.S. and U.K. Governments are authorised to reproduce and distribute reprints for Government purposes notwithstanding any copyright notation hereon.

References

1. Bornea, M., Dolby, J., Fokoue, A., Kementsietsidis, A., Srinivas, K.: An offline optimal sparql query planning approach,
   http://researcher.watson.ibm.com/researcher/files/us-achille/techreport.pdf
2. Bornea, M., Dolby, J., Kementsietsidis, A., Srinivas, K., Dantressangle, P., Udrea, O., Bishwaranjan, B.: Building an efficient rdf store over a relational database. In: Proceedings of the ACM SIGMOD Conference, SIGMOD 2013 (2013)
3. Chaudhuri, S.: An overview of query optimization in relational systems. In: SIGACT-SIGMOD-SIGART, pp. 34–43 (1998)
4. Graefe, G.: The cascades framework for query optimization. Data Engineering Bulletin 18 (1995)
5. Graefe, G., DeWitt, D.J.: The exodus optimizer generator. SIGMOD Record, 160–172 (1987)
6. Guo, Y., Pan, Z., Heflin, J.: LUBM: A benchmark for OWL knowledge base systems. Journal of Web Semantics 3(2-3), 158–182 (2005)
7. Haas, L.M., Freytag, J.C., Lohman, G.M., Pirahesh, H.: Extensible query processing in starburst. SIGMOD Record, 377–388 (1989)

8. Hartig, O., Heese, R.: The SPARQL query graph model for query optimization. In: Franconi, E., Kifer, M., May, W. (eds.) ESWC 2007. LNCS, vol. 4519, pp. 564–578. Springer, Heidelberg (2007)

9. Ibaraki, T., Kameda, T.: On the optimal nesting order for computing n-relational joins. ACM Trans. Database Syst. 9(3), 482–502 (1984), http://doi.acm.org/10.1145/1270.1498

10. Ioannidis, Y.E.: Query optimization. In: The Computer Science and Engineering Handbook, pp. 1038–1057 (1997)

11. Jarke, M., Koch, J.: Query optimization in database systems. ACM Comput. Surv., 111–152 (1984)

12. Ma, L., Yang, Y., Qiu, Z., Xie, G., Pan, Y., Liu, S.: Towards a complete owl ontology benchmark. In: Sure, Y., Domingue, J. (eds.) ESWC 2006. LNCS, vol. 4011, pp. 125–139. Springer, Heidelberg (2006), http://dx.doi.org/10.1007/11762256_12

13. Maduko, A., Anyanwu, K., Sheth, A., Schliekelman, P.: Estimating the cardinality of rdf graph patterns. In: WWW, pp. 1233–1234 (2007)

14. Morsey, M., Lehmann, J., Auer, S., Ngonga Ngomo, A.-C.: DBpedia SPARQL Benchmark – Performance Assessment with Real Queries on Real Data. In: Aroyo, L., Welty, C., Alani, H., Taylor, J., Bernstein, A., Kagal, L., Noy, N., Blomqvist, E. (eds.) ISWC 2011, Part I. LNCS, vol. 7031, pp. 454–469. Springer, Heidelberg (2011)

15. Muralikrishna, M., DeWitt, D.J.: Equi-depth histograms for estimating selectivity factors for multi-dimensional queries. In: SIGMOD, pp. 28–36 (1988)

16. Neumann, T., Weikum, G.: The RDF-3X engine for scalable management of RDF data. The VLDB Journal 19(1), 91–113 (2010)

17. Poosala, V., Ioannidis, Y.E., Haas, P.J., Shekita, E.J.: Improved histograms for selectivity estimation of range predicates. In: SIGMOD, pp. 294–305 (1996)

18. Schmidt, M., Hornung, T., Lausen, G., Pinkel, C.: SP2Bench: A SPARQL Performance Benchmark. CoRR abs/0806.4627 (2008)

19. Selinger, P.G., Astrahan, M.M., Chamberlin, D.D., Lorie, R.A., Price, T.G.: Access path selection in a relational database management system. In: SIGMOD (1979)

20. Stocker, M., Seaborne, A., Bernstein, A., Kiefer, C., Reynolds, D.: SPARQL basic graph pattern optimization using selectivity estimation. In: WWW (2008)

21. Tsialiamanis, P., Sidirourgos, L., Fundulaki, I., Christophides, V., Boncz, P.: Heuristics-based query optimisation for SPARQL. In: EDBT, pp. 324–335 (2012)

Agents, Models and Semantic Integration in Support of Personal eHealth Knowledge Spaces

Haridimos Kondylakis[1], Dimitris Plexousakis[1], Vedran Hrgovcic[2], Robert Woitsch[2], Marc Premm[3], and Michael Schuele[3]

[1] Institute of Computer Science, FORTH, N. Plastira 100, Heraklion, Greece
{kondylak,dp}@ics.forth.gr
[2] BOC Asset Management GmbH, Operngasse 20B, Vienna, 1040, Austria
{vedran.hrgovcic,robert.woitsch}@boc-eu.com
[3] Universität Hohenheim, Information Systems 2 (530 D), 70593 Stuttgart, Germany
{marc.premm,michael.schuele}@uni-hohenheim.de

Abstract. The advancements in healthcare practice have brought to the fore the need for flexible access to health-related information and created an ever-growing demand for the design, development and management of personalized knowledge spaces. In this paper, we present a web-based platform that generates a Personal eHealth Knowledge Space as an aggregation of several knowledge sources relevant for the provision of individualized personal services. To this end, novel technologies are exploited, such as *knowledge on demand* to lower the information overload for the end-users, *agent-based communication and reasoning* to support cooperation and decision making, and *semantic integration* to provide uniform access to heterogeneous information. All three different technologies are combined to create a novel web-based platform allowing seamless user interaction through a portal that supports personalized, granular and secure access to relevant information.

Keywords: Web-based Information Systems, Knowledge on Demand, Semantic Integration, Agent-based Systems, Meta Modelling.

1 Introduction

Medicine is undergoing a revolution that is transforming the nature of healthcare from reactive to preventive. The changes are catalyzed by a new systems approach which focuses on integrated diagnosis, treatment and prevention of disease in individuals. This will replace the current practice of medicine over the coming years with a personalized predictive treatment. While the goal is clear, the path is fraught with challenges [1].

The fragmentation of knowledge about personal risk factors hinders the assessment of disease risks. In order to decide on preventive or therapeutic actions, physicians, patients and caregivers are required to obtain all relevant user-specific knowledge. Relevant knowledge sources include health records, patient records, databases on medical literature or environmental information, wearable or portable devices for

B. Benatallah et al. (Eds.): WISE 2014, Part I, LNCS 8786, pp. 496–511, 2014.
© Springer International Publishing Switzerland 2014

health monitoring, and common ubiquitous internet services (including user generated information). Currently petabytes of data are produced every day, which generate the following problems: (1) inability to access correct data – e.g. by not being able to select the right knowledge source within an available timeframe; (2) inability to process the relevant data with available resources; and (3) inability to easily downsize and save the selected data so that they become portable–creating thus personal knowledge spaces [2].

The eHealthMonitor (http://www.ehealthmonitor.eu/)project aspires to create a platform that generates a Personal eHealth Knowledge Space (PeKS) as an aggregation of all relevant sources (e.g., EHRs, PHRs, medical sensors, weather services etc.) relevant for the provision of individualized personal eHealth Services. The technological challenges for such a goal are the following:

1. The first challenge in providing such services is *the identification of the available knowledge and data sources* concerning a *particular domain* or part thereof in a specific *timeframe* for each specific *user* type, i.e. the knowledge space [3].
2. Having identified all data sources we need to *establish then a seamless, transparent mechanism to access and query them.* For example, a weather service can be provided through a web service with an XML output, a medical device can provide a custom API whereas EHR/PHR systems might store data in a relational database.
3. However, even when we have uniform access to the information under consideration *different privacy requirements and conflicting interests might exist among the participating entities.* For example, an insurance company might have an interest in accessing the entire medical history of patients, whereas, for the patients themselves, it would be better if the insurance company had access only to specific data that their contract requires to share. Moreover, personal medical information that is known to a patient and a physician should not be shared with any other participant that is not explicitly authorized to access this information.

Although good solutions exist for addressing each one of those challenges independently, to the best of our knowledge there is no solution addressing all those challenges simultaneously. To achieve that, in this paper, we present a platform thatuses models, agents and semantic integration technologies to address these challenges. We make advancements both in combining these technologies as in each of these fields independently. More specifically our contributions are following:

- A novel platform configured on demand for each type of users, created thought models, using the Knowledge on Demand approach [4].This approach defines a set of procedures and measures to provide knowledge required at the right time, in a satisfying granularity and content level. This is implemented by *using hybrid modelling languages and their extensions* towards dynamic knowledge spaces [2].
- Data integration methodologies are extended to support uniform access to semi-structured and fully-structured databases, as well as to the results of web service execution, using a *novel ontology* as global schemaextending TMO/TMKB [5].
- Since our platform will be used as a continuous companion of a citizen, it should allow knowledge to evolve. Novel mechanisms allow *data integration under evolving ontologies* that are used as global schema [6].

- Then, *software agents* are used, one agent for each participating actor *implementing her/his interests*, mapping their mutual relationships to a multi-agent organisation [7].
- *Adaptive coordination methods* among several agents, are used then to analyse the provision of personal guidance services for cooperative decision making in eHealth service networks [8].

The combination of all these technologies allows the personalization of information presented to the user in a comprehensible, uniform and secure manner. The remaining of this paper is structured as follows: In Section 2 we present the building blocks of our architecture and we describe the front-end of the platform. Then in Sections 3, 4 and 5 we present the Knowledge on Demand, the Semantic and the Multi-Agent components respectively. Section 6 reviews other related projects with similar goals and Section 7 summarizes and presents an outlook for further work.

2   Conceptual and Technical Architecture

In order to derive the requirements for the architecture and to define appropriate models and methods to address the challenges imposed by the need to support the PeKS design, execution and development, a series of interviews and modelling workshops was conducted with the end-users coming from the Dementia, Cardiovascular and Chronic Obstructive Pulmonary Disease. The end-users of the platform were doctors, patients and informal or formal caregivers. As an outcome (details in [9]), a set of 78 interaction processes was identified, carried out by 26 different stakeholders, whereas the total number of significant knowledge resources was 137. In the next step these insights were used as a starting base to define a total of 82 functional and 50 non-functional requirements that influenced the selection of appropriate tools and methods to address the relevant challenges. The conceptual architecture, depicted in Fig. 1, follows a three-tier approach:

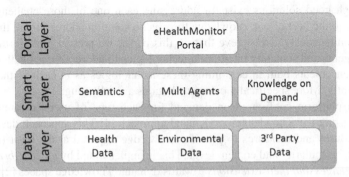

Fig. 1. High-Level Architecture

- (1) *The Presentation Layer* with the Portal Component.
- (2) *The Smart Layer* with Semantic, Knowledge on Demand and Multi-Agent components.

- (3) *The Data Layer* where large datasets consisting of different sources are accessed such as EHR/PHR, sensor data streams, environmental readings, clinical guidelines, databases.

The *portal* has the role of a single access point for all stakeholders. It is used for "pull" as well as "push" communication channel to support the cooperative decision making between, for example, the informal caregiver and medical professionals. Given the diversity of the stakeholder skills and technological environment they may have access to, when processing the knowledge space, the information visualization and delivery functionality has to be highly adaptable.

As a running example consider a patient named David suffering from Chronic Obstructive Pulmonary Disease who is traveling from Erlangen, Germany to Heraklion, Greece. Based on his current location the information (changes in the diagnosis of the patient, weather changes etc.) is delivered within a portal-app installed on the patient smartphone. In case no internet access is available, the portal may provide the information through more simple means of transport such as SMS, in a push message fashion. On the other hand, as he is moving from city to city, different online web services might be used to identify weather conditions in each location in order to send to him the proper alerts. Besides this, consider also Jim, a dementia patient, whose skills may change dynamically as his disease is progressing. In this case, based on specific thresholds for his state, the medium of visualization may have to change. This means that the single portal – based on the current conditions – may need to switch from a standard UI toward single yes/no button interfaces and still be able to provide the full functionality. In addition, consider an insurance company requiring access to some of the patient information that is dictated by his contract. However, the patient wants to share exactly that information and no more information.

Based on the dynamic environment in which the platform is used and the required modularity of the underlying services, the implementation of the components followed the service-oriented architecture (SOA) approach with a strong focus on the model-driven basis of the Knowledge on Demand component. In the following sections, we will focus on the Smart Layer[1] highlighting the novel aspects of our work.

3   Knowledge on Demand for Personal Knowledge Spaces

The Knowledge on Demand approach – defined as a set of procedures and measures to provide knowledge required by the stakeholders at the right time and in the right processable context – is used in this component to achieve the following goals of the project: (1) to enable a model-driven management approach for patient empowerment; (2) to support the cooperative decision making by calculating correlation between stakeholder preference dimensions (e.g. technology, time, location). In order to apply the Knowledge on Demand approach the following challenges have to be

---

[1] Functionalities concerning the Data layer integration are described in the Semantic Layer whereas description of data sources has been omitted due to space constraints. All descriptions can be retrieved from D2.2 [11] and D4.3 [12]

considered: (1) the modelling has to be simplified and adapted both to the stake-holders (skill aspects) as well as to the technology used (mobile aspects) and (2) an effective and efficient way of reducing the semantic distance had to be applied to enable application of identified languages and tools as well as to guarantee the exten-sibility of the approach. From the Knowledge on Demand point of view we distin-guish between the following stakeholder groups involved in Personal Knowledge Spaces: (1) medical professionals and patients/non-professional caregivers, (2) know-ledge engineers and (3) service providers.

Based on the requirements identified for those stakeholders we have derived a three-layer architecture, namely the domain the knowledge and the technical layers. In each of these layers several modelling toolkits (MTK) are used by a group of seven specific MTKs to allow them specify the rules for the automatic configuration of our platform. In the following, we describea subset:

- on the domain layer – the medical professionals and other domain experts – use the so-called eSpace MTK to semi-formalise use cases from everyday situations (e.g. CT examination) by applying simplified notation and implicit modelling guidelines – thus reducing the burden to apply graphical modelling to unskilled users.
- on the knowledge layer – knowledge engineers use Multi-Agent MTK (for the machine interpretation) of PROMOTE MTK (for the human interpretation) to de-scribe the knowledge space using an applicable modelling language.
- on the technical layer – service providers use e.g. BPMN MTK to define abstract and executable workflows [10].

These tools reflect a minimum working set required to configure an instance of an eHealthMonitor based knowledge platform. The aforementioned working set was based on a small subset of possible scenarios within the eHealth domain that primarily aims to address those requirements. However, it also covers generic aspects that are highly relevant to other parts of the domain.

In the following, we will briefly describe a scenario applying a subset of available modelling toolkits for the three stakeholder groups to configure an instance of an eHealthMonitor-like knowledge platform supporting the COPD patient David. In the first step of such a scenario a healthcare professional selects an eSpace pattern. This is a model type of the eSpace MTK (see [2] for details) that closely resembles a specific health scenario concerned with change of the location. The complexity within the pattern itself has been abstracted for the end user and transferred to mechanisms that are triggered based on the users positioning of specific elements. Such elements in-clude: (1) the Actors – such as Patient, MD, Nurse; (2) the eSpaces – represent an aggregation of specific environment such as hospital, home environment; and (3) the Interaction Processes – such as making an appointment in the current city. In the next step, the healthcare professional,may further update or extend the pattern – to person-alize the use case based on the actual situation (e.g. patient may not be tech-savvy therefore communication is either through appointment or by phone) and may also request support from the eHealthMonitor platform services (determining the appropri-ate means of communication, e.g. if abroad SMS is preferred). Similarly, a healthcare professional can configure the platform for Jimto switch from normal UI to sing-leyes/no buttons if hisdementia stage goes beyond a specific value.

Theseconfigurations are achieved in the modelling environment by simply marking activities, within the interaction process, that should be supported by the Smart Layer services. After performing this task the following are automatically generated: (1) the Knowledge Workbenches that act as a portal toward the Personal Knowledge Space for all involved actors – based on the access rights there may be one or many such workbenches – all configured using the Knowledge Workbench modelling toolkit; and (2) the specific agent acting on behalf of the user, deployed on the Multi-Agent Network.In our example the generated Agent for David, will be configured automatically to act proactively by requesting weather updates based on his location and will inform David by SMS if certain conditions are met. All configurations generated using the aforementioned modelling toolkits can be updated or extended and this will result in the reconfiguration of the agents and the portal.

After (1) and (2) have been generated, human stakeholders can interact with e.g. Mobile Apps required to process the relevant parts of the knowledge space – e.g. to decide to do a surgery or to apply "wait and see" strategy, and agents can reason on the accessible knowledge sources. The available knowledge sources have been described using PROMOTE by knowledge engineers and provide data for the decision making process. Apps that a stakeholder may use range from functionality point of view from atomic apps toward orchestrated sets of apps to provide the required functionality. The formalization and orchestration of the apps is done by a Service Provider stakeholder using the App Orchestration MTK.

This scenario as outlined previously is strongly influenced by the identified challenges and more specifically by the necessity to provide a simple modelling approach. This is crucial as the stakeholders in the eHealthMonitor scenarios (based on the interviews with the actual users) tend to reject the tool if found to be too complex. Therefore the complexity that is required to e.g. to generate the knowledge workbenches or perform the multi-agent network formation has to be hidden in the platform and be configurable by a user using a tablet or a mobile phone – through touch and swipe actions. Furthermore, the requirement to provide a model-based environment to configure such complex scenarios requires minimizing the semantic distance between the stakeholders to a minimum; and additionally to be open for extending the model basis by introducing new meta-models – that may become relevant in future or due to different scenarios. This has been achieved by applying the Hybrid Modelling [13]. The main advantage of Hybrid Modelling is that the meta-models are amalgamated instead rather than tightly integrated and that many different scenarios as depicted in [2] can be applied allowing interactions with toolkits that enable the aforementioned simplified modelling. This, on the other hand, allows inclusion of the patients to the cooperative decision making by enabling processing with information found in previously unused knowledge sources.

4   Semantic Integration Component

Besides selecting the appropriate knowledge source, information should also be presented to the end-user, or fed to an agent, in a uniform, transparent way. This is the task of the semantic integration component. Semantic integration is the process of interrelating information from diverse sources [14] and has to resolve several hetero-geneity problems since the sources might use different modelling choices to model data and different structures to store them. During the last 15 years, numerous systems have been developed to tackle semantic integration problems. The main approaches [9] are either centralized – e.g. data warehouses, where data is stored locally – or federated – where data is left at the sources and accessed on demand. The selection of either approach depends on the type of solution to be deployed. Data ware-housesprovide better efficiency and allow tighter control on data but they face issues with outdated data. Federated approaches always access updated data, allow partial and non-managed data connections, but suffer from efficiency issues.

In eHealthMonitor, since most of the data sources to be integrated are not directly managed by the consortium, we adopt a federated semantic integration solution also known as *query translation approach*. More specifically the system that we devel-oped and extended is the *exelixis* system [15] that accepts SPARQL queries and re-writes the query to the subsequent sources to be answered as shown in Fig. 2.

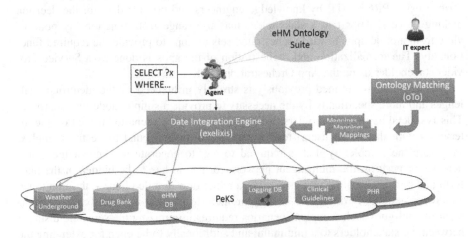

Fig. 2. Integration of several data sources using *exelixis*

In order to formulate the SPARQL queries, an ontology named eHealthMonitor Ontology Suite (eHMOSuite)[16] is developed extending TMO/TMKB.TMO/TMKB [5] integrates a wide range of about 23 high-quality biomedical ontologies modelling chemical, genomic, proteomic, disease and treatment data and is extended by eHMO-Suite to model also environmental conditions, patient symptoms and family health history.

When using ontologies to integrate structured and semi-structured data, one is re-quired to produce mappingsto the sources, i.e. to link similar concepts or relationships from the ontology/ies to the sources by way of an equivalence relation. This is the *mapping definition process*[6] and the output of this task is the *mapping*, i.e., a collec-tion of mapping rules. In practice, this process is done manually with the help of graphical user interfaces and it is a *time-consuming*, *labour-intensive* and *error-prone* activity [6]. In our case, to help the mapping process the OtO ontology matching system is used [17] which employs a set of state-of-the-art schema and instance matching algorithms to identify similar concepts between ontology and source schemata. Then, based on these matchings the appropriate Global as View mappings are created which are being used by the query rewriting algorithm of our engine to rewrite queries to the sources. Query rewriting algorithm uses the Ontop system (http://ontop.inf.unibz.it/) and is based on description logic.

In our running example, consider that the agent acting on behalf of David identifies that his location has changed and that he is now located in Heraklion, Greece. Automatically the agent issues proactively the query shown ontop of Fig. 3 to identify the weather conditions there and to issue the necessary alerts if required. To be able to answer this query the semantic integration engine should already be linked with the *weather underground*(http://www.wunderground.com/) web service by means of mappings. Some example mappings for linking the property "*hasTemperature*" in the ontology and the "*temp_c*" field of the weather underground result is shown in Fig. 3. Due to lack of space we omit the description of the mechanisms that allows the semantic integration engine to view XML pages as relational data and to query them on the on-the-fly.

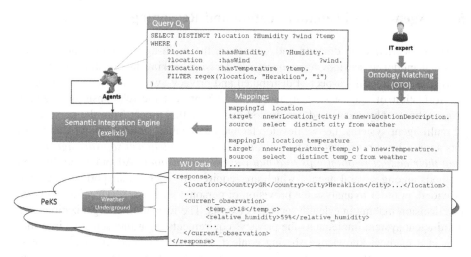

Fig. 3. Example scenario for retrieving weather in heraklion

When designing a system that is going to be a life-time companion to the citizens, updates on the knowledge are expected to occur. Due to the rapid development of research, ontologies are frequently changed to depict the new knowledge that is acquired. The problem that occurs is the following: when ontologies change, the mappings may become invalid and should somehow be updated or adapted. In our running example, let's assume that an ontology expert works on the eHM Ontology Suite and generates the next version v2. In the new version the property "hasTemperature" is renamed to "hasCelciusTemperature" since now the ontology also models the temperature in Fahrenheit. As a result the previously generated mapping "location temperature" is invalidated and should be updated. However, this specific property might be involved in other mappings as well that might need to be updated similarly. Moreover, besides simple renaming other more complex changes might also occur such as split, merge, pull up, delete etc. [6].

To face these difficult problems *exelixis* incorporates algorithms that handle ontology evolution efficiently and effectively [6]. The system, instead of invalidating the mappings and then forcing the experts to correct them, automatically identifies the changes in the ontology and uses those changes to rewrite input queries to the correct ontology version. For example the query Q_1, when issued by an agent, is automatically rewritten into Q_0(shown in Fig. 3) exploiting the identified renaming in the ontology [6].

Q_1: SELECT ?location ?humidity ?wind ?tempWHERE {

?location	:hasHumidity	?humidity.
?location	:hasWind	?wind.
?location	:hasCelciusTemperature	?temp.

FILTER regex(?location, "Heraklion", "i") }

5   Agent-Based Communication and Reasoning

Next to knowledge on demand and semantics, the multi-agent component provides personal guidance services for cooperative decision making in eHealth service networks. We address the aspect of distributed and privacy-aware knowledge sharing by the formation of agent-based organizations to represent the relationships of actors (e.g., patients, physicians and caregivers) and study this aspect from the perspective of multi-agent systems; i.e. we develop technology enabled collaboration solutions.In this approach *we represent the participating actors by intelligent software agents and map their mutual relationships to a multi-agent organization*. Adaptive coordination methods among several agents, which are required for decision support services, are provided, in order to analyse the provision of personal guidance services for cooperative decision making in eHealth service networks. The formation of organizations in a multi-agent system implements the privacy requirements regarding sensitive information in a medical knowledge sharing context. Personal medical information that is known to a patient and a physician should not be shared with any other participant that is not explicitly authorized to access it.Interaction protocols for distributed, adaptive knowledge sharing are developed. They construct the interaction between the agents in the system and develop technology enabled collaboration solutions from portal registrations to the execution phase. The privacy-related coordination mechanism is provided as a distributed concept itself. The control of sharing sensitive information is local and the communicated informed is limited to the minimal set of data.

Following organization theory, the concepts of roles and structures constitute the two major measures to be applied to multi-agent systems [13]. Ferber et al. [18] present concepts and implementation guidelines describing these roles, structures, and interactions within a multi-agent system. An organization-centered multi-agent system (OCMAS) is a multi-agent system whose foundation lies in particular organizational concepts such as groups, communities, and roles [18]. An OCMAS extends the more familiar notion of an agent-centered multi-agent system (ACMAS), which solely deals with the individual agents' mental states. As the classical ACMAS approach reveals a number of drawbacks on the organizational level, Ferber et al. [18] focus on the design of multi-agent systems using organizational concepts only and so propose a framework where agents with different cognitive abilities interact with each other. Based on their organizational framework they provide a generic organizational model called AGR (Agent/Group/Role) and demonstrate the usefulness of these concepts by means of an illustrating example. We adapt this meta-model to suit our purpose with the following basic elements shown inFig. 4:

- **Agent:** An agent represents/acts on behalf of /supports a real world actor. An agent is able to hold multiple roles and may be member of several organizations.
- **Role:** A role describes a position in an organization.
- **Organization:** An organization consists of a set of agents that share one service. Agent communication is restricted to agents that belong to the same organization with the exception of communication that is part of the formation of an organization. An agent may belong to several organizations at the same time.
- **Organizational structure:** The organizational structure maps services to organizations and defines which roles form which organizational structure. An agent has to play a role to be part of an organization. Some roles may be played by several agents, e.g. in a patient-centric care giving scenario several care givers attend one patient.
- **Function:** A role has a function, which describes the task of the role. In this special context, we introduced by following the "knowledge on demand" vision, the functions (1) Knowledge Provider and (2) Knowledge Consumer, which are required as a general pattern for the knowledge sharing activities. Knowledge Consumers are limited to supply the process with knowledge, while Knowledge Providers are able to analyse knowledge in order to gain new knowledge for the purpose of provision.

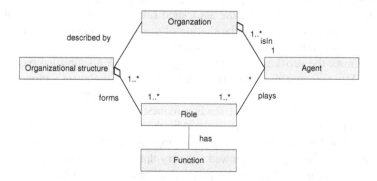

Fig. 4. Meta-Model of multi-agent organizations

In order to base the formation of an organization on the described model and to guarantee the restricted communication, an organizational reasoning mechanism is developed. This accounts for the following two steps: (1) the presented meta-model is converted to an ontology, added to the eHM Ontology Suite, which represents the organizational concepts, and (2) the agents are enabled to execute reasoning on an ontology.

The integration of semantic web technologies into multi-agent systems, proposed in previous work [19]is based on the concept of deliberative Beliefs-Desires-Intentions (BDI) agents [20]. The BDI paradigm specifies explicitly (1) the current facts about the world (beliefs), (2) the motivational attitude that form concrete goals (desires), and (3) the appropriate actions to achieve the given goals (intentions).

The standard BDI concept has been integrated with explicit semantics: the agent's beliefs, stored in the agent's semantic belief base. Conceptual definitions of the organizations are given using the aforementioned sub-ontology.

Every agent holds their own copy of this predefined ontology. In the case of the formation of a new organization or change of an existing organization the new organizational information is inserted as instances into the knowledge base/ontology, which is automatically enriched using description logicreasoning. Agents can then retrieve the results of reasoning via the beliefs. An embedded OWL reasoning engine provides the core semantic functions inside the agents. TheOWL engine is connected to the BDI agent via the beliefbase. The BDI related practical reasoning mechanism [20] is enhanced by the mechanism of reasoning on structured semantic data. The implemented BDI agent plans add or modify facts in the semantic database, which may activate other rules through reasoning inside the semantic core.

When reasoning changes the semantic representation of the agent beliefs inside the semantic core, it can trigger a goal via the BDI beliefs. When goals activate selected plans, the semantic core is updated. Two different prototype implementations for the embedded lightweight semantic core have been analyzed:Using (1) the Jena SW toolkit, (2) Jena built-in rules and Pellet OWL reasoner (applicable as SWRL rule engine as well). Both solutions provided a small and effective extension to our BDI agents [21]. We use the Jena SW toolkit for reasoning on organization ontologies

The organizational reasoning itself gives information about the organizations in which the agent is member and the associated role and function. As a result, active and passive communication and the sharing of sensitive information can be restricted to those agents that belong to the same organizations as the agent itself.

Fig. 5shows the verification of access rights based on a simplified request procedure as UML-sequence diagram. An external trigger or the agent's proactive behaviour determines aninteraction demand regarding another agent that is member in one of his organizations. So the agent sends a request message to the other agent. After receiving this message, it has to be verified by the receiving agent. The request procedure is extended by an organizational reasoning check. The organizational reasoning checks if the sending agent is allowed to query that request. This verification mechanism prevents misuse and guarantees privacy of sensitive information.

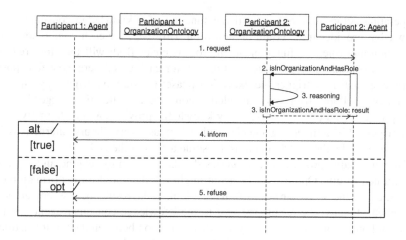

Fig. 5. UML-sequence diagram of access rights verification

The verification maybe also executed optionally on the sender side before sending a request. In the case of a positive verified request the agent reacts according to the content; otherwise the agent refuses the request.

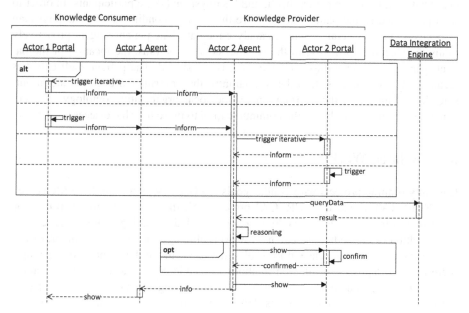

Fig. 6. UML-sequence diagram of execution phase

The formation phase of an organization consists of several steps, which are based on previous work [7]. After forming different organizations, the actors will actually work together in the execution phase.Fig. 6shows the process as a UML-sequence diagram and the corresponding web services that are involved in the process. The two cooperating actors may either fulfil the function of a knowledge consumer or of a knowledge provider in a certain organization. However, for each combination of organization and role, the function of this role is clear. Both will therefore reason on their organization ontology to get its function for each case.There exist four possible alternatives that may trigger the execution phase: (1) Iterative trigger by knowledge consumer, (2) Portal trigger by knowledge consumer, (3) Iterative trigger by knowledge provider, and (4) Portal trigger by knowledge provider.While alternatives 1 and 3 are triggered by a timer, e.g. check weather forecast every 2hours, alternatives 2 and 4 are triggered by the portal, and thus, by some action of the end-user.

The knowledge provider agent queries the data integration engine and reasons on the returned results. Optionally, the agent might ask the portal for confirmation of the results, e.g. if the physician wants to have a look at the results before they are sent to the patient. Finally, the results are sent to the knowledge consumer agent and to the portal of both actors.In this context, the differentiation between data, information and knowledge is important, because even though the knowledge consumer may deliver information to the knowledge provider, the information can only be processed reasonably with the knowledge of the knowledge provider.

Referring to the running example, the information David is sharing is limited to the minimal set of data. In the service network, which is represented by the multi-agent organization, David, the patient itself, and his physician(s) are participants. In order to send proper alerts, the patient has to share the weather conditions with the physician respectively the physicians agents to analyse the data. The patient's agent proactively limits the shared information to the weather conditions, thus, the location of the patient, which is considered as sensitive data, is not shared. The patient itself executes location specific requests. In order to guarantee that information does not attain outside the organization, the verification by the local organization ontology of every communication activity limits the communication to those how have access.

6   Related Work

Currently running projects with similar goals to eHealthMonitor are the p-Medicine[2], the INTEGRATE[3] and the EURECA[4] projects. All these projects aspire to develop innovative infrastructures to enable data and knowledge sharing to foster large-scale collaboration in biomedical research. However, although the goals are similar, different strategies have been selected to resolve the problems in the domain. For example there are no adaptive coordination methods to enable collaborative decision making and no knowledge on demand mechanisms for generating personal knowledge spaces. On the other hand all these projects define a methodology for semantic integration. However in the aforementioned projects data are extracted from their sources, transformed and then loaded to a central data-warehouse. To store the data in the data warehouse, ontologies are used as schemata and the information needs to be updated

---

[2] http://www.p-medicine.eu/
[3] http://www.fp7-integrate.eu/
[4] http://eurecaproject.eu/

every time something is changed in the data sources. In our case however, we avoid the constant update of a central repository since we do not store data to a central repository but we directly link to them.

Other projects in the area employee as well semantic integration methodologies, such as MyHealthAvatar[5], MD-Paedigree[6] and CARE[7] employ ontologies for this purpose. There exists also a set of initiatives and personal health record systems concerned with the patient empowered knowledge spaces, such asMicrosoft HealthVault, Web MD Health Manager, NoMoreClipboard, PatientsLikeMe, Patient Ally , Patient Fusion MyOscar, myMediConnect PHR, eclinicalWorks Patient Portal, MedHelp PHR, MyALERT, CareZone PHR, Indivo-X, Epic MyChart, 911 Medical ID, zweena PHR, MedicAlert, Tolven, HealtheTracks, LifeLedger, OpenMRS, KIS PHR, MedicKey PHR, Dossia and Minerva Health Manager (see [22] for a comparison) but as opposed to the work presented here they do not proactively support the cooperative decision making process (e.g. by dynamically enlarging the decision set), but act as a static knowledge spaces – providing storage and role based access to the knowledge resources. Finally, the project Commodity12[8] provides a system that builds a multi-layered multi-parametric infrastructure for continuous monitoring of diabetes type 1 and 2. They integrated a multi-agent system (MAS) where patients are represented as agents that monitor data to produce alerts based on anomalous temporal patterns in the patients' physiological data using event calculus. To handle a larger number of patients the agents are saved in a database from which the agent's state is resumed. However, an agent is just active if the patient is logged in.

To the best of our knowledge eHealthMonitor is the first project to combine knowledge on demand methodology, semantic integration technology and agents.

7   Conclusions

This paper presents a novel platform providing individualized personal eHealth services. Novel technologies such as knowledge on demand, semantic integration and multi-agent systems are cooperating to enable distributed, adaptive knowledge sharing coordination methods.

A key next step is to produce the reasoning mechanisms that will enable decision support, to efficiently and correctly locate and interpret knowledge and to provide this knowledge to the interested parties. We also plan to evaluate the whole platform in a real-world context in three scenarios (Dementia, Cardiovascular and Chronic Obstructive Pulmonary Disease) in three European countries (Germany, Poland and Greece). Preliminary evaluation performed provided initial evidences about the added value and the usability of our approach which will be extensively reported in a follow-up paper.Without a doubt this is an important area for healthcare delivery that will only become more critical as healthcare delivery continues to grapple with current challenges.

---

[5] http://www.myhealthavatar.eu/
[6] http://www.md-paedigree.eu/partners/
[7] http://www.carre-project.eu/
[8] http://www.commodity12.eu/

Acknowledgement. This work has been supported by the eHealthMonitor Project (www.ehealthmonitor.eu) and has been partly funded by the European Commission under the contract: FP7-287509.

References

1. Kondylakis, H., Koumakis, L., Tsiknakis, M., et al.: Smart recommendation services in support of patient empowerment and personalized medicine. In: Multimedia Services in Intelligent Environments – Recommendation Services, vol 25, pp. 39–61. Springer, Heidelberg (2013)
2. Hrgovcic, V., Woitsch, R.: Evolution of eHealth Knowledge Spaces: Meta Model Based Approach for Semantic Lifting. In: eChallenges e-2013 Conference (2013)
3. Karagiannis, D., Woitsch, R.: Knowledge Engineering in Business Process Management. In: Handbook on Business Process Management 2, International Handbooks on Information Systems, pp. 463–485. Springer (2010)
4. Hrgovcic, V., Utz, W., Woitsch, R.: Knowledge Engineering in Future Internet. In: Karagiannis, D., Jin, Z. (eds.) KSEM 2009. LNCS, vol. 5914, pp. 100–109. Springer, Heidelberg (2009)
5. Luciano, J.S., Andersson, B.: The Translational Medicine Ontology and Knowledge Base: Driving personalized medicine by bridging the gap between bench and bedside. Journal of Biomedical Semantics 2(suppl. 2), S1 (2011)
6. Kondylakis, H., Plexousakis, D.: Ontology evolution without tears. In: Web Semantics: Science, Services and Agents on the World Wide Web, vol. 19, pp. 42–58 (2013)
7. Schuele, M., Widmer, T., Premm, M., Criegee-Rieck, M., Wickramasinghe, N.: Improving Knowledge Provision for Shared Decision Making in Patient-Physician Relationships - A Multiagent Organizational Approach. In: HICSS Conference, pp. 646–655 (2014)
8. Widmer, T., Premm, M., Karaenke, P.: Sourcing Strategies for Energy-Efficient Virtual Organizations in Cloud Computing. In: Business Informatics (CBI) Conference, pp. 159–166 (2013)
9. eHealthMonitor Consortium: D2.3: Analysis of Models and Methods (2012), http://goo.gl/9mPNT4
10. Hrgovcic, V., Woitsch, R., Karagiannis, D.: Hybrid Service Modeling in Enterprise Computing. In: 3rd IEEE Conference on Commerce and Enterprise Computing (CEC), pp. 207–211 (2011)
11. eHealthMonitor Consortium: D2.2: Technical Analysis and Requirement Specification (2012), http://goo.gl/dwbC7X
12. eHealthMonitor Consortium: D4.3 Semantic Integration Methodology (2013), http://goo.gl/OcjwtC
13. Kirn, S., Herzog, O., Lockemann, P., Spaniol, O. (eds.): Multiagent Engineering, Theory and Applications in Enterprises, Springer (2006)
14. Wikipedia Article: http://en.wikipedia.org/wiki/Semantic_integration (last visited November 14, 2013)
15. Kondylakis, H., Dimitris, P.: Exelixis: Evolving Ontology-Based Data Integration System. In: SIGMOD/PODS, pp. 1283–1286 (2011)
16. eHealthMonitor Consortium, D4.1: Analysis of Models and Methods (2012), http://goo.gl/aDqx4f
17. Daskalaki, E., Plexousakis, D.: OtO Matching System: A Multi-strategy Approach to Instance Matching. In: CAiSE, pp. 286–300 (2012)

18. Ferber, J., Gutknecht, O., Michel, F.: From Agents to Organizations: an Organizational View of Multi-Agent Systems. In: Giorgini, P., Müller, J.P., Odell, J.J. (eds.) AOSE 2003. LNCS, vol. 2935, pp. 214–230. Springer, Heidelberg (2004)
19. Karaenke, P., Schuele, M., Micsik, A., Kipp, A.: Inter-organizational Interoperability through Integration of Multiagent, Web Service, and Semantic Web Technologies. In: Agent-based Technologies and Applications for Enterprise Interoperability, vol. 98, pp. 55–75 (2013)
20. Bratman, M.E., Israel, J., Pollack, M.E.: Plans and resource-bounded practical reasoning. In: Computional Intelligence, vol. 4, pp. 349–355 (1988)
21. Karaenke, P., Leukel, J., Sugumaran, V.: Ontology-based QoS aggregation for composite web services. In: WI 2013 Conference, pp. 1343–1357 (2013)
22. Genitsaridi, I., Kondylakis, H., Koumakis, L., Marias, K., Tsiknakis, M.: Towards Intelligent Personal Health Record Systems: Review, Criteria and Extensions. In: Procedia Computer Science, vol. 21, pp. 327–334 (2013)

Probabilistic Associations as a Proxy
for Semantic Relatedness

Shahida Jabeen, Xiaoying Gao, and Peter Andreae

School of Engineering and Computer Science
Victoria University of Wellington, P.O. Box 600, Wellington, New Zealand
{shahidarao,xgao,peter.andreae}@ecs.vuw.ac.nz

Abstract. Semantic relatedness computation is a well known problem
with multidisciplinary applications. Existing approaches to computing
semantic relatedness ignore the asymmetric associations of words. In the
absence of an explicit topical context, these asymmetric associations can
be effectively used to represent the relation of words in directional contexts. Motivated by the idea of word associations, this paper presents
a new approach to computing semantic relatedness using asymmetric
association based probabilities of words extracted from the directional
contexts of words based on the Wikipedia corpus. The performance evaluation of the proposed approach on a variety of publicly available benchmark datasets shows that the asymmetric association based measures
outperformed not only the baseline symmetric measures but also most
of the state-of-art approaches.

Keywords: Semantic relatedness, Free word associations, Asymmetric
word associations, Wikipedia corpus, Directional context.

1 Introduction

Semantic relatedness computation is the task of measuring the degree of association of two words. It is a core component of many computational linguistics
applications such as word sense disambiguation, keyword extraction, document
classification, Microblog clustering, machine translation and spelling corrections.
In conventional approaches to semantic relatedness computation, the relatedness
of two terms is assumed to be symmetric, such that $rel(a, b) = rel(b, a)$, where
a and b are two given terms. This assumption essentially disregards the contexts
in which these terms could be associated, hence considers their relation in a
context-independent way. In real scenarios, such as in web document retrieval
and microblog clustering, where the context is quite diverse and changes rapidly,
computing realistic scores for a word pair according to the context in which it
appears, is critical. Hence, context consideration is important for computing
semantic relatedness.

There are three types of contexts: *local context, topical context* and *directional
context*. A set of words that occur in the proximity of a given word is called its
local context. This *local context* is taken into account at document level. There

B. Benatallah et al. (Eds.): WISE 2014, Part I, LNCS 8786, pp. 512–522, 2014.
© Springer International Publishing Switzerland 2014

is a stream of research on semantic relatedness which used this local context for relating two terms [1,2,3,4]. The second type of context, *topical context*, refers to an explicit and wider context in which the strength of association of two given words is computed. For instance, in the context of *study*, the association strength of a word pair book and library is higher than the word pair book and bookshop, whereas in the context of *publishing and marketing*, the association strength of a word pair book and bookshop is higher than the word pair book and library. The third type of context is directional context, which assumes that the topical context is determined by the first word of a given word pair in the absence of an explicit topical context.

The task of measuring word relatedness takes a word pair without any explicit topical context as the input and produces a relatedness score as the output. Thus, following free word association task, we use the individual words of a given word pair as stimulus for determining the context, in the absence of a given topical context. This idea of directional context is derived from the free word association task, where given a *cue* or *stimulus* word, the goal is to find out the most strongly associated *response* words. In this task, the cue word provides the context in which the association of a cue-response pair is computed. Following [5], this paper presents a novel approach to computing semantic relatedness guided by the directional contexts.

2   Related Work

Statistical approaches to computing the semantic relatedness are based on three main assumptions: *topicality, proximity* and *parallelism*.

According to the *topicality* assumption words found in similar context tend to be semantically similar and two words that are similar are likely to be used in similar contexts. Examples of topicality-based approaches include LSA [2], Random Indexing [1] and second order co-occurrence vector-based approaches such as [3,4].

The *Proximity* assumption, like the *topicality* assumption, relies on the word co-occurrences within a proximity window but the way it measures the co-occurrences is different from the *topicality* assumption. It states that two semantically related words that are semantically associated tend to co-occur near each other rather than having similar context words. Examples of approaches based on *proximity* assumption include PMI (Pointwise Mutual information) [6], PMI-IR [7] and LC-IR [8]. Proximity assumption applies to the semantic relatedness as well as semantic similarity.

The final assumption is based on grammatical *parallelism*, in which similar words tend to have similar grammatical frames. This assumption considers the frequency with which words are linked to other words by grammatical relations. Such approaches are based on the grammatical functions such as (subject-verb) and (verb-object) as well as selectional properties of verbs. Approaches based on this assumption include [9,10].

This paper is also based on proximity assumption but the specifics differ from the conventional proximity assumption-based approaches in using asymmetric associations of words occurring near each other. The idea of guiding relatedness computation based on asymmetric word associations is originally introduced in [5]. The aim of this paper is to extend the idea of computing asymmetric associations based relatedness by considering the context of each word at corpus level.

3   Asymmetry-Based Probabilistic Relatedness Measure

Asymmetry-based probabilistic associations refer to the co-occurrence-based probabilities of word associations computed in the context of each word. These probabilistic associations for computing relatedness are extracted from unstructured text corpora. Hence, the proposed approach uses the Wikipedia corpus to extract the co-occurrence-based associative probabilities of words. These probabilities are then used to compute two types of relatedness measures: asymmetric association based relatedness measures and symmetric relatedness measures. To use Wikipedia as an unstructured text corpus, an inverted index of Wikipedia articles is constructed.

Asymmetry-based Probabilistic Relatedness Measure. APRM measure is based on computing and combining the directional association strengths of a term pair to get their relatedness score. When the association strength of a given term pair (T_i, T_j) is determined in the context of first term T_i, it is referred to as *forward association strength* and is denoted by $Association(T_i \to T_j)$. Similarly, when the association strength of the same term pair is computed in the context of the second term T_j, it is called *backward association strength*, denoted by $Association(T_j \to T_i)$. APRM combines these asymmetric association strengths into a symmetric relatedness score.

Given a term pair (T_i, T_j), the context of each term is extracted from the inverted index. Context of a term refers to a set of those Wikipedia articles in which that term appears. The directional association strength of the pair of input terms is computed as the probability of co-occurrence of both terms within a proximity window of size 2w+1 in the context of each term.

$$Association(T_i \to T_j) = \frac{p\,(T_i\ near\ T_j)}{p\,(T_i)}$$

where $p(T_i\ near\ T_j)$ is the fraction of Wikipedia articles having given two terms with the proximity window and $p(T_i)$ is the fraction of Wikipedia articles containing the term T_i. The asymmetric forward and backward association strengths of a given term pair (T_i, T_j) are linearly combined to generate the final symmetric relatedness scores as follows:

$$APRM(T_i, T_j) = (1 - \lambda) \times Association(T_i \to T_j) + \lambda \times Association(T_j \to T_i)$$

where λ is the coefficient of association and is set to 0.5 to give equal importance to both directional association strengths.

APRM is a proximity assumption based measure to capture and use asymmetric word associations for estimating the strength of their relatedness. However, there are certain cases where methods based on proximity assumption fail to cope effectively. For instance, if an input term pair consists of synonymous words such as (construct,build) then it is less likely that these synonyms co-occur in close proximity very often in the corpus. In such cases, synonymous word pairs get lower scores than collocational words pairs. APRM, being a co-occurrence based measure also suffers from the same limitation. To cope with this, the second component of the semantic association computation approach exploits the knowledge source aspect of Wikipedia. Two structural features of Wikipedia are used as explicit indicators of synonymy: *redirects* and *senses*.

For a given term pair, both terms are matched with corresponding Wikipedia articles. For each matched article, its redirects are collected. Redirects are various surface forms of an article title that are used to refer to the same article and signify the synonyms of that article. For a given term pair, if a match is found in the redirect sets of both terms then the term pair is considered as a synonymous pair and is assigned maximum score. If either term does not match with a corresponding Wikipedia article then all senses of its Wikipedia label are retrieved and searched for a match, as in the case of redirects. If there is a synonymy relation between given terms, either match results in producing maximum score for a term pair.

3.1 Symmetry-Based Relatedness Measures

In this research, four baseline symmetric relatedness measures based on co-occurrence probabilities are implemented and compared with APRM.

The first symmetric measure is *Adapted Normalized Google Distance* (ANGD). Cilibrasi *et al.* [11] used the tendency of two terms to co-occur in web pages by proposing NGD as a measure of semantic distance of words and phrases, given as,

$$NGD = \frac{max(logf(x), logf(y)) - logf(x,y)}{logM - min(logf(x), logf(y))}$$

where $f(x)$ denoted the number of web pages containing word x and $f(x,y)$ represents the number of web pages containing both words x and y. Gracia and Mena [12] noted that NGD is a generalized measure that could be used with any web search engine. They transformed the NGD formula so that the distance scores it computes are bounded in a new range of [0-1] rather than the original range [0-∞]. Following their transformation, ANGD is given as,

$$ANGD(T_i, T_j) = e^{-2NGD(T_i, T_j)}$$

The second symmetric measure is *Dice coefficient*, which is a well known measure used in information retrieval. Given the two terms x and y, the Dice coefficient is computed as,

$$Dice(x, y) = \frac{2 \times f(x,y)}{f(x) + f(y)}$$

where $f(x)$ denotes the number of Wikipedia articles containing word x and $f(x,y)$ denotes the number of articles containing both words x and y within a proximity window.

The *Simpson coefficient*, often called *Overlap coefficient* is another symmetric relatedness measure that computes the overlap between two sets. The formula for the Simpson coefficient is given as,

$$Simpson(x,y) = \frac{log(f(x,y))}{min(log(f(x)), log(f(y)))}$$

The document overlap set of two words is normalized by the size of the smaller set to avoid any bias introduced by the larger set size.

The last symmetric relatedness measure is *Adapted-PMI* (APMI), which is inspired by the PMI measure [6] and is computed as follows:

$$APMI(x,y) = \sqrt{\frac{log(f(x,y))}{log(f(x)) * log(f(y)))}}$$

Both asymmetric and symmetric relatedness measures are implemented using the same experimental settings. The underlying corpus for all measures is Wikipedia and synonymy matching is used by all measures to augment their scores.

4   Experimental Results and Discussion

The direct evaluation of the APRM measure is based on computing and comparing the ranks of automatically computed scores with that of manual judgments using Spearman's rank order correlation coefficient. Details of the benchmark datasets used for evaluation of semantic similarity and relatedness approaches are summarized in Table 1.

The experiment used three versions of APRM and baseline measures on eight datasets and nine different window sizes (ranging from 1 to 20). Results of the

Table 1. Benchmark datasets used in the experiments

Dataset	Year	Language	No. of Pairs	Part of Speech	No. of Subjects	Range	Annotator Agreement
R&G	1965	English	65	N	51	[0-4]	0.85
M&C	1991	English	30	N	38	[0-4]	.90
WS-353	2002	English	353	N,V,A	13-16	[0-10]	0.87
YP-130	2005	English	130	V	6	[0-4]	.76 and .79
WS-Similarity	2009	English	203	N,V,A	2	[0-10]	.80
WS-Relatedness	2009	English	252	N,V,A	2	[0-10]	.80
MTURK-287	2011	English	287	N,V,A	23	[0-5]	-
MTURK-771	2012	English	771	N,V,A	20	[0-5]	0.89

experiment were evaluated by measuring the correlation of automatically computed scores with the human judgments using Spearman's correlation coefficient. Finally, the results of APRM with the optimal version and window size are compared with other state-of-art measures.

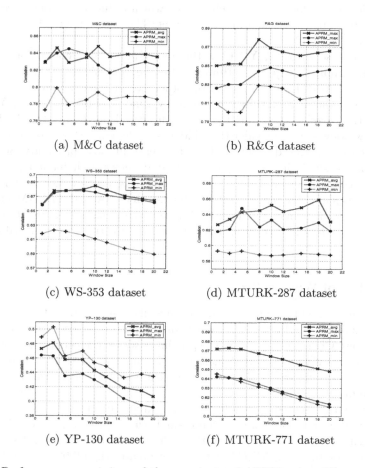

(a) M&C dataset

(b) R&G dataset

(c) WS-353 dataset

(d) MTURK-287 dataset

(e) YP-130 dataset

(f) MTURK-771 dataset

Fig. 1. Performance comparison of three variants of APRM using different window sizes on all datasets

4.1 Combining Directional Association Strengths

To analyze the impact of combining directional association strengths on the overall performance, three combining methods were explored. APRM_max combines the directional association strengths by selecting the maximum of the two association strengths for each term pair. APRM_avg linearly combines the two association strengths using a coefficient of association λ. APRM_min takes the minimum of the two association strengths into account.

A comparison of these three combinations on various datasets is shown in Figure 1. Overall, APRM_avg performed better than the other two methods. Hence, we opt to use it in later comparisons with other state-of-art and baselines measures. There was no optimal window size on which each version always performed the best. The highest correlation was achieved by each version of APRM using different window sizes on each dataset which is due to the difference in the nature of term pairs included in each dataset. This factor is investigated in the next section in detail.

On all datasets (except YP-130), the lowest performing combining method was APRM_min, as expected. However, on the verb similarity dataset (YP-130 dataset), APRM_min surpassed the other two methods on all window sizes and achieved the highest value on proximity window with $w = 3$. On YP-130 dataset, the correlation is found to have an overall decreasing trend with increasing window size. The wider the window size the lower the correlation value. Consequently, on this dataset, looking for verbs in close proximity produced more realistic similarity scores that correlated better with human judgments than the scores of wider window sizes. The results also revealed that humans assigned low similarity scores to the verb pairs that is why the APRM_min correlated well with human judgments. However, the maximum correlation value achieved on verb similarity dataset is still the lowest correlation achieved on any dataset.

4.2   Effect of Window Size

Both symmetric and asymmetric approaches discussed in the previous section used a proximity window of size $2w + 1$ for computing the co-occurrence based probabilities. Changing the size of the proximity window affects the performance of the proximity window based measures. To understand the behavior of the APRM measure on various window sizes, nine different windows sizes were tested as shown in the Figure 2.

APRM is represented by a solid line in each graph. Figure 2 shows that the difference in correlation values is quite small which proves that APRM is not very sensitive to the window size: the difference of maximum and minimum correlation produced by APRM_avg was not greater than 0.026 on all datasets except YP-130 datasets, where this difference increased to 0.066. Overall, using a proximity window with $w = 10$ produced consistently good results. The Spearman's correlation of APRM_avg using $w = 10$ on M&C and R&G datasets is 0.85 and 0.87 respectively. However, on verb similarity dataset, it was 0.44 which means that APRM_avg is not a measure of choice for computing verb similarity. On MTURK-287, which is a complicated relatedness-based dataset, APRM_avg performed quite well with a Spearman's correlation of 0.65. The following section compares APRM using proximity window with $w = 10$ with state-of-art approaches on various datasets.

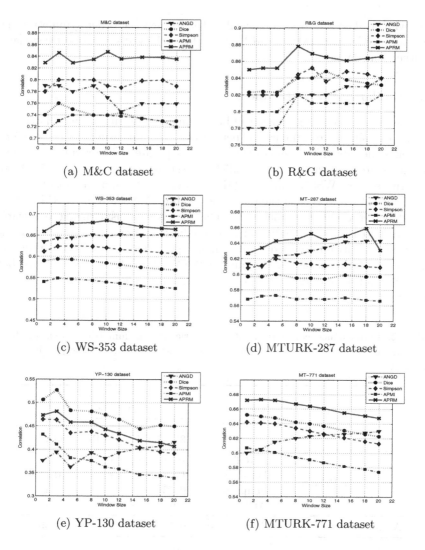

Fig. 2. Effect of changing window size on the performance of various relatedness measures on all datasets.

4.3 Symmetry vs. Asymmetry

Figure 2 also compares the performance of APRM_avg with that of four symmetric measures: Adapted Normalized Google distance (ANGD), the Dice coefficient, the Simpson coefficient and Adapted PMI (APMI). It is clear from the figure that APRM_avg surpassed all the symmetric measures on all datasets except YP-130 dataset. The other asymmetric measure APRM_max also outperformed the symmetric measures on three dataset and performed comaprable to the best symmetric measure on two datasets. Among symmetric measures, there is no clear winner. The Simpson coefficient outperformed the other symmetric measures on two datasets (M&C and R&G), while achieved second best correlation on other four datasets.

4.4 Similarity vs. Relatedness

Since the WS-353 dataset consists of both semantically similar and related word pairs, it is desirable to compare the performance of all measures on the similarity and relatedness-based subsets of this dataset separately. The results on similarity and relatedness subsets of WS-353 dataset are shown in Table 2. On the similarity subset, WS353-Sim, APRM_avg preformed comparable to the Simpson coefficient. However, on the relatedness subset, WS353-Rel, APRM_avg performed better than other measures. This shows that asymmetric association based APRM is a better indicator of relatedness and can be a good choice for detecting semantically related words but still performs as well as the other measures on similarity datasets. Overall, on WS353-All dataset consisting of both similarity and relatedness based term pairs, APRM_avg outperformed all other measures by a good margin.

Table 2. Performance comparison of symmetric and asymmetric association-based measures on two subsets of WS-353: WS353-Sim and WS353-Rel using proximity window w=10

Aspect	Measure	Datasets		
		WS353-Sim	WS353-Rel	WS353-All
Symmetric Relatedness	ANGD	0.72	0.64	0.62
	Dice	0.70	0.61	0.64
	Simpson	**0.73**	0.63	0.63
	APMI	0.65	0.58	0.59
Asymmetric Associations	APRM_avg	**0.73**	**0.65**	**0.69**

Note: Correlation is significant at $\alpha = 0.05$ level.

4.5 Comparison with State-of-Art Approaches

Table 3 compares the performance of APRM_avg with five state-of-art approaches on MTURK-771 dataset: DS (Distributional Similarity); ESA (Explicit Semantic Analysis); TSA (Temporal Semantic Analysis); LDA (Laten Dirichlet Allocation); and CLEAR (Constrained LEArning of Relatedness). The details of these approaches listed in Table 3 were originally reported in [13]. MTURK-771 is the largest and the most recent of all similarity and relatedness datasets, hence not reported much in the literature. Moreover, MTURK-771 is a relatedness-based dataset.

Table 3. Performance comparison of APRM_avg measure (using a proximity window with $w = 10$) with existing state-of-art approaches on MTURK-771 dataset

Measures	Source	Datasets
		MTURK-771
DS	Corpus	0.57
ESA	Wikipedia	0.60
TSA	Wikipedia & Corpus	0.60
LDA	Corpus	0.61
CLEAR	WordNet & Corpus	**0.72**
APRM_avg (w=10)	Wikipedia & Corpus	0.65[*]
APRM_avg (best)	Wikipedia & Corpus	0.66[*]

[*] Correlation is significant at $\alpha = 0.05$ level.

On the MTURK-771 dataset, APRM_avg surpassed all other approaches except CLEAR which produced highest correlation on this dataset. CLEAR combines the distributional statistics from multiple informal text corpora and learns the relatedness of words constrained by the known related word pairs extracted from WordNet synsets. However, two aspects of the MTURK-771 dataset construction are worth mentioning: first, this dataset is based on extracting related word pairs, which are linked with graph distances between 1-4 in the WordNet graph; and second, the word pairs used in this dataset are all nouns. Clearly, the first point indicates a bias in CLEAR as it learns its relatedness constrained by the related word pairs extracted from WordNet which is the same knowledge source used to create the MTURK-771 dataset. Although CLEAR achieved the highest performance, it depends on preprocessing of three huge text corpora to generate and use training data for learning and optimization of model parameters, which is the downside of the approach. On the other hand, APRM_avg, without requiring huge computational resources or training data, outperformed most of the state-of-art approaches in predicting relatedness scores. The good performance of APRM_avg on the MTURK-771 dataset conforms that it is quite good at detecting the relation of noun-noun term pairs.

5   Conclusion and Future Work

This chapter presented a new measure of semantic associations, APRM, using associative probabilities based on the idea of directional contexts. The evaluation of APRM on all the publicly available benchmark datasets of term similarity and relatedness demonstrated that its performance is superior to most other existing state-of-art approaches. The APRM measure is found to be particularly useful in predicting relatedness of noun term pairs. However, it did not perform well on verb relatedness dataset. The reason of this low performance is partly the nature of the underlying knowledge source, Wikipedia, which being an encyclopedia does not focus on verbs as much as on nouns and partly because humans estimates of verb similarity are generally lower than the noun similarity. The future work involves conducting a task-driven indirect evaluation for investigating the performance of APRM measure in an NLP application.

References

1. Sahlgren, M.: Vector-based semantic analysis: Representing word meanings based on random labels. In: Proceedings of ESSLI Workshop on Semantic Knowledge Acquistion and Categorization. Kluwer Academic Publishers (2001)
2. Landauer, T.K., Dumais, S.T.: A solution to plato's problem: The latent semantic analysis theory of acquisition, induction, and representation of knowledge. Psychological Review, 211–240 (1997)
3. Islam, A., Inkpen, D.: Second order co-occurrence pmi for determining the semantic similarity of words. In: Proceedings of the International Conference on Language Resources and Evaluation (LREC 2006), pp. 1033–1038 (2006)
4. Liu, H., Bao, H., Xu, D.: Concept vector for semantic similarity and relatedness based on wordnet structure. Journal of Systems and Softwares 85, 370–381 (2012)
5. Jabeen, S., Gao, X., Andreae, P.: Directional Context Helps: Guiding Semantic Relatedness Computation by Asymmetric Word Associations. In: Lin, X., Manolopoulos, Y., Srivastava, D., Huang, G. (eds.) WISE 2013, Part I. LNCS, vol. 8180, pp. 92–101. Springer, Heidelberg (2013)
6. Church, K.W., Hanks, P.: Word association norms, mutual information, and lexicography. Comput. Linguist. 16, 22–29 (1990)
7. Turney, P.D.: Mining the web for synonyms: Pmi-ir versus lsa on toefl. In: Proceedings of the 12th European Conference on Machine Learning, EMCL 2001, pp. 491–502 (2001)
8. Higgins, D.: Which statistics reflect semantics? rethinking synonymy and word similarity. In: Proceedings of International Conference on Linguistic Evidence, pp. 265–284 (2004)
9. Lin, D.: An information-theoretic definition of similarity. In: Proceedings of 15th International Conference on Machine Learning (ICML1998), pp. 296–304 (1998)
10. Bollegala, D., Matsuo, Y., Ishizuka, M.: A web search engine-based approach to measure semantic similarity between words. IEEE Trans. on Knowl. and Data Eng. 23(7), 977–990 (2011)
11. Cilibrasi, R.L., Vitanyi, P.M.B.: The google similarity distance. IEEE Trans. on Knowl. and Data Eng. 19(3), 370–383 (2007)
12. Gracia, J.L., Mena, E.: Web-based measure of semantic relatedness. In: Bailey, J., Maier, D., Schewe, K.-D., Thalheim, B., Wang, X.S. (eds.) WISE 2008. LNCS, vol. 5175, pp. 136–150. Springer, Heidelberg (2008)
13. Halawi, G., Dror, G., Gabrilovich, E., Koren, Y.: Large-scale learning of word relatedness with constraints. In: Proceedings of the 18th ACM SIGKDD International Conference on Knowledge Discovery and Data Mining, KDD 2012, pp. 1406–1414 (2012)

A Hybrid Model for Learning Semantic Relatedness Using Wikipedia-Based Features

Shahida Jabeen, Xiaoying Gao, and Peter Andreae

School of Engineering and Computer Science
Victoria University of Wellington, P.O. Box 600, Wellington, New Zealand
{shahidarao,xgao,peter.andreae}@ecs.vuw.ac.nz

Abstract. Semantic relatedness computation is the task of quantifying the degree of relatedness of two concepts. The performance of existing approaches to computing semantic relatedness is highly dependent on particular aspects of relatedness. For instance, taxonomy-based approaches aim at computing similarity, which is a special case of semantic relatedness. On the other hand, corpus-based approaches focus on the associative relations of words by taking their distributional features into account. Based on the assumption that different aspects of knowledge sources cover different kinds of semantic relations, this paper presents a hybrid model for computing semantic relatedness of words using new features extracted from various aspects of Wikipedia. The focus of this paper is on finding the optimal feature combination(s) that enhance the performance of the hybrid model. The empirical evaluation on benchmark datasets has shown that hybrid features perform better than single features by providing a complementary coverage of semantic relations, leading to improved correlation with human judgments.

Keywords: Semantic relatedness, Wikipedia hyperlinks, Asymmetric associations, Machine learning, Regression model, Cosine similarity.

1 Introduction

Semantic relatedness computation has multidisciplinary applications such as information retrieval [1], word sense disambiguation [2], topic identification [3] and document clustering [4]. Existing approaches to computing semantic relatedness focus on certain aspects of corpora or knowledge source(s), thus leading to better predictions of some semantic relations but performing poor on others. Such approaches have their own strengths as well as limitations. For instance, taxonomy or structure-based approaches are good at computing the similarity between two concepts that exist in the same hierarchy and are connected through a path. Such measures are good at finding semantically similar concepts but are limited by the coverage and structure of the underlying taxonomy. On the other hand, approaches based on the distributional properties of two concepts focus more on the usage-based aspects of closeness of two concepts. Such approaches are good at judging the relatedness of concepts which co-exist frequently but

B. Benatallah et al. (Eds.): WISE 2014, Part I, LNCS 8786, pp. 523–533, 2014.
© Springer International Publishing Switzerland 2014

might not be strongly related. For example, distributional approaches assign a relatedness score higher than synonyms to the word pair (bread,butter) due to a common phrase *bread and butter*, as indicated by [5]. Content-based approaches focus on concepts rather than mere words. Such approaches make use of informative-content or glosses of concepts and use various *Vector Space Models* (VSM) to compute relatedness scores. Such approaches are biased towards the size of informative-content and suffer from limited knowledge coverage of the underlying knowledge source(s).

A recent trend in semantic relatedness computation research is to exploit multiple aspects of knowledge source(s) for computing the semantic relatedness. Such approaches combined various aspects of knowledge source(s) and proved that a relatedness measure utilizing multiple aspects of knowledge source(s) perform better than those which rely on individual aspects [6,7,5]. This paper proposes a hybrid model based on three new features generated from hyperlink structure and informative-content of Wikipedia and proves that learning semantic relatedness using features based on multiple aspects shows significant improvement over that using features based on individual aspects.

2   Related Work

According to the nature of various aspects of underlying knowledge sources, the research in relatedness computation literature is broadly categorized into three main streams: *structure-based approaches, content-based approaches* and *hybrid approaches*. Each research direction has its own advantages and limitations.

The *structure-based approaches* rely on well structured knowledge sources, such as taxonomies, thesauri and lexical databases, which organize the human intellect into semantically rich and well defined semantic networks or graphs. Literature review indicates that there are two research directions of structure-based approaches.

First research direction includes the approaches which are based on predefined structures or taxonomies of existing knowledge sources such as the IS-A hierarchy of WordNet or hyperlink structure of Wikipedia. Since, these structures are predefined, their strengths lie in the huge coverage of knowledge in a structured way and explicit semantic encoding in their structure. On the other hand, their downside is the structural inflexibility of the predefined knowledge sources in explicitly controlling the semantics. Examples of such approaches include the work of [3,8].

Second research direction consists of approaches that automatically generate huge semantic networks or graphs from various existing knowledge sources. These approaches have an edge of controlling flexibility of the way the semantics are encoded in their structure but are computationally expensive to generate and are static in nature, hence suffer from scalability issues. Some worth mentioning approaches in this research direction are [9,10].

Content-based approaches make an effective use of informative-content of articles or documents derived from web, corpus or knowledge sources. Such approaches usually consider words as lexemes rather than concepts and rely on their distributional properties or statistical features. Usually, such approaches are some variants of *Vector Space Models* (VSM) and are known to perform better than structure-based approaches but are still biased towards more frequently occurring words. Example of web-based approaches of relatedness computation include [11,12]. Some worth mentioning relatedness computation approaches based on corpus statistics are [13,14,15].

In *hybrid approaches*, multiple aspects of knowledge source(s) are combined to extract semantics. One research stream consists of hybrid approaches that combine multiple aspects of the same knowledge source such as [6,10,16]. In the other research stream, hybrid approaches combine various aspects of multiple knowledge sources such as [17,7,5]. Following hybrid approaches, this paper proposes a new model for computing semantic relatedness of concepts by mining various features from Wikipedia.

3   A Hybrid Model of Semantic Relatedness

This paper presents a hybrid model of computing semantic relatedness that generates three new Wikipedia-based features and combines them using a regression function to predict the final relatedness score.

3.1   Feature Generation

When asked to relate two terms, human intrinsically relate their corresponding concepts in a wider context rather than merely relating two lexical forms of those concepts. Hence, the features presented in this paper are based on representing the terms by Wikipedia articles. In Wikipedia, each article is dedicated to one particular concept[1] and is connected to many other concepts through hyperlinks. Structural semantics mining is done by exploiting the hyperlink structure of Wikipedia concepts which intrinsically encode a variety of semantic and lexical relations among Wikipedia concepts.

Feature 1: *Structure-based Relatedness (SRel)* is the first feature that takes into account the proportion of links shared by two concepts and is computed as:

$$SRel(t_1, t_2) = \frac{1}{2} \times [link\_overlap(t_1, t_2) + OverlapStrength(t_1, t_2)] \quad (1)$$

here $link\_overlap(t_1, t_2)$ is computed as follows:

$$link\_overlap(t_1, t_2) = \begin{cases} \frac{|LS_1 \cap LS_2|}{\min(|LS1|, |LS2|)} & \text{if } |LS_1 \cap LS_2| \neq 0 \\ 0 & \text{otherwise} \end{cases} \quad (2)$$

where LS_1 and LS_2 are the Wikipedia link sets corresponding to the input terms and $|LS_1 \cap LS_2|$ is the set of all links shared by both link sets. $SRel(t_1, t_2)$ com-

---

[1] The terms *article* and *concept* are used interchangeably.

bines the $link\_overlap(t_1, t_2)$ with the $OverlapStrength(t_1, t_2)$, which is computed as follows:

$$OverlapStrength(t_1, t_2) = \frac{\sum_{a_s \in |C_s|} SR(a_s, article_1) \times SR(a_s, article_2)}{|L|} \quad (3)$$

where $|L|$ represents the total number of links of both concepts and $|C_s|$ is the total number of shared links. The association strength of a shared link is computed as the product of its *Semantic Relatedness* (SR) with the corresponding article of each term. It is computed using Article comparer of Wikipedia Miner Toolkit [18].

Feature 2: *Context Profile based Relatedness (CPRel)* is the second feature based on informative-content of Wikipedia concepts. A context profile of each input term is computed by finding Wikipedia labels in the informative-content of the concepts and discarding the labels having *Link Probability* (LP)[2] values below a certain cutoff threshold α.

To compute statistics for weighting schemes, filtered labels are stemmed. The weight of each label in the context profile is computed as follows:

$$W(r) = \frac{2 \times LE(r) \times NTF(r)}{LE(r) + NTF(r)} \quad (4)$$

Normalized Term Frequency (NTF) is computed as the ratio of sum of frequencies of all the inflectional forms of a root word to the sum of weights of all the root words in an article and is given as:

$$NTF(r) = \frac{\sum_{i=1}^{k} TF(d_i)}{\sum_{j=1}^{m} TF(r_j)} \quad (5)$$

where k represents the total number of derivational forms d_i of a root word r and m represents the total number of root words in a Wikipedia article.

Link Estimation (LE) is the probability of a root word being used as a link in the whole corpus and is defined as the ratio of number of documents, where each inflectional form di of a root word occurs as a link $(Dlink_{d_i})$ to the number of documents where each inflectional word occurs at all (Dw_{d_i}). *Link estimation* is computed as below:

$$LE(r) = \frac{\sum_{i=1}^{k} count(Dlink_{d_i})}{\sum_{i=1}^{k} count(Dw_{d_i})} \quad (6)$$

where k represents the number of inflectional forms d_i of a root word r.

The weighted context profiles are then compared using Cosine similarity to compute final semantic relatedness score of a term pair.

---

[2] *Link Probability* (LP) [19] is a proven measure to signify keyphraseness of a term.

Feature 3: *Directional Context based Relatedness Measure (DCRM)* is based on the semantics mined from Wikipedia hyperlink structure as well as informative-content of Wikipedia concepts. It is inspired by the idea of directional context borrowed from free word associations [20]. For a given term pair, the context of each term is derived from Wikipedia hyperlink structure. A subset of this context is then selected based on the co-occurrence of both input terms within a window of reference of size $\omega = 20$ in the informative-content of them. The directional association strength of a given term pair is computed in this subset of context of each term individually. For a given term pair (t_i, t_j), the relatedness score based on the directional association of input terms in two different contexts is computed as follows:

$$DCRM(t_1, t_2) = \lambda \times Association\_Strength(t_i, t_j)$$
$$+ (1 - \lambda) \times Association\_Strength(t_j, t_i) \quad (7)$$

where λ is the asymmetric coefficient and its value is set to 0.5, so as to give equal importance to both forward and backward association strengths. The association strength of an input term pair is computed by the following formula:

$$Association\_Strength(t_i, t_j) = \frac{\sum_{a \in C_s} rel(a, a_i) \times rel(a, a_j)}{|C_i|} \quad (8)$$

where C_i is the context vector of the first term and $rel(a, a_i)$ and $rel(a, a_j)$ are computed using *Article Comparer* of Wikipedia Miner Toolkit [8].

3.2   Validation and Learning of Semantic Relatedness

The model automatically computes the relatedness score of each test bed term pair using individual features and combines them using regression. To generate optimal supervised regression model, four different classifiers are used: Gaussian Process (GP) with the Pearson VII function-based universal kernel (PUK), which implements Gaussian processes for regression without hyper-parameter-tuning; SMOreg (SMO) with the Pearson VII function-based universal kernel (PUK), which implements the Support Vector Machine for non-linear regression; Linear Regression (LR); and Multi-Layer Perceptrons (MLP), which uses back propagation to classify instances. All classifiers are used with default parameter settings. Each classifier learns how to most effectively combine the features to maximize the correlation of the predicted scores with the human judgments.

To investigate the performance of features combinations, three single features were used to generate various feature combinations and the effect of each feature combination on the performance of semantic association computation was analyzed. In the following experiments, the individual features SRel, DCRM and CPRel are referred to as f_1, f_2 and f_3. Based on these three single features three 2-feature combinations, (f_{12}), (f_{13}) and (f_{23}) and one 3-feature combination (f_{123}) were generated, resulting in seven features in total.

Since the datasets are small in size, cross validation is used to avoid over fitting. The learning process is validated using two different classifier evaluation algorithms: 10-fold cross validation and leave-one-out cross validation. Leave-one-out cross validation is the same as a K-fold cross-validation with K representing the number of observations in the original dataset.

4 Evaluation

To evaluate the hybrid model, the automatically computed scores are compared with manual judgments using two correlation metrics: Pearson's correlation (γ) and Spearman's correlation (ρ). To get the manual judgments, three datasets are used: M&C dataset, consisting of 30 noun term pairs rated by 38 human judges on a scale of 0-4; R&G dataset, consisting of 65 noun term pairs rated by 51 human judges on a scale of 0-4; and WS-353, consisting of 353 term pairs rated by 13-16 judges on a scale of 0-10.

5 Results and Discussion

The results in Table 1 shows the correlation-based performance comparison of the hybrid model using both leave-one-out and 10-fold cross validations on all datasets using the 3-feature combination (f_{123}).

Table 1. Performance comparison of four classifiers based on the 3-feature combination

Classifier	Correl.	L-1-O CV			10-fold CV		
		M&C	R&G	WS-353	M&C	R&G	WS-353
SMOReg	ρ	0.76[0.55-0.8]	**0.89**[0.82-0.93]	0.67[0.60-0.72]	0.73[0.50-0.86]	0.83[0.73-0.89]	0.66[0.59-0.71]
	γ	**0.85**[0.70-0.92]	0.89[0.82-0.93]	0.68[0.61-0.73]	**0.83**[0.67-0.91]	0.85[0.76-0.90]	0.66[0.59-0.71]
GP	ρ	**0.77**[0.56-0.88]	**0.89**[0.82-0.93]	**0.71**[0.64-0.74]	**0.74**[0.51-0.86]	**0.84**[0.74-0.90]	**0.70**[0.63-0.74]
	γ	**0.87**[0.70-0.92]	**0.90**[0.84-0.93]	**0.70**[0.64-0.74]	0.82[0.65-0.91]	**0.87**[0.79-0.92]	**0.68**[0.61-0.73]
LR	ρ	0.72[0.48-0.85]	0.82[0.72-0.88]	0.64[0.57-0.69]	0.72[0.48-0.85]	0.80[0.69-0.87]	0.63[0.56-0.68]
	γ	0.78[0.58-0.89]	0.84[0.74-0.89]	0.53[0.45-0.60]	0.74[0.51-0.86]	0.81[0.70-0.88]	0.51[0.42-0.58]
MLP	ρ	0.72[0.48-0.85]	0.82[0.72-0.88]	0.64[0.57-0.69]	0.65[0.37-0.81]	0.83[0.73-0.89]	0.65[0.58-0.70]
	γ	**0.85**[0.70-0.92]	0.87[0.79-0.91]	0.64[0.57-0.69]	0.67[0.40-0.82]	**0.87**[0.79-0.91]	0.64[0.57-0.69]

Note: Correlation is significant at $\alpha = 0.05$ level.

The reported results are statistically significant at 95% confidence level with $\alpha = 0.05$. The correlation values are accompanied by their corresponding confidence intervals. Overall, the performance of GP regression model was consistently good across all datasets. Hence, this classifier is used to compare the performance of the hybrid model with the existing approaches. The Spearman's correlation based performance of leave-one-out cross validation surpassed the 10-fold cross validation because of more training instances but this difference diminished on the largest dataset, WS-353.

5.1   Performance Comparison of Features

To further analyze the individual performance of the features and their combinations, their averaged feature ranks were computed to pick the overall best
performing feature(s). All features were sorted and ranked according to their
correlation values on each dataset. Low ranks correspond to high correlation
values and vice versa. Ranks of each feature on all datasets were averaged to get
its averaged_rank score. The features are sorted and ranked again according to
their averaged_rank scores. The feature(s) with lowest rank are selected as the
best performing feature(s).

A comparison of averaged feature ranking is presented in Table 2. The aim of
this experiment is to analyze the overall rank based performance of each feature.
The best feature on leave-one-out cross validation was f_{123} and on 10-fold cross
validation was f_{13}. It is intuitive to note that the first single feature f_1, which is a
combination of two structure-based measures, outperformed the other two single
features, while the third single feature f_3, which is the informative-content based
association measure produced the lowest correlation of all. However, when the
structure-based measure is combined with the informative-content based measure using a regression function, this new feature f_{13} surpassed all other feature
combinations. These results demonstrate that distinct features complement each
other in various association scenarios, resulting in significant performance improvement. That is why the feature combination f_{13} outperformed the individual
features f_1 and f_3.

Table 2.   A comparison of averaged feature ranking on all datasets. Highest rank
correspond to lowest correlation-based performance of a feature and vice versa.

Feature	Rank Number			
	L1O	L1O	10fold	10fold
	(ρ)	(γ)	(ρ)	(γ)
f_1	3	2	1	3
f_2	5	5	5	5
f_3	6	6	6	6
f_{12}	3	3	3	3
f_{13}	2	1	1	1
f_{23}	4	4	4	4
f_{123}	1	1	2	2

5.2   Hybrid Model with Maximum Function

To gain further insight into the behavior of the hybrid model, the regression
function is replaced by the maximum function. In this experiment, given three
features for a term pair, the hybrid model picks the maximum of the three as
the final relatedness score. The results of this experiment are reported in the
Table 3 and Table 4 of the following section.

It is surprising to discover that the results of hybrid model with maximum function were superior to more sophisticated supervised regression learning using the Pearson's correlation (in all cases) and on the largest dataset, WS-353, using the Spearman's correlation. Section 3 of the paper indicated that the three features are based on different aspects of Wikipedia. Consequently, these features are good at detecting different kinds of term relations. For instance, if the structure-based feature is unable to find out a strong relation of two terms in the Wikipedia structure, it does not mean that the relation does not exist at all. It could be the inability of that feature to detect the relation of the given term pair due to its underlying aspect—Wikipedia hyperlink structure. Hence, if the informative-content based feature finds out a stronger association of the same term pair then the low associations predicted by other features should be overshadowed by this feature's indication of a stronger relation. The maximum function represents these semantics by selecting the kind of relation that is strongest. Surprisingly, the supervised regression appears to be unable to represent and learn these semantics better than the maximum_function based hybrid_model. This experiment also shows the poor ability of the classifiers used in the supervised regression learning to cope with such scenarios due to lack of their internal mechanism in handling the maximum likelihood of individual features while converging to the final regression function.

Table 3. Performance comparison of the proposed hybrid model-based approach with existing approaches on M&C dataset

| | | M&C |
Method	Source	(γ)
Strube and Ponzetto [16]	Wikipedia	0.46
Sahami and Heilman [12]	Web	0.58
Gladson [11]	Web	0.55
Wu and Palmer [11]	WordNet	0.78
Resnik [11]	WordNet	0.74
Leacock [11]	WordNet	0.82
Lin [11]	WordNet	0.82
Jiang and Conarth [11]	WordNet	0.84
Bollegala et al. [11]	Web	0.87
Jarmasz [14]	Thesaurus	0.87
Agirre et al. [7]	WordNet+Corpus	**0.93**
Hassan et al. [13]	Wikipedia	0.87
Hybrid_Model (max)	Wikipedia	0.85[*]
Hybrid_Model (GP)	Wikipedia	0.87[*]

[*]Correlation is significant at $\alpha = 0.05$ level.

5.3   Comparison With Existing Approaches

The results of the proposed hybrid model are compared with existing state-of-art approaches on three benchmark datasets. The knowledge source used by each

Table 4. Performance comparison of the hybrid model-based approaches with existing approaches on two benchmark datasets: R&G and WS-353

Method	Source	R&G (γ)	WS-353 (γ)	R&G (ρ)	WS-353 (ρ)
Strube and Ponzetto [16]	Wikipedia	0.52	0.49	–	–
Milne and Witten [8]	Wikipedia	–	–	0.64	0.69
Launder et al. [21]	Corpus	0.64	0.56	0.60	0.58
Gabrilovich and Markovich [15]	Wikipedia	–	0.50	–	0.75
Jermasz [14]	Corpus	0.81	0.53	0.80	0.41
Agirre et al.(supervised) [7]	WordNet+Corpus	–	–	**0.96**	0.78
Agirre et al.(unsupervised) [7]	WordNet+Corpus	–	–	0.88	0.66
Hassan and Mihalceae [13]	Corpus	0.80	0.67	0.79	0.71
Lushan et al. [22]	Corpus	0.81	0.62	0.84	0.66
Bollegala et al. [11]	Web	–	–	0.86	0.74
Navigli and Ponzetto [9]	Wikipedia	–	0.59	–	0.65
Hybrid_Model (max)	Wikipedia	**0.91**[*]	**0.79**[*]	0.86[*]	**0.81**[*]
Hybrid_Model (GP)	Wikipedia	0.90[*]	0.70[*]	0.89[*]	0.71[*]

[*] Correlation is significant at $\alpha = 0.05$ level.

approach is also mentioned with the results. Table 3 indicates the comparison of the proposed hybrid model based approaches with the existing state-of-art approaches on M&C dataset. The results of the existing approaches on M&C dataset are generally reported using Pearson's correlation. Hence, following [11], this paper also reports the results on M&C dataset in a separate table. On this dataset, the GP-based hybrid_model produced the second highest correlation while outperforming many of the existing approaches. Also, the hybrid_model based on maximum function also performed quite well on this dataset.

The results on R&G and WS-353 datasets are reported in Table 4. The maximum function-based Hybrid_Model surpassed all other approaches on WS-353 dataset using both Spearman and Pearson's correlation and on R&G dataset using Pearson's correlation. Moreover, on both R&G and WS-353 datasets, the GP-based hybrid_model outperformed the existing approaches using Pearson's correlation and produced second best results after maximum function-based Hybrid_Model.

On the smaller datasets, M&C and R&G, which are solely similarity-based dataset, Agirre et al. performed the best while the Hybrid_Models were strongly competitive. However, on the largest dataset WS-353, which represents a variety of word relations, the maximum function-based Hybrid_Model performed the best by surpassing all existing approaches including Agirre et al. as well.

Agirre et al. outperformed other approaches on M&C dataset (using Pearson's correlation) and R&G dataset (using Spearman's correlation) using their supervised approach, which is based on combining multiple features from two different knowledge sources. However, the performance of their best unsupervised approach was not even close to the unsupervised Hybrid_Model presented in this paper. Overall, both supervised and unsupervised Hybrid_Model based approaches performed quite well by surpassing most of the existing approaches on all datasets.

6   Conclusion and Future Work

This paper addressed the problem of semantic association computation using a supervised machine learning based hybrid model. The research contributions of the paper are two-fold: first, it presented a new hybrid model based on three new features generated from multiple aspects of Wikipedia for learning semantic association computation; second, it used a correlation-based feature ranking to select an optimal feature combination(s). The experiments demonstrated the effectiveness of the hybrid model on computing semantic associations of words. The paper investigated the impact of individual features and their combinations on learning semantic association computation and empirically showed that the best performing feature combinations are the ones that are based on multiple aspects of Wikipedia. Empirical results have also shown that the maximum function based hybrid model performed quite well, exceeding the GP-based hybrid model on all datasets using the Spearman's correlation. The reason was the preferential selection of a feature with maximum association score over those features that yielded poor estimates of semantic associations. The performance of presented approach could be enhanced further by improving the context. This could be done by considering features from other knowledge sources such as WordNet and Wiktionary. This will be an intuitive avenue to explore in future.

References

1. Hofmann, T.: Probabilistic latent semantic indexing. In: Proceedings of the 22nd Annual International ACM SIGIR Conference on Research and Development in Information Retrieval (SIGIR 1999), pp. 50–57 (1999)
2. Patwardhan, S., Banerjee, S., Pedersen, T.: Using measures of semantic relatedness for word sense disambiguation. In: Gelbukh, A. (ed.) CICLing 2003. LNCS, vol. 2588, pp. 241–257. Springer, Heidelberg (2003)
3. Schonhofen, P.: Identifying document topics using the wikipedia category network. In: Proceedings of the International Conference on Web Intelligence (WI 2006), pp. 456–462. IEEE Computer Society (2006)
4. Huang, A., Milne, D., Frank, E., Witten, I.H.: Clustering documents using a wikipedia-based concept representation. In: Theeramunkong, T., Kijsirikul, B., Cercone, N., Ho, T.-B. (eds.) PAKDD 2009. LNCS, vol. 5476, pp. 628–636. Springer, Heidelberg (2009)
5. Yih, W., Qazvinian, V.: Measuring word relatedness using heterogeneous vector space models. In: Proceedings of the 2012 Conference of the North American Chapter of the Association for Computational Linguistics: Human Language Technologies (NAACL HLT 2012), pp. 616–620 (2012)
6. Budanitsky, A., Hirst, G.: Evaluating wordnet-based measures of lexical semantic relatedness. Comput. Linguist. 32, 13–47 (2006)

7. Agirre, E., Alfonseca, E., Hall, K., Kravalova, J., Paşca, M., Soroa, A.: A study on similarity and relatedness using distributional and wordnet-based approaches. In: Proceedings of Human Language Technologies: The 2009 Annual Conference of the North American Chapter of the Association for Computational Linguistics (NAACL 2009), pp. 19–27 (2009)

8. Milne, D., Witten, I.H.: An effective, low-cost measure of semantic relatedness obtained from wikipedia links. In: Proceeding of AAAI Workshop on Wikipedia and Artificial Intelligence: an Evolving Synergy, pp. 25–30 (2008)

9. Navigli, R., Ponzetto, S.P.: Babelrelate! a joint multilingual approach to computing semantic relatedness. In: Proceedings of the Twenty-Sixth Conference on Artificial Intelligence, AAAI 2012 (2012)

10. Yazdani, M., Popescu-Belis, A.: Computing text semantic relatedness using the contents and links of a hypertext encyclopedia. Artif. Intell. 194, 176–202 (2013)

11. Bollegala, D., Matsuo, Y., Ishizuka, M.: A web search engine-based approach to measure semantic similarity between words. IEEE Trans. on Knowl. and Data Eng. 23(7), 977–990 (2011)

12. Sahami, M., Heilman, T.D.: A web-based kernel function for measuring the similarity of short text snippets. In: Proceedings of the 15th International Conference on World Wide Web (WWW 2006), pp. 377–386 (2006)

13. Hassan, S., Banea, C., Mihalcea, R.: Measuring semantic relatedness using multilingual representations. In: Proceedings of the First Joint Conference on Lexical and Computational Semantics (SemEval 2012), pp. 20–29 (2012)

14. Jarmasz, M., Szpakowicz, S.: Roget's thesaurus: a lexical resource to treasure. CoRR (2012)

15. Gabrilovich, E., Markovitch, S.: Computing semantic relatedness using wikipedia-based explicit semantic analysis. In: Proceedings of the 20th International Joint Conference on Artificial Intelligence (IJCAI 2007), pp. 1606–1611 (2007)

16. Ponzetto, S.P., Strube, M.: Knowledge derived from wikipedia for computing semantic relatedness. J. Artif. Intell. Res. (JAIR) 30, 181–212 (2007)

17. Mihalcea, R., Corley, C., Strapparava, C.: Corpus-based and knowledge-based measures of text semantic similarity. In: Proceedings of the Association for the Advancement of Artificial Intelligence (AAAI 2006), pp. 775–780 (2006)

18. Milne, D., Witten, I.H.: An open-source toolkit for mining wikipedia. Artificial Intelligence 194, 222–239 (2013); Artificial Intelligence, Wikipedia and Semi-Structured Resources.

19. Mihalcea, R., Csomai, A.: Wikify!: linking documents to encyclopedic knowledge. In: Proceedings of the Sixteenth ACM Conference on Information and Knowledge Management (CIKM 2007), pp. 233–242 (2007)

20. Jabeen, S., Gao, X., Andreae, P.: Directional Context Helps: Guiding Semantic Relatedness Computation by Asymmetric Word Associations. In: Lin, X., Manolopoulos, Y., Srivastava, D., Huang, G. (eds.) WISE 2013, Part I. LNCS, vol. 8180, pp. 92–101. Springer, Heidelberg (2013)

21. Landauer, T.K., Dumais, S.T.: A solution to plato's problem: The latent semantic analysis theory of acquisition, induction, and representation of knowledge. Psychological Review, 211–240 (1997)

22. Han, L., Finin, T., McNamee, P., Joshi, A., Yesha, Y.: Improving word similarity by augmenting pmi with estimates of word polysemy. IEEE Trans. Knowl. Data Eng. 25(6), 1307–1322 (2013)

An Ontology-Based Approach
for Product Entity Resolution on the Web

Raymond Vermaas, Damir Vandic, and Flavius Frasincar

Erasmus University Rotterdam
P.O. Box 1738, 3000 DR, Rotterdam, The Netherlands
research@raymondvermaas.nl, {vandic,frasincar}@ese.eur.nl

Abstract. Product entity resolution is an important part of online product search, where product entities coming from different websites need to be aggregated in the search results. In this paper, we propose an approach to product entity resolution using the descriptive power of an ontology. In our algorithm, we use similarity measures that are defined specifically for each type of product feature and learn the feature weights by means of a genetic algorithm. In the evaluation of our algorithm, we obtain F_1-measures of 59% and 72% for two product classes that we consider. The obtained results are significantly better than those obtained from a state-of-the-art product entity resolution algorithm.

Keywords: entity resolution, product ontologoy, product features.

1 Introduction

With the current increase in Web sales [5], online product search has become an important tool for consumers on the Web. The main reasons for its popularity are simple. First, it allows consumers to search for products they prefer. Compared to traditional shopping, online shopping makes it easier for consumers to find their desired product using information search facilities. Second, by comparing several Web shop offers, consumers have a good overview of the available prices, which allows them to choose the best bargain.

However, online product search providers, like Google Product Search and Shopping.com, are suffering from several issues. One of these issues is that the automatic retrieval of products is not easy, as there is some manual work involved if one wants to display the correct product information. The reason for this is that most products are available on more than one website, which results in retrieving products multiple times from different Web shops. In this case, one needs an approach to automatic product entity resolution, i.e., automatically finding duplicate products. In other words, in order to allow for comparison of the product offers from different Web shops, it is necessary to map all the offers found to their corresponding product. One solution to this problem is to use a universal product domain ontology to describe products in Web shops. Although this is an interesting approach, it falls outside the scope and goal of this paper.

B. Benatallah et al. (Eds.): WISE 2014, Part I, LNCS 8786, pp. 534–543, 2014.
© Springer International Publishing Switzerland 2014

Recent semi-automatic approaches for product entity resolution are able to obtain an F_1-measure between 45% and 81% [12,13]. In order to improve these results, in this paper, we consider the use of an ontology-based product entity resolution approach that automatically takes into account the product specific properties. Because product features are heterogeneous and can describe various aspects of products, we hypothesize that a feature-based approach for product entity resolution can improve the current state-of-the-art approaches. For describing products and their features, we make use of a product ontology that refines an existing upper-level e-commerce ontology.

The structure of the paper is defined as follows. Section 2 explains the related work. In Sect. 3, we describe the various aspects of our product entity resolution framework. This is followed by the evaluation of our method in Sect. 4. Last, we conclude our work and give directions for future research in Sect. 5.

2   Related Work

Elmagarmid et al. [8] survey the domain of entity resolution and identify several important steps that are involved in this process. The first step mainly focuses on the extraction, transformation, loading (ETL), and standardization of data. The next step is field matching, which deals with the methods used to compare the fields of two selected records. In this step, three types of field matching methods are identified: character-based similarity (e.g., Levenstein), token-based similarity (e.g., n-grams), and phonetic similarity (e.g., Soundex). The next step describes the different duplicate detection techniques in which a set of records is processed. The authors divided the techniques into the following types: probabilistic matching, supervised learning, active learning, distance-based techniques, rule-based techniques, and unsupervised learning. In the rest of this section, we present a few instances of such entity resolution algorithms.

Köpcke et al. [12] perform entity resolution on product offers. This is the same domain as we use, only instead of using all the product features, the authors of this paper use only the title of product offers. From the title they extract the product features, the brand, and the model. Next, an adaptive learning strategy using three string measures is applied and a support vector machine for the entity resolution is used. For product matching, the authors two different approaches. The Universal Product Code (UPC) is used as a reference mapping for the matching. This gives an F_1-measure of 55% for the product category *non-accessories* and 45% for the product category *accessories*. For the category *TV's*, the authors achieve a F_1-measure of 69% and for the category *digital cameras* their approach achieved a F_1-measure of 81%. While the authors of [12] focus on only title of a Web page, we focus on the entire set with product features. Furthermore, because our goal is to have a fully automatic approach, we do not consider comparing our approach to this solution, as it requires to manually create regular expressions.

In [22], the authors propose a method that determines whether two Web pages are referring to the same person. They assess the available features, like the

URL of the Web page, most frequent names, and other text on the Web pages, using one or more similarity measures like the cosine similarity, string similarity, number of overlaps, Pearson's correlation, and extended Jaccard similarity. As this method focuses on persons and unstructured data, it is less relevant for our approach, where we focus on products and a structured ontology.

Lee et al. [15] use relational evidence to perform Web scale entity resolution. They divide the relational evidence in two types of evidence, negative evidence and positive evidence. The negative evidence relies on the *birds of a feather* principle, which states that entities that are not similar do not share the same properties. Unfortunately, the *birds of a feather* principle does not hold when performing entity resolution in the product domain, since (different) product instances in the same product class have many properties in common (e.g., display size and display standard in the TV product class). For positive evidence the authors make use of Wikipedia. Since Wikipedia does not describe products with detailed feature information, we cannot use this approach.

In [3] the authors investigate entity resolution in a customer database. The authors propose three algorithms for the propose of pairwise entity resolution. The first is G-Swoosh, a generic brute force algorithm. The second is R-Swoosh, where the source records are immediately discarded if a match between records is found. The last algorithm is F-Swoosh, which uses a cache of the compared data features to perform entity resolution. The authors report that G-Swoosh is extremely expensive and not practical when many matches occur. The authors also report that F-Swoosh is between 1.1 and 11.4 times faster than R-Swoosh. The focus of the three algorithms is on entity resolution in an arbitrary domain. This results in a general purpose matching and merging functions, which can be implemented in any domain. We use R-Swoosh as a starting point for our research, due to its scalability and extensibility properties.

3   Product Entity Resolution Platform

As explained in Sect. 1, Web shops use various ways to display their product information. If this information is aggregated into one product ontology, it might occur that the ontology contains some products more than once. With the here proposed product entity resolution framework, we aim to solve this problem. In order to reach this goal, we need to (i) find an efficient way to compare products, (ii) use the extra information available from the product ontology when matching products, and (iii) merge the matching products without losing information. The process of ontology population is outside the scope of this paper. For this task we refer the reader to existing literature [16,18].

3.1   Overview of the Proposed Approach

Our proposed solution is called Product Entity Resolution Platform (PERP) and can be divided into two parts. The first part concerns the domain ontology, which contains the data and the structure of the product information. The second part covers the algorithm that employs novel matching and merging routines.

In the first part of the approach, i.e., the domain ontology, we use OWL, the standard Web ontology language. An OWL ontology offers more expressivity than a relational database, which is normally used in the field of entity resolution [8]. Compared to relational databases, this increased expressivity allows us to create a more precise representation of a product. For example, entity resolution in a relational database is restricted to the predefined structure of a row, while in an ontology one can specify different types of relations between complex objects. The latter description makes it easier to differentiate between products. Various synonyms for products and product features are stored in our lexicalized domain ontology. This simplifies the instantiation of products from the Web by covering a large set of lexical representations. Although the actual instantiation of an ontology falls out the scope of this research, the structural description of products using domain ontologies can be exploited in the process of entity resolution, as we shall demonstrate.

In this paper, an extension of the GoodRelations [10] Consumer Electronics ontology (CEO) [4] is used as the domain ontology. This ontology describes electronic products, like televisions and digital audio players. It also defines several property types (e.g., qualitative properties and quantitative properties) that can be used to describe products. We extend the CEO ontology by adding missing products, like Router and Laptop, and extending the already described products with more product-specific properties, like the number of HDMI ports on a television or the number of headphone ports on a digital audio player. These extensions make the ontology more suitable for entity resolution, because of the extra information gained by using these new features. We also extended the Consumer Electronics ontology to support units of measurement, which is achieved using the Unified Code for Units of Measure ontology (UCUM) [19]. This allowed us to add units of measurements to product properties and identify the relations between the units of measurements (e.g., meter vs. centimeter and inch vs. centimeter). Figure 1(a) shows an excerpt of the domain ontology from the UCUM ontology and Fig. 1(b) shows how for 'Voltage' the UCUM ontology and our product ontology are linked. As we can see, this is done using the 'Quality value' of UCUM and the 'Quantitative value float' from the CEO ontology.

The most specific aspects of our platform are the product matching and merging routines. The foundation that underlies our approach is the R-Swoosh algorithm [3]. The reason for choosing R-Swoosh is that the algorithm offers an efficient alternative to a brute-force approach in pairwise entity resolution by decreasing the number of comparisons needed. It is an abstract algorithm, which allows us to use custom match and merge functions. We did not choose for the more scalable F-Swoosh, because we experiment on a relatively small dataset for which caching is not needed. We use the R-Swoosh implementation of the Stanford Entity Resolution Framework (SERF) [2].

3.2 Matching

In this section we describe the matching function of PERP. We start with an explanation of the different types of data and the corresponding distance measures. This is followed by a motivation for the use of product features weights. Last, we present the algorithm that employs the matching function.

Matching Functions. Product features can be roughly divided in three categories. First, there are the qualitative properties. These properties contain mainly categorical data, for example, the different supported data formats in digital audio players. The second category are the quantitative properties, which contain quantitative data that differs per product instance, for example, the number of HDMI ports on a television. Quantitative properties can also be expressed in a particular unit of measurement, for example, the weight of a product in kilograms. The last category are the data type properties, which contain product features that do not fit in the two previous categories. These are mainly Boolean features (e.g., has remote control) and complex features (e.g., contrast ratio). For each of these categories we have different matching functions. The matching functions always return a Boolean value that indicates if two product instances match for a particular category. Matching functions can consist of an exact match or an approximate match between features, where the latter is achieved by using one or more distance measures and a threshold.

The matching function of the qualitative properties is rather simple, it is an exact match between two product features. Such a precise comparison is needed, because these values are predefined in the domain ontology, so there is no room for their approximation.

The quantitative values are handled differently. These are processed using two distance measures. The first is the numerical distance measure, which is defined as the relative difference between two values. If this distance is below a certain threshold, the values are considered similar. In case the quantitative value is a

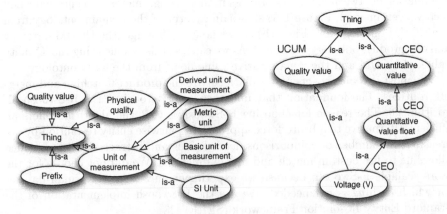

(a) An excerpt of the UCUM ontology

(b) An example of the link between UCUM and part of the CEO product ontology

Fig. 1. The UCUM ontology and the CEO product ontology

conjunction between a numerical value and a unit of measurement, we use the numerical distance measure for the numerical part and an exact match for the unit of measurement. The units of measurement are of the same magnitude in our representation.

The data type property values are also processed using two types of distance measures. Boolean variables are compared using the exact match from the qualitative values and complex values (represented as strings) are compared using the Level 2 Jaro Winkler distance [11,17,21]. We have chosen to use this distance because it showed to perform well compared to other distance measures when used for string comparisons [6].

Weights. Because products usually have many features, an interesting issue arises, i.e., among the products within a certain product type, some features are far more important than others. For example, all digital audio players are able to play MP3 files. This does not really help in the process of entity resolution, so we need to focus on the more distinct product features for entity resolution. To address the previous problem, we propose to use weights for the different product features. The weights are different for each product type, since some features (like weight) are used in multiple product types and can have a different importance depending on the product type. The weights are decimal values between 0.0 and 10.0. We determine the weights using a genetic algorithm built on the Watchmaker Framework [7], which is also used in the literature [9,14] and in the Apache Mahout project [1]. Our fitness function is the F_1-measure for a set of merged products. We use the genetic algorithm to determine the weights for one product type at a time. In order to run the genetic algorithm we split the data in a training and a test set, and run the genetic algorithm on the training set.

Matching Algorithm. The matching functions and feature weights are used in the matching algorithm. The algorithm executes a pairwise comparison on two records $R1$ and $R2$ containing product features that represent the same product type. In order to run this comparison, a set of keys of product features in the records is required, where a key is the name of the feature in the current concept. Each key has a similarity function that operates on the corresponding data type and a weight, as described previously. Another variable required for the algorithm is the matching threshold. This is the minimum percentage of matching product features and it is determined using a hill climbing procedure on the training dataset.

The algorithm iterates through all the possible keys (product attributes) of the product type in question. For each key, the similarity is computed for all possible pairs of values between the two records for that key. This is necessary, because a record can have multiple values for a certain key (e.g., weight in ounce and grams). Next the algorithm loops through all the results, computes the percentage of matching values, and compares this percentage to a threshold. Two products are matching if the percentage of matching values is higher than the threshold.

3.3 Merging

The merge algorithm merges two records for a given set of keys. It starts by checking if the two records are exactly the same. If that is the case, one of the two records is returned. If that is not the case, the algorithm iterates through all the keys and adds for each key all the non-matching values of $R2$ to $R1$. The reason to add only non-matching values is that one does not want to have the same values twice in a record.

We chose to also include the different values for matching keys, since we do not want to lose product relevant information. For example, if a product has a height in centimeters and in inches, we include both measurements in the new product. Also, in the case of contradicting product feature values, we keep both values. For example, one Web shop might suggest that a particular television has one HDMI port, while another Web shop might claim that it has two HDMI ports. Subsequently processing this data, an expert could decide which one of the conflicting values is the correct one, although this step is not required by our platform.

4   Method Evaluation

In this section we evaluate our proposed approach. In Sect. 3.1, we mentioned that our domain ontology was based on the GoodRelations Consumer Electronics ontology [4]. This ontology does not contain any product instances, so we had to instantiate those ourselves. As we mentioned previously, in a production environment, ontology population should be automated [16,18]. However, in order to perform a fair evaluation of the considered algorithms, we performed the instantiation manually. Having relatively detailed product descriptions in our data set, the instantiation and verification of this data turned out to be a very time consuming task, sometimes taking up to 5 hours per product instance. Consequently, we obtained a relatively small dataset, nevertheless of similar size as reported in related work for similar reasons [20].

We gathered data for only two product classes, i.e., *TV's* and *Digital Audio Players*, due to the tremendous effort involved in the manual instantiations. We chose for TV's and Digital Audio Players because of the variation in the granularity of the available information for the two product categories. Whereas televisions are usually described in a lot of detail, the description of Digital Audio Players is often limited to some basic features, like color and memory capacity. In order to ensure that we would have enough duplicates to conduct our experiments on, we used two Web shops (BestBuy.com and NewEgg.com) to gather the data. We chose for these websites, since they both have a wide assortment in consumer electronics, which made it easy to find duplicates.

We instantiated in total 49 products for two product types using the data from the two Web shops. We aimed to create a fair product distribution between BestBuy.com and NewEgg.com, so that we would not create a bias in our results, while at the same time have enough duplicate records for our product entity resolution algorithm. We also aimed to get a fair distribution between duplicate and non-duplicate records (or unique records) in order to prevent overfitting.

In order to compensate for the small dataset, we ran our experiments 100 times. In each run, we randomly selected a training and a test set. The distribution of unique and duplicate records is the same for the training and test set and proportionately depends on the product distribution for each product class. In each run 60% of the products are randomly assigned to the training set and 40% of the products are randomly assigned to the test set, while fulfilling the above distribution constraint.

The genetic algorithm is run for 100 generations for each random training set. This is only used for our own method. The method of [3] does not use weights, so there is also no need for a genetic algorithm. The same test set is used for both methods in each run. We tested different configurations of the crossover rate, mutation rate, and selection method for the genetic algorithm. We varied the crossover rate randomly from 10% up to 50% during one generation of the algorithm. The mutation rate was set to 15% and we used the Roulette Wheel selection as selection technique. These values were determined by using a hill-climbing procedure on the training dataset.

4.1 Results

We use a two-sample two-tailed paired t-test with a 95% confidence interval to compare our method to the benchmark method of [3] based on the F_1-measure from each run. The null hypothesis is $H_0 : \mu_{our} = \mu_{ref}$ and the alternative hypothesis is $H_A : \mu_{our} \neq \mu_{ref}$. The μ_{our} is the F_1 average of our method and μ_{ref} is the F_1 average of the reference method.

From the parameter optimization process described earlier, we used the best parameters for both our own method and those for the benchmark method. For our own method, the best set of parameters are a match threshold of 0.8, a threshold for the Level 2 JaroWinkler distance of 0.9 and a threshold for the numerical comparison 0.9 (so a maximum difference of 10% between two numerical values). The benchmark method has a match threshold of 0.9 and a threshold for the Jaccard distance of 0.75. These parameters are applied for both product classes.

The average F_1-measure over 100 runs for Digital Audio Players and Televisions is 0.721 and 0.595 for our approach, respectively. Compared to the benchmark method (0.177 and 0.509), our approach scores significantly better at a significance level of 5%.

5 Conclusion and Future Work

In this paper we investigated the problem of product entity resolution using a domain ontology-driven approach. Most solutions in the literature use databases for entity resolution, while employing the extra descriptive power of ontologies can help to obtain better results. We created a domain ontology for electronic devices that is based on the GoodRelations [10] Consumer Electronics Ontology [4]. Our approach is based on the state-of-the-art R-Swoosh algorithm [3], as this approach is both extensible and scalable. We propose a novel matching algorithm that, differently than the matching algorithm from R-Swoosh, employs product feature type-specific similarity measures, determined by the type information stored in the product ontology. Furthermore, the proposed matching algorithm uses weights for the duplicate detection task, in order to account for the difference in importance among the product features.

We evaluated our algorithm on a dataset that contains two different product classes: televisions and digital audio players. We obtained a F_1-measure on the merging of 59% on the television product class and 72% on the digital audio player product class. These results are significantly higher than the results obtained with the reference method [3]. From this, we can conclude that the use of product feature type-specific similarity measures and product feature weights improves the overall performance of product entity resolution.

In future work we would like to further exploit the descriptive power of an ontology by employing negative relations to emphasize that a certain product does not have a certain feature. Also we plan to make use of the part-whole relations, e.g., a memory chip is part of a laptop rather than a simple property and thus should have a higher importance in the product matching process.

Acknowledgment. Damir Vandic is supported by an NWO Mozaiek scholarship; project 017.007.142, *Semantic Web Enhanced Product Search (SWEPS)*.

References

1. Apache: Apache Mahout (2014), http://mahout.apache.org/
2. Benjelloun, O., Garcia-Molina, H., Kawai, H., Larson, T.E., Menestrina, D., Su, Q., Thavisomboon, S., Widom, J.: Generic Entity Resolution in the SERF Project. Bulletin, 1–9 (June 2006), http://goo.gl/rOhCFh
3. Benjelloun, O., Garcia-Molina, H., Menestrina, D., Su, Q., Whang, S.E., Widom, J.: Swoosh: a Generic Approach to Entity Resolution. The VLDB Journal 18(1), 255–276 (2008)
4. Brandl, X., Deckert, C., Frommer, F., Karl, D., Koslowski, D., Schley, D., Sonntag, D., Wechselberger, A., Hepp, M.: CEO: Consumer Electronics Ontology - An Ontology for Consumer Electronics Products and Services (2014), http://goo.gl/eFOMV1
5. Carini, A., Sehgal, V., Freeman Evans, P., Roberge, D.: European Online Retail Forecast, 2010 To 2015, Tech rep., Forrester Research, Inc. (2011), http://goo.gl/mxxD5J
6. Cohen, W.W., Ravikumar, P.D., Fienberg, S.E.: A Comparison of String Distance Metrics for Name-Matching Tasks. In: IJCAI 2003 Workshop on Information Integration on the Web (IIWeb 2003), pp. 73–78 (2003)
7. Dyer, D.W.: Watchmaker Framework (2014), http://goo.gl/CZjwVg
8. Elmagarmid, A., Ipeirotis, P., Verykios, V.: Duplicate Record Detection: A Survey. IEEE Transactions on Knowledge and Data Engineering 19(1), 1–16 (2007)

9. Halalai, R., Lemnaru, C., Potolea, R.: Distributed Community Detection in Social Networks with Genetic Algorithms. In: 6th IEEE International Conference on Intelligent Computer Communication and Processing (ICCP 2010), pp. 35–41. IEEE Computer Society (2010)
10. Hepp, M.: GoodRelations: An Ontology for Describing Products and Services Offers on the Web. In: Gangemi, A., Euzenat, J. (eds.) EKAW 2008. LNCS (LNAI), vol. 5268, pp. 329–346. Springer, Heidelberg (2008)
11. Jaro, M.A.: Advances in Record-Linkage Methodology as Applied to Matching the 1985 Census of Tampa, Florida. Journal of the American Statistical Association 84(406), 414–420 (1989)
12. Köpcke, H., Thor, A., Thomas, S.: Tailoring Entity Resolution for Matching Product Offers. In: 15th International Conference on Extending Database Technology (EDBT 2012), pp. 545–550. ACM (2012)
13. Köpcke, H., Thor, A., Rahm, E.: Evaluation of Entity Resolution Approaches on Real-World Match Problems. VLDB Endowment 3, 484–493 (2010)
14. Kryvyy, R., Tkachenko, S., Karkuljovskyy, V.: Analysis of Frameworks for Developing Genetic Algorithms. In: 7th International Conference on Perspective Technologies and Methods in MEMS Design (MEMSTECH 2011), pp. 209–210. IEEE Computer Society (2011)
15. Lee, T., Wang, Z., Wang, H.: Web Scale Entity Resolution using Relational Evidence, Microsoft Research, Technical Report MSR-TR-2011-30 (2011), http://goo.gl/OhNU3A
16. McDowell, L.K., Cafarella, M.: Ontology-Driven, Unsupervised Instance Population. Journal of Web Semantics: Science, Services and Agents on the World Wide Web 6(3), 218–236 (2012)
17. Monge, A., Elkan, C.: An Efficient Domain-Independent Algorithm for Detecting Approximately Duplicate Database Records. In: SIGMOD Workshop on Data Mining and Knowledge Discovery (DMKD 1997). ACM (1997)
18. Petasis, G., Karkaletsis, V., Paliouras, G., Krithara, A., Zavitsanos, E.: Ontology Population and Enrichment: State of the Art. In: Paliouras, G., Spyropoulos, C.D., Tsatsaronis, G. (eds.) Multimedia Information Extraction. LNCS, vol. 6050, pp. 134–166. Springer, Heidelberg (2011)
19. Polo, L., Berrueta, D.: Measurement Units Ontology (2008), http://goo.gl/DMQUEJ
20. Singla, P., Domingos, P.: Entity Resolution with Markov Logic. In: Sixth International Conference on Data Mining (ICDM 2006), pp. 572–582. IEEE (2006)
21. Winkler, W.E.: Using the EM algorithm for weight computation in the fellegisunter model of record linkage. In: Section on Survey Research Methods. pp. 354–359. American Statistical Association (1990)
22. Yerva, S.R., Miklós, Z., Aberer, K.: Towards Better Entity Resolution Techniques for Web Document Collections. In: 1st International Workshop on Data Engineering meets the Semantic Web (DESWeb 2010), pp. 209–214 (2010)

Author Index

Aberer, Karl I-276
Achilleos, Achilleas P. II-304
Agathangelou, Pantelis I-47
Agrawal, Shradha I-135
Aksoy, Cem I-448
Alabau, Vicent II-460
Alba, Alfredo II-17
Albitar, Shereen I-105
Aldalur, Iñigo I-293
Almulla, Mohammed II-32
Altingovde, Ismail Sengor II-78
Andreae, Peter I-512, I-523
Arellano, Cristóbal I-293

Bai, Changqing II-381
Bass, Len II-425
Bell, David A. I-115
Berberich, Klaus I-156
Bernardino B. de Campos, Gilda Helena
  II-351
Bianchini, Devis I-218
Bieliková, Mária I-372
Blazquez, Desamparados II-435
Bochmann, Gregor V. II-365
Boon, Ferry I-433
Bornea, Mihaela I-480
Bouzidi, Sabri I-433
Brunk, Sören II-521
Burger, Roman I-372

Cao, Longbing I-1
Cao, Yu II-246
Casanova, Marco Antonio I-324, II-351
Ceroni, Andrea II-90
Chaudry, Rabia II-425
Chawda, Bhupesh II-278
Chen, Fengjiao I-95
Chen, Wei I-170
Chen, Xi II-336
Chowdhury, Israt J. I-146
Chung, Tonglee I-308

Dao, Bo I-398
Dass, Ananya I-448
De Antonellis, Valeria I-218

de Boer, Nienke I-357
de Lara, Juan II-505
Dey, Akon II-262
Díaz, Oscar I-293
Dietze, Stefan I-324
Dimitriou, Aggeliki I-448
Ding, ZhiJun I-79
Di Pietro, Roberto I-15
Djafari Naini, Kaweh II-90
Dolby, Julian I-480
Domenech, Josep II-435
Drews, Clemens II-17
Duong, Thi II-474
Dustdar, Schahram II-415
D'yakonov, Alexander II-541

Enoki, Miki II-395
Erdmann, Maike II-109
Espinasse, Bernard I-105

Fei, Jinlong II-336
Feigenbutz, Florian II-294
Fekete, Alan II-319, II-425
Fetahu, Besnik II-351
Firmenich, Sergio I-293
Fisichella, Marco II-90
Fokoue, Achille I-480
Fournier, Sébastien I-105
Frasincar, Flavius I-357, I-418, I-433,
  I-534

Gan, Zaobin I-63
Gao, Xiaoying I-512, I-523
Gao, Yang I-186
Georgescu, Mihai II-90
Geva, Shlomo I-125
Giannopoulos, Giorgos II-189
Gil, Jose A. II-435
Gong, Zhiguo II-62
Gruhl, Daniel II-17
Gu, Jun II-246
Guabtni, Adnene II-425
Gunopulos, Dimitrios II-178
Guo, Xiaohui I-464
Gupta, Himanshu II-278

Hattori, Gen   II-109
Hirzel, Martin   II-395
Hogenboom, Frederik   I-418
Holzmann, Helge   II-47
Hong, Jun   I-115
Horii, Hiroshi   II-395
Hrgovcic, Vedran   I-496
Hu, Weishu   II-62
Husmann, Maria   II-199

IJntema, Wouter   I-418
Ikeda, Kazushi   II-109
Ilic, Alexander   II-531
Ishizaki, Hiromi   II-109

Jabeen, Shahida   I-512, I-523
Jiajie, Xu   I-170
Jin, Cheqing   II-489
Jourdan, Guy-Vincent   II-365
Joy, Mike   II-158

Kang, Qiangqiang   II-489
Kapitsaki, Georgia M.   II-304
Katakis, Ioannis   I-47
Kau, Chris   II-17
Kawase, Ricardo   II-351
Keller, Christine   II-521
Kementsietsidis, Anastasios   I-480
Kermarrec, Anne-Marie   I-276
Kliem, Andreas   II-294
Kokkoras, Fotios   I-47
Kompatsiaris, Ioannis   II-168
Kondylakis, Haridimos   I-496
Koniaris, Marios   II-189

Lee, Kevin   II-319
Leiva, Luis A.   II-460
Lewis, Neal   II-17
Li, Fangfang   I-1
Li, Guohui   I-244
Li, Jianjun   I-244
Li, Kan   I-31, I-95
Li, Victor O.K.   II-125
Li, Xue   II-1
Li, Xueming   II-1
Li, Yuefeng   I-186, I-408
Liu, An   II-141
Liu, Anna   II-319, II-425
Liu, Guanfeng   II-141
Liu, Shengli   II-336
Liu, Xudong   I-464

Liu, Yongbin   I-308
Lofi, Christoph   I-340
Long, Yi   II-125
Lu, Hongwei   I-63
Luo, Changyin   I-244
Luo, Wei   I-266

Ma, Jiangang   I-256
Ma, Yuanchao   I-388
Maamar, Zakaria   II-32
Macha, Meghanath   I-135
Medina, Haritz   I-293
Melchiori, Michele   I-218
Mendes, Pablo N.   II-17
Miller, James   II-445
Mirtaheri, Seyed M.   II-365
Mora Segura, Ángel   II-505
Móro, Róbert   I-372

Nagarajan, Meena   II-17
Nayak, Richi   I-125, I-146
Nebeling, Michael   II-199
Nguyen, Thin   I-266, I-398, II-474
Nieke, Christian   I-340
Niu, Guolin   II-125
Norrie, Moira C.   II-199
Ntonas, Konstantinos   I-47

Olteanu, Alexandra   I-276
Onal, Kezban Dilek   II-78
Onut, Iosif-Viorel   II-365, II-445
Ozsoy, Makbule Gulcin   II-78

P. Paes Leme, Luiz André   I-324
Pai, Deepak   I-135
Panev, Kiril   I-156
Panwar, Abhimanyu   II-445
Papadopoulos, Apostolos   II-541
Papadopoulos, Symeon   II-168
Peng, Shu   II-246
Pereira Nunes, Bernardo   I-324, II-351
Petrocchi, Marinella   I-15
Phung, Dinh   I-266, I-398, II-474
Plexousakis, Dimitris   I-496
Pongelli, Stefano   II-199
Pont, Ana   II-435
Premm, Marc   I-496
Puurula, Antti   II-541

Qadan Al Fayez, Reem   II-158
Qian, Weining   I-234, II-541

Rabello Lopes, Giseli   I-324
Rajani, Meena   II-262
Rao, Weixiong   II-246
Read, Jesse   II-541
Risse, Thomas   II-47
Roels, Reinout   II-215
Röhm, Uwe   II-262

Sánchez Cuadrado, Jesús   II-505
Saravanou, Antonia   II-178
Scekic, Ognjen   II-415
Schlegel, Thomas   II-521
Schouten, Kim   I-357
Schuele, Michael   I-496
Sellis, Timos   II-189
Semenov, Stanislav   II-541
Seyfi, Majid   I-125
Sharaf, Mohamed   II-1
Sharang, Abhijit   I-135
Shemshadi, Ali   I-202, I-256
Sheng, Quan Z.   I-202, I-256, II-405
Shuai, Kaiyan   I-308
Signer, Beat   II-215, II-231
Siméon, Jérôme   II-395
Sladescu, Matthew   II-319
Spognardi, Angelo   I-15
Srinivas, Kavitha   I-480
Stanik, Alexander   II-294
Sun, HaiChun   I-79
Sun, Hailong   I-464
Švec, Jan   II-541
Szabo, Claudia   II-405

Takishima, Yasuhiro   II-109
Tao, Han   II-336
Tayeh, Ahmed A.O.   II-231
Theodoratos, Dimitri   I-448
Tian, Nan   I-408
Tran, Truyen   I-266
Tran, Tuan   II-90
Truong, Hong-Linh   II-415
Tsakalidis, Adam   II-168
Tsoumakas, Grigorios   II-541

Unankard, Sayan   II-1

Valkanas, George   II-178
Vandic, Damir   I-357, I-418, I-433, I-534
van Leeuwen, Marijtje   I-357
van Luijk, Ruud   I-357
Vassiliou, Yiannis   II-189
Venkatesh, Svetha   I-266, I-398, II-474
Vermaas, Raymond   I-433, I-534
Vologiannidis, Stavros   II-541

Wang, PengWei   I-79
Wang, Ting   II-381
Wang, X. Sean   II-246
Wang, Yanghao   I-464
Welch, Steve   II-17
Weng, Daiyue   I-115
Woitsch, Robert   I-496
Wu, Chao   I-388

Xu, Bin   I-308, I-388
Xu, Chen   I-234
Xu, Guandong   I-1
Xu, Runhua   II-531
Xu, Yong   I-256
Xu, Yue   I-186, I-408

Yahyaoui, Hamdi   II-32
Yang, Min   II-246
Yao, Lina   I-256
Yuan, Bozhi   I-308, I-388
Yuan, Ji   I-464
Yuefei, Zhu   II-336

Zhang, Liang   I-234
Zhang, Richong   I-464
Zhang, Wei Emma   I-202
Zhang, Yihong   II-405
Zhang, Zhao   II-489
Zhao, Lei   I-170, II-141
Zhao, Qian   I-63
Zheng, Kai   I-170
Zhong, Jiang   II-1
Zhong, MingJie   I-79
Zhou, Aoying   I-234, II-489
Zhou, Xiaofang   I-170, II-141
Zhu, Feng   II-141
Zhu, Guangyao   I-31